10th International Conference on Lead and Zinc Processing (Lead-Zinc 2023)

AA007514

Journal of Physics: Conference Series Volume 2738

Changsha, China
17 - 20 October 2023

ISBN: 978-1-7138-9728-6
ISSN: 1742-6588

Printed from e-media with permission by:

Curran Associates, Inc.
57 Morehouse Lane
Red Hook, NY 12571

Some format issues inherent in the e-media version may also appear in this print version.

Institute of Physics
Dirac House, Temple Back
Bristol BS1 6BE UK

Phone: 44 1 17 929 7481
Fax: 44 1 17 920 0979

techtracking@iop.org

Additional copies of this publication are available from:

Curran Associates, Inc.
57 Morehouse Lane
Red Hook, NY 12571 USA
Phone: 845-758-0400
Fax: 845-758-2633
Email: curran@proceedings.com
Web: www.proceedings.com

TABLE OF CONTENTS

PbZn 2023

10th International Symposium on Lead and Zinc Processing

15 – 22 October 2023

Changsha, China,

www.leadzinc2023.com

Organizing Societies:

GDMB Gesellschaft der Metallurgen und Bergleute e.V. · MetSoc · MMIJ · NFSOC The Nonferrous Metals Society of China · TMS

Sponsorships:

NFSOC The Nonferrous Metals Society of China · Central South University · ENFI 中国恩菲 · 湖南省科学技术协会 Hunan Association for Science and Technology

Preface

The Lead-Zinc conference series, initiated by the American Institute of Mining, Metallurgical, and Petroleum Engineers (AIME) in 1970, has evolved into a global collaboration of metallurgical societies. This conference series is held every three or five years. It is heartening to acknowledge the significant role played by international organizations including the Minerals, Metals & Materials Society (TMS), Metallurgy & Materials Society of the Canadian Institute of Mining, Metallurgy and Petroleum (MetSoc), the Mining and Materials Processing Institute of Japan (MMIJ), the Society of Metallurgists and Miners (GDMB, Germany), and the Nonferrous Metals Society of China (NFSoc) in fostering knowledge exchange and advancements in lead and zinc industries.

The 10th International Symposium on Lead and Zinc Processing (PbZn 2023) was held in Changsha, China, organized by NFSoc and co-sponsored by Central South University (CSU), China ENFI Engineering Corporation (ENFI), and Hunan Science and Technology Association (HSTA). PbZn 2023 had more than 800 attendees, 230 presentations from over 20 countries. This proceedings volume is the result of over a year of collaborative effort, involving authors, editors, and volunteers. Comprising contributions from 20 countries, the volume reflects the global nature of lead and zinc industries, covering business trends, plant operations, fundamental developments, emerging technologies, and environmental considerations. We hope that this compilation will serve as a valuable record of the Lead-Zinc 2023 symposium and become a standard reference for the processing of lead and zinc.

We extend our heartfelt gratitude to all the contributors, reviewers, and volunteers who have dedicated their time and expertise to make this proceedings volume a reality.

Dr. Wei Sun

Executive Chairman
Professor, School of Minerals Processing and Bioengineering,
Central South University, China

International organization

Organizing Societies

Nonferrous Metals Society of China (NFSoc)

The Minerals, Metals and Materials Society (TMS)

Metallurgy & Materials Society - Canadian Institute of Mining, Metallurgy and Petroleum (MetSoc)

Society of Metallurgists and Miners (GDMB)

The Mining and Materials Processing Institute of Japan (MMIJ)

Organizing Chairs

Mingxing Jia, NFSoc

Yuehua Hu, Central South University

Cheng Liu, ENFI

Executive Chairman

Wei Sun, Central South University

Co-organizing Societies

Nonferrous Metals Society of China

Central South University

China ENFI Engineering Corporation

Hunan Science and Technology Association

Co-sponsors

PROCESS INTERFACE CHEMISTRY COMMITTEE

CHINA NONFERROUS METALS SOCIETY NONFERROUS METALLURGICAL RESOURCES COMPREHENSIVE UTILIZATION COMMITTEE

NATIONAL KEY LABORATORY OF NEW TECHNOLOGY OF

NONFERROUS METAL STRENGTHENING METALLURGY

NATIONAL HEAVY METAL POLLUTION PREVENTION AND CONTROL ENGINEERING TECHNOLOGY RESEARCH CENTER

MINISTRY OF EDUCATION ENGINEERING RESEARCH CENTER FOR DEVELOPMENT AND UTILIZATION OF METAL RESOURCES CARBON EMISSION REDUCTION

BGRIMM TECHNOLOGY GROUP

GRINM GROUP

CHINA NERIN ENGINEERING CO., LTD

CINF ENGINEERING CO., LTD

KUNMING UNIVERSITY OF SCIENCE AND TECHNOLOGY

JIANGXI UNIVERSITY OF SCIENCE AND TECHNOLOGY

HUNAN NONFERROUS METALS HOLDING GROUP CO., LTD

ZHUZHOU SMELTER GROUP CO., LTD

HENAN YUGUANG GOLD LEAD GROUP CO., LTD

SHENZHEN ZHONGJIN LINGNAN NONFERROUS METALS CO., LTD.FANKOU LEAD-ZINC MINE

BEIJING CHINA NON-FERROUS METAL CO., LTD

HUNAN NONFERROUS METALS INVESTMENT CO., LTD

ZIJIN MINING GROUP COMPANY LIMITED

CHINA NONFERROUS MINING GROUP CO., LTD

WESTERN MINING CO., LTD

JIANGXI COPPER LEAD&ZINC METAL CO., LTD

CHINA NONFERROUS METALS SOCIETY MINING AND METALLURGY

MITSUI MINING & SMELTING CO., LTD

DOWA METALS & MINING CO., LTD

TOHO ZINC CO., LTD

International Organizing Committee

Journal of Physics: Conference Series **2738** (2024) 011001 doi:10.1088/1742-6596/2738/1/011001

Editorial Committee

Chairman

Wei Sun, Central South University

Vice-Chairman

Huanzhi Gao, NFSoc

Secretary-general

Haisheng Han, Central South University

Honghu Tang, Central South University

Shihong Xu, Central South University

Assistant Secretary-general

Weiping Liu, Central South University

Steven Pooley, Central South University

Qingjun Guan, Hunan University of Science & Technology

Yongjie Bu, Hunan University of Science & Technology

Tong Yue, Central South University

Bing Li, China ENFI Engineering Corporation

Yu Zhang, Central South University

Zhongxi Jiang, Central South University

Hepeng Zhou, Jiangxi University of Science and Technology

Yan Liu, Northeastern University

Jialiang Zhang, University of Science and Technology Beijing

Journal of Physics: Conference Series **2738** (2024) 011001 doi:10.1088/1742-6596/2738/1/011001

Fen Jiao, Central South University

Degang Xu, Central South University

Ailiang Chen, Central South University

Huahao Xiang, BGRIMM Technology Group

Wenjun Wang, Central South University

Xianhui Qiu, Jiangxi University of Science and Technology

Huashan Yan, Jiangxi University of Science and Technology

Ye Chen, Guangxi University

Qicheng Feng, Kunming University of Science and Technology

Runqing Liu, Central South University

Feng Jiang, Central South University

Yunfeng Song, Central South University

Jianjun Wang, Central South University

Jian Cao, Central South University

Shangyong Lin, Central South University

Chenyang Zhang, Central South University

Yubiao Li, Wuhan University of Technology

Hao Zhang, University of Science and Technology Beijing

Yongxing Zheng, Kunming University of Science and Technology

Qing Shi, Central South University

Junwei Han, Central South University

Bei Sun, Central South University

Xu Yan, Central South University

Weifeng Liu, Central South University

Journal of Physics: Conference Series **2738** (2024) 011002 doi:10.1088/1742-6596/2738/1/011002

Peer Review Statement

All papers published in this volume have been reviewed through processes administered by the Editors. Reviews were conducted by expert referees to the professional and scientific standards expected of a proceedings journal published by IOP Publishing.

- **Type of peer review:** Single Anonymous
- **Conference submission management system:** Morressier
- **Number of submissions received:** 53
- **Number of submissions sent for review:** 51
- **Number of submissions accepted:** 50
- **Acceptance Rate (Submissions Accepted / Submissions Received × 100):** 94.3
- **Average number of reviews per paper:** 1.96
- **Total number of reviewers involved:** 24
- **Contact person for queries:**
  Name: Honghu Tang
  Email: honghu.tang@csu.edu.cn
  Affiliation: Central South University

Drillability classification and identification for rock mass based on machine learning

Yumeng WU[1], Shaofeng WANG[1, a]

[1]School of Resources and Safety Engineering, Central South University, Changsha 410083, China

[a]sf.wang@csu.edu.cn (S. WANG)

Abstract. In order to better apply the drilling method to underground mines, rock drillability classification and identification in situ by drilling process monitoring technology is a convenient and effective method to achieve the rock mass drillability. In this study, a database was established based on 188 groups of drilling parameters, drillability parameters and rock mechanics parameters. By analyzing the correlation between mechanical parameters and drillability parameters, rock drillability was classified using the TOPSIS-RMR method. Then, drilling force (*F*), torque (*T*), rotation speed (*N*), rate of penetration (*V*), specific energy (SE) and drillability index (I_d) were used as machine learning input variables to predict drillability grades. Finally, the machine learning classification models include SVM, ELM, BPNN, RBF, RF and LSTM are compared to select the optimal model. The efforts and results can be used to evaluate the rock mass drillability and provide support for the design optimization of drilling and blasting method. It can effectively protect the safety and improve efficiency of underground mining.

Keywords. lead-zinc ore mining, monitoring while drilling, TOPSIS-RSR method, machine learning, drillability classification and identification

1. Background

China possesses significant deposits of lead and zinc minerals, making it the second-largest producer globally, surpassed only by Australia. Given the rapid pace of industrialization, modernization, and urbanization, it becomes imperative to exploit vast quantities of lead and zinc ore to sustain these developments. The rapid development of lead and zinc industry has caused an impact on the ecological environment, so it is necessary to make strict requirements from the mining process to achieve sustainable development. In addition, Deep mine development will become the new norm, and deep rocks are in a complex environment of high situ stress, high temperature, high water pressure and strong excavation disturbance [1]. In order to improve resource utilization and reduce the risk of disaster, drilling and blasting methods must be improved.Therefore, rock drillability classification and identification in situ by drilling process monitoring technology is proposed.

The performance of drilling is intrinsically dependent on the rock mass drillability, which is a parameter that reflects the interaction between the drill bit and the rock, and is influenced by a combination of various factors. These factors can be categorized into two groups: controllable and uncontrollable [2]. Controllable factors include machine and drilling parameters such as drill bit type and diameter, drilling force and rotation speed, while uncontrollable factors include rock characteristics that cannot be directly controlled, such as rock composition, texture, and brittleness [3]. In short, there are many factors affecting drillability classification, complex classification system, single drilling performance evaluation index and other deficiencies, which make mine drilling and blasting construction explosive consumption and overall construction risk factor large.

At present, neural networks are often introduced to predict or classify traditional rock mechanical properties. For example, Sarkar et al. [4] uses feedforward and backpropagation neural network technology to predict the compressive strength of surrounding rock by using wave velocity, point load and density of rock mass. Ocak et al. [5] used multi-layer perceptron neural networks (MLPNN) to predict the elastic modulus of intact rocks.

In order to understand the distribution of drillability among different rock strata, researchers have established numerous empirical relationships through laboratory tests and field engineering experiments. Including schmidt hammer number, point load index, shore hardness, taber abrasion, rock brittleness, equivalent quartz content (EQC), drilling rate index (DRI), bit wear index (BWI), coefficient of rock strength (CRS), impact strength index (ISI), cerchar abrasion index (CAI), cerchar hardness index (CHI), specific energy (SE), texture coefficient (TC), among others [6]. Therefore, based on the aforementioned indices, a novel method for classifying the drillability grades of the rock mass has been proposed. In this method, mechanical parameters of rock mass are used to classify different grades, and machine learning method is used to evaluate the drillability grade of rock mass in real time during drilling.

2. Method

In order to realize the rock mass drillability classification and identification, TOPSIS-RMR method and machine learning classification model were selected in this study, and the optimal model of identification was determined by comparing different models. The research route of this study is shown in Figure 1. The classification method and machine learning model are introduced below.

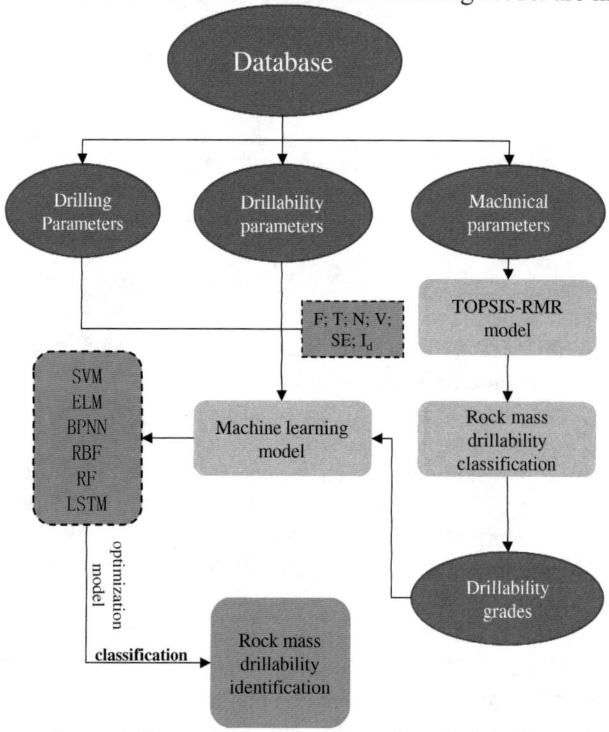

Figure 1. Schematic diagram of research methods in this study

2.1. TOPSIS-RMR method

The TOPSIS method is typically utilized for multi-objective decision making. It involves cotrend processing and normalization matrix to identify the optimal and worst targets among multiple targets. The Euclidean distance is then calculated and the proximity between each index value and the ideal solution is obtained and sorted as the basis for evaluating the quality of the target.

The Rank-sum ratio method (RSR) is a well-known comprehensive evaluation technique. It involves obtaining dimensionless RSR values by sorting indexes in the original data matrix, which can be used to directly rank or grade evaluation targets.

The combination of TOPSIS and RSR methods can overcome the limitations of using a single method and obtain more accurate and comprehensive results. These two methods provide different perspectives for analyzing and evaluating the targets. The RSR method focuses on sorting while the TOPSIS method provides a grading perspective. This combined approach can identify and alleviate issues related to data dispersion and information loss [7]. The specific steps are as follows:

(1) Construct a standardized decision matrix

Assuming that the drillability grade of m rock samples with n characteristic variables is to be divided, the original data matrix is:

$$X=\begin{bmatrix} x_{11} & x_{12} & \cdots & x_{1n} \\ x_{21} & x_{22} & \cdots & x_{2n} \\ \vdots & \vdots & \vdots & \vdots \\ x_{m1} & x_{m2} & \cdots & x_{mn} \end{bmatrix}_{m\times n} \tag{1}$$

The original data are treated with the cotrend processing and normalization. The data in this study are normalized by $Y_{ij}=\frac{x_j^+-x_{ij}}{x_j^+-x_j^-}$ (*i=1, 2, ..., m*; *j=1, 2, ..., n*), where $x_j^+=\max\left\{x_{1j},x_{2j},\cdots,x_{mj}\right\}$, $x_j^-=\min\left\{x_{1j},x_{2j},\cdots,x_{mj}\right\}$. The calculated standard matrix is:

$$Y=\begin{bmatrix} y_{11} & y_{12} & \cdots & y_{1n} \\ y_{21} & y_{22} & \cdots & y_{2n} \\ \vdots & \vdots & \vdots & \vdots \\ y_{m1} & y_{m2} & \cdots & y_{mn} \end{bmatrix}_{m\times n} \tag{2}$$

(2) Calculate relative proximity D_i

The optimal vector Y^+ and the worst vector Y^- are obtained according to the standard matrix.

$$Y^+=(y_1^+,y_2^+,\cdots,y_n^+) \tag{3}$$

$$Y^-=(y_1^-,y_2^-,\cdots,y_n^-) \tag{4}$$

where, $y_j^+=\max\left\{y_{1j},y_{2j},\cdots,y_{mj}\right\}$, $y_j^+=\min\left\{y_{1j},y_{2j},\cdots,y_{mj}\right\}$, (*j=1, 2, ..., n*).

Build the weight index $W=[W_1, W_2, ..., W_n]$ based on the correlation between rock machnical parameters and rock drillability parameters. And W is considered in the calculation of the distance B_i^+ and B_i^- between each evaluation object and the optimal vector and the worst vector.

Based on the correlation between rock machnical parameters and drillability parameters, a weight index $W=[W_1, W_2, ..., W_n]$ is constructed, and W is taken into account when calculating B_i^+ and B_i^- by equation (5) and equation (6).

$$B_i^+=\sqrt{\sum_{j=1}^{n}W_j(y_{ij}-y_j^+)^2}\ \ (i=1,\ 2,\ \ldots,\ m) \tag{5}$$

$$B_i^- = \sqrt{\sum_{j=1}^{n} W_j (y_{ij} - y_j^-)^2} \quad (i = 1,\ 2,\ \ldots,\ m) \tag{6}$$

where B_i^+ and B_i^- respectively represent the distance between the ith rock sample and the optimal vector and the worst vector, and y_{ij} represent the value of the jth index of the ith rock sample.

The relative proximity D_i between each index and the optimal value is calculated comprehensively:

$$D_i = \frac{B_i^-}{B_i^+ + B_i^-} \quad (i = 1,\ 2,\ \ldots,\ m) \tag{7}$$

(3) Determine D_i distribution

D_i value is used as RMR value to sort its size; At the same time, the frequency (f) and the cumulative frequency ($\sum f$) of each rock sample are listed; Determine the rank (R) and the average rank (R^*); Calculate the ranking cumulative frequency $p=R^*/n \times 100\%$ (the last item is modified with $(1-1/4n) \times 100\%$); Query 'Percentage and probability unit conversion Table' according to the ranking cumulative frequency to get the value of Probit.

(4) Calculate regression equation

As shown in equation (8), Probit was taken as the independent variable and D_i value as the dependent variable to calculate the regression equation.

$$D_i = \alpha + \beta \times Probit \tag{8}$$

(5) Grading of rock drillability

According to the estimated value of D_i derived from regression equation, the drillability of rock samples is classified and sorted.

2.2. Machine classification model

This study adopts the following machine learning classification models: Support Vector Machines (SVM), Extreme Learning Machines (ELM), Backpropagation Neural Networks (BPNN), Radial Basis Neural Network (RBNN), Random Forests (RF), and Long Short-Term Memory Networks (LSTM). Here's a brief introduction to each of the machine learning classification model:

The SVM is a powerful algorithm that is effective in handling high-dimensional data. It works by mapping the data to a higher dimensional space where a hyperplane is used to separate the data into different classes [8]. The ELM is a fast and efficient algorithm that uses a randomly generated input layer and a single hidden layer to establish a nonlinear mapping between inputs and outputs. The output layer weights can be easily calculated through a linear regression [9]. The BPNN is capable of learning complex relationships between inputs and outputs, and can adapt well to new data. BP algorithms use a supervised learning approach to adjust the weights between the neurons in each layer during training [10]. The RBNN is capable of capturing non-linear patterns in data. The hidden layer in RBNN networks uses radial basis functions to model the data and the output layer weights are calculated through a linear regression [11]. The RF is a robust algorithm that is capable of handling noise and outliers in data. RF uses an ensemble learning approach by constructing multiple decision trees and combining them to reduce the risk of overfitting [12].The LSTM is a type of recurrent neural network that can be used for sequential data. LSTM networks use a memory cell to remember long-term dependencies in the data and process input sequences in a sequential manner [13].

Overall, the machine learning classification model includes a wide range of algorithms that can handle various types of data and achieve high accuracy and efficiency in classification tasks.

3. Rock mass drillability classification and identification

Through data collection, a total of 188 sets of diamond coring bit data consisting of drilling parameters, drillability parameters, and mechanical parameters were gathered [14]. The drilling parameters include drilling force (F), torque (T), rotation speed (N), rate of penetration (V), The drillability parameters include specific energy (SE) and drillability index (I_d) and drilling rate index (DRI), and the mechanical parameters include uniaxial compressive strength (UCS), tensile strength (TS), Cerchar abrasion index (CAI) and Leeb hardness (HL). These parameters are closely linked to the interaction between rock mass and drill bit, which can well reflect rock mass drillability [15]. The specific data distribution is shown in Table 1.

Table 1. Statistical distribution of parameters in the database

Variables	Size	Maximum	Minimum	Mean	St. D
UCS/MPa	188	148	56	94.138	24.863
TS/MPa	188	10.9	2.9	7.919	2.216
SE/103MPa	188	89.877	2.77	19.9	15.86
Id	188	0.039	0.0026	0.0103	0.0058
DRI	188	86	32	58.872	13.83
CAI	188	3.78	0.8	2.175	1.024
ESD	188	848	548	650.692	80.41
F/kN	188	1.1	0.5	0.806	0.222
T/N·m	188	275.053	35.102	115.917	56.172
N/r·min-1	188	470	210	336.702	106.530
V/mm·min-1	188	90.5	3	21.535	15.594

3.1. Rock mass drillability classification

TOPSIS-RMR method is used to classify the rock mass drillability. The drillability grade of rocks is determined by analyzing the correlation between rock mechanical parameters and drillability parameters. The drillability grades thus obtained served as the basis for training a machine learning classification model to identify the drillability grade in real time. To optimize the classification accuracy of the model, various models were trained and evaluated, and the most accurate model was selected. This comprehensive approach provides a reliable and efficient tool for assessing the drillability grade of a rock mass during drilling operations.

In order to avoid the randomness of a single drilling performance evaluation index, the correlation between UCS, TS, CAI, HL and *V*, SE, DRI was calculated, and the coefficient of determination R^2 was used as the evaluation index to evaluate the degree of the correlation. Then, the mean value of R^2 was calculated and normalized to obtain the weight index *W*. Table 2 shows the value of R^2 and weight index *W*.

Table 2. The mean and weight of R^2

Variable	UCS	BTS	CAI	HL
V	0.08	0.367	0.017	0.179
SE	0.064	0.068	0.024	0.246
DRI	0.121	0.681	0.048	0.393
Mean	**0.0883**	**0.372**	**0.02967**	**0.2727**
Weight	0.1158	0.4878	0.0389	0.3575

The standardized decision matrix is constructed successively according to the data. Table 3 shows the relative proximity C_i calculated after cotrend processing and normalization and the ranking situation. The frequency *f*, cumulative frequency $\sum f$, rank *R*, average rank R^* of each group and the corresponding Probit value are shown in Table 4.

Table 3. TOPSIS evaluation results

Number	Positive ideal solution B_i^+	Negative ideal solution B_i^-	Relative proximity D_i	Rank
1	0.605	0.560	0.481	7
2	0.477	0.563	0.541	5
3	0.963	0.093	0.088	16

4	0.746	0.297	0.285	15
5	0.193	0.952	0.832	2
6	0.626	0.401	0.391	14
7	0.105	0.950	0.900	1
8	0.627	0.477	0.432	12
9	0.728	0.517	0.415	13
10	0.579	0.482	0.454	11
11	0.369	0.676	0.647	4
12	0.645	0.543	0.457	10
13	0.370	0.752	0.670	3
14	0.527	0.551	0.511	6
15	0.544	0.494	0.476	8
16	0.605	0.560	0.481	9

Table 4. Distribution of relative proximity(D_i)

Relative proximity D_i	Frequency f	Cumulative frequency $\sum f$	Mean rank R^*	R^*/n*100%	Probit
0.088	1	1	1.0	6.3	3.466
0.285	1	2	2.0	12.5	3.85
0.391	1	3	3.0	18.8	4.113
0.415	1	4	4.0	25.0	4.326
0.432	1	5	5.0	31.3	4.511
0.454	1	6	6.0	37.5	4.681
0.457	1	7	7.0	43.8	4.843
0.458	1	8	8.0	50.0	5
0.476	1	9	9.0	56.3	5.157
0.481	1	10	10.0	62.5	5.319
0.511	1	11	11.0	68.8	5.489
0.541	1	12	12.0	75.0	5.674
0.647	1	13	13.0	81.3	5.887
0.67	1	14	14.0	87.5	6.15
0.832	1	15	15.0	93.8	6.534
0.9	1	16	16.0	98.4	7.154

The relative proximity C_i in Table 4 was used as the dependent variable and the value of Probit as the independent variable to conduct linear regression. The regression equation was obtained as follows:

$$D_i = -0.463 + 0.188 Probit$$

The D_i critical value is calculated according to the fixed Probit critical value. The Probit critical value varies according to the number of levels. In this study, the drillability is divided into four levels, and the critical value of Probit is 3.5, 5 and 6.5 respectively. The calculated results of regression equation D_i were 0.195, 0.477 and 0.759 respectively, and the drillability classification was carried out accordingly, as shown in the following Table 5. Among them, the higher the grade, the worse the drillability.

Journal of Physics: Conference Series **2738** (2024) 012001 doi:10.1088/1742-6596/2738/1/012001

Table 5. Drillability grade of rock samples

Grade	Percentile threshold	Probit threhold	D_i fitting value	Result
4	~6.681	~3.5	~0.195	3
3	6.681~	3.5~5	0.195~0.477	4, 6, 8, 9, 10, 12, 15, 16
2	50~	5~6.5	0.477~0.759	1, 2, 11, 13, 14
1	93.319~	6.5~	0.759~	5, 7

3.2. Drillability grades identification

In order to determine the most suitable machine learning classification model for drillability grade identification, SVM, ELM, BPNN, RBNN, RF and LSTM were selected for drillability grade identification. Take *F*, *T*, *N*, *V*, SE and I_d from the database as input and the corresponding drillability grade as output. In a unified manner, 80% of the data is randomly selected to form training set, and the remaining 20% constitutes testing set.

The SVM model uses linear kernel function for classification, with penalty factor c=0.1. ELM sets the number of neurons in the hidden layer as 20, and uses sigmoid function to activate. BPNN sets the number of hidden layers as 15; The radial basis function expansion speed of RBNN model is set to 100. RF model sets the number of decision trees as 10 and the minimum number of leaves as 5. LSTM Sets the maximum number of iterations to 500. The confusion matrix and identification results for testing set of different machine learning classification models are shown in Figure 2 and Figure 3 respectively. The accuracy rates of SVM, ELM, BPNN, RBNN, RF and LSTM were 64.86%, 67.57%, 72.97%, 81.08%, 78.38% and 72.97% respectively.

According to the application results of classification model, RBNN model has the highest identification accuracy, and it can be considered that RBNN algorithm is the optimal algorithm for rock drillability classification perception and identification.

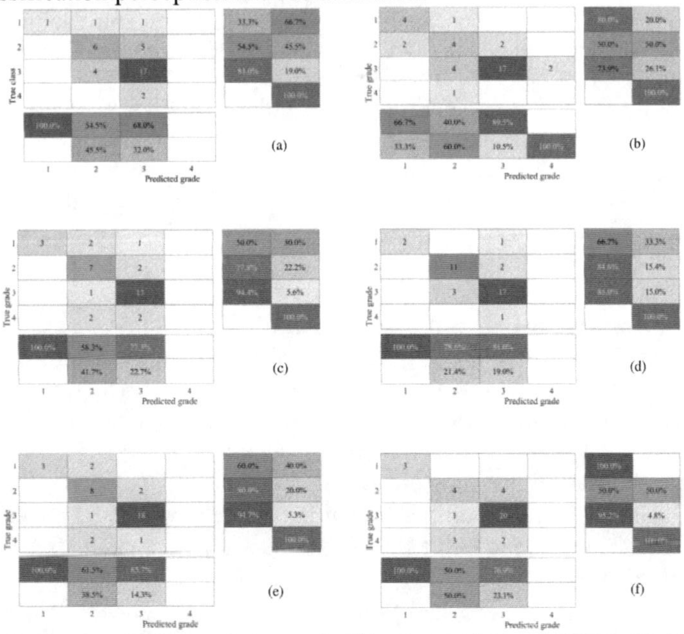

Figure 2. The confusion matrix for testing set of different models: (a) SVM; (b) ELM;(c) BPNN; (d) RBNN; (e) RF; (f) LSTM

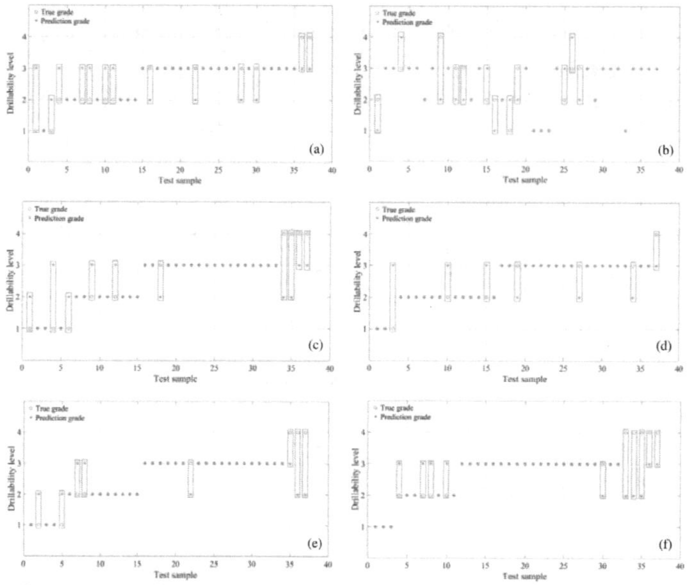

Figure 3. Identification results of different models: (a) SVM; (b) ELM;(c) BPNN; (d) RBNN; (e) RF; (f) LSTM

4. Discussion

The size of the dataset is a crucial factor in limiting performance of classification. Since the values of drilling force (*F*) and rotation speed (*N*) in the current database are fixed, the overall accuracy of identification is not an ideal level. Therefore, to enhance the reliability of classification, future research should focus on collecting a larger amount of data for the database.

The existing rock mass mechanical parameters used for drillability classification are insufficient, as factors such as groundwater state, rock joint conditions, and RQD can also affect drillability. The current study only considers UCS, TS, CAI, and HL for drillability classification. Therefore, further research is necessary to incorporate various geological conditions into the classification process.

Although the aforementioned study indicates that TOPSIS-RMR method and RBNN model can achieve drillability classification and identification of rock mass. they can not provide a comprehensive assessment for different engineering projects. In future research, it is recommended to include practical engineering as the research subjects to verify the applicability of the classification method.

5. Conclusions

The drillability classification is based on the TOPSIS-RMR method to classify the drillability of each rock sample. Additionally, the drillability grade was identified by inputting the drilling force (*F*), torque (*T*), rotation speed (*N*), rate of penetration (*V*), specific energy (SE), and drillability index (I_d) into the machine learning classification model. Finally, the optimal drillability grade classification model is determined by comparing identification accuracy of different models. The main achievements and conclusions of this study are as follows:

(1) Based on TOPSIS-RMR method, the drillability of rock samples is successfully divided into four grades. The higher the grade, the worse the drillability.

(2) Compared with SVM, ELM, BPNN, RBNN, RF and LSTM models, the classification accuracy of drillability is 64.86%, 67.57%, 72.97%, 81.08%, 78.38% and 72.97% respectively. The RBNN algorithm with the highest accuracy is the best model for rock drillability identification.

(3) A rapid rock mass drillability classification method is established, and in-situ drillability recognition is realized based on monitoring while drilling technology and machine learning algorithm, which is helpful to improve the safety and efficiency of underground mining.

Acknowledgements

The authors gratefully acknowledge the edited assist from professor Zilong Zhou and the financial support from the Independent Exploration and Innovation Project for Graduate Students of Central South University, China (2023ZZTS0497).

References

[1] Li P and Cai M 2021 Challenges and new insights for exploitation of deep underground metal mineral resources *Transactions of Nonferrous Metals Society of China* **31** 3478–505

[2] Yarali O and Kahraman S 2011 The drillability assessment of rocks using the different brittleness values *Tunnelling and Underground Space Technology* **26** 406–14

[3] Shao-feng W, Yu T and Shan-yong W 2021 Influence of brittleness and confining stress on rock cuttability based on rock indentation tests *J. Cent. South Univ.* **28** 2786–800

[4] Sarkar K, Tiwary A and Singh T N 2010 Estimation of strength parameters of rock using artificial neural networks *Bull Eng Geol Environ* **69** 599–606

[5] Ocak I and Seker S E 2012 Estimation of Elastic Modulus of Intact Rocks by Artificial Neural Network *Rock Mech Rock Eng* **45** 1047–54

[6] Altindag R 2010 Assessment of some brittleness indexes in rock-drilling efficiency *Rock Mech Rock Eng* **43** 361–70

[7] Zhao C, Liu B, Li J, Li S, Liu Y, Guo Y and Zhang X 2022 Evaluation of Laboratory Management Based on a Combination of TOPSIS and RSR Methods: A Study in 7 Provincial Laboratories of China *Front. Public Health* **10** 883551

[8] Wu X, Kumar V, Ross Quinlan J, Ghosh J, Yang Q, Motoda H, McLachlan G J, Ng A, Liu B, Yu P S, Zhou Z-H, Steinbach M, Hand D J and Steinberg D 2008 Top 10 algorithms in data mining *Knowl Inf Syst* **14** 1–37

[9] Huang G-B, Zhou H, Ding X and Zhang R 2012 Extreme Learning Machine for Regression and Multiclass Classification *IEEE Transactions on Systems, Man, and Cybernetics, Part B (Cybernetics)* **42** 513–29

[10] Zhang J-R, Zhang J, Lok T-M and Lyu M R 2007 A hybrid particle swarm optimization–back-propagation algorithm for feedforward neural network training *Applied Mathematics and Computation* **185** 1026–37

[11] Firat M, Yurdusev M A and Turan M E 2009 Evaluation of Artificial Neural Network Techniques for Municipal Water Consumption Modeling *Water Resour Manage* **23** 617–32

[12] Breiman L 2001 Random Forests *Machine Learning* **45** 5–32

[13] Greff K, Srivastava R K, Koutnik J, Steunebrink B R and Schmidhuber J 2017 LSTM: A Search Space Odyssey *IEEE Trans. Neural Netw. Learning Syst.* **28** 2222–32

[14] Sakız U, Aydın H and Yaralı O 2021 Investigation of the rock drilling performance of rotary core drilling *Bull Eng Geol Environ* **81** 24

[15] Wang S, Tang Y, Cao R, Zhou Z and Cai X 2022 Regressive and Big-Data-Based Analyses of Rock Drillability Based on Drilling Process Monitoring (DPM) Parameters *Mathematics* **10** 628

Journal of Physics: Conference Series **2738** (2024) 012002 doi:10.1088/1742-6596/2738/1/012002

Numerical study on fractal characteristic of blast-induced cracks in deep rock mass

Peiyu Wang[1], Zilong Zhou[1*], Jiale Fan[1], Ruishan Cheng[2], Xin Cai[1], Cuigang Chen[1] and Wencan Jiao[3]

[1]School of Resources and Safety Engineering, Central South University, Changsha 410010, China
[2]Center for Infrastructural Monitoring and Protection, School of Civil and Mechanical Engineering, Curtin University, Bentley, Australia
[3]Guangxi Xinfazhan Commun Grp Co Ltd, Nanning 530029, China

*Corresponding author: Zilong Zhou (zlzhou@csu.edu.cn)

Abstract. The quantitative representation on the characteristics of blasting-induced cracks in rock mass is of great significance to guide the fine blasting design in practice. However, the quantitative relation between blasting-induced cracks of rock mass and in-situ stresses is currently lacking. To provide the reference for the fine blasting design to excavate rock mass with different in-situ stresses, the present study adopts the fractal theory to quantitatively represent the characteristics of blasting-induced cracks in rock mass. The effects of in-situ stress levels and lateral coefficients on the modes and fractal dimensions of blasting-induced cracks are investigated by numerical simulation combined with image processing. The results show that with the increase of in-situ stresses and lateral coefficients, the fractal dimensions of blast-induced cracks gradually decrease. In addition, quantitative relations between the fractal dimensions of blasting-induced cracks and in-situ stresses are established to provide the reference for the design of rock blasting.

1. Introduction

The demands on mineral resources increasingly grow and significantly promotes the development of underground mining toward the deep earth [1]. However, the high in-situ stresses in deep earth greatly hinders the connection of the blast-induced cracks between different blasting boreholes, thereby easy to cause poor fragmentation of excavated rock [2,3]. Therefore, it is essential to investigate the effects of in-situ stresses on the blasting-induced cracks to give the reference for the blasting design in deep mining and improve the mining efficiency in deep rock mass.

Many studies [4–7] have examined the actions of in-situ stresses on the initiation, propagation and coalescence of blasting-induced cracks. For instance, Xie et al. [8] numerically investigated the effects of in-situ stresses/confining stresses on the forming results of cracks induced by cut blasting in deep rock mass. The results showed that the lengths of blasting-induced cracks decreased with increasing hydrostatic stress. It is noted that previous studies mainly focused on the qualitative discussion on the effects of in-situ stresses on blasting-induced cracks of rock mass. Very limited study has provided quantitative relation between in-situ stresses and blasting-induced crack levels.

To quantitatively represent the levels of blasting-induced cracks, the fractal theory has been widely used in existing studies [9]. For example, Yang et al. [10,11] calculated the fractal dimensions of blasting-induced cracks with varying the charge weight and eccentric charge blasting. The results indicated the fractal dimensions well characterized the levels of blasting-induced cracks. In addition, Ding et al. [12,13] experimentally investigated the effects of decoupled coefficients, coupled media, and uniaxial stress on the fractal dimensions of blast-induced cracks. However, the uniaxial stress cannot indicate the real in-situ stress states of deep rock mass. Currently, the effects of in-situ stresses on the characteristics of blasting-induced cracks has not been well investigated. The quantitative relation between them is still lacking.

In this study, The fractal characteristics of blast-induced cracks under different in-situ stresses are investigated using LS-DYNA. Firstly, the numerical model of rock mass with a single borehole is established and is calibrated based on the results of a small-scale blasting test. Then, the modes of blast-induced cracks under different confining stresses (i.e., in-situ stresses) and lateral stress coefficients are examined. The fractal dimensions of blasting-induced cracks of rock mass are quantitively compared. Finally, a quantitative relation between fractal dimensions of the blast-induced cracks and in-situ stresses is established.

2. Material model and validation

In LS-DYNA, many material models can be used to simulate the behavior of rock mass subjected to blasting loading. Amongst, the RHT model is popular one [3,8], which is used in this study. The details of RHT model can refer to Borrvall and Riedel [14]. In this study, a small-scale blasting test of cylinder rock subjected to internal blasting loads conducted by Banadaki [15] is utilized to calibrate the model. The rock mass employed in the test is a cylindrical granite sample with a diameter of 144 mm. The PETN explosive with a diameter of 1.65 mm is filled at the center of the granite sample. A two-dimensional numerical model with the same size as the granite sample in the test is established in this study. The parameters of the RHT model to simulate the granite are given in Table 1. *EOS_JWL and *MAT_HIGH_EXPLOSIVE_BUREN are respectively used as the equation of state and material models of the PETN explosive. Its density, velocity of detonation and bursting pressure are determined as 1320 $kg \cdot m^{-3}$, 6690 $m \cdot s^{-1}$ and 16 GPa, respectively. *EOS_JWL parameters of the explosive, i.e., A=586 GPa, B=21.6 GPa, R_1=5.81, R_2=1.77, ω=0.282, E_0=7.38 GPa are mainly determined based on Banadaki [15]. Figure 1(a) and (b) respectively show the measured and processed results of blast-induced cracks modes. Figure 1(c) shows the simulated results. It can be found that the distributions of simulated blast-induced cracks have good agreements with the experimental results. Therefore, it is concluded that the RHT model can accurately predict the blasting-induced response of rock mass.

Table 1. The parameters of RHT model for granite.

Parameter	Value	Parameter	Value
Mass density (kg/m^3)	2660	Break compressive strain rate	3×10^{25}
Elastic shear modulus (GPa)	21.9	Break tensile strain rate	3×10^{25}
Relative shear strength	0.18	Lode angle dependence factor Q0	0.68
Relative tensile strength	0.04	Lode angle dependence factor B	0.01
Parameter for polynomial EOS T1 (GPa)	35.27	Compressive yield surface parameter	0.53
Parameter for polynomial EOS T2 (GPa)	0	Tensile yield surface parameter	0.70
Damage parameter D1	0.04	Crush pressure (MPa)	125
Damage parameter D2	1.0	Compaction pressure (GPa)	6
Hugoniot polynomial coefficient A1 (GPa)	35.27	Shear modulus reduction factor	0.5
Hugoniot polynomial coefficient A2 (GPa)	39.58	Eroding plastic strain	2.0
Hugoniot polynomial coefficient A3 (GPa)	9.04	Minimum damaged residual strain	0.01
Failure surface parameter A	1.60	Porosity exponent	3.0
Failure surface parameter N	0.61	Initial porosity	1.0
Residual surface parameter AF	1.60	Pressure influence on plastic flow in	0.001

Journal of Physics: Conference Series **2738** (2024) 012002 doi:10.1088/1742-6596/2738/1/012002

		tension	
Residual surface parameter NF	0.61	Tensile strain rate dependence exponent	0.036
Parameter for polynomial EOS B0	1.22	Compressive strength (MPa)	167.80
Parameter for polynomial EOS B1	1.22	Compressive strain rate dependence exponent	0.032
Reference compressive strain rate	3×10^{-5}	Gruneisen gamma	0
Reference tensile strain rate	3×10^{-6}		

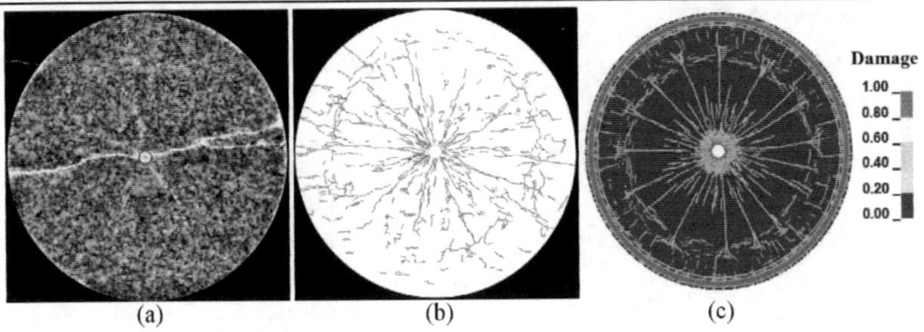

Figure 1. Measured and simulated blasting-induced cracks: (a) measured result [15], (b) processed test result [15] and (c) simulated cracks.

3. Numerical model and result discussion

3.1. Numerical model

With the calibrated model, a two-dimensional numerical model of rock mass with a single borehole is established in this section, as shown in Figure 2. The size of the numerical model is 5 m in width and 5 m in height. According to the specification of drilling machine, the diameter of the borehole located in the centre of the model is 42 mm. The explosive filled in the borehole has a diameter of 32 mm. The solid elements with gradually increased sizes from 2 mm to 20 mm are used to mesh the numerical model. The fluid-structure coupling algorithm is adopted to control the interaction of explosive, air and rock. The partial rock domain overlap with air domain to ensure the blasting loads to pass into rock domain. Non-reflective boundary conditions are applied to the model boundaries using the keyword *BOUNDARY_NON_REFLECTING. A quasi-static method [16] is utilized to apply in-situ stresses to the numerical model. The application time of the pre-stress is determined to be 80 ms, and the explosive is detonated at 90 ms.

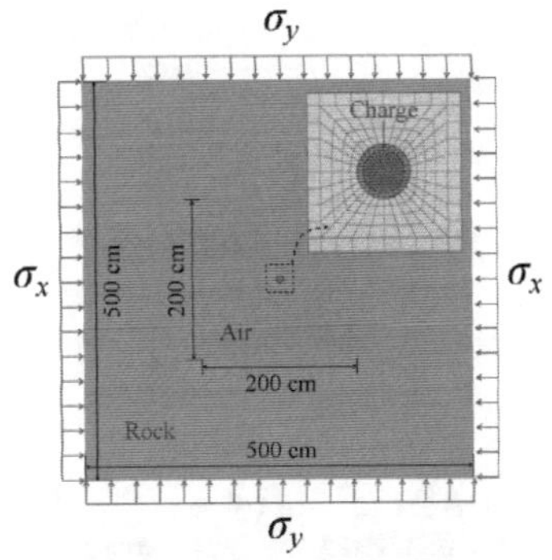

Figure 2. Configurations of the numerical model.

3.2. Modes of blast-induced cracks under different in-situ stresses

Figure 3 shows the modes of blast-induced cracks under confining stress of 0 MPa, 10 MPa, 20 MPa, 30 MPa and 40 MPa. It can be seen from that the confining stress significantly affects the length and distribution of the blast-induced cracks. With the increase of confining stress, the lengths of blasting-induced cracks decrease in vertical and horizontal directions. In addition, the lengths of the blast-induced cracks in the two directions are similar in the same case. It may be because the circumferential stress around the borehole is isotropic under equal bidirectional stress. The results indicate that the increasing the confining stress is disadvantageous to expand blasting-induced cracks.

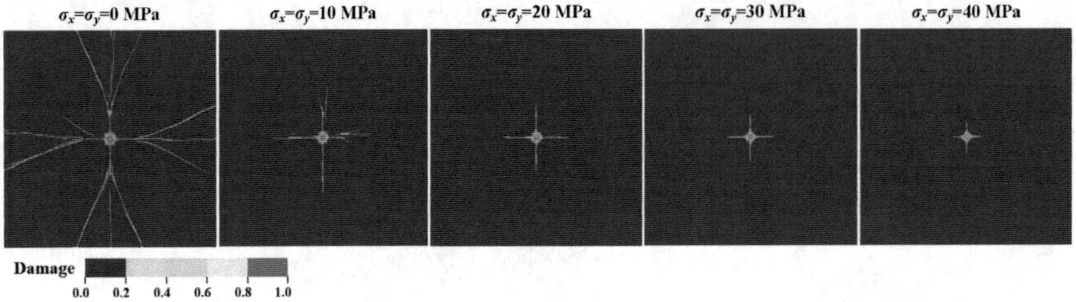

Figure 3. Modes of blast-induced cracks under different confining stresses.

The modes of blast-induced cracks of rock mass under same vertical in-situ stress (i.e., 20 MPa) but different lateral stress coefficients (i.e., 0.25, 0.5, 1.0, 1.5, and 2.0) are shown in Figure 4. It can be seen that the constraint action of in-situ stresses on the levels of blasting-induced cracks is enhanced as the lateral stress coefficient increases from 0.25 to 2.0, i.e., increased horizontal in-situ stress. In addition, the blasting-induced cracks are easier to propagate along the direction of the maximum principal stress. It is because that the constraints of in-situ stresses on the blasting-induced cracks are smallest along the direction of maximum principal stress when compared other directions.

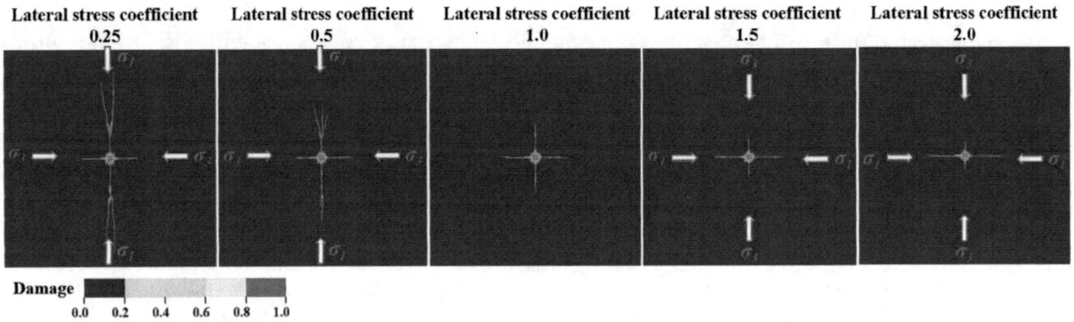

Figure 4. Modes of blast-induced cracks under different lateral stress coefficients.

3.3. *Fractal characteristic of blast-induced cracks under different in-situ stresses*

To quantify the blasting-induced cracks, the box-counting fractal dimension approach is further introduced to describe the cracks. Simulated blast-induced cracks in Sections 3.1 and 3.2 are binarized by image processing and the results are shown in Figure 5.

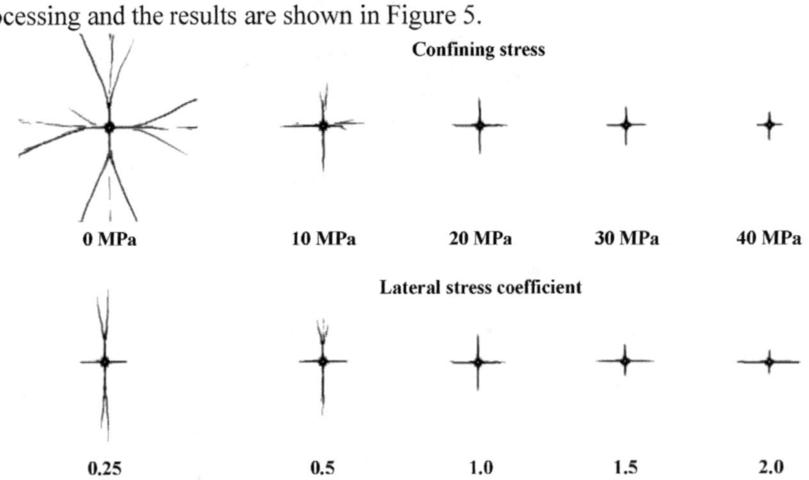

Figure 5. Binary image of the blast-induced cracks under different confining stresses and lateral

stress coefficients.

To improve the accuracy of fractal dimensions, the resolution of binary image in Figure 5 is enhanced to 6181px × 6181px. The box-counting fractal dimension can be calculated as

$$D = \lim_{k \to \infty} \frac{\lg N(\delta_k)}{-\lg \delta_k} \tag{1}$$

where D is the box-counting fractal dimension, δ_k is the side length of the square box, $N(\delta_k)$ is the minimum number of square boxes covering blasting-induced cracks. Figure 6(a) and (b) show the fractal dimensions of blast-induced crack under different confining stresses and lateral stress coefficients. It can be seen that the fractal dimensions of the blast-induced cracks change with the variations of confining stresses and lateral stress coefficients. The fractal dimensions of blasting-induced cracks nonlinearly decrease with the increase of confining stresses and lateral stress coefficients. Meanwhile, the quantitative relation between the fractal dimensions of blasting-induced cracks and in-situ stresses are established as given in Figure 6(a) and (b).

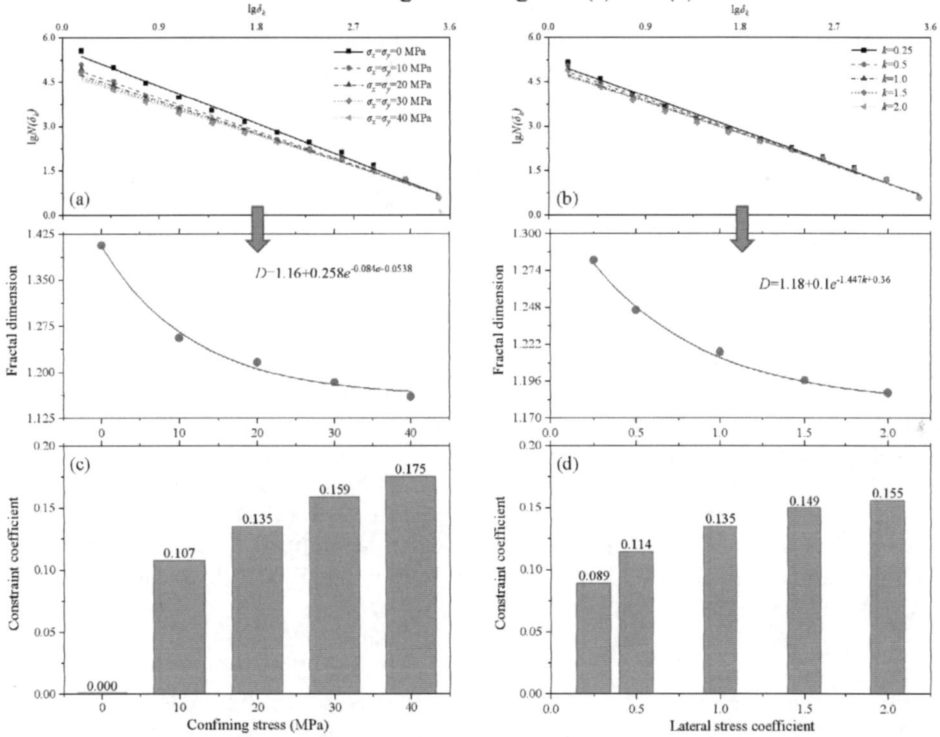

Figure 6. Fractal dimensions under different (a) confining stresses and (b) lateral stress coefficients, constraint coefficient under different (c) confining stresses and (d) lateral stress coefficients.

To quantify the constraint of in-situ stress on the blast-induced cracks, the constraint coefficient is defined as

$$C_{clamp} = \frac{D_0 - D_{stress}}{D_0} \tag{2}$$

where C_{clamp} is the constraint coefficient, D_0 is the fractal dimension of blast-induced cracks without in situ stress, D_{stress} is the fractal dimension of blast-induced cracks with given in-situ stress. The constraint coefficients under different confining stresses and lateral stress coefficients are calculated using Eq. (2) and the results are shown in Figure 6(c) and (d). It can be found that from Figure 6(c) and (d) the freezing effect for the blast-induced cracks becomes more obvious as the increase of in-situ stress. The quantitative relations between in-situ stresses and fractal dimensions of blasting-induced cracks and the defined constraint coefficient can give the reference for the fine blasting design to excavate the deep rock mass.

4. Conclusion

The present study investigates the modes and fractal dimension characteristics of blasting-induced cracks in rock mass. The effects of in-situ stress levels and lateral stress coefficients on the characteristics of blasting-induced cracks are examined. The main conclusions are summarized as follows.

(1) Blasting-induced crack levels of rock mass reduce with the increase of confining stresses. Cracks are prone to develop towards the direction of maximum in-situ stresses.

(2) The quantitative relations between the fractal dimensions of blast-induced cracks and the in-situ stresses are established, which can be used to guide blasting design in practice.

Acknowledgments

This research was funded by the Project (2020-24) supported by the Key science and technology project of Guangxi transportation industry and the Project (2023ZZTS0516) supported by the Fundamental Research Funds for the Central Universities of Central South University.

References

[1] Zhou Z, Wang P, Cai X, Cao W 2023 Influence of Water Content on Energy Partition and Release in Rock Failure: Implications for Water-Weakening on Rock-Burst Proneness *Rock Mech. Rock Eng.* **56** 6189–6205

[2] Wang H, Wang Z, Wang J, Wang S, Wang H, Yin Y and Li F 2021 Effect of confining pressure on damage accumulation of rock under repeated blast loading *Int. J. Impact Eng.* **156** 103961

[3] Li X, Liu K, Yang J and Song R 2022 Numerical study on blast-induced fragmentation in deep rock mass *Int. J. Impact Eng.* **170** 104367

[4] Kutter H K and Fairhurst C 1971 On the fracture process in blasting *Int. J. Rock Mech. Min. Sci.* **8** 181–202

[5] Rossmanith H P, Knasmillner R E, Daehnke A and Mishnaevsky L 1996 Wave propagation, damage evolution, and dynamic fracture extension. Part II. Blasting *Mater. Sci.* **32** 403–10

[6] Yi C, Johansson D and Greberg J 2018 Effects of in-situ stresses on the fracturing of rock by blasting *Comput. Geotech.* **104** 321–30

[7] Li X, Zhu Z, Wang M, Wan D, Zhou L and Liu R 2021 Numerical study on the behavior of blasting in deep rock masses *Tunn. Undergr. Space Technol.* **113** 103968

[8] Xie L X, Lu W B, Zhang Q B, Jiang Q H, Chen M and Zhao J 2017 Analysis of damage mechanisms and optimization of cut blasting design under high in-situ stresses *Tunn. Undergr. Space Technol.* **66** 19–33

[9] Xie H 1996 *Introduction to fractal-rock mechanics* (Science Press)

[10] Yang R, Xiao C, Chen C, Ding C, Zheng C, Zhao Y and Zhao Z 2020 Experimental study on blasting dynamic caustics of different charges based on fractal theory *Vib. Shock Chin.* **39** 80-86+93

[11] Yang R, Xiao C, Li Y, Xu P, Ding C, Zheng C and Kang Y 2020 Fractal study on blasting damage of uncoupled eccentric charging structures *Vib. Shock Chin.* **39** 129–34

[12] Ding C, Yang R, Lei Z, Wang M, Zhao Y and Lin H 2021 Fractal damage and crack propagation in decoupled charge blasting *Soil Dyn. Earthq. Eng.* **141** 106503

[13] Ding C, Yang R and Yang L 2021 Experimental results of blast-induced cracking fractal characteristics and propagation behavior in deep rock mass *Int. J. Rock Mech. Min. Sci.* **142** 104772

[14] Borrvall T and Riedel W 2011 The RHT concrete model in LS-DYNA *Proc. 8th European LS-DYNA Users Conference*

[15] Dehghan Banadaki M M and Mohanty B 2012 Numerical simulation of stress wave induced fractures in rock *Int. J. Impact Eng.* **40–41** 16–25

[16] Yang J, Liu K, Li X and Liu Z 2020 Stress initialization methods for dynamic numerical simulation of rock mass with high in-situ stress *J. Cent. South Univ.* **27** 3149–62

Rockburst classification based on cross reconstruction learning under small-sample condition

Xin Cai * & Liye Chen & Zilong Zhou & Hang Yuan & Peiyu Wang & Ao Zheng
School of Resources and Safety Engineering, Central South University, Changsha 410083, China

*Corresponding author
Email: xincai@csu.edu.cn

Abstract: Rockburst is a prevalent geological hazard in deep geotechnical engineering, and its accurate prognostication is vital for prevention measures. Consequently, this research proffers a pioneering classification prediction methodology, namely Cross Reconstruction Learning (CR), underpinned by conventional machine learning algorithms and metric learning strategies. Initially, this technique partitions and restructures the original dataset, where each sample feature intersects and reconfigures with features from other samples within the set. During this amalgamation, new samples are assigned labels based on the degree of divergence or congruity between two sets of sample labels, thereby forming a new set of samples. Subsequently, an array of machine learning algorithms is utilized to train and test this new dataset. Ultimately, employing a universal class voting mechanism and decoding test set results through probability assignment, the predicted labels are converted back into rock burst outcomes, thereby determining the final prediction classification. The proposed model was trained on a database encompassing 239 instance samples, and its performance was validated against the currently proficient models (KNN, XGBoost, and Random Forest algorithms) employed in rock burst prediction. The outcome revealed a decline in the performance metrics of all three machine learning algorithms when interfaced with the Cross Reconstruction learning method, particularly the KNN algorithm, owing to the doubled feature dimensions in the combined dataset. However, the metrics of ensemble models, XGBoost and Random Forest, exhibited a notable improvement compared to the original classification models. On comparing multiple performance metrics, it was discovered that the CR-XGBoost model outperformed others across all evaluations, thereby offering significant guidance for practical engineering applications.

Keywords: Rockburst classification; Small-Sample Dataset; Machine learning; XGBoost; Deep mining

1 Introduction

Rockburst is characterized as a dynamic instability catastrophe wherein, under substantial geostress conditions in subterranean engineering, rocks abruptly discharge stored elastic strain energy due to excavation or alternative load disturbances, culminating in rock exfoliation, fragmentation, and ejection[1]. Incidences of rockburst have been documented extensively in numerous lead-zinc mines across China. For instance, with the escalation in mining depth at the Huize Lead-Zinc Mine, geostress-related calamities such as rockbursts have proliferated, inflicting conspicuous equipment damage[2]. During the deep stope backfilling procedure at the Fankou Lead-Zinc Mine, the rock mass under significant geostress was perturbed by blasting, thereby exacerbating rockburst manifestations and gravely impairing operational safety[3]. As shallow mineral resources are being depleted, deep mining is progressively emerging as the mainstay in mineral resource exploitation[4], thereby imposing severe challenges on rockburst mitigation[5]. Therefore, accomplishing precise predictions of rockbursts becomes indispensable for the advancement of future subterranean spatial engineering.

At present, numerous scholars have employed machine learning methodologies for rockburst prediction. For instance, Zhou et al. [6] constructed a database rooted in 132 rockburst instance data and utilized the Particle Swarm Optimization (PSO) algorithm to expedite the optimization velocity of SVM algorithm hyperparameters, discerning that the hybrid model possessed robust performance. X. Yin et al. [7] harnessed data mining techniques to attenuate dimensions, identify outliers, and replace outliers in the gathered 246 sets of rockburst data. Subsequently, they employed the Stacking ensemble method for the amalgamation and comparative analysis of disparate models. The outcomes indicated that the Stacking ensemble method possesses distinct benefits when engaging with unbalanced data. The accuracy and dependability of rockburst prediction methodologies predicated on machine learning predominantly hinge on the quantity and quality of rockburst instances[8]. However, the majority of existing research is grounded in one to two hundred data sets, thereby generating less reliable prediction models. Given that the number of rockburst instances is unlikely to experience a significant surge in the short term, augmenting the prediction accuracy of models based on the existing modest volume of data emerges as an urgent issue in contemporary research.

To mitigate the aforementioned challenge, this investigation introduces a unique prediction methodology - Cross Reconstruction Learning (CR), which is rooted in metric learning strategies[9] and conventional machine learning algorithms. This investigation deploys models (KNN, XGBoost, and Random Forest algorithms) that have evidenced efficacy in prior rockburst prediction research [10-12], integrating them with the proposed Cross Reconstruction technique to fabricate a novel hybrid prediction model. Assessment and comparison of assorted prediction schemes suggest that the proposed Cross Reconstruction strategy efficaciously amplifies the predictive precision of the model.

2 Dataset description

Predicated on the rockburst instances gathered by J. Zhou et al. [26], discrepancies and missing data were rectified and augmented, culminating in a total of 239 comprehensive rockburst cases. These served to construct a rockburst class prediction database. The collected rockburst instances are classified into four classes - none rockburst (39 cases), light rockburst (71 cases), moderate rockburst (97 cases), and strong rockburst (32 cases) - contingent upon the severity of the occurrence and the destruction characteristics. The dataset also manifests a class imbalance phenomenon.

The initial database is partitioned into a training set (comprising 80% of the samples) and a test set (encompassing 20% of the samples) for model instruction and performance validation. To address the issue of class sample imbalance, stratified sampling techniques are employed to segregate the original database. A semi-violin plot is utilized to analyze each feature variable in the training and testing sets, as illustrated in Figure 1. The figure manifests the scatter distribution, kernel density distribution, and additional information of each feature variable in the training set and test set separated from the original database. Notably, the data distribution disparities of each feature variable in the segregated training and test sets are comparatively minor. Considering that anomalies can transpire in actual operational conditions, this paper utilizes data inclusive of outliers for model training and prediction.

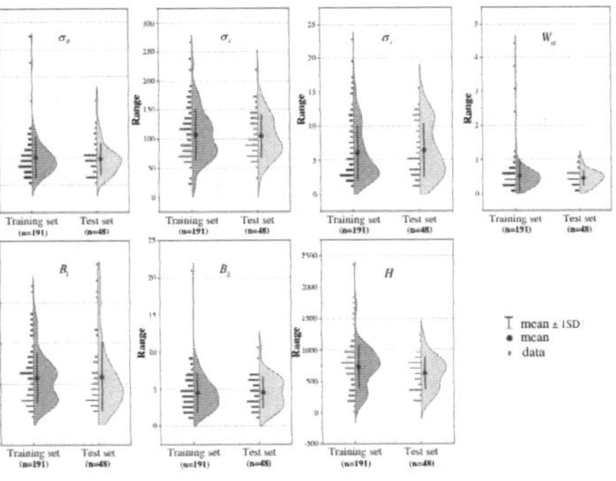

Figure 1. Comparative distribution of each feature variable in the training and test Sets.

3 Methods

3.1 Cross reconstruction learning

Cross reconstruction learning method. The principle is shown in Figure 2, with detailed introduction as follows:

(1) Construct combined training set

Firstly, the training set is duplicated, and the feature variable names in the original traini ng set and the duplicated training set are modified to (H_1,$\sigma_{\theta1}$,σ_{c1},σ_{t1},W_{et1},B_{11},B_{21})and(H_2,$\sigma_{\theta2}$,σ_{c2},σ_{t2},W_{et2},B_{12},B_{22}), respectively. The samples in the original training set and the duplicated training set undergo cross-reconstruction, with the class encoding of the new samples determin ed based on the similarity or discrepancy in the amalgamated sample categories, if the two sa mple classes belong to the same class, the new sample class label is designated as 1, whereas if the two sample classes pertain to distinct categories, the label is set as 0. Consequently, e ach new sample data comprises (H_1,$\sigma_{\theta1}$,σ_{c1},σ_{t1},W_{et1},B_{11},B_{21},H_2,$\sigma_{\theta2}$,σ_{c2},σ_{t2},W_{et2},B_{12},B_{22},Y), w here Y can assume a value of 0 or 1. Given that the research is a multi-classification proble m, the quantity of new samples with class 0 will considerably surpass the quantity of new sa mples with class 1. Random undersampling techniques are employed to randomly eliminate sa

Journal of Physics: Conference Series **2738** (2024) 012003 doi:10.1088/1742-6596/2738/1/012003

mples from the larger class. The residual new samples collectively formulate the combined trai ning set.

(2) Construct combined test set.

Initially, the feature variable names of the original test set are altered to (H_2,$\sigma_{\theta 2}$,σ_{c2},σ_{t2}, W_{et2},B_{12},B_{22}). If the original training set exhibits class imbalance whilst constructing the com bined test set, it can influence subsequent decoding work. Therefore, it is requisite to apply ra ndom undersampling to the original training set to ensure an equitable quantity of samples acr oss all categories. Subsequently, the processed original training set and original test set sample s are cross-reconstructed individually. The new samples thus generated collectively constitute t he combined test set, and the class labels of the new samples are sourced from ensuing mode l predictions.

(3) Decoding of predicted class labels in the combined test set

The samples from the combined test set and their corresponding prediction outcomes are merged to constitute the query test set. For a sample in the original test set, all samples that serve as the (H_2,$\sigma_{\theta 2}$,σ_{c2},σ_{t2},W_{et2},B_{12},B_{22}) component in the query test set are retrieved. Subse quently, the corresponding categories in the original training set are tallied for each sample (H_1,$\sigma_{\theta 1}$,σ_{c1},σ_{t1},W_{et1},B_{11},B_{21}) when the predicted class Y=1. The frequencies of each class are computed using formulas (1)~(2), and the class with the highest frequency is selected as the r ockburst classification outcome for the test set sample. This process is iterated until all predict ion outcomes for the original test set are procured.

$$P_i = \frac{r_{(i=k|Y=1)}}{t} \text{ (k=0,1,2,3)} \tag{1}$$

$$y_{pre} = y_{max(P_0,P_1,P_2,P_3)} \tag{2}$$

where, P_i is the probability of class ;r is the frequency of class appearing in the new samples where(H_1,$\sigma_{\theta 1}$,σ_{c1},σ_{t1},W_{et1},B_{11},B_{21}) part has a prediction result of 1; k is the class label; Y is the predicted class; t is the quantity of each class sample in the original training set after undersampling; y_{pre} is the rock burst prediction class for sample x; y_{max} is the rockburst class corresponding to the maximum value among P_0~P_3.

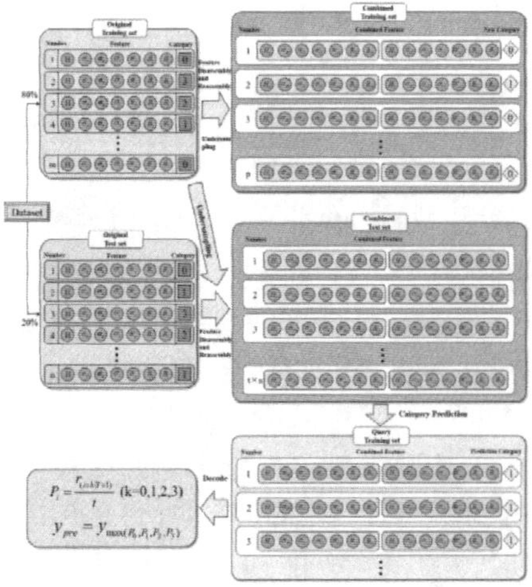

Figure 2. Schematic diagram of the principle of cross-reconstruction learning.

3.2 Model assessment indicators

To ensure objectivity in the ultimate prediction result comparison, all models predict categories from the original test set, which remains entirely independent throughout the model training procedures. Several evaluation metrics are selected for comparative assessment, including Accuracy, Precision, Recall, and F1-score. Given the existence of sample imbalances among various categories in the original test set, Precision, Recall, and F1 scores are evaluated using the Macro-average method to appraise prediction outcomes.

4 Results and discussion

The results of each evaluation metric are delineated in Table 1. It is discernible that, upon the implementation of the cross-reconstruction learning methodology, the predictive performance of the KNN model has experienced a minor decline. The cause for this can be attributed to the fact that subsequent to the construction of the combined training set, the feature dimensionality has been amplified twofold, engendering a larger distance between each sample in the high-dimensional space. This is not conducive to the KNN model's training and prediction of test set class labels through the measurement of distance between samples. The values of each evaluation metric for the ensemble models XGBoost and RF have seen considerable augmentation, signifying that the cross-reconstruction learning methodology has performed more proficiently for models predicated on decision tree ensembles, thereby validating the efficacy of the data combination methodology in enhancing model accuracy. Among them, the data combination classification prediction scheme predicated on the XGBoost model performed optimally in all metric evaluations (ACC=0.81; PRE_M=0.86; REC_M=0.75; $F1_M$=0.78).

Table 1. Performance evaluation of models.

Model	ACC	PRE_M	REC_M	$F1_M$
KNN	0.71	0.82	0.67	0.71
CR-KNN	0.69	0.77	0.65	0.68
XGBoost	0.75	0.80	0.71	0.73
CR-XGBoost	0.81	0.86	0.75	0.78
RF	0.71	0.72	0.67	0.68
CR-RF	0.75	0.81	0.69	0.70

5 Conclusions

This manuscript proposes a classification prediction methodology—Cross-Reconstruction Learning—that effectively augments the prediction precision of models dealing with datasets possessing small sample sizes. Through the comparison and analysis of prediction outcomes from various model prediction schemes, it was observed that the utilization of the Cross Reconstruction Learning methodology resulted in a performance decrement in the KNN model due to an amplification in feature dimensions. Conversely, the ensemble models XGBoost and RF demonstrated significant enhancements in all prediction assessment results, thereby validating the stability and efficacy of the proposed Cross Reconstruction Learning methodology. Furthermore, following a multi-metric comparison, the hybrid CR-XGBoost model exhibits the most favorable prediction outcomes (ACC=0.81; PRE_M=0.86; REC_M=0.75; $F1_M$=0.78), thereby providing invaluable guidance for practical rockburst prediction in engineering undertakings.

References

[1] WANG J A, PARK H D. Comprehensive prediction of rockburst based on analysis of strain energy in rocks[J]. Tunnelling and Underground Space Technology, 2001, 16(1): 49–57.

[2] Yongfu Qiu. Construction technology of ore chute under complex conditions in deep well [J]. World Nonferrous Metals, 2016(22): 121-123. (in Chinese)

[3] Guiqiang Yuan. Analysis and prevention of rock burst geological factors in deep part of Fankou Lead-Zinc Mine [J]. Metallurgical Collection, 2016, No.223(03): 37-42. (in Chinese)

[4] Wu, S., Wu, Z., Zhang, C., 2019. Rock burst prediction probability model based on case analysis. Tunnelling and Underground Space Technology 93, 103069.

[5] Feng, X., Zhao, H., 2002. Prediction of rockburst using support vector machine (in Chinese). J. Northeastern Univ. (Natural Science) 1, 57–59.

[6] ZHOU J, LI X B, SHI X Z. Long-term prediction model of rockburst in underground openings using heuristic algorithms and support vector machines[J]. Safety Science, 2012, 50(4): 629–644.

[7] YIN X, LIU Q S, PAN Y C, et al. Strength of stacking technique of ensemble learning in rockburst prediction with imbalanced data: comparison of eight single and ensemble models[J]. Natural Resources Research, 2021, 30(2): 1 795–1 815.

[8] LI N, JIMENEZ R, FENG X D. The influence of Bayesian networks structure on rock burst hazard prediction with incomplete data[J]. Procedia Engineer, 2017, 191: 206–214.

[9] Yuanyuan Shen, Yan Yan, Hanzi Wang. Progress in supervised distance metric learning algorithms [J]. Journal of Automation, 2014, 40(12): 2673-2686. (in Chinese)

[10] TANG Zhili, XU Qianjun. Research on rockburst prediction based on 9 machine learning algorithms[J]. Chinese Journal of Rock Mechanics and Engineering, 2020, 39(4): 773–781.(in Chinese)

[11] Wenkan Tan, Nanyan Hu, Yicheng Ye, et al. Rock burst intensity grading prediction based on four major integrated learning [J]. Chinese Journal of Rock Mechanics and Engineering, 2022, 41(S2): 3250-3259. (in Chinese)

[12] ZHOU J, LI X B, MITRI H S. Classification of rockburst in underground projects: comparison of ten supervised learning methods[J]. Journal of Computing in Civil Engineering, 2016, 30(5): 04016003.

Effects of seawater on the adsorption of xanthate onto galena and sphalerite

Ningbo Song [1, 2, 3], **Jin Yao** [1, 2,*] and **Wanzhong Yin** [1, 2]

1 School of Resources and Civil Engineering, Northeastern University, Shenyang 110819, China

2 State Key Laboratory of Mineral Processing, Beijing 102628, China

3 BGRIMM Technology Group，Beijing 100160, China

*Corresponding author: Jin Yao, Email: yaojin@mail.neu.edu.cn

ABSTRACT. Seawater contains divalent calcium and magnesium cations. Under alkaline conditions, calcium and magnesium ions react with hydroxide ions to form insoluble hydroxyl complexes or hydroxide precipitates. The hydrophilic substances that may be adsorbed on the mineral surface during the flotation process hinder the adsorption of the collector, affecting mineral hydrophobicity, and thus reducing the floatability of the mineral. In this study, the effects of seawater on the adsorption of xanthate onto galena and sphalerite were investigated. The results show that under strong alkaline conditions, seawater has significant and slight adverse effects on sphalerite and galena, respectively. Flotation regulators such as ethylenediamine tetraacetic acid, sodium hexametaphosphate, and sodium silicate can eliminate the adverse effect on galena and sphalerite flotation to a certain extent. The mechanisms were revealed through microflotation experiments, contact angle measurements, bubble-particle attachment tests, zeta potential measurements, and XPS analysis.

Keywords: Seawater flotation; galena; sphalerite; hydroxyl complexes; regulator

1. Introduction

Pb and Zn are widely used in metallurgy, national defense, automotive, and electronics [1, 2]. Galena and sphalerite are the main sources of Pb and Zn extracted by these industries and often coexists in nature [3, 4]. Galena is naturally hydrophobic, making it suitable for flotation separation [5-10]. The floatability of sphalerite, on the other hand, depends on its impurities, particularly the iron content. Generally, sphalerite is a sulfide mineral that usually does not exhibit flotation. The high floatability of sphalerite is attributed to the natural activation of heavy metal ions, especially copper ions, or mineral oxidation [11-14].

In the industrial practice of separating Pb and Zn through flotation, lime is commonly used as a pH regulator and inhibitor of sphalerite because the calcium in lime forms hydrophilic hydroxyl complexes that are adsorbed on the sphalerite surface under alkaline conditions, reducing its floatability. Seawater contains significant amounts of Mg^{2+} and Ca^{2+} [15-19], which can act as natural and inexhaustible inhibitors of sphalerite. Seawater flotation of Pb and Zn can conserve freshwater resources and reduce the amount of lime required. Therefore, it has attracted considerable research interest.

To analyze the effect of seawater on the flotation of galena and sphalerite, microflotation test was initially conducted using DI water and then repeated using seawater employing the same process and flotation reagent conditions. Three reagents, namely ethylenediamine tetraacetic acid (EDTA), sodium hexametaphosphate (SHMP) and sodium silicate (SS), were used to regulate Pb and Zn flotation in seawater. Among these reagents, EDTA was a chelating agent of calcium and magnesium ions [20, 21], while SHMP [22-24] and SS [25, 26] were the dispersants. By comparing the test results, the flotation mechanism was discussed through microflotation tests, contact angle measurements, bubble-particle attachment tests, XPS analysis, and zeta potential measurements.

2. Materials and methods

2.1. Materials

The pure galena and sphalerite used in the experiments were obtained from Guilin, Guangxi and Taolin, Hunan, China. After crushing and screening, the ore sample was prepared into particle sizes of -74+38 μm and -38 μm. Samples with a particle size of -74+38 μm were individually packed into small vacuum bags, each weighing 20.0 g, for X-ray diffraction (XRD), XPS analysis, and microflotation tests. Furthermore, the -38 μm sample was further ground to -5 μm and used for zeta potential and FTIR measurements. An XRD analysis was performed to determine the chemical composition and purity of the sample. The results are presented in Figure 1 and Table 1. Artificially simulated seawater was prepared refer to the findings of Kester et al. [27]. The concentrations of the major ions are listed in Table 2.

Table 1. Pure galena and sphalerite compositions analysis (wt. %).

Contents	Pb	Zn	S	Else
Galena	85.40	–	13.20	1.40
Sphalerite	–	65.80	32.85	1.35

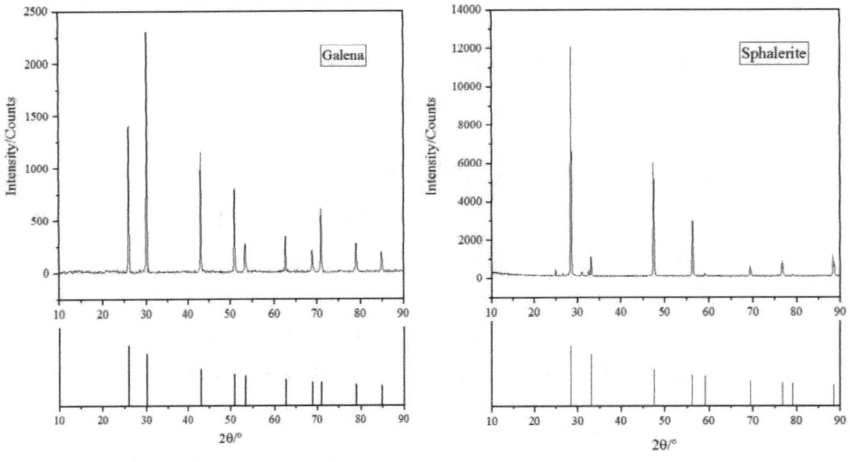

Figure 1. XRD pattern of pure galena and sphalerite.

Table 2. The concentrations of main ions in artificial simulated seawater (g/L).

Ions	Na^+	Mg^{2+}	K^+	Ca^{2+}	Cl^-	SO_4^{2-}	HCO_3^-
Concentration	11.60	1.31	0.50	0.41	19.58	2.62	0.15

2.2. Microflotation tests

Flotation test was carried on a XFG model machine produced by Jilin Exploration Machinery Factory, China. The spindle speed set to 1992 RPM. Potassium butyl xanthate (PBX) was used as collector and methyl isobutyl carbinol (MIBC) was used as frother. HCl and NaOH were used as pH-adjusting agents. EDTA, SHMP and SS were used as regulators. Pure samples (2.0 g) were added to a 30.0 ml flotation tank, and the pulp within the flotation tank was continuously stirred to ensure uniform dispersion of the samples in the solution. After one minute of stirring, the pH value of the pulp was adjusted, followed by the addition of the corresponding regulator at two-minute intervals. Next, PBX (stirring for 3 min) and MIBC (stirring for 1 min) were added, and the bubbles were manually scraped for 3 min. The concentrate and tailings were collected, dried, weighed, and sequentially tested to calculate the recovery. The test process is illustrated in Figure 2.

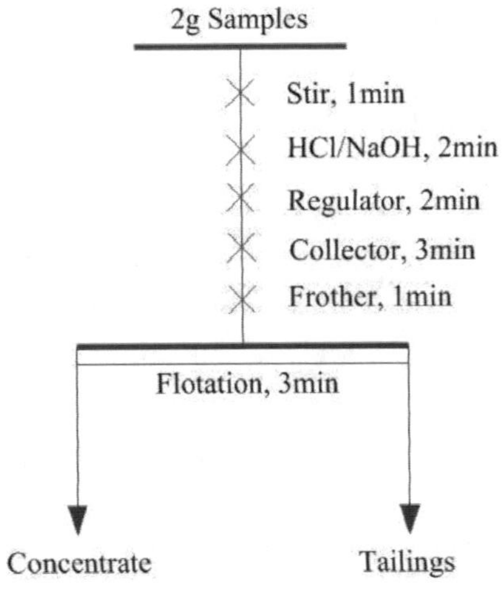

Figure 2. Flowchart of micro-flotation test.

2.3. Contact angle measurements

Contact angle measurements are commonly used to determine the surface wettability of minerals. The most frequently-used sample preparation method for measuring contact angles of powder samples is the Powder Tablet Method [28-30]. The contact angles of the galena and sphalerite were measured in DI water, seawater, and seawater containing EDTA, SHMP, and SS. For each measurement, adding 2.0 g pure mineral to a 30.0 ml solution. After treatment like those in the microflotation test, sample was dried in a vacuum oven at 40℃ for 24h and then pressed into 1-2 mm wafer under 10.0 MPa pressure for two minutes. A contact angle measuring instrument (JC2000A, Powereach, Shanghai, China) was used to measure the contact angle. Water with a particle diameter of approximately 2–3 mm was dropped onto the sample surface using a micro syringe. Each measurement was performed three times and the mean and standard deviation values were calculated [31, 32].

2.4. Bubble-particle attachment tests

A self-constructed visual measurement system was used to measure the bubble–particle attachment. The test system was primarily composed of a high-speed camera, vibration exciter, light source, lifting table, and computer. The samples were prepared in a square glass tank measuring 5 × 5 × 5 cm^3. The preparation process and dosage were based on the flotation tests. After preparation, the samples were transferred to a lifting table, and the height was adjusted for subsequent tests. During the test, one end of the capillary tube was connected to the micro syringe, while the other end was placed below the liquid level of the square glass tank. Bubbles with a diameter of 2 mm were created using a micro syringe. The bubbles moved downward in each test, making contact with the mineral particle bed, and then returned to their initial position. The above was a complete round of testing. The device displacement was adjusted to ensure that each bubble was in contact with the sample particle bed, and different dwell times were set for each test. The process of mineral particles adhering to the bubbles was observed using a high-speed dynamic camera. Each experiment was repeated three times, and representative images were selected for analysis.

2.5. Zeta potential distribution measurements

A Nano ZS90 nanometer particle potentiometer (Malvern Company) was used to measure the surface potential of the minerals before and after interaction with the reagents. The specific operation steps were as follows: For each test, a 30.0 mg sample with a particle size of -5 μm was placed in 30.0 ml KCl solution with a concentration of 1×10^{-3} mol/L, and the container was placed on a magnetic stirrer for continuous agitation. The test process and dosing system were consistent with the flotation test. After each test, the solution in the container was allowed to stand for 10 min, and the upper clarified solution was extracted for zeta potential analysis. Each group of samples was measured thrice, and the average value and standard deviation were calculated as the final test results.

2.6. X-ray photoelectron spectroscopy (XPS) analysis test

The processing method of XPS samples were referred to flotation test. After sufficient interaction with flotation agents according to the flotation process, the ore samples were filtered washed and vacuum dried for XPS testing. XPS analysis was performed using the Thermo KAlpha (ESCALAB 250xi) X-ray photoelectron spectrometer. The size of Al Kα microfocus monochromatic source was 5 μm, and the data acquisition power was 72 W. The vacuum degree in the analysis chamber should not exceed 2×10^{-7} mbar. In the full-spectrum analysis, the pass energy and energy scan step size were 100 eV and 1 eV, respectively, and in the narrow-spectrum analysis, they were 50 eV and 0.1 eV. The acquired spectra were calibrated using standard C1s peaks (284.8eV), and XPS measurement data were analyzed using Thermo Vantage software.

3. Results and discussions

3.1. Microflotation results

Figure 3 shows that in deionized water, the galena recovery remained stable above 90% within the pulp pH value range of 2–8, with the highest value of 94.65% (pH = 5.98). In addition, sphalerite recovery was low, with a maximum of 72.40% at pH 5.99, after which it began to decline. In seawater, galena recovery reached its highest value of 96.95% at a pH of 5.98, but sharply dropped when the pH exceeded 10.0. The sphalerite recovery was relatively low, with the maximum value of only 60.50% (pH = 5.97). It is evident that seawater has an inhibitory effect on both galena and sphalerite flotation, particularly on the latter.

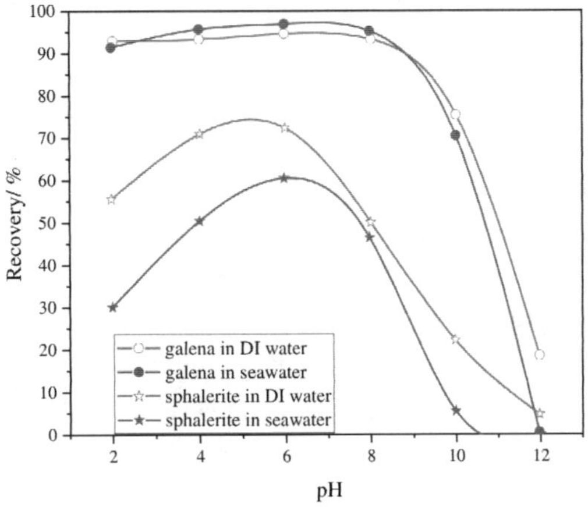

Figure 3. Effect of pH value on galena and sphalerite flotation (C_{PBX}=50mg/L, C_{MIBC}=50 mg/L).

The separation of Pb and Zn is usually performed under alkaline conditions, which adversely affect the floatability of sulfide minerals in seawater. Therefore, this study investigated the changes in galena and sphalerite recovery after adding three regulators (EDTA, SHMP, and SS) to both DI water and seawater at a pH of 10.0, as shown in Figure 4. After adding three reagents, the recoveries of galena and sphalerite were increased to different extent. The variation range of galena is small while that of sphalerite is large. The results show that the three reagents are effective in eliminating the adverse effects of seawater on galena and sphalerite flotation.

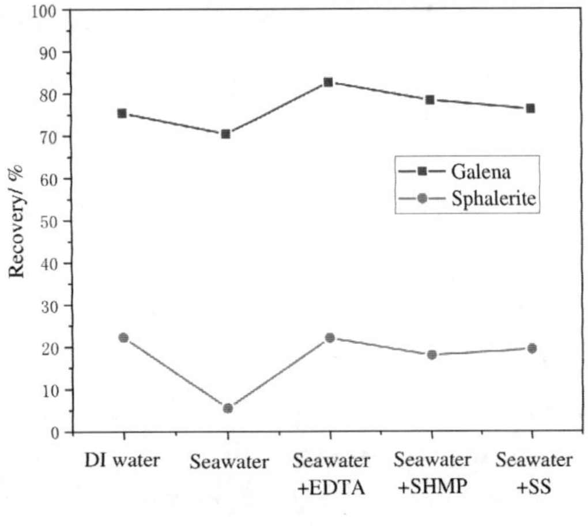

Figure 4. Effect of regulator on galena and sphalerite in seawater flotation (pH=10.0 , C_{PBX}=50mg/L, C_{MIBC}=50 mg/L, C_{EDTA}=60mg/L, C_{SHMP}=60 mg/L C_{SS}=60mg/L).

3.2. Contact angle determination

The contact angle is an important parameter for measuring the wettability of a liquid on a mineral surface. It reflects the floatability of the minerals to a certain extent [30]. Figure 5 shows the contact angles of galena and sphalerite in DI water, seawater, and seawater with EDTA, SHMP, and SS before and after interaction with the collector PBX.

Compared with DI water, the contact angles in seawater were smaller, indicating that hydrophilic substances were adsorbed onto the mineral surface. The contact angle of galena in seawater decreased to 66.53°. However, after the addition of EDTA, SHMP, and SS, the contact angles became 63.85°, 66.57°, and 69.85°, respectively, indicating that these three regulators had minute influence on the contact angle of galena in seawater. The contact angle of the sphalerite in seawater was reduced to 40.67°. However, after the addition of EDTA, SHMP, and SS, the contact angles changed to 61.02°, 65.41°, and 62.57°, respectively, indicating that the three regulators significantly increased the contact angle of sphalerite in seawater. The collector can increase mineral hydrophobicity, leading to an increase in the contact angle of the mineral. In seawater, the change in the galena contact angle was significantly greater than that of sphalerite after the addition of the collector PBX, indicating a higher adsorption amount of the collector PBX on the galena surface compared to that on sphalerite.

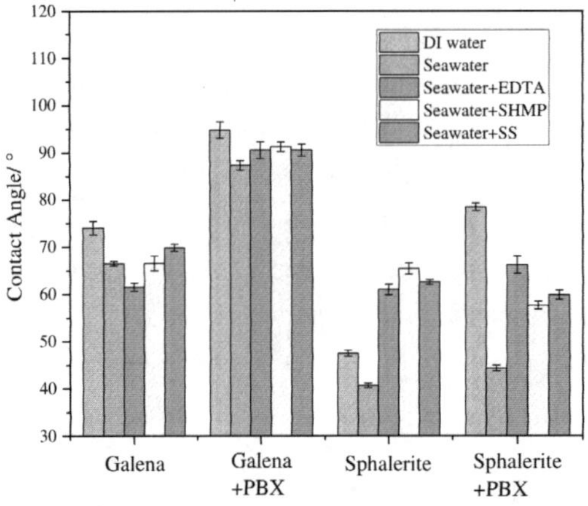

Figure 5. Contact angles of galena and sphalerite in DI water, seawater, seawater with EDTA, seawater with SHMP and seawater with SS. (pH=10.0, C_{PBX}=50 mg/L, C_{EDTA}=60mg/L, C_{SHMP}=60 mg/L C_{SS}=60mg/L).

3.3. Bubble-particle attachment analysis

Contact angle measurements and bubble-particle attachment tests are commonly used to characterize mineral hydrophobicity [33, 34]. Among these tests, the bubble-particle attachment test is a kinetic measure of hydrophobicity, making it more suitable for describing flotation behavior [35-38]. Figures 6 and 7 show the bubble-particle attachment under the three systems at dwell time 10, 20 and 30 ms.

Mineral particle adhesion on the bubble increased with a longer dwell time. In addition, at the same dwell time, galena particles have similar adhesion with bubbles in different systems. However, the adhesion of sphalerite particles to bubbles in seawater was significantly lower than that in DI water and was significantly improved when EDTA was added to seawater, indicating that seawater has a stronger inhibition on the adhesion of sphalerite particles to bubbles.

Figure 6. The bubble-galena particle attachment at different dwell time in DI water (a, b, c), seawater (d, e, f), seawater with EDTA (g, h, i) (C_{PBX}=50 mg/L, C_{EDTA}=60mg/L, pH=10.0).

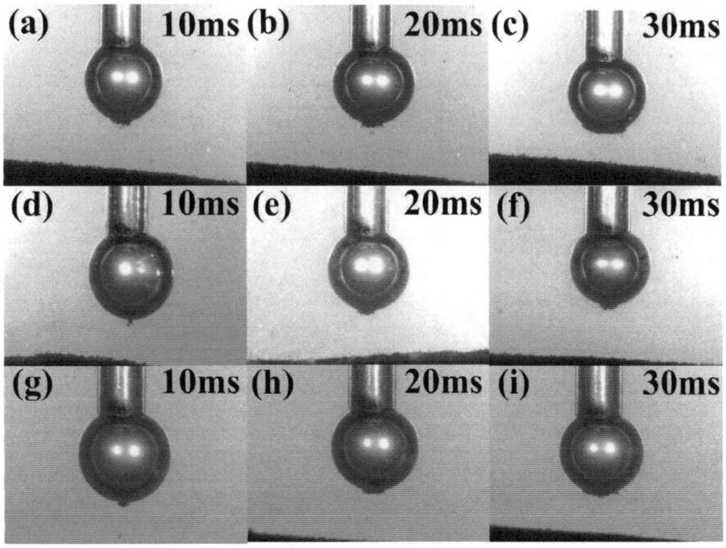

Figure 7. The bubble-sphalerite particle attachment at different dwell time in DI water (a, b, c), seawater (d, e, f), seawater with EDTA (g, h, i) (C_{PBX}=50 mg/L, C_{EDTA}=60mg/L, pH=10.0).

3.4. Zeta potential distribution analysis

The adsorption of chemical substances on the mineral interface can affect their electrochemical properties, which is directly manifested as the alteration of zeta potential [39-42]. These substances include hydroxyl complexes, ions, and other molecules. Figure 8 shows the zeta potentials of galena and sphalerite in different systems at a pH of 10.0.

In seawater, the zeta potentials of the galena and sphalerite shifted positively, indicating adsorption of positively charged substances onto the surface of the minerals. The zeta potential of galena increased from -17.67 mV to -11.27 mV, with an increase of 6.40 mV. Similarly, the zeta potential of sphalerite increased from -25.48 mV to -12.89 mV, with an increase of 12.59 mV, which was significant higher than that of galena, indicating higher adsorption amount of positively charged substances on the sphalerite surface compared to galena. However, after the addition of EDTA, SHMP and SS, the Zeta potentials of both minerals shifted negatively, indicating that the addition of the three regulators reduced the adsorption of positively charged substances on the mineral surface, but did not return to the value of DI water. When the PBX collector was added to the DI water system, the zeta potentials of both minerals decreased significantly, indicating that the anionic PBX collector was strongly adsorbed onto the mineral surface. In the seawater system, after the addition of the collector, the galena zeta potential decreased from -11.27 mV to -20.74 mV (reduction was 9.47 mV), and the sphalerite zeta potential decreased from -12.89 mV to -18.93 mV (reduction was 6.04 mV), which was lower than that of galena. Therefore, the adsorption capacity of PBX on the sphalerite surface was much lower than that of galena in seawater. This is consistent with the aforementioned conclusion regarding contact angle.

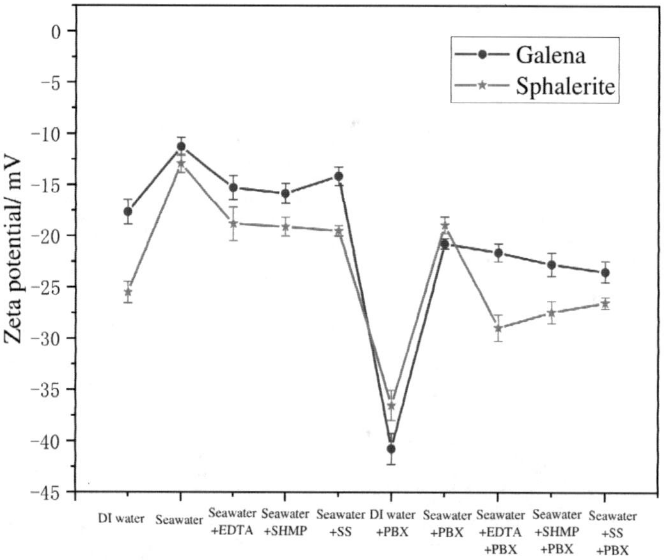

Figure 8. Zeta potential of galena and sphalerite in different flotation systems when pH=10.0.

3.5. XPS analysis

Tables 3 and 4 show the surface element contents of galena and sphalerite in the different systems, respectively. Notably, the Mg content of galena in seawater was 0.43%, and the O content increased from 17.95% to 24.15%, indicating that small amounts of substances containing Mg and O were adsorbed on the galena surface. Furthermore, the Mg content increased to 0.89% after the addition of EDTA. Additionally, the increase in the N content indicated that the EDTA-complexes were adsorbed on the galena surface. However, after the addition of SHMP and SS, the Mg content decreased to 0.24% and 0.31%, respectively, indicating that the addition of SHMP and SS reduced the adsorption of Mg-containing substances.

Similarly, the Mg content of the sphalerite in the seawater was 0.52%, and the O content increased from 10.62% to 14.77%, indicating that substances containing Mg and O were adsorbed on the sphalerite surface. In addition, the Mg content on the sphalerite surface was higher than that of galena, indicating that more hydroxyl complexes containing Mg were adsorbed on the sphalerite surface. Furthermore, the Mg content increased to 0.89% after the addition of EDTA. Similarly, the increase in the N content indicated that the EDTA-complexes were adsorbed on the sphalerite surface. After adding SHMP and SS, the Mg content decreased to 0.25% and 0.38%, respectively, indicating that the addition of SHMP and SS reduced the adsorption of Mg-containing substances. Notably, the decrease in the C and S contents on the surfaces of both sulfide minerals in seawater may be owing to the inhibition of the PBX adsorption on their surfaces.

Table 3. Atomic concentration on the galena surface.

Samples	Atomic concentration/%						
	Pb 4f	S 2p	C 1s	O 1s	Mg 1s	Ca 2p	N 1s
DI water+PBX	20.59	23.87	37.59	17.95	–	–	–
Seawater+PBX	18.20	22.66	34.46	24.15	0.43	0.10	–
Seawater+EDTA+PBX	14.02	12.66	44.90	25.57	0.89	0.16	1.80
Seawater+SHMP+PBX	21.13	22.37	38.49	17.75	0.24	0.02	–
Seawater+SS+PBX	19.97	23.63	37.16	18.86	0.31	0.07	–

Table 4. Atomic concentration on the sphalerite surface.

Samples	Atomic concentration/%						
	Zn 2p	S 2p	C 1s	O 1s	Mg 1s	Ca 2p	N 1s
DI water+PBX	30.26	24.61	34.51	10.62	–	–	–
Seawater+PBX	28.15	22.90	33.62	14.77	0.52	0.04	–
Seawater+EDTA+PBX	22.26	17.30	40.00	15.32	0.66	0.13	4.33
Seawater+SHMP+PBX	29.13	22.34	35.49	12.76	0.25	0.03	–
Seawater+SS+PBX	27.97	23.63	35.78	12.19	0.38	0.05	–

3.6. Discussion

PBX is an anionic heteropolar-sulfhydryl collector. When $C_5H_9CSS^-$ dissociates in water, it reacts with Pb atoms on the galena surface, producing lead xanthate through a process known as chemisorption [43]. However, several studies have suggested that the primary mechanism of xanthate-adsorbing sphalerite is the formation of double xanthate on its surface. At pH 10, the seawater mainly generated

$Mg(OH)^+$ complexes, including a small amount of $Ca(OH)^+$ complexes and hydroxide precipitation. The coatings of such hydrophilic substances on the mineral surface prevent adsorption of the collector, thereby reducing mineral recovery. The addition of EDTA can produce complex reactions with calcium and magnesium ions to reduce the ion content in seawater. The addition of SHMP and SS can produce precipitate reactions with calcium and magnesium ions. On the other hand, SHMP and SS can disperse particles, weaken the attraction between complexes and mineral particles, and reduce their adsorption onto mineral surfaces [44, 45]. These agents can reduce the adverse effects of seawater on the adsorption of xanthate on both galena and sphalerite surfaces. The contact Angle of sphalerite after adding PBX in seawater is much smaller than that in DI water, indicating that seawater greatly affects the adsorption of PBX on sphalerite surface. Bubble particle attachment analysis can more directly reflect the decrease degree of hydrophobicity of galena and sphalerite in seawater. Zeta potential distribution analysis showed that the potential drop of galena was more obvious when PBX was added in seawater. XPS analysis shows that there is more calcium and magnesium content on sphalerite surface in seawater. The above studies show that the adsorption of hydroxyl complexes on galena surface was less than that on sphalerite surface. Accordingly, the adverse effect of seawater on galena flotation was slighter than that of sphalerite.

Galena and sphalerite are the most representative sulfide minerals of Pb and Zn, respectively. Flotation separation of galena and sphalerite is a common process in industry, which generally adopts preferential flotation, that is, flotation of galena first, then flotation of sphalerite. The application of seawater in the flotation separation of galena and sphalerite can reduce the consumption of fresh water resources. This study shows that seawater inhibits galena and sphalerite to different extents. On this basis, the effect of seawater on the flotation separation of galena and sphalerite is worthy of further research.

4. Conclusions

This study investigated the effect of seawater on galena and sphalerite flotation.

- Seawater has a significantly adverse effect on sphalerite flotation and a slightly adverse effect on galena flotation.
- The adverse effect is mainly due to the formation of hydroxyl complexes on the surface of minerals under alkaline conditions, which affects the adsorption of the collector PBX on the galena and sphalerite surfaces.
- The slight effect of seawater on galena flotation is due to the weak adsorption of hydrophilic substances on the surface of galena, and the characteristics of seawater to promote foaming make up for the loss of galena hydrophobicity.
- The adverse effects can be eliminated to some extent by the use of EDTA, SHMP, and SS, although their mechanisms differ. EDTA mainly reduces the content of magnesium and calcium ions in seawater flotation through chelation. Conversely, SHMP and SS mainly alter the forces of the hydroxyl complex and mineral particles, thereby reducing the adsorption of complexes on mineral particles.

Reference:

[1] Y. Chen, B. Feng, Y. Guo, T. Wang, L. Zhang, C. Zhong, H. Wang, The role of oxidizer in the flotation separation of chalcopyrite and galena using sodium lignosulfonate as a depressant, *Minerals Engineering*, 172 (2021) 107160.

[2] Y. Miao, S. Wen, Z. Shen, Q. Feng, Q. Zhang, Flotation separation of chalcopyrite from galena using locust bean gum as a selective and eco-friendly depressant, *Separation and Purification Technology*, 283 (2022) 120173.

[3] J. Li, K. Song, D. Liu, X. Zhang, Z. Lan, Y. Sun, S. Wen, Hydrolyzation and adsorption behaviors of SPH and SCT used as combined depressants in the selective flotation of galena from sphalerite, *J. Mol. Liq.*, 231 (2017) 485-490.

[4] H. Zhu, B. Yang, R. Martin, H. Zhang, D. He, H. Luo, Flotation separation of galena from sphalerite using hyaluronic acid (HA) as an environmental-friendly sphalerite depressant, *Minerals Engineering*, 187 (2022) 107771.

[5] W. Qin, Q. Wei, F. Jiao, C. Yang, R. Liu, P. Wang, L. Ke, Utilization of polysaccharides as depressants for the flotation separation of copper/lead concentrate, *International Journal of Mining Science and Technology*, 23 (2013) 179-186.

[6] B. Yang, H. Yan, M. Zeng, P. Huang, F. Jia, A. Teng, A novel copper depressant for selective flotation of chalcopyrite and molybdenite, *Minerals Engineering*, 151 (2020) 106309.

[7] X. Zhang, L. Lu, H. Zeng, Z. Hu, Y. Zhu, L. Han, A macromolecular depressant for galena and its flotation behavior in the separation from molybdenite, *Minerals Engineering*, 157 (2020) 106576.

[8] R. Liu, H. Lu, Z. Xu, C. Wang, W. Sun, M. Wu, Y. Dong, L. Bai, New insights into the reagent-removal mechanism of sodium sulfide in chalcopyrite and galena bulk flotation: a combined experimental and computational study, *Journal of Materials Research and Technology*, 9 (2020) 5352-5363.

[9] C. Su, J. Cai, X. Yu, R. Peng, Q. Zheng, Y. Ma, R. Liu, P. Shen, D. Liu, Effect of pre-oxidation on galena in the flotation separation of chalcopyrite from galena with calcium lignosulfonate as depressant, *Minerals Engineering*, 182 (2022) 107520.

[10] X.R. Zhang, Y.G. Zhu, G.B. Zheng, L. Han, B. McFadzean, Z.B. Qian, Y.C. Piao, C. O'Connor, An investigation into the selective separation and adsorption mechanism of a macromolecular depressant in the galena-chalcopyrite system, *Minerals Engineering*, 134 (2019) 291-299.

[11] A.P. Chandra, A.R. Gerson, A review of the fundamental studies of the copper activation mechanisms for selective flotation of the sulfide minerals, sphalerite and pyrite, *Advances in Colloid and Interface Science*, 145 (2009) 97-110.

[12] B. Xu, J. Wu, Z. Dong, T. Jiang, Q. Li, Y. Yang, Flotation performance, structure–activity relationship and adsorption mechanism of a newly-synthesized collector for copper sulfide minerals in Gacun polymetallic ore, *Applied Surface Science*, 551 (2021) 149420.

[13] J.S. Laskowski, Q. Liu, Y. Zhan, Sphalerite activation: Flotation and electrokinetic studies, *Minerals Engineering*, 10 (1997) 787-802.

[14] J. Liu, M. Ejtemaei, A.V. Nguyen, S. Wen, Y. Zeng, Surface chemistry of Pb-activated sphalerite, *Minerals Engineering*, 145 (2020) 106058.

[15] Y. Chen, X. Chen, Y. Peng, The depression of molybdenite flotation by sodium metabisulphite in fresh water and seawater, *Minerals Engineering*, 168 (2021) 106939.

[16] G. Ai, K. Huang, C. Liu, S. Yang, Exploration of amino trimethylene phosphonic acid to eliminate the adverse effect of seawater in molybdenite flotation, *International Journal of Mining Science and Technology*, 31 (2021) 1129-1134.

[17] A. Ramirez, L. Gutierrez, J.S. Laskowski, Use of "oily bubbles" and dispersants in flotation of molybdenite in fresh and seawater, *Minerals Engineering*, 148 (2020).

[18] Y.B.A. Li, W.Q. Li, Q. Xiao, N. He, Z.J. Ren, C. Lartey, A.R. Gerson, The Influence of Common Monovalent and Divalent Chlorides on Chalcopyrite Flotation, *Minerals*, 7 (2017).

[19] W. Li, Y. Li, Improved understanding of chalcopyrite flotation in seawater using sodium hexametaphosphate, *Minerals Engineering*, 134 (2019) 269-274.

[20] W. Xiao, C. Fang, J. Wang, Q. Liang, P. Cao, X. Wang, L. Zhang, G. Qiu, J. Hu, The role of EDTA on rutile flotation using Al(3+) ions as an activator, *RSC Adv*, 8 (2018) 4872-4880.

[21] J. Tian, L. Xu, W. Sun, H. Han, X. Zeng, S. Fang, K. Hong, Y. Hu, The selective flotation separation of celestite from fluorite and calcite using a novel depressant EDTA, *Powder Technology*, 352 (2019) 62-71.

[22] L.Z.Z. Wang, Y.B. Li, R.H. Fan, R. Fan, Influencing mechanisms of sodium hexametaphosphate on molybdenite flotation using sea water, *Physicochemical Problems of Mineral Processing*, 55 (2019) 1091-1098.

[23] W. Li, Y. Li, Q. Xiao, Z. Wei, S. Song, The Influencing Mechanisms of Sodium Hexametaphosphate on Chalcopyrite Flotation in the Presence of MgCl2 and CaCl2, *Minerals*, 8 (2018).

[24] W. Li, Y. Li, Z. Wei, Q. Xiao, S. Song, Fundamental Studies of SHMP in Reducing Negative Effects of Divalent Ions on Molybdenite Flotation, *Minerals*, 8 (2018).

[25] A. Ramirez, A. Rojas, L. Gutierrez, J.S. Laskowski, Sodium hexametaphosphate and sodium silicate as dispersants to reduce the negative effect of kaolinite on the flotation of chalcopyrite in seawater, *Minerals Engineering*, 125 (2018) 10-14.

[26] E. Rebolledo, J.S. Laskowski, L. Gutierrez, S. Castro, Use of dispersants in flotation of molybdenite in seawater, *Minerals Engineering*, 100 (2017) 71-74.

[27] D.R. KESTER, I.W. DUEDALL, D.N. CONNORS, R.M. PYTKOWICZ, PREPARATION OF ARTIFICIAL SEAWATER1, 12 (1967) 176-179.

[28] Z. Zhu, W. Yin, D. Wang, H. Sun, K. Chen, B. Yang, The role of surface roughness in the wettability and floatability of quartz particles, *Applied Surface Science*, 527 (2020) 146799.

[29] H. Sun, B. Yang, Z. Zhu, W. Yin, Q. Sheng, Y. Hou, J. Yao, New insights into selective-depression mechanism of novel depressant EDTMPS on magnesite and quartz surfaces: Adsorption mechanism, DFT calculations, and adsorption model, *Minerals Engineering*, 160 (2021) 106660.

[30] N.B. Song, C.Y. Sun, W.Z. Yin, J. Yao, B. Yang, Exploration on flotation behavior of galena in seawater and related mechanism, *Physicochemical Problems of Mineral Processing*, 58 (2022).

[31] B. Yang, Z. Zhu, H. Sun, W. Yin, J. Hong, S. Cao, Y. Tang, C. Zhao, J. Yao, Improving flotation separation of apatite from dolomite using PAMS as a novel eco-friendly depressant, *Minerals*

Engineering, 156 (2020) 106492.

[32] Q. Sheng, W. Yin, B. Yang, S. Cao, H. Sun, Y. Ma, K. Chen, Improving surface sulfidization of azurite with ammonium bisulfate and its contribution to sulfidization flotation, *Minerals Engineering*, 171 (2021) 107072.

[33] Y. Ye, S.M. Khandrika, J.D. Miller, Induction-time measurements at a particle bed, *International Journal of Mineral Processing*, 25 (1989) 221-240.

[34] J. Drelich, J.D. Miller, R.J. Good, The effect of drop (bubble) size on advancing and receding contact angles for heterogeneous and rough solid surfaces as observed with sessile-drop and captive-bubble techniques, *Journal of Colloid and Interface Science*, 179 (1996) 37-50.

[35] D. Fuerstenau, J. Rosenbaum, J. Laskowski, Effect of surface functional groups on the flotation of coal, *Colloids and Surfaces*, 8 (1983) 153–173.

[36] J. Finch, G.W. Smith, DYNAMIC SURFACE TENSION OF ALKALINE DODECYLAMINE ACETATE SOLUTIONS IN OXIDE FLOTATION, Transactions of the Institution of Mining and Metallurgy, Section C: *Mineral Processing and Extractive Metallurgy*, 81 (1972) c213-c218.

[37] Q. Cao, X. Wang, J.D. Miller, F. Cheng, Y. Jiao, Bubble attachment time and FTIR analysis of water structure in the flotation of sylvite, bischofite and carnallite, *Minerals Engineering*, 24 (2011) 108-114.

[38] R. Smith, R. Lai, On the relationship between contact angle and flotation behavior, Trans. *AIME*, 235 (1966) 413-418.

[39] X. Wang, B. Zhao, J. Liu, Y. Zhu, Y. Han, Dithiouracil, a highly efficient depressant for the selective separation of molybdenite from chalcopyrite by flotation: Applications and mechanism, *Minerals Engineering*, 175 (2022) 107287.

[40] C. Oliveira, J. Rubio, Zeta potential of single and polymer-coated microbubbles using an adapted microelectrophoresis technique, *International Journal of Mineral Processing,* 98 (2011) 118-123.

[41] H. Han, W. Yin, B. Yang, D. Wang, J. Yao, Z. Zhu, Adsorption behavior of sodium oleate on iron minerals and its effect on flotation kinetics, *Colloids and Surfaces A*: Physicochemical and Engineering Aspects, 647 (2022).

[42] D.W. Fuerstenau, Correlation of Contact Angles, Adsorption Density, Zeta Potentials and Flotation Rate, *Trans AIME*, 208 (1957) 1365 ~ 1367.

[43] J. Su, Study of adsorption behavior and mechanism of heteropolar sulfhydryl collectors on galena, *Central South University*, 2012.

[44] Y. Li, H. Zhu, W. Li, Y. Zhu, A fundamental study of chalcopyrite flotation in sea water using sodium silicate, *Minerals Engineering*, 139 (2019) 105862.

[45] W.Q. Li, Y.B. Li, Improved understanding of chalcopyrite flotation in seawater using sodium hexametaphosphate, *Minerals Engineering*, 134 (2019) 269-274.

Journal of Physics: Conference Series **2738** (2024) 012005 doi:10.1088/1742-6596/2738/1/012005

Influences of coordination structures on the electronic properties of Zn ion and the adsorption of O-containing and S-containing molecules

Chongjun Liu ,Guiye Wu ,Huinan Liu ,Yanhong Ma ,Tong Lu ,Yangge Zhu*

State Key Laboratory of Mineral Processing, BGRIMM Technology Group, Beijing, China, 102628.

*Email: zhuyangge@bgrimm.com

Abstract. Ligands in minerals have an important effect on the chemical properties of metal ions. The electronic properties of Zn ions formed by O and S ligands have been studied using density functional theory (DFT), and the interaction strength between O and S-containing molecules and Zn ions has been analyzed. The results show that the electronic properties of Zn ions may be influenced by the type of ligands, the number of ligands, and the distance between ligands and Zn ions. The adsorption capacity of zinc ions decreased with an the in increase ligand coordination number, but increased with an increase in the distance between the ligand and zinc ion. The adsorption of O- and S-containing molecules on sphalerite smithsonite and hemimorphite was then studied. It was indicated that O-containing molecules had a strong collecting ability for sphalerite, smithsonite and hemimorphite, but the adsorption capacity of S-containing molecules was weaker than that of O-containing molecules. The influence of water molecules on the adsorption behavior of O- and S-containing molecules was studied. The results of the calculations show that the relationship between H2O and O-containing sulfur-containing molecules is competitive adsorption on the surface of sphalerite smithsonite and hemimorphite. Adsorption of water molecules can reduce the adsorption energy of O-containing and S-containing molecules on mineral surfaces.

1. Introduction

Zinc is an important non-ferrous metal that primarily exists in nature in the forms of sphalerite, smithsonite, and hemimorphite. To extract and concentrate zinc resources, the flotation method is typically employed. Flotation is a physicochemical process that occurs at the surface of minerals[1]. The primary mechanism in the flotation process is the selective adsorption of flotation reagents on the mineral surface. The metal ions on the mineral surface act as active sites for the interaction with flotation reagents. Understanding the flotation process thoroughly and improving production parameters based on this understanding need extensive research into the mechanism of interaction between metal ions on the mineral surface and flotation reagents. To establish regularities, mineral processing experts have put up a variety of theoretical ideas.

As early as the 1930s, the solubility product hypothesis was proposed in the literature[2], which explained the interaction between collectors and metal ions on the mineral surface from a chemical reaction perspective. It suggested that the smaller the solubility product of the collector and metal ion, the easier the collector would interact with the mineral surface. In 1950, the molecular adsorption hypothesis was proposed in the literature[3], which explained the relationship between different thiol collectors and the dissociation constant in sphalerite flotation[4]. In the mid-1950s[5-6], the electrostatic interaction model was proposed in the literature, taking into account the relationship between the electrical properties of the mineral surface and the adsorption of reagents, where cationic collectors adsorb on negatively charged mineral surfaces, while anionic collectors adsorb on positively charged mineral surfaces. The literature also discovered the relationship between sulfide flotation and electrochemistry[7], leading to the development of the sulfide flotation electrochemical theory, which has been a research hotspot for half a century. It explains the role of oxygen in flotation and the electron transfer mechanism between reagents and the mineral surface.

The above theories can explain some flotation phenomena and contribute to the understanding of the flotation process. However, they still simplify the model of metal ions on the mineral surface into free ions in classical chemistry and fail to fully consider the limitations of mineral crystal chemistry and other properties. They do not provide a thorough understanding of the micro-mechanism of the interaction between flotation reagents and the mineral surface. The fundamental question of the existence state of metal ions on the mineral surface and their influence on the adsorption of flotation reagents has not been clearly understood, significantly limiting the understanding of mineral flotation mechanisms and the development of efficient reagents.

In fact, the presence and properties of metal ions on the mineral surface are different significantly from their free ion states due to the influence of surrounding ions[8-11]. For example, both Sphalerite and smithsonite surfaces contain Zn ions, but their floatability is entirely different. This indicates that the coordinating atoms surrounding the metal ions in the mineral crystal structure have a significant constraining effect, resulting in differences in the floatability of different minerals. It can be seen that it is very important to study the effect of ligands on the chemical properties of metal ions in the crystal structure of minerals to understand the difference in the adsorption properties of flotation agents on different mineral surfaces.

In recent years, the development of density functional theory in flotation has provided possibilities for exploring the influence of surface environment on the properties of metal ions. The literature has used density functional theory to study the effects of lattice defects and surface spatial structure on the properties of sulfide mineral surfaces and the adsorption of reagents[12-18]. Additionally, the hydrated environment on the mineral surface also has a significant impact on the adsorption of flotation reagents[19-20]. The aforementioned results all indicate that the interaction between metal ions on the mineral surface and reagent molecules is not only related to the properties of the metal ions themselves but also influenced by the surrounding environment.

In order to study the difference in chemical properties of Zn ions on the surface of different kinds of zinc-containing minerals, the effect of ligand structure on the interaction between Zn ions and collectors was analyzed. In this study, first-principles calculations were employed to discuss the effects of factors such as the types of coordinating atoms, the number of ligands, and the distance between ligands on the technical charge and electrophilicity of zinc ions. The differences in the adsorption of different types of collectors on the mineral surface and the influence of water molecules on the adsorption of flotation reagents were investigated. The research results provide important references and insights for theoretical studies on mineral flotation and the design of reagent molecules.

2. Computational models

Density functional theory (DFT) was used to investigate Zn oxide and sulfide minerals. The CASTEP module was utilized for electronic and structural calculations, while the DMol3 module was employed for adsorption simulations. In order to investigate the effects of the types of coordination atoms, the number of ligands and the distance between ligands and metal ions on the chemical properties of metal ions, the effects of different factors on the chemical properties of metal ions were investigated.

Mineral models of MeO and MeS were constructed by coordinating O and S ligands to simulate oxide and sulfide minerals, respectively.

Then the adsorption of collector molecules containing O and S on these minerals was simulated. In the CASTEP calculations, ultrasoft pseudopotentials were employed to evaluate the interaction between

Journal of Physics: Conference Series **2738** (2024) 012005 doi:10.1088/1742-6596/2738/1/012005

valence electrons and ions. The self-consistent field convergence accuracy was set to 2.0×10^{-6}eV/atom, and the displacement between two atoms was constrained to be less than 2.0×10^{-3}Å. Convergence thresholds for maximum energy change, force, and stress during geometry optimization were set to 2×10^{-5}eV/atom, 0.05 eV/Å, and 0.1 GPa, respectively.

For the DMol3 calculations, convergence thresholds for energy change, maximum force, maximum displacement, maximum step size, and self-consistent field were set to 1.0×10^{-5}Ha, 2.0×10^{-3}Ha·Å^{-1}, 5.0×10^{-3}Å, 0.3 Å, and 1.0×10^{-6}eV/atom, respectively. We utilized the DNP atomic orbital basis set, including all electrons in the calculation. The GGA-PBE exchange-correlation function was employed for both CASTEP and DMol3 calculations.

3. Results and discussion

3.1. Effects of O and S on the properties of Zn ion

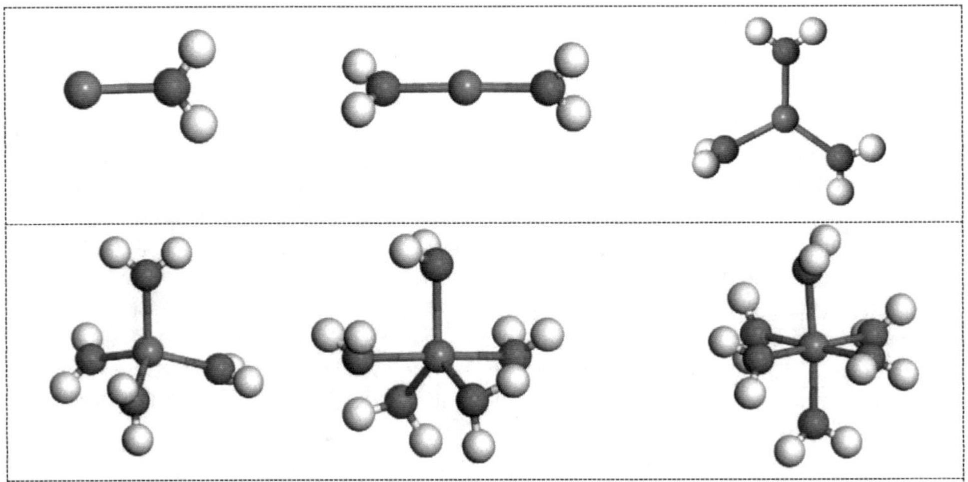

Figure 1. Models of Zn oxide minerals formed by coordinated O.

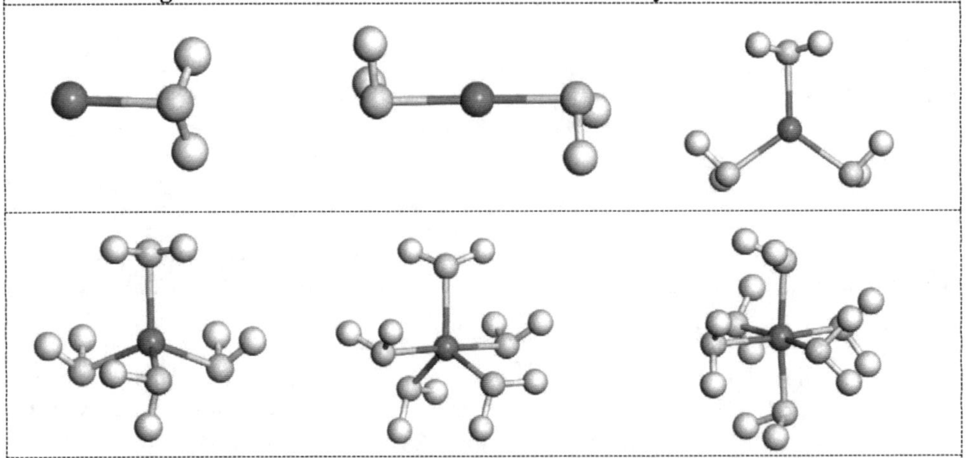

Figure 2. Models of Zn sulfide minerals formed by coordinated S.

Table 1. Parameters in Different Coordination Models of Zinc Ions.

	Charge(e)	Bond length(Å)		Charge(e)	Bond length(Å)
Zn-O	1.720	1.860	Zn-S	1.186	2.281
Zn-2O	1.600	1.894	Zn-2S	0.705	2.336
Zn-3O	1.155	1.963	Zn-3S	0.552	2.338
Zn-4O	1.056	2.018	Zn-4S	0.450	2.347
Zn-5O	0.868	2.135	Zn-5S	0.241	2.407

Table 2. Electrophilicity of Zn Ion under Different Constraint Conditions.

	Fukui		Fukui
Zn-O	0.816	Zn-S	0.627
Zn-2O	0.712	Zn-2S	0.438
Zn-3O	0.624	Zn-3S	0.365
Zn-4O	0.465	Zn-4S	0.286
Zn-5O	0.328	Zn-5S	0.072

Table 3. Electrophilicity of Zn Ion at Different Constraint Distances.

O-Bond length(Å)	Charge(e)	Fukui	S-Bond length(Å)	Charge(e)	Fukui
2.0	0.821	0.266	2.0	0.032	-0.026
2.2	0.839	0.276	2.2	0.063	-0.024
2.4	0.853	0.283	2.4	0.092	-0.020
2.6	0.862	0.287	2.6	0.118	-0.013
2.8	0.868	0.290	2.8	0.141	-0.004

The electrophilicity of a metal ion represents the strength of its interaction with negatively charged groups. Based on the electrophilicity of metal ions, the action of ligands can be discussed. The electrophilicity of metal ions can be characterized using the Fukui index. A larger value of $f+$ (M) indicates that the metal ion is more prone to attack by ligands, indicating higher electrophilicity.

Table 2 presents the calculated electrophilic indices of Zn at different coordination numbers using first-principles theoretical calculations. The electrophilicity at coordination number 0 represents the electrophilicity of the free ion. From Table 2, it can be observed that as the coordination number increases, the electrophilicity of the Zn ion decreases, and the influence of ligand O on the electrophilicity of the metal ion is weaker compared to ligand S.

Furthermore, Table 3 demonstrates that with an increase in the coordination distance, the electrophilicity of the Zn ion gradually increases, and again, the influence of the ligand O on the electrophilicity of the metal ion is weaker compared to the ligand S.

3.2. Adsorption of O- and S-containing collectors on Zn ions

For the present study, the S-containing collector(ethylxanthate (CH_3CH_2-OCSSH)) and the O-containing collector(fatty acid (CH_3CH_2-COOH)) are chosen as the adsorbent representatives. Tables 4 and 5 show the adsorption energies of S- and O-containing collector molecules.

Table 4. Electrophilicity of Zn Ion at Different Constraint Distances.

O	Adsorption energy of ethylxanthate	Adsorption energy of fatty acid	S	Adsorption energy of ethylxanthate	Adsorption energy of fatty acid
1	-1455.40	-1494.44	1	-1368.93	-1382.83
2	-1191.95	-1250.03	2	-1183.42	-1075.51
3	-1008.85	-1020.65	3	-1040.47	-958.21
4	-931.97	-953.76	4	-978.41	-969.10
5	-923.32	-894.51	5	-872.67	-892.55

Table 5. Adsorption energy of O- and S-containing collectors in the model at different constraint distances, kJ/mol.

O-Bond length(Å)	Adsorption energy of ethylxanthate	Adsorption energy of fatty acid	S-Bond length(Å)	Adsorption energy of ethylxanthate	Adsorption energy of fatty acid
2.00	-956.78	-945.02	2.00	-978.34	-1005.28
2.20	-962.78	-963.56	2.20	-990.28	-1011.27
2.40	-1002.80	-981.79	2.40	-1019.95	-1042.51
2.60	-1005.64	-1028.24	2.60	-1030.57	-1049.78
2.80	-1013.21	-1067.63	2.80	-1035.81	-1060.25

This demonstrates that O- and S-containing collectors have different binding energies with Zn ions under different coordination environments. In each coordination environment, the binding energy between fatty acids and Zn ions is greater than that between xanthate and Zn ions, indicating a stronger binding ability between fatty acids and Zn ions. Table 4 shows that as the number of ligands increases, the binding energy between O- and S-containing collectors and Zn ions decreases. This indicates that both O and S ligands reduce the electrophilic ability of Zn ions, thereby lowering the binding energy between Zn ions and O- and S-containing collectors. Moreover, as shown in Table 4, the binding energy between O- and S-containing collectors and Zn ions increases with an increase in the relative distance between ligands. This suggests that both O and S ligands, when their relative distance from Zn ions gradually increases, have a diminishing effect on Zn ions, thereby enhancing the binding energy between Zn ions and O- and S-containing collectors.
To investigate the difference in binding energy between O- and S-containing collectors and Zn ions, the adsorption configurations of Zn and O- and S-containing collectors were studied.

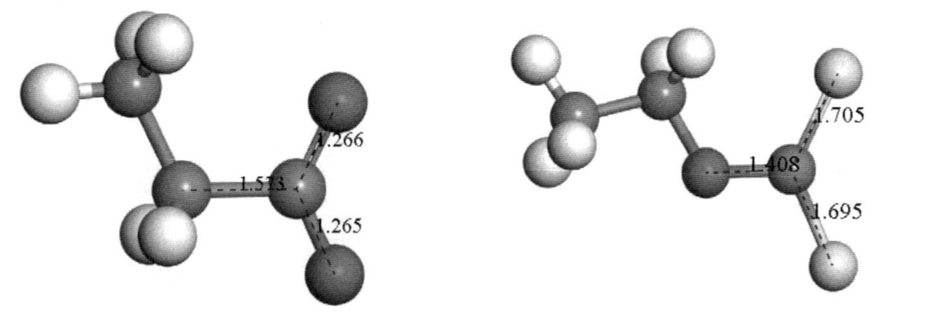

Figure 3. Models of O- and S-containing collectors.

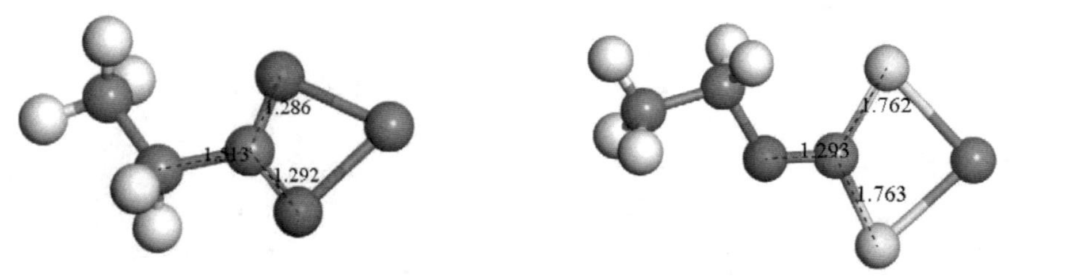

Figure 4. The adsorption models of O- and S-containing collectors with Zn ions.

Figures 3 and 4 depict the molecular configurations of the fatty acid and xanthate, which are O- and S-containing collectors, respectively, as well as the binding models of O- and S-containing collectors with Zn ions. From the figures, it can be observed that the C-O bond lengths in the fatty acid are 1.265 Å and 1.266 Å, and after adsorption onto Zn ions, the C-O bond lengths become 1.286 Å and 1.292 Å, indicating a negligible change in the molecular structure of the fatty acid upon adsorption. The positive and negative values of COOP are bonding and anti-bonding states, respectively. The value of overlap population between carbon-oxygen has been computed via integration of bonding state to Fermi level. COOP analysis reveals that the COOP value for the C-O bond before adsorption is 0.47, which decreases to 0.43 after adsorption. This suggests that there is no significant feedback effect from Zn ions on the fatty acid, as the molecular orbitals exhibit minor changes. In contrast, the C-S bond lengths in xanthate are 1.705Å and 1.695Å, and upon adsorption of Zn ions, they increase to 1.762Å and 1.763Å, indicating a noticeable change in the molecular structure of xanthate upon adsorption. The COOP value for the C-S bond before adsorption is 0.44, which decreases to 0.36 after adsorption, indicating the presence of a feedback π -bonding interaction between Zn ions and xanthate, which can reduce the strength of the C-S bond.

3.3. Adsorption of O- and S-containing collectors on oxide and sulfide minerals

Figures 5, 6, and 7 illustrate the adsorption configurations of O- and S-containing collector and water on the surfaces of sphalerite, smithsonite, and hemimorphite, respectively. The calculated adsorption energies of O- and S-containing collectors and water on the surfaces are as follows: -405.992 kJ/mol, -313.61 kJ/mol, and -324.02 kJ/mol for sphalerite; -386.46 kJ/mol, -298.67 kJ/mol and -344.65 kJ/mol for smithsonite; and -352.43 kJ/mol, -299.71kJ/mol and -311.89 kJ/mol for hemimorphite. From this result, it can be seen that the adsorption energy of water is more than the adsorption energy of S molecules on the surface of the three zinc-containing minerals and that the adsorption energy of O molecules is greater than the adsorption energy of water molecules. It is evident that the interaction between mineral surface activity sites and collector is affected by the coordination structure of mineral surface activity sites

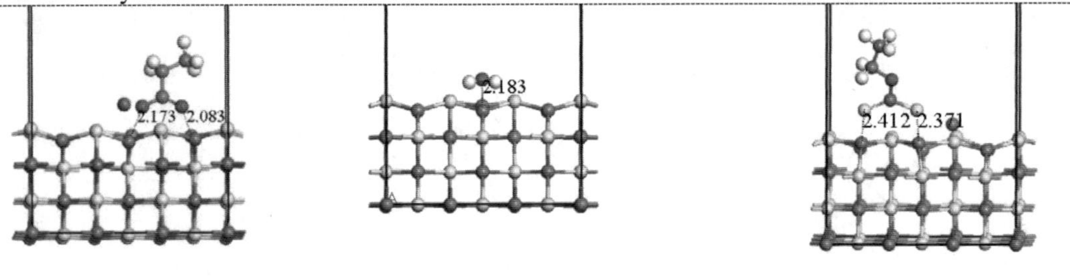

Figure 5. The adsorption configurations of O- and S-containing collectors and water on the surface of sphalerite.

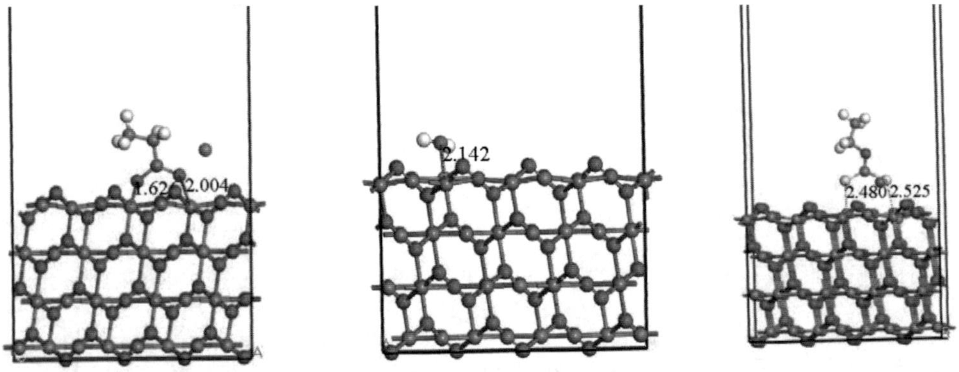

Figure 6. The adsorption configurations of O- and S-containing collectors and water on the surface of smithsonite.

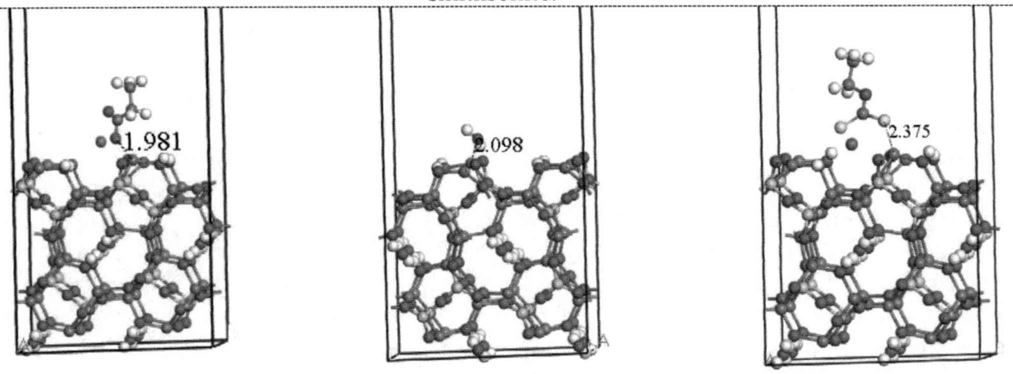

Figure 7. The adsorption configurations of O- and S-containing collectors and water on the surface of hemimorphite.

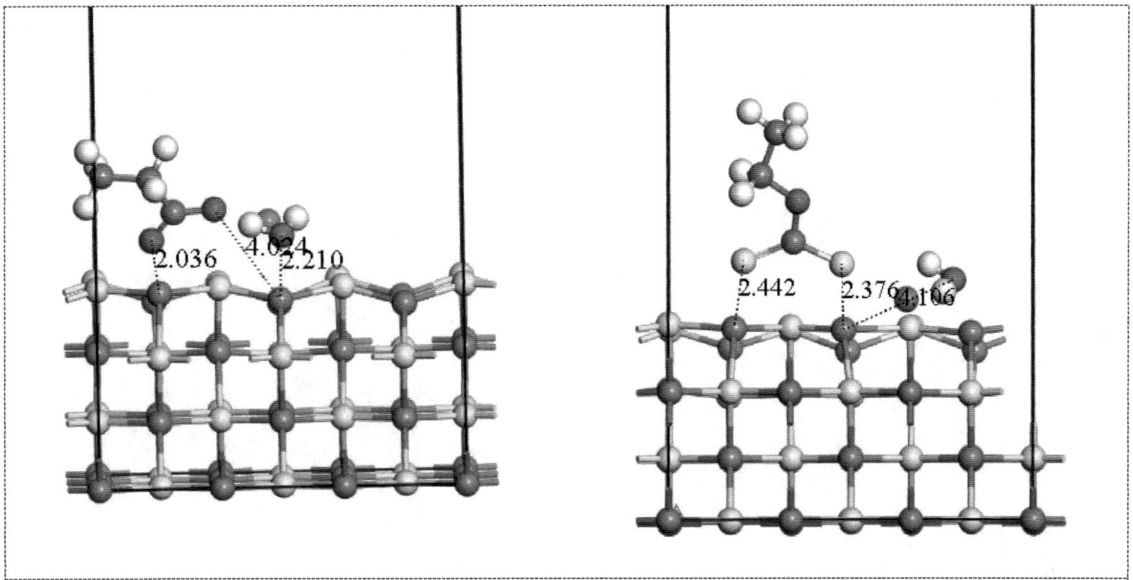

Figure 8. Competitive adsorption model of O- and S-containing collectors on sphalerite.

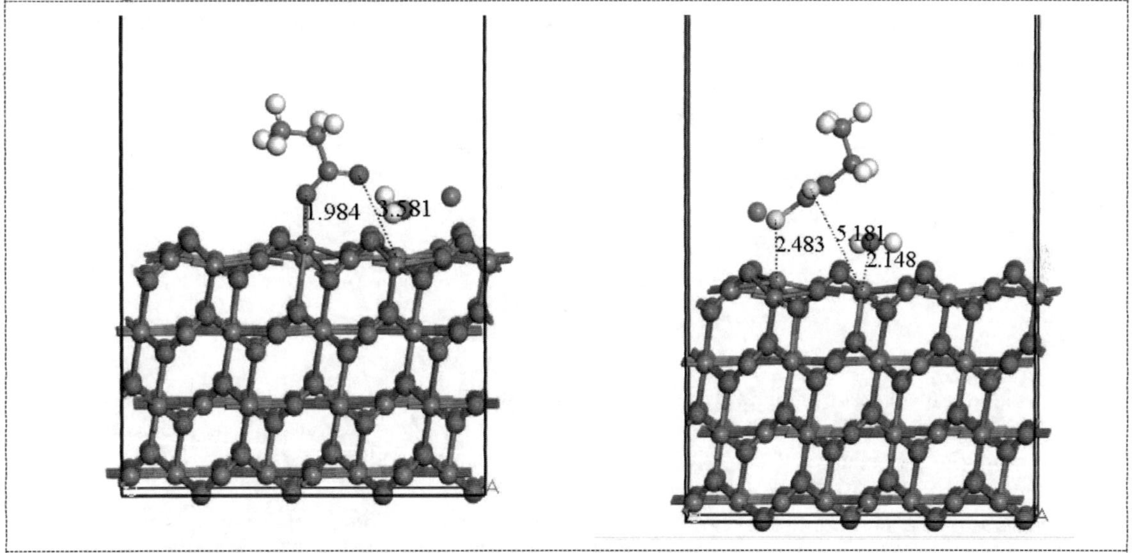

Figure 9. Competitive adsorption model of O- and S-containing collectors on smithsonite.

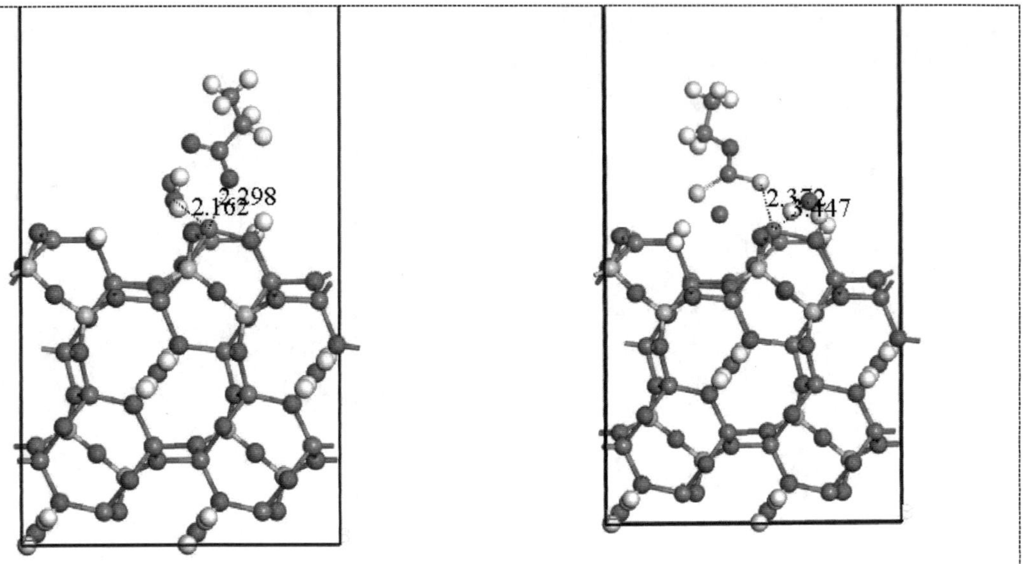

Figure 10. Competitive adsorption model of O- and S-containing collectors on hemimorphite.

Water-competitive adsorption models on the surfaces of sphalerite, smithsonite, and hemimorphite are shown in Figures 8, 9, and 10. It is known that the adsorption energies of O- and S-containing collectors on the surfaces of sphalerite and smithsonite, respectively, can be -360.99kJ/mol and -262.02kJ/mol and -341.27kJ/mol and -280.26kJ/mol, respectively, in the presence of water molecules. The adsorption energies of an O- and S-containing collector on hemimorphite's surface are -328.43 kJ/mol and -251.87 kJ/mol, respectively. As can be shown, water molecules can make O and S molecules less likely to adhere to the surfaces of three zinc minerals.

4. Conclusions

In this study, the effect of O ligands and S ligands on the electronic properties of Zn ions was investigated using density functional theory. The calculated results show that Zn ions coordinated with O have more positive charges than Zn ions coordinated with S. In addition, an analysis of the Fukui index indicates that Zn ions coordinated with O are more likely to exhibit stronger electron absorption characteristics. The increase in the number of ligands or the distance between the ligands and the central ion could enhance the electron absorption characteristics of Zn ions. This indicates that the properties of ligands significantly affect the adsorption capacity of zinc ions. The S-containing collector (ethyl xanthate, CH_3CH_2-OCSSH) and the O-containing collector (fatty acid, CH_3CH_2-COOH) were selected to conduct adsorption simulations in order to investigate the influence of ligands on the adsorption characteristics of zinc ions. The results revealed that S-containing molecules can form π backbonding with zinc ions, whereas the O-containing molecules cannot. With an increase in the coordination number of ligands, the adsorption capacity of zinc ions decreases. However, with an increase in the distance between the ligands and zinc ions, the adsorption capacity of zinc ions increases.

The adsorption of O- and S-containing molecules on sphalerite, smithsonite, and hemimorphite was investigated. The findings showed that the adsorption capability of molecules containing sulfur (S) was slightly weaker than that of molecules containing oxygen (O) towards sphalerite, smithsonite, and hemimorphite.

Additionally, the investigation explores how water molecules affect the adsorption of molecules containing O and S. The results of simulations showed that H_2O and molecules containing O- or S- compete for adsorption on the surfaces of sphalerite, smithsonite, and hemimorphite. The adsorption energy for both molecules containing O and molecules containing S on the mineral surfaces decreased in the presence of water molecules.

Acknowledgments

The authors are grateful for the financial support provided by the National Natural Science Foundation of China (NSFC) (grant number U20A20269) and the Fundamental Research Funds for the BGRIMM, China (02-2274).

References

[1] SRDJAN M B.Handbook of Flotation Reagents[M].Amsterdam: Elsevier Science&Technology Publisers,2007:4.
[2] Taggart A F, Ziehl O A, del Giudice G R M. The case for the chemical theory of flotation[M]. 1934. 4 ,112:348-381.
[3] Cook M A, Nixon J C. The Theory of Water-Repellent Films on Solids Formed by Adsorption from Aqueous Solutions of Heteropolar Compounds[J]. The journal of physical chemistry, 1950, 54(4): 445-459.
[4] Steininger J. Collector ionization in sphalerite flotation with sulfhydryl compounds[J]. Trans. Aime, 1967, 238: 251.
[5] Gaudin A M, Fuerstenau D W. Streaming potential studies. Quartz flotation with cationic collectors[J]. Mining Eng., 1955, 7:958-962.
[6] Fuerstenau D W. Correlation of contact angles, adsorption density, zeta potentials, and flotation rate[J]. Trans AIME 1957, 208(7): 1365-1367.
[7] Salamy S G, Nixon J C. Reaction between a mercury surface and some flotation reagents: and electochemical study. I. Polarization curves[J]. Australian journal of chemistry, 1954, 7(2): 146-156.
[8] Hu H, Li M, Li L, et al. Improving bubble-particle attachment during the flotation of low rank coal by surface modification[J]. International Journal of Mining Science and Technology, 2020, 30(2): 217-223.
[9] Nasirimoghaddam S, Mohebbi A, Karimi M, et al. Assessment of pH-responsive nanoparticles performance on laboratory column flotation cell applying a real ore feed[J]. International Journal of Mining Science and Technology, 2020, 30(2): 197-205.
[10] YU Hongdong, SUN Chuanyao. Study of differences Of Floatability and physical Property of copyright from Different geogenetic deposits [J]. Journal Of China University of Mining&Technology, 2010, 39 (5):758-764.
[11] CAO Shiming, CAOYijun, MAZilong, etal.The Floatation separation of fine pyrite locked in coking coal[J]. Journal of China University of Mining&Technology, 2019, 48 (6): 1366-1374.
[12] Chen J, Chen Y. A first-principle study of the effect of vacancy defects and impurities on the adsorption of O2 on sphalerite surfaces[J]. Colloids and Surfaces A: Physicochemical and Engineering Aspects, 2010, 363(1-3): 56-63.
[13] Chen Y, Chen J, Guo J. A DFT study on the effect of lattice impurities on the electronic structures and floatability of sphalerite[J]. Minerals engineering, 2010, 23(14): 1120-1130.
[14] Chen J, Wang L, Chen Y, et al. A DFT study of the effect of natural impurities on the electronic structure of galena[J]. International Journal of Mineral Processing, 2011, 98(3-4): 132-136.
[15] Chen J, Ye C, Li Y. Quantum-mechanical study of effect of lattice defects on surface properties and copper activation of sphalerite surface[J]. Transactions of Nonferrous Metals Society of China, 2010, 20(6): 1121-1130.
[16] Chen J, Wang J, Li Y, et al. Effects of surface spatial structures and electronic properties of

chalcopyrite and pyrite on Z-200 selectivity[J]. Minerals Engineering, 2021, 163: 106803.
[17] Chen Y, Liu X, Chen J. Steric hindrance effect on adsorption of xanthate on sphalerite surface: A DFT study[J]. Minerals Engineering, 2021, 165: 106834.
[18] Chen J, Long X, Chen Y. Comparison of multilayer water adsorption on the hydrophobic galena (PbS) and hydrophilic pyrite (FeS2) surfaces: a DFT study[J]. The Journal of Physical Chemistry C, 2014, 118(22): 11657-11665.
[19] Long X, Chen Y, Chen J, et al. The effect of water molecules on the thiol collector interaction on the galena (PbS) and sphalerite (ZnS) surfaces: A DFT study[J]. Applied Surface Science, 2016, 389: 103-111.
[20] Long X, Chen J, Chen Y. Adsorption of ethyl xanthate on ZnS (110) surface in the presence of water molecules: A DFT study[J]. Applied Surface Science, 2016, 370: 11-18.

Journal of Physics: Conference Series **2738** (2024) 012006 doi:10.1088/1742-6596/2738/1/012006

Beneficiation process reengineering under the background of green development-- take the full utilization of a low-grade lead-zinc ore as an example

Xiaoxing Hu[1], Yangge Zhu[1,*], Zhenguo Song[1], Kangkang Chen[1]

1. State Key Laboratory of Mineral Processing, BGRIMM Technology Group, Beijing, China, 102628.

*Email: zhuyangge@bgrimm.com

Abstract： A zero-waste mine is an advanced mine development and resource utilization strategy that aims to encourage the reduction of mine solid waste at source and the utilization of all its components. A low-grade lead-zinc ore in Guangxi, China, contains valuable components like galena, sphalerite, pyrite, barite, etc. and the gangue mineral is primarily dolomite. This paper proposes a flow of "waste pre-discarding by dense medium cyclone - sequential priority flotation of lead and zinc recovery - flotation and gravity combined process for barite recovery" to realize full component utilization. In this technical approach, 38.81% of the waste rock that can be utilized as concrete aggregate was removed from the feed ore through pre-discarding prior to grinding, after that recovers lead and zinc by the sequential priority flotation method with yield of Pb concentrate and Zn concentrate at 7.61%. In order to more efficiently recover barite and pyrite from lead-zinc flotation tailings, the gravity equipment of reflux classifier (RC) was used, which obtained barite concentrate with a yield of 7.97% and pyrite concentrate with a yield of 2.28%. The remaining yield of 43.34% tailings can be used for filling after the recovery of barite concentrate. Utilizing waste rock, lead, zinc, barite, and other resources might help the mine achieve the dynamic balance of "mining-discarding-beneficiation-filling" and realize zero tailings discharge.

Keywords: zero-waste mine, lead-zinc sulfide ore, comprehensive utilization, process flowsheet, pre-discarding

1. Introduction

The lead-zinc sulfide ores in China are typically characterized by large deposit scale, low valuable components, and challenging comprehensive recovery of associated resources.The exploitation of these resources often results in the generation of significant amounts of tailings. As China places increasing emphasis on safety, environmental concerns, and the conservation of water resources, the admission criteria for tailings ponds have undergone significant enhancements, resulting in a general reduction rather than an increase in the number of tailings ponds. Therefore, the issue of tailings disposal has become a key constraint on the survival and development of mines.A low-grade lead-zinc ore in Guangxi, China, currently only recovers lead and zinc through the sequential priority flotation method, despite containing valuable components such as barite, pyrite, etc., and with the primary gangue mineral being dolomite. This paper is dedicated to the development of technology for the full utilization of all components in this ore, aiming to achieve the goal of a zero-waste mine.

A zero-waste mine is an advanced mine development and resource utilization strategy that aims to reduce mine solid waste at its source and maximize the utilization of all components. In recent years, pre-discarding technology prior to grinding has garnered increasing attention due to its ability to reduce solid waste generation in mines from the source. The application of dense medium separation technology in nonferrous and nonmetal mineral processing is gradually on the rise (Lv et al., 2019; Luo and Tian, 2004; Zhou and Yu, 2019; Luo et al., 2015).When valuable minerals in low-grade lead-zinc sulfide ores are embedded within aggregates, a significant amount of individual gangue materials is produced after intermediate crushing. These gangue materials can be efficiently removed in advance through dense medium separation technology to prevent them from entering the grinding operation. Based on the characteristics of the waste rocks generated by pre-discarding, the development of resource utilization technologies capable of absorbing a large quantity of waste rocks may significantly extend the service life of the tailings pond.Considering the recovery characteristics of various accompanying elements, developing beneficiation techniques that do not affect the recovery of the main target minerals is another way to address the current dilemma. Currently, in the recovery of barite from sulphide ore flotation tailings, most existing research has primarily employed flotation methods (Xiao and Dong, 2017; Zhang, 2017; Cui et al., 2011; Liu et al., 2019). However, the flotation method can lead to mutual interference between sulphide ore and non-sulphide ore backwater, limiting the application range of hydrophobic barite concentrate and presenting other challenges. Therefore, there is an urgent need to develop non-flotation recovery technologies.

By incorporating techniques such as multistage waste pre-discarding and the comprehensive recovery of co-associated resources, this paper reconstructs the complete component utilization process for the comprehensive utilization of waste rock, lead, zinc, barite, and other resources. This approach is anticipated to facilitate the sustainable development of the mine.

2. Experimental

2.1. Samples

The low-grade lead-zinc ore used for this investigation was obtained from Guangxi province, China. The chemical composition of the feed ore is provided in Table 1. As shown in Table 1, the feed sample contains 0.84% Pb and 3.56% Zn, which are the main target elements for recovery. Additionally, the content of BaSO4 is 17.09%, making it suitable for comprehensive recovery. The mineralogical composition of the sample, as determined by QEMSCAM, is presented in Table 2. Dolomite is the primary gangue mineral, constituting 61.44% of the sample, and is rarely associated with sulfide minerals.

Table 1. Chemical composition of the feed sample (mass fraction, %)

Composition	Pb	Zn	$BaSO_4$	Fe	S	Cu	Ag*	SiO_2	Mn	LOI
Content (wt%)	0.84	3.56	17.09	3.72	8.80	<0.005	9	4.31	0.31	25.53
Composition	Na_2O	Al_2O_3	CaO	MgO	K_2O	C	As	P	Cd	
Content (wt%)	0.32	1.17	18.70	12.37	0.34	7.47	0.083	0.014	0.012	

*The unit of silver content is grams per ton (g/t).

Table 2. QEMSCAM quantitative analysis showing the mineralogical composition (mass fraction, %)

Mineral	Formula	(wt%)
Sphalerite	ZnS	5.31
Galena	PbS	0.77
Geocronite	$Pb(Sb,As)_2S_8$	0.21
Pyrite	FeS_2	7.97
Dolomite	$CaMg(CO_3)_2$	61.44
Barite	$BaSO_4$	17.09
Albite	$Na_2O\cdot Al_2O_3\cdot 6SiO_2$	2.28
Sericite	$KAl_2(AlSi_3O_{10})(OH)_2$	1.36
Quartz	SiO_2	1.14
Calcite	$CaCO_3$	1.62
others		0.81

2.2. Methodology

The current process at the lead-zinc mine only involves the sequential priority flotation method for recovering lead and zinc. The yield of lead and zinc concentrates is approximately 8%, with 45% of the tailings can be used for filling. However, the remaining 47% of tailings remain untreated. The aim of reengineering the beneficiation process is to address the issue of disposing of this portion of tailings.

The main valuable minerals in the ore feed are galena (with a density ranging from 7.4 to 7.6 g/cm^3), sphalerite (with a density of 3.5 to 4.2 g/cm^3), barite (with a density of 4.0 to 4.6 g/cm^3), and pyrite (with a density of 5.1 g/cm^3), all falling within a density range of 3.5 to 7.6 g/cm^3. In contrast, the primary gangue minerals have a density of 2.8 to 2.9 g/cm^3.The significant density difference between valuable minerals and gangue minerals allows for the effective removal of gangue minerals prior to grinding through the application of dense medium separation technology.The use of dense medium cyclones in the processing of nonferrous and nonmetallic minerals is steadily increasing. This technology offers advantages such as high separation precision, a wide range of particle sizes, efficient gangue removal in the pre-discarding stage, and a sewage-free separation process.In this experiment, dense medium cyclones will be employed for pre-discarding operations to eliminate the majority of the dolomite gangue. The use of this gangue as concrete aggregate will substantially reduce the volume of tailings that pose treatment challenges.

Recycling a portion of barite from the lead-zinc flotation tailings and enriching it to meet marketable standards represent an additional strategy for dealing with challenging-to-dispose tailings. However, the value of barite is relatively low, and its comprehensive recovery should not interfere with the main lead and zinc recovery process. In light of this prerequisite, it becomes imperative to explore the development of a cost-effective, gravity-based process with minimal impact on process water circulation.Furthermore, given that the value of lead in the ore is significantly lower than that of zinc, minimizing zinc losses during the lead recovery process is another prerequisite for process reengineering.Therefore, maintaining the existing sequential priority process remains the preferred approach for lead and zinc recovery.

Based on these considerations, the full component utilization technology route for this lead-zinc ore was developed, as shown in Figure 1. This innovative technology comprises three main stages: pre-discarding before grinding, lead-zinc flotation, and the recovery of barite and pyrite from flotation tailings.

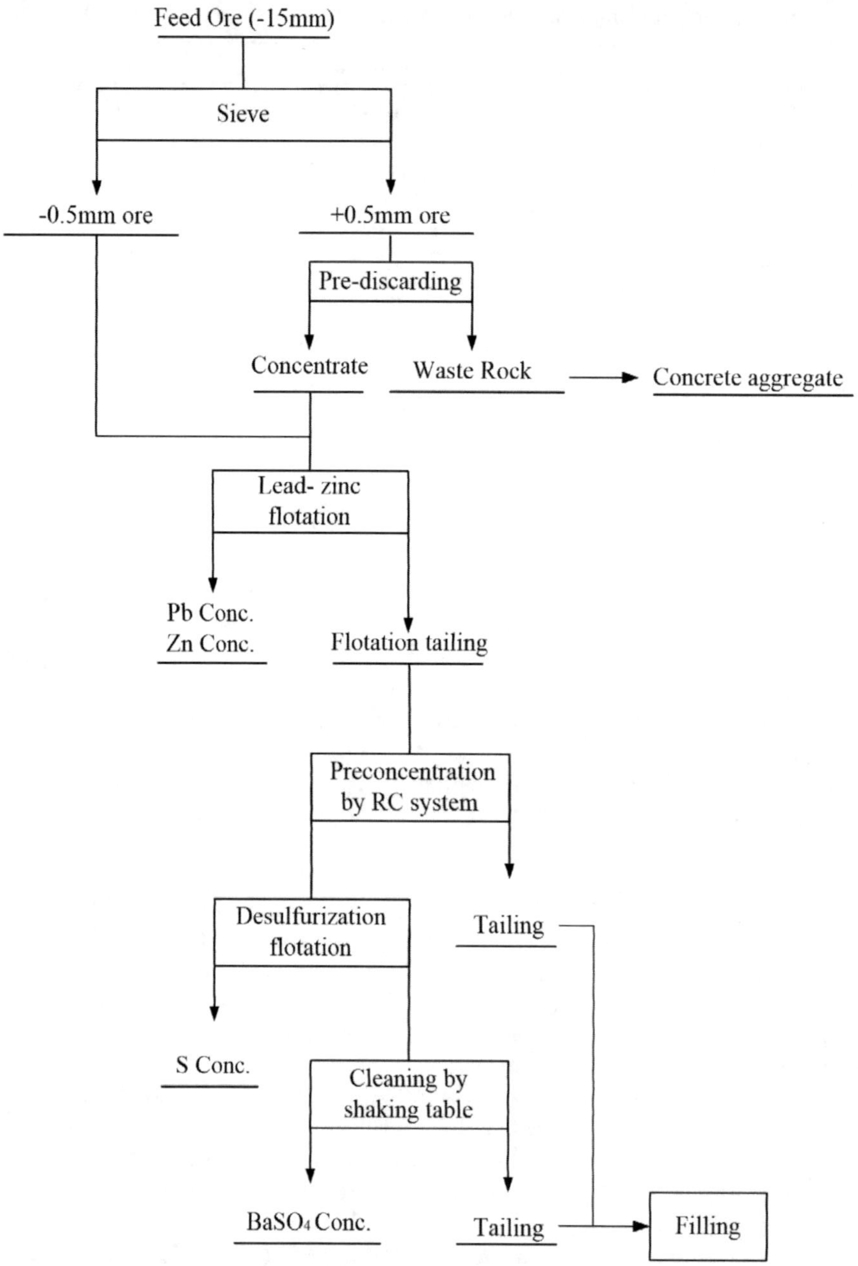

Figure 1. Technology route for full component utilization of the lead-zinc ore

2.2.1. Pre-discarding tests by dense medium cyclone. Pre-discarding tests by dense medium cyclone were conducted on the size fraction -15+0.5 mm of raw ore to remove gangue minerals, primarily dolomite, from the low-grade lead-zinc ores. The equipment flowsheet for the dense medium cyclone system is shown in Figure 2, and the pilot-scale cyclone used in the experiment was manufactured by Weihai Haiwang, China.

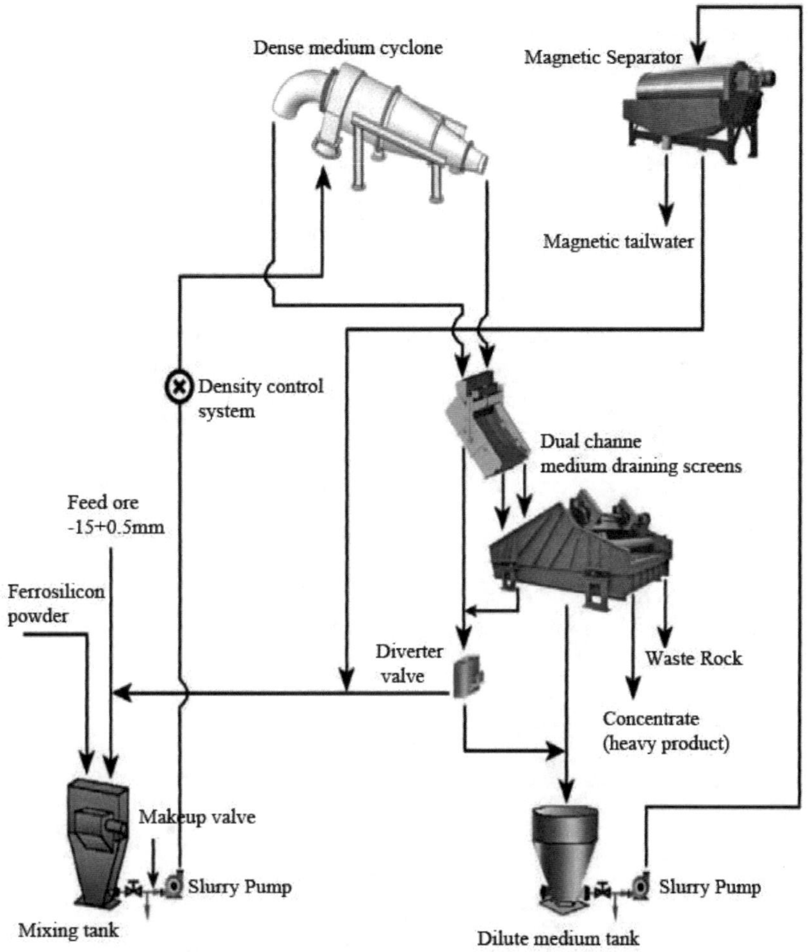

Figure 2. Equipment flowsheet for pre-discarding tests

2.2.2. Lead-zinc flotation tests. For the recovery of lead and zinc, a sequential priority flotation process is adopted, and corresponding closed-circuit tests are conducted. As a result, the flotation test comprises various stages including roughing, scavenging, and cleaning. The flotation tests were conducted using an XFG series flotation machine (Jilin Prospecting Machinery Factory, China) equipped with plexiglass flotation cells of various volumes to accommodate different flotation operations.The flotation feed consists of two parts: the pre-discarding concentrate and the -0.5mm size fraction of the raw ore (mixed proportionally by yield). The mixed feed sample was crushed, then rotary-split into 1 kg portions, and subsequently wet grinded to achieve a 50% solids content with 70% passing through a 74 μm sieve in a mild steel ball mill. The resulting slurry was transferred to a 3-litre flotation cell for roughing and scavenging, and the pulp concentration was adjusted to 27%. Multiple flotation cells with different volumes were used for cleaning.The impeller speed was set to 1750 rpm before reagent addition and flotation. The flotation reagents involved in this study are all industrial grade. After the flotation process, the products were collected, dried, and weighed. Finally, the recovery rate was calculated based on the dry weights and grades of the products.

2.2.3. Reflux classifier tests. This study employs a laboratory-scale reflux classifier, as illustrated in Figure 3, for preconcentrating barite in lead-zinc flotation tailings. The reflux classifier, recognized as an innovative pre-discarding technology (Chu et al., 2019; Liu et al., 2021; Zeng et al., 2017), was utilized to reduce throughput and enhance the separation efficiency in the subsequent "reverse flotation for desulfurization - shaking table for cleaning" process.As depicted in Figure 3, prior to commencing the experiment, a 15 kg ore sample is prepared to achieve a specific slurry mass concentration. At the outset of the test, the ⑦ -agitator motor is activated, adjusted to the required test speed, and the ⑤ - pump is set to deliver water at the desired test flow rate. The adjusted pulp is then transferred to the ① - agitation tank for thorough mixing, ensuring even feeding into the sorting column. The valve of the ② - agitation tank is opened, and the pulp flow rates of both the ③ - feeding pump and the ⑥ - underflow pump are adjusted. After 45 minutes of stable sorting, samples are collected from both the overflow and underflow. Each sampling duration lasts 60 seconds, repeated three times under identical conditions, with a 24-minute interval between each sampling cycle.Subsequent to drying, the samples are weighed, analyzed for quality and grade, and various performance indicators such as yield, recovery rate, and sorting efficiency are calculated.

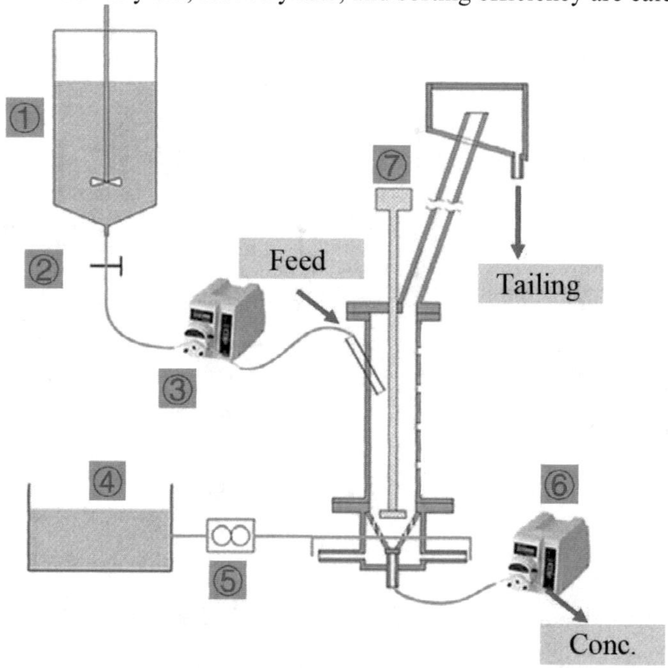

①-Agitation tanks; ②-Valve; ③-Feed pump; ④-Water tank; ⑤-Pump for rising water; ⑥-Underflow pump; ⑦-Agitator motor

Figure 3. Operation process diagram of reflux classifier

2.2.4. Shaking tabling tests. The shaking table, Ly-1100x500, was used to conduct a cleaning test on the lead-zinc tailings after preconcentration and desulfurization in order to obtain a high-grade barite concentrate.

3. Results and discussion

3.1. Results of sieve analysis

Under laboratory conditions, sieve analysis was performed on the raw ore crushed to below 15mm, and the results are shown in Table 3. In the sieve analysis, the -0.5mm fraction exhibited a $BaSO_4$ content of 28.52%, with a $BaSO_4$ distribution rate of 37.71%. Compared to the +0.5mm fraction, the -0.5mm fraction had a higher $BaSO_4$ content but lower Pb and Zn content.

Table 3. Results of sieve analysis for raw ore(-15mm)

Size,mm	Yield，%	Grade，%			Distribution rate，%		
		Pb	Zn	$BaSO_4$	Pb	Zn	$BaSO_4$
+0.5	77.38	0.87	3.71	13.77	80.37	80.51	62.29
-0.5	22.62	0.73	3.07	28.52	19.63	19.49	37.71
Total	100.00	0.84	3.56	17.11	100.00	100.00	100.00

3.2. Results of pre-discarding tests by dense medium cyclone

Perform pre-discarding tests on the +0.5mm fraction of raw ore. The results of these tests at different specific gravities of the dense medium are presented in Figure 4. According to the results, the discard rate increases with the increase of the dense medium specific gravity, leading to a corresponding increase in the recovery rate of Pb and Zn lost in the waste rock. It is considered appropriate to conduct pre-discarding tests when the specific gravity of the dense medium is set to approximately 2.55. Under this condition, the discard rate of the pre-discarding process is approximately 50.16%, and the yield of waste rock to the raw ore is 38.81%. The Pb+Zn grade in the waste rock is 0.31%, with the recovery rates of Pb and Zn in the process being 3.16% and 3.38%, respectively.

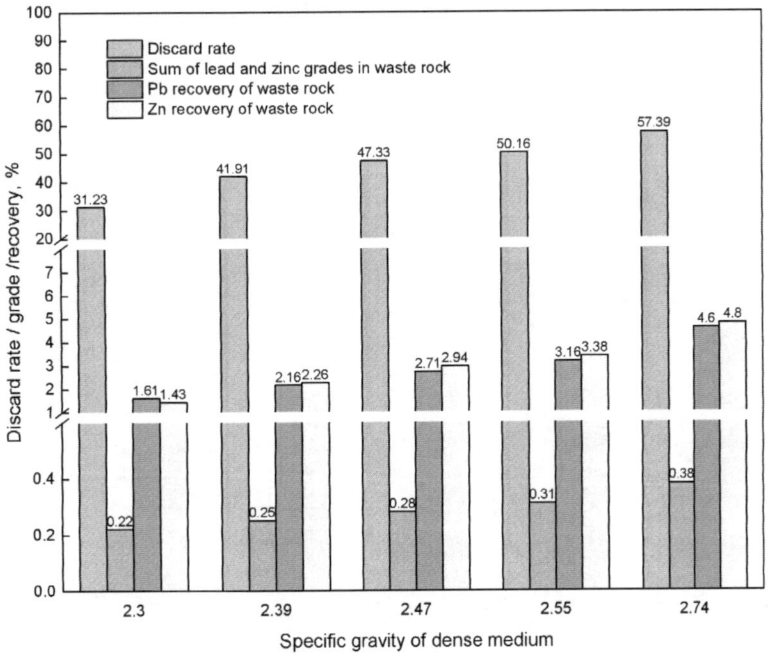

Figure 4. Pre-discarding tests results at different dense medium specific gravities

The flotation feed for lead and zinc recovery consists of two sample components: the -0.5mm raw ore and the concentrate from the pre discarding test. Pre-discarding tests are conducted under optimal dense medium specific gravity conditions to obtain the feed for subsequent flotation tests. Through pre-discarding, waste rock with a yield of 38.81% can be removed from the raw ore before the grinding process, significantly reducing the consumption of milling energy and flotation reagents during the subsequent lead-zinc recovery process. Table 4 presents the grade and recovery indicators of the flotation feed, which has a yield of 61.19%, and the Pb and Zn recovery rates are both above 97%, while the $BaSO_4$ recovery rate exceeds 98%.

Table 4. Indicators of the flotation feed obtained after pre-discarding

Feed	Yield, %	Grade, %			Recovery, %		
		Pb	Zn	$BaSO_4$	Pb	Zn	$BaSO_4$
-0.5mm ore	22.62	0.73	3.07	28.52	19.63	19.49	37.71
Conc. for pre-discarding	38.57	1.70	7.19	26.85	77.83	77.79	60.54
Flotation feed	**61.19**	**1.34**	**5.66**	**27.47**	**97.46**	**97.28**	**98.25**
Waste Rock	38.81	0.06	0.25	0.77	2.54	2.72	1.75
Raw ore	100.00	0.84	3.56	17.11	100.00	100.00	100.00

3.3. *Results of flotation test for lead and zinc recovery*

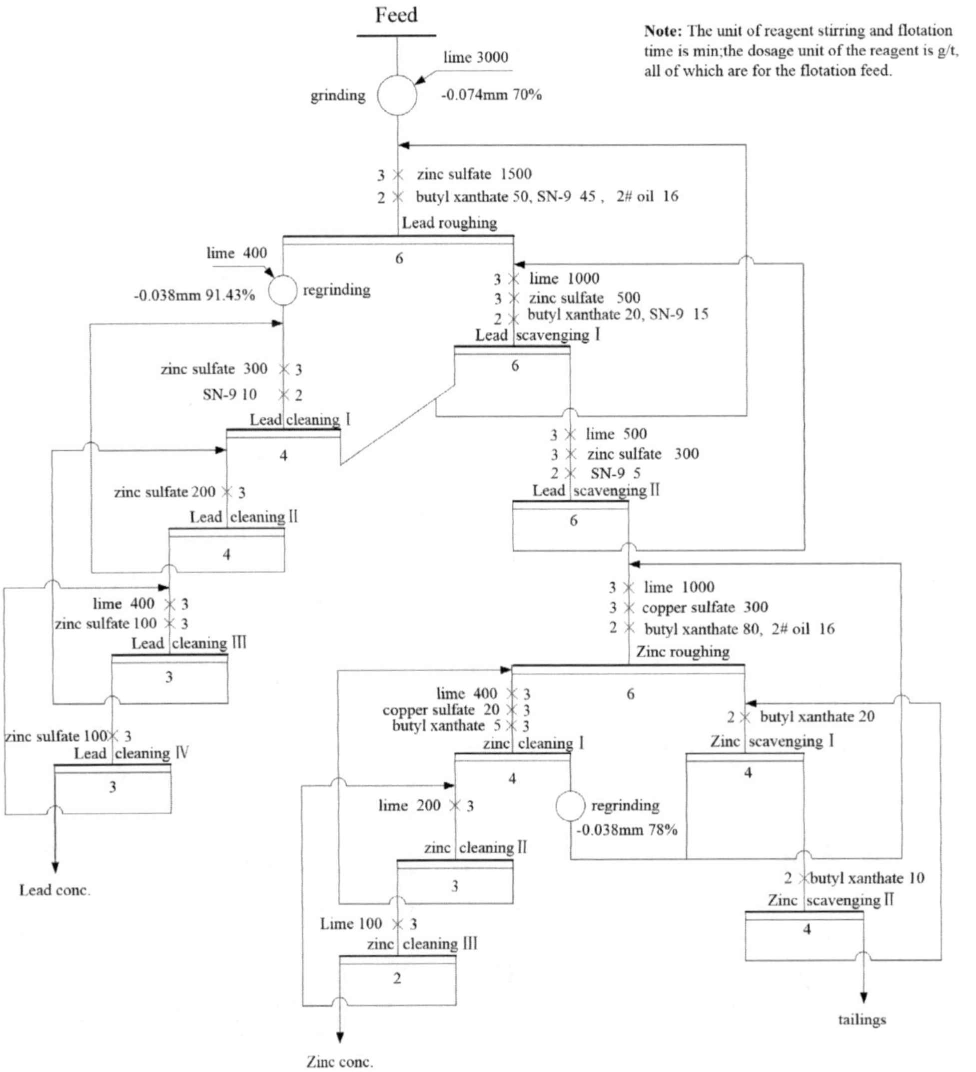

Figure 5. The flowsheet of locked circuit flotation tests

Conduct locked circuit flotation tests on the flotation feed presented in Table 4, following the process flowsheet for "lead-zinc sequential priority flotation" as illustrated in Figure 5, and the results are shown in Table 5. The flotation process and reagent regimen for the flotation feed obtained after pre-discarding are essentially identical to those used for direct flotation of the raw ore.

The experimental results indicate that even after pre-discarding the flotation feed that has lost some of the lead and zinc minerals, the lead and zinc concentrates can still achieve high recovery rates for Pb and Zn (64.17% and 92.77%, respectively). While, the recovery rates of Pb and Zn to the raw ore are 62.55% and 92.24%, respectively, which are essentially consistent with the indicators obtained through direct flotation of the raw ore.

Table 5. The results of locked circuit flotation tests

Product	Yield, %	Grade, %		Recovery for flotation feed, %		Recovery for raw ore, %	
		Pb	Zn	Pb	Zn	Pb	Zn
Lead Concentrate	1.01	52.12	6.33	64.17	1.84	62.55	1.79
Zinc Concentrate	6.60	1.57	48.75	12.63	92.77	12.31	90.24
Flotation tailings	53.58	0.35	0.35	23.20	5.39	22.61	5.24
Feed	61.19	1.34	5.66	100.00	100.00	97.46	97.28

3.4. *Results of barite recovery by combined technology of flotation-gravity separation*

A novel technology has been developed for the recovery of barite from lead-zinc flotation tailings, as depicted in Figure 6. This innovative approach incorporates a reflux classifier (RC) for preconcentration, which reduces the throughput of subsequent operations such as “reverse flotation for desulfurization” and “shaking table for cleaning”, thereby shortening the barite recovery process. Due to the fine particle size of the lead-zinc flotation tailings, a hydrocyclone is employed for desliming to ensure that the ore sample entering the reflux classifier meets equipment requirements. Following preconcentration, reverse flotation is conducted to effectively remove pyrite, which shares a similar density with barite, from the preconcentrated concentrate, referred to as RC concentrate. The desulfurization process utilizes oxalic acid as an activator and butyl xanthate as a collector. The resulting sulfur concentrate can be sold or disposed of, and the backwater generated during desulfurization closely resembles that of the lead-zinc flotation process. Finally, the rough concentrate obtained after desulfurization undergoes multi-stage shaking table cleaning to yield qualified barite concentrate.

Preconcentration is a crucial operation in the new barite recovery technology; therefore, the effects of barite preconcentration using the reflux classifier were investigated under different concentrate operation yields, as shown in Figure 7. The results indicate that when the RC concentrate operation yield is 45.54%, both the $BaSO_4$ grade and operation recovery rate of the RC concentrate are favorable, measuring 48.77% and 72.91%, respectively.Under this condition, subsequent tests involving "reverse flotation for desulfurization" and "shaking table for cleaning" were conducted. Finally, as shown in Table 6, through the combined flotation-gravity separation process, a barite concentrate with a $BaSO_4$ grade of 93.20% and a $BaSO_4$ recovery of 44.99% was obtained. In addition, a sulfur concentrate with an Fe grade of 37.02% and a S grade of 42.31% was also obtained.

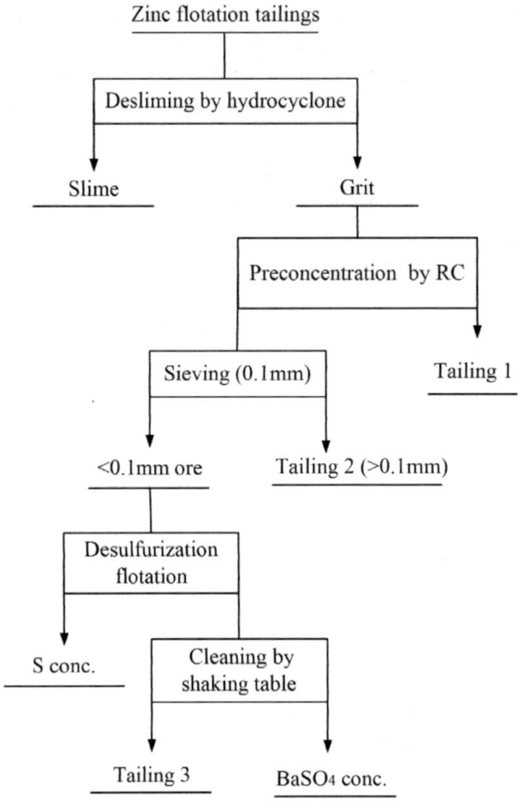

Figure 6. Flowsheetof flotation-gravity separation combined technology

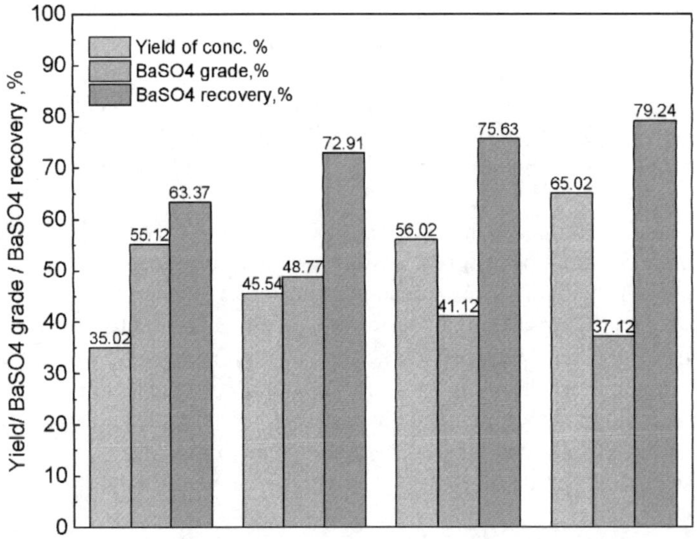

Figure 7. The effect of barite preconcentration with RC under different concentrate operation yields

Table 6. The results of flotation-gravity separation combined tests

Product	Yield for Zn tailing, %	Grade, %		Operation recovery, %	
		$BaSO_4$	Fe*	$BaSO_4$	Fe
Slime	25.87	31.79	5.59	26.70	29.21
Tailing 1	40.37	15.15	3.82	19.85	31.15
Tailing 2	4.70	3.95	3.97	0.60	3.77
Tailing 3	9.93	22.33	1.73	7.20	3.47
S Conc.	4.25	4.76	37.02	0.66	31.78
$BaSO_4$ Conc.	14.87	93.20	0.20	44.99	0.61
Zn flotation tailing	100.00	30.81	4.95	100.00	100.00

*Due to the presence of barite, it is more appropriate to use the content of Fe to characterize the content of pyrite.

3.5. *Results of the complete flowsheet test*

By combining waste pre-discarding technology, comprehensive recovery technology for tailings and the existing lead-zinc flotation technology, a new full component utilization process has been developed, referred to as the "waste pre-discarding by dense medium cyclone - sequential priority flotation of lead and zinc recovery - flotation and gravity combined process for barite recovery".The test results obtained using this new process are shown in Table 7.

The results show that waste rock with yield of 38.81% can be removed from the feeding through pre-discarding prior to grinding, while the overall yield of Pb concentrate and Zn concentrate remains at 7.61%, and the recovery rates of Pb and Zn are essentially the same as those achieved through direct flotation. After recovering a barite concentrate with a yield of 7.97% and a sulfur concentrate with a yield of 2.28% from the lead-zinc flotation tailings, the remaining tailings with a yield of 43.34% can be utilized for filling.

Additionally, research was conducted on the utilization of waste rock as a construction material. The results demonstrated that waste rock can be used as concrete aggregate to produce C30, C35, C40, and C45 concrete. The slump, expansion, density, and compressive strength of the concrete at different ages all met the corresponding strength grade requirements for concrete.

Table 7. The results of the complete flowsheet test

Product	Yield, %	Grade, %				Recovery, %			
		Pb	Zn	$BaSO_4$	Fe	Pb	Zn	$BaSO_4$	Fe
Pb Conc.	1.01	52.12	6.33	1.41	6.24	62.55	1.79	0.08	1.69
Zn Conc.	6.60	1.57	48.75	4.35	3.45	12.31	90.24	1.68	6.12
S Conc.	2.28	1.30	0.66	4.76	37.02	3.52	0.42	0.63	22.67
$BaSO_4$ Conc.	7.97	0.04	0.06	93.20	0.20	0.38	0.13	43.41	0.44
Tailing	43.34	0.36	0.39	20.70	4.14	18.71	4.68	52.45	48.21
Waste Rock	38.81	0.06	0.25	0.77	2.00	2.54	2.72	1.75	20.88
Raw ore	100.00	0.84	3.56	17.11	3.72	100.00	100.00	100.00	100.00

4. Conclusion

This paper reconstructed a full component utilization process flow of "waste pre-discarding by dense medium cyclone - sequential priority flotation of lead and zinc recovery - flotation and gravity combined process for barite recovery" by adding techniques like multistage waste pre-discarding and comprehensive recovery of co-associated resources.

In this technical approach, 38.81% of the waste rock that can be utilized as concrete aggregate was removed from the feed through pre-discarding prior to grinding, significantly reducing the consumption of milling energy and flotation reagents in the subsequent lead-zinc flotation process. The Pb concentrate and Zn concentrate are obtained with an overall yield of 7.61% through the sequential priority flotation method. A simplified process for recovering barite from lead-zinc flotation tailings has been developed, which can recover barite concentrate with the yield of 7.97% and sulfur concentrate with the yield of 2.28%. The remaining yield of 43.34% tailings can be consumed through filling. The utilization of waste rock, lead, zinc, barite, and other resources may help the mine achieve a dynamic balance in the "mining-discarding-beneficiation-filling" cycle and realize zero tailings discharge.

Acknowledgments
This work is financially supported by the Natural Science Foundation of China under grants: U20A20269.

References

[1] Lv C, Liang Y Q, Zhao X, et al. Study on the separation of a lead-zinc mine in the southwestern area of the heavy medium-flotation combined process 2019 *Mining Research and Development* **39(3)** 33-36
[2] Luo X M, Tian S H. Research on the beneficiation process of a heavy-medium cyclone in a lead-zinc mine 2004 *Hunan Nonferrous Metals* **20(4)** 7-10
[3] Zhou F, Yu X. Research on the design of pre-selection and throwing tailing process of a tungsten mine in Kazakhstan 2019 *Non-ferrous Metals (Mineral Processing section)* **(6)** 43-49
[4] Luo H H, Zhai D J, Miao H J, et al. Yichang low-grade collophane ore screening and classification micro-differential density heavy-medium separation technology 2015 *Journal of Wuhan Institute of Technology* **37(9)** 35-38
[5] Xiao J, Dong Y H. Experimental study on beneficiation of barite from lead-zinc tailings in Panlong lead-zinc mine 2017 *Industrial Minerals & Processing* **46(11)** 14-18,21
[6] Zhang H Y. Study on the recovery of barite from a lead zinc tailings 2017 *Materials research and application* **11(03)** 187-191
[7] Cui C Z, Gou M L, Sun Y, et al. Application research on the recovery of barite from lead-zinc tailings 2011 *Multipurpose Utilization of Mineral Resources* **(03)** 47-49
[8] Liu C, Chen Z Q, Luo C S , et al. Beneficiation test on recovery of barite by combined magnetic separation-flotation process 2019 *Industrial Minerals & Processing* **48(10)** 40-42+45
[9] Chu H R, Wang Y H, Zeng F S , et al. Experimental study of reflux classifier for pre-enrichment of fine antimony oxide tailings 2019 *Mining and metallurgical engineering* **39(2)** 45-48
[10] Liu Z Q, Lu D F, Zeng F S , et al. Flow field simulation and experimental study of self-made reflux classifier for pre-concentration of low-grade cervantite ore 2021 *Mining and metallurgical engineering* **41(2)** 39-43
[11] Zeng F S, Wang Y H, Lu D F, et al. Application of self-made reflux classifier in gravity preconcentration of lateritic nickel and flow field simulation thereof 2017 *Mining and metallurgical engineering* **37(1)** 52-56

Recent University Research, Implementation and Education in Support of Pyrometallurgical Lead Processing

Evgueni Jak [*,1], Maksym Shevchenko [1], Denis Shishin [1], Evgenii Nekhoroshev [1], Roman Starykh [1], Jiang Chen and Peter Hayes[1]

Pyrometallurgy Innovation Centre, School of Chemical Engineering, University of Queensland, Brisbane, Queensland, Australia, 4072

e.jak@uq.edu.au

Abstract. The increasing chemical complexity of lead process streams encountered in industrial high temperature processing operations, as the result of declining primary resources, increased metal recycling and increased overall range of metals in modern devices has highlighted the urgent need for new predictive tools, fundamental phase equilibria and thermodynamic information and thermodynamic models to characterise the chemical behaviour of these systems. The paper examines recent progress in experimental and thermodynamic modelling research on process fundamentals, the availability of advanced, predictive computer-based tools and the implementation of the research outcomes into industrial practice. A wide range of chemical systems and phase assemblages have been studied. Some examples are taken from the current research program at PYROSEARCH, which involves the characterisation of multi-component, multi-phase gas-slag-matte-speiss-metal-solids systems with the PbO-ZnO-"Cu_2O"-FeO-Fe_2O_3-CaO-Al_2O_3-MgO-SiO_2-S as major and As-Sn-Sb-Bi-Ag-Au-Ni-Co-Cr-Na as minor elements with focus on systems directly relevant to lead primary and recycling pyrometallurgical processes. Examples of the application of advanced analytical techniques to fundamental and applied industrial research are also given. The implementation of new research outcomes into industrial practice depends critically on commitments by research staff as well as industry management and the availability of well-trained metallurgical engineers. We examine the current status of research implementation, university research, metallurgical engineering education and the availability of suitable educational pathways and initiatives that can be taken to increase undergraduate enrolments. Active engagement and support by industry is critical in ensuring the continuation of academic programs and advanced technical skills required by the industry.

1. Introduction

As a society we face critical global challenges associated with sustainable energy and technological developments. We are using more metals, more different metals and more combinations of metals than ever before in the production of the new advanced materials needed to address these challenges. As a result, the chemical compositions of the process streams in primary and secondary metal production, refining and recycling processes, which are necessary to support these technological developments, are becoming more and more complex and with increased variability. These are significant and increasing technical and societal problems with major implications for industrial and economic sustainability, and environment impact.

In addition to providing new scientific information, predictive tools and technical innovation to address these issues, we need to develop and grow a skilled workforce to be able to implement these advances in industrial practice. In the present paper we review recent advances in characterisation of pyro-metallurgical systems provided through experimental and thermodynamic modelling studies, as well as recent trend in University research and education.

Present paper presents only individual views of the authors based on personal experience and does not represent the official University position.

1.1 Stages of process optimisation

For existing pyrometallurgical operations, improvements in the form of increased efficiency, technical and economic productivity can be achieved in several ways, for example, by increased process stability, process optimisation, adoption of feed-forward process control, increased recoveries and forward planning. Generally, three stages of process optimisation must be done (see Figure 1) – i) ***measurement*** of all factors influencing the process output, ii) ***control*** of those factors and finally iii) ***optimisation*** of the process to achieve and maintain optimal process output.

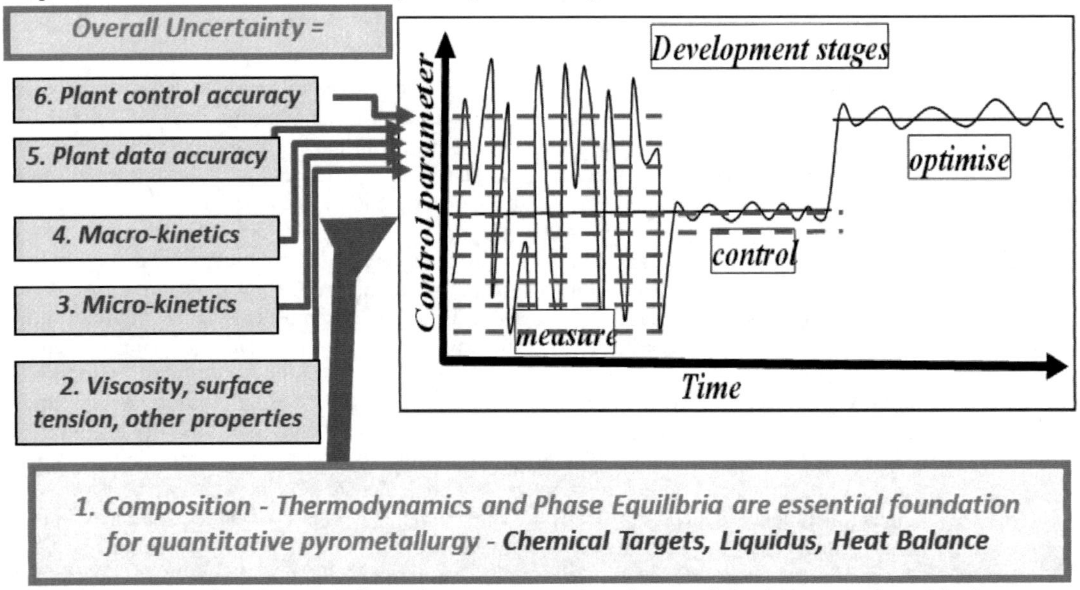

Figure 1. Some of the factors influencing process uncertainty and the key steps in achieving process optimisation – measurement, control and optimisation

1.2 Predictive tools

Critical for the metallurgical process optimisation of complex metallurgical processes is the availability of not only accurate data on the process performance and process conditions but also accurate predictive tools. Improved instrumentation and data storage on computer systems can provide the ability to gather key information about the process performance and changes in real time. Appropriate models and predictive tools can now be developed with the significantly increased computer speed and capacity. One approach is to analyse these data to determine trends, patterns, and correlations between measurements, currently referred to as data mining. These methods are useful to understand and quantify behaviour within a defined range of conditions, that is, within defined compositional and process conditions. However, these empirical correlations cannot be used to confidently predict behaviour outside these current conditions. Typically, chemical and physical properties in metallurgical systems do not vary in continuous or linear relations with key process variables. These non-linear relationships arise from the formation of new phases within the system, which change the compositions and proportions of all other phases present. For the reasons cited above, process development involving the use of novel chemical compositions or process conditions is preferably undertaken with the support of accurate fundamentally based models or tools.

In the case of pyrometallurgical processing, many factors can influence the outcomes of pyrometallurgical processes. Some of these factors are summarized in Figure 1. The major controlling input parameters for pyrometallurgical processes are the amounts and compositions of the materials introduced into the reactor. These input compositions control three key output process characteristics (see Table 1) – product compositions (***chemical targets***), ***heat balances*** and ***phase equilibria*** (liquidus temperatures or proportion of solids). All these three factors are directly determined by the thermodynamic and phase equilibria properties of the system (see examples given in Table 2).

Fundamental research can deliver fundamental scientific information on the heat capacities of elements and compounds, and enthalpies of reaction, to enable the calculation and prediction of mass and heat balances in pyrometallurgical reaction systems. It is now possible to better understand the relative stabilities of elements and compounds, and the conditions for chemical equilibrium, through the development of thermodynamic theories, such as the concepts of Gibbs free energies and Gibbs energy minimisation. The application of these fundamental scientific concepts would not be possible without quantitative descriptions of these systems, for example, thermodynamic data on enthalpies and entropies of all components, and potential reactants and products, in these systems. The development of phase diagrams has enabled detailed thermodynamic information on the relative stabilities of phases and phase changes under different process conditions to be summarised and translated into useful graphical and mathematical forms. The use of phase diagrams has provided industry professionals with reliable methods of predicting flux requirements to achieve smooth and efficient operation of smelting and refining systems. The increased use of complex ores, end-of-life materials and process wastes has led to the increased chemical complexity of metallurgical process streams. Phase diagrams of binary and ternary systems are no longer sufficient to design and select the optimum compositions and process conditions in these industrial process streams.

Table 1. Typical input, control and target parameters used in pyrometallurgical processing

Main control input pyrometallurgical reactor parameters	Kinetic "calibration" parameter	Target output conditions / parameters
Oxygen coefficient (O_2/feed ratio) Duration of blow and rate	Oxygen efficiency	*Chemical* – e.g. matte grade, Cu or Pb in slag, %S in blister
O_2 enrichment, fuel (coal, old), dust and reverts, feed rate, composition (e.g. Cu/S/Fe) and mineralogy, electric power	Heat loss	*Heat balance* – temperature
SiO_2 and CaO fluxes, slagging impurity levels (e.g. Al_2O_3, CaO, MgO)	Flux utilization	*Phase equilibria* – Liquidus, % solids, freeze-lining thickness, frequently controlled by Fe/SiO_2, CaO/SiO_2 in slag

Table 2. Examples of the application of fundamental thermodynamic research to industrial pyrometallurgical practice.

Fundamental data	Industrial Applications
Heat capacities of elements and compounds, enthalpies of reaction	Mass and heat balances
Enthalpies and entropies of compounds	Prediction of compound stabilities and operating temperatures
Phase diagrams	Liquidus determination and fluxing strategies
Component activities in solutions, Elemental distributions	Reaction equilibria, Partitioning of elements between phases
Gibbs free energies of all components and thermodynamic models	Predictive tools for process optimisation of multiphase systems
Advanced computer based databases and computer platforms	Virtual reactors for reactor and process design

Advanced computer-based predictive models are now needed in order to provide both improved accuracy and to undertake the mathematical operations necessary to determine chemical equilibria in multicomponent, multiphase systems. These thermodynamic predictive tools can be used to develop process predictive tools (virtual reactors) and be incorporated into systems similar to the GPS systems used to define directions of travel (see Figure 2). Input to the computer program interface is in the form of feed/phase compositions, temperature, pressure and solid phase characteristics. The outputs can be presented in a wide variety of formats providing quantitative data on the amounts of individual phases and their compositions, and the overall reaction enthalpies or heat changes associated with the chemical reactions. These advanced tools can be used to not only determine process outcomes form individual reactors but also can be applied to all process streams in an operation, making it possible to optimize multi-step processes and metal recycle streams within the operations. These then are powerful tools that have the potential to significantly enhance overall technical and economic performance.

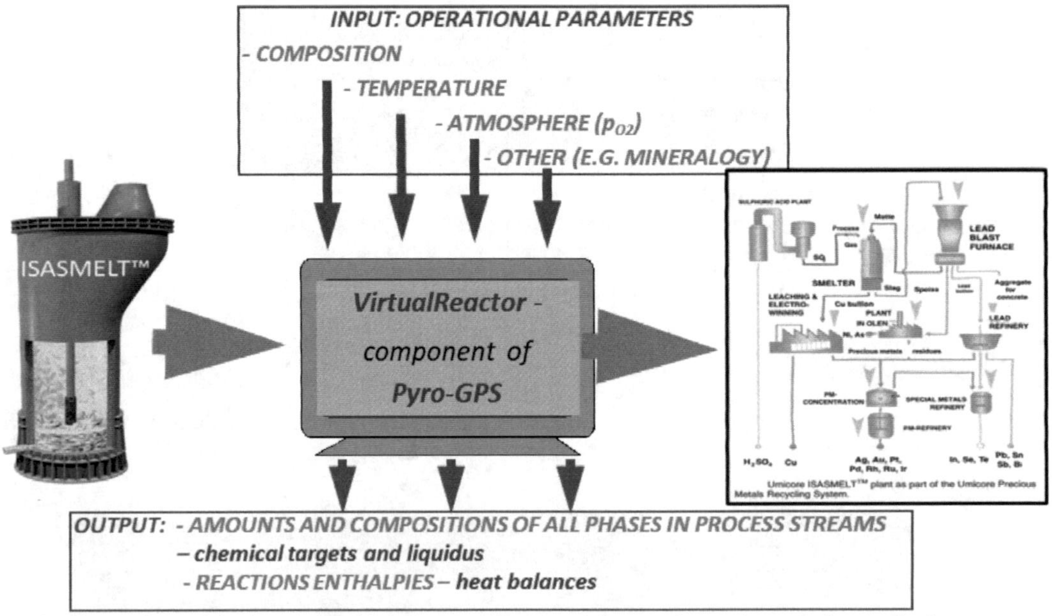

Figure 2. Illustration of the key role of fundamentally based thermodynamic tools in predicting process outcomes and assisting in process optimisation.

1.3 Thermodynamic databases
Since the thermodynamic databases that are at the heart of these predictive tools describe the interactions of individual elements and the relative stabilities of a wide range of compounds and phases, they are more reliable predictors of reactions outside the current industrial practice. These thermodynamically based process tools also have the advantage that the predictions are independent of the technologies being used and therefore can be used to define and explore all combinations of process variables within the range of systems that have been optimised in the database.

Significant gaps in knowledge on slag properties still exist – these are due to difficulties associated with high temperature research. *The demand* for the accurate fundamental information on phase equilibria, thermodynamic and physicochemical properties of the complex multi-component systems from metallurgical and recycling industries is growing due to a) stronger economic competition, b) stricter environmental regulations, and c) better equipment and options in process control. Many other factors also contribute, including increasing complexity of ores, the number of available smelting technology options. *The supply* of the needed fundamental data on the chemistry of the processes is becoming possible since a) new experimental techniques are becoming available due to the developments of modern advanced analytical techniques, dramatic improvement of their capabilities and availability, and b) new theoretical modelling approaches due to the significantly increased computer power. This demand / supply combination is the basis for *the renaissance in research on the high temperature chemistry of metallurgical systems.*

1.4 Fundamental research developments
An overview of earlier research on the phase chemistry relevant to lead/zinc smelting has been provided in 2010 Lead-zinc conference [1]. References to subsequent fundamental research on thermodynamics and phase equilibria of lead-containing systems directly relevant to primary and secondary processing are provided in the present paper. They are grouped in the following categories: *Smelting* [2-21], *Reduction* [21-29], *Refining* [30-43] *and Elemental Distributions* [44-47].

The increased number of different elements present in the system results in an increase in the number of chemical components and species interacting simultaneously. Many of these chemical interactions, and the chemical subsystems defining these interactions, have not previously been studied or accurately characterised. To be able to accurately describe reactions in these systems it is necessary to undertake further experimental measurements and incorporate this information into suitable mathematical forms using thermodynamic models and appropriate model parameters. The unambiguous determination of the unique set of binary and ternary model parameters accurately describing a whole range of conditions from low to high order systems is critical for underpinning the predictive power of the tool, and requires experimental data for all binary and ternary subsystems, and for selected quaternary subsystems. Accurate predictive models cannot be developed if accurate, fundamental, experimentally determined information is not available.

2. PYROSEARCH's Integrated Experimental and Thermodynamic Modelling Approach

2.1. Integration of experimental and thermodynamic modelling research
At the Pyrometallurgy Innovation Centre an integrated approach has been developed that involves the parallel and complementary characterisation of the thermodynamic properties of the systems using experimental and thermodynamic model and database development.

Experimental and thermodynamic modelling studies are fully integrated in the present research program. Thermodynamic assessment is used to identify discrepancies and to plan experiments. Experimental results are then used to fix model parameters to improve model accuracy. If there are no experimental data – there can be no reliable thermodynamic parameters, and consequently no reliable model. The thermodynamic parameters in the present study therefore are determined by targeted experiments specifically planned and undertaken to determine the parameters rather – this is different from studies fixing thermodynamic parameters only from the data available from literature. A brief summary of the experimental and modelling techniques, and their application is provided in the following paragraphs.

2.2. Experimental methodology

The accurate measurements of fundamental physical and chemical properties of lead/zinc containing systems pose particular experimental difficulties due to the high vapour pressures of the metal species and the highly reactive/corrosive properties of liquid phases in these systems. Conventional experimental techniques, such as thermal analysis can be used to determine liquidus temperatures and phase diagrams at low temperatures and in relatively simple low or silica-free systems containing eutectic reactions and stoichiometric compounds. However, these approaches cannot provide the accuracy required for thermodynamic database development at high temperatures, particularly for systems containing peritectic reactions, solid solutions and multicomponent systems.

The equilibration/quench/microanalysis methodology developed by PYROSEARCH has overcome all of these limitations and it is now the preferred approach to phase equilibria determination in these complex systems. The technique has been first discussed in 1995 [48] (Figure 3a). Essentially the methodology involves equilibration of small masses (typically less than 0.5g of material) under controlled process conditions at temperature. The sample is then rapidly quenched to room temperature thus retaining the phase assemblage and phase compositions present at the equilibration temperature. The phases and the phase compositions present in the samples are determined through the use of advanced microanalytical techniques including: i) electron probe X-ray microanalysis (EPMA), and ii) laser ablation inductively coupled plasma mass spectrometry (LA-ICPMS). The compositions and process conditions are deliberately selected so that multi-phase materials are formed in the equilibrated products. In these conditions' changes, so long as the phases present remain the same, any change to the bulk composition of the sample only changes the proportions of the phases, not their compositions. The methodology then has the important advantage that the results do not depend on small bulk composition changes that may take place during equilibration. Recent developments in experimental techniques mean that phase equilibrium measurements and elemental distributions between phases can now be obtained in slags/mattes/metal alloys/speisses/solid solutions in closed and open systems. Contamination from crucible materials is avoided by using synthetic substrates of the primary phase material. A range of metal substrates (Pt, Pd, Ir, Re, Rh/Pt, Au, Mo, W, Fe, Co, Ni, Cu) have also been used for specific systems and process conditions. Successful equilibrium experiments have been carried at temperatures up to 1740°C. In all cases, proof of equilibrium is established through the "4-point" test protocol, which requires that the results are independent of 1. reaction time and 2. the starting conditions of the system. 3. the outcomes should be not restricted by the reaction mechanism and reaction pathway and 4. the compositions of the phases should be homogeneous and not vary with position. Overviews of the latest advances in the integrated experimental and modelling research approach to the thermodynamic database development for these pyrometallurgical systems has been provided in [49-51]. The use of these microanalysis techniques has provided breakthrough capabilities in phase equilibrium studies for pyrometallurgical applications, greatly extending the range of elements, bulk compositions and conditions that can be characterised.

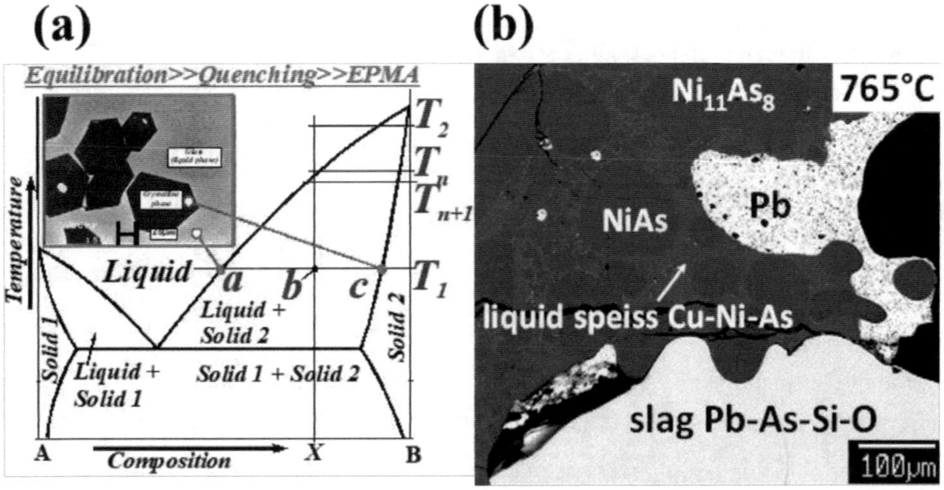

Figure 3. a) Illustration of the PYROSEARCH approach to high temperature phase equlbria determination, enabling accurate measurement of phase compositions in multi-phase systems under defined process conditions, b) An example of a complex multiphase, multicomponent system at equilibrium.

2.3. Thermodynamic models and databases

PYROSEARCH researchers use the FactSage computer platform and the associated thermodynamic models to develop the new thermodynamic database. For every portion of the database, the methodology used involves first constructing preliminary models using experimental thermodynamic and phase equilibria data available in the literature data, the identification of key areas of disagreement or thermodynamic inconsistency and the selection of critical experiments required to resolve these inconsistencies or omissions. The database is developed so as to achieve a self-consistent set of parameters for the thermodynamic models, accurately describing phase equilibria and all other thermodynamic properties for all phases over the complete range of conditions in terms of temperature, composition and atmospheric conditions.

The modified quasichemical thermodynamic model (MQM) is available through the FactSage software [52]. Better predictive capacity and flexibility of MQM in multicomponent systems is achieved by using pair (or quadruplet) fractions for polynomial expansion of Gibbs energy change in quasichemical reactions [53], by using composition-dependent coordination numbers [53], by combining quasichemical and Bragg-Williams random mixing contribution to the excess Gibbs energy [54], by using different geometric extrapolation methods of binary model parameters into multicomponent space [54, 55], and by using quadruplet formalism for ionic liquids with several possible cations and anions [56]. Recent examples of application of MQM for different cases are provided below: slags [57-61], mattes and metals [40-42, 62], slags in equilibrium with mattes [63-65] and/or metals [66].

In addition, there have been major revisions to the database in the form of implementation of an improved methodology to accurately describe the heat capacities of elements and compounds, corrections to the heat capacity and enthalpy of melting of SiO_2 [20] and a number of other pure end members, and the introduction of parameters that accurately describe the intermediate silica concentrations with maximum ordering (near orthosilicate composition) as well as high silica slags. All of these improvements have been critical in obtaining improved thermodynamic description of high and low silica systems, the accurate prediction of liquid immiscibilities and the complete revision of the all subsystems in this 20-component system, including the FeO-Fe_2O_3-CaO-SiO_2 subsystem, which is now being prepared for publication.

A major issue emerging in the development of multicomponent databases is the exponential increase in chemical interactions to be described with the increasing number of components. Adding a single component, or even incorporating new experimental data into the database, requires the iterative re-optimisation of all low-order and corresponding high-order systems; a significant task when handling a 20 component system that has over 1000 binary and ternary sub-systems.

To tackle this problem and to be able to efficiently update the databases, a new optimisation methodology has been implemented by the PYROSEARCH team. The key points in this new methodology are as follows. The sets of target experimental points to be described and corresponding weights are selected based on the experimental information available. A matrix of first derivatives showing the sensitivity of each target point to each possible model parameter is calculated using traditional Gibbs energy minimization calculations, which is a relatively slow step. The initial slow Gibbs energy minimisation calculations are then replaced by a fast analytical approach with linear extrapolation of the existing values through matrix multiplications of the form: $\Delta\overline{y} = \mathbf{A}\cdot\Delta\overline{x}$, where **A** is the matrix of first-order derivatives (*n* target values by *k* model parameters), $\Delta\overline{x} = \overline{x} - \overline{x}_0$ is the difference between the final set of model parameters and their initial approximation, and $\Delta\overline{y} = \overline{y} - \overline{y}_0$ is the difference between the final model predictions and model predictions at $\overline{x}_0$. The optimum values of the model parameters $\overline{x}_{optimal}$ are obtained by solving the relationship: $\overline{x}_{optimal} = \overline{x}_0 + \left(\mathbf{A}^T\cdot\mathbf{A}\right)^{-1}\cdot\mathbf{A}^T\cdot\left(\overline{y}_{\text{target}} - \overline{y}_0\right)$ for the condition $\overline{y} \rightarrow \overline{y}_{\text{target}}$. This non-iterative analytical (rather than numerical) optimization approach is orders of magnitude faster than the combination of the thermodynamic calculations using Gibbs energy minimisation and numerical non-linear minimisation.

The first-derivative-based linear extrapolation approach also enables a) the *real-time* graphical presentation of predicted and target points as well as b) the *real-time* systematic tabular presentation of the statistical analysis of agreement between predictions and target values, which makes the optimisation process truly interactive. Also, the new *formalised and semi-automated* methodology makes it possible to increase the efficiency and flexibility of collaborative work between researchers by organising parallel simultaneous optimisations by several researchers, thus distributing the database development efforts between the research team: once the procedure to optimise the model parameters is formalised and semi-automated, the group of researchers each will contribute to the thermodynamic parameters optimisation by planning and undertaking new experiments, adding corresponding target points and correcting weights, rather than "manually" optimising model parameters. Thus the discrepancies and conflicts within the system will be resolved by the formalised semi-automated system significantly more efficiently and with less oversight. Only periodically will the re-calculations using Gibbs energy minimisations be needed to update the matrix of derivatives. The new methodology enables researchers undertaking experimental work on a particular sub-system to personally contribute to the thermodynamic optimisation of that system and to select further experimental target points further increasing productivity. Optimisation is undertaken in iterative cycles, the major discrepancies are identified at each step, and new experiments conducted to specifically resolve discrepancies within time intervals from several days to several weeks rather than months and years using the more traditional approaches.

3. Fundamental research outcomes

Examples of the experimental results on chemical systems relevant to lead smelting, reduction and refining obtained during recent studies are provided in the following figures. These demonstrate the number of different phases, both solid and liquid, that can form in these polymetallic systems in the presence of minor elements, as exemplified in the examples given in Figure 4. The predicted phase diagrams show the influence of bulk composition on the liquidus and primary phases formed, as illustrated in Figure 5. The application of the database to the prediction of phase equilibria in complex multicomponent systems is illustrated by the fluxing diagram reproduced in Figure 6, which clearly shows the effect of CaO/SiO_2 ratio on slag liquidus and primary phase formation. Once the database has been established the program can be used to systematically quantify the effects of changing composition or process conditions.

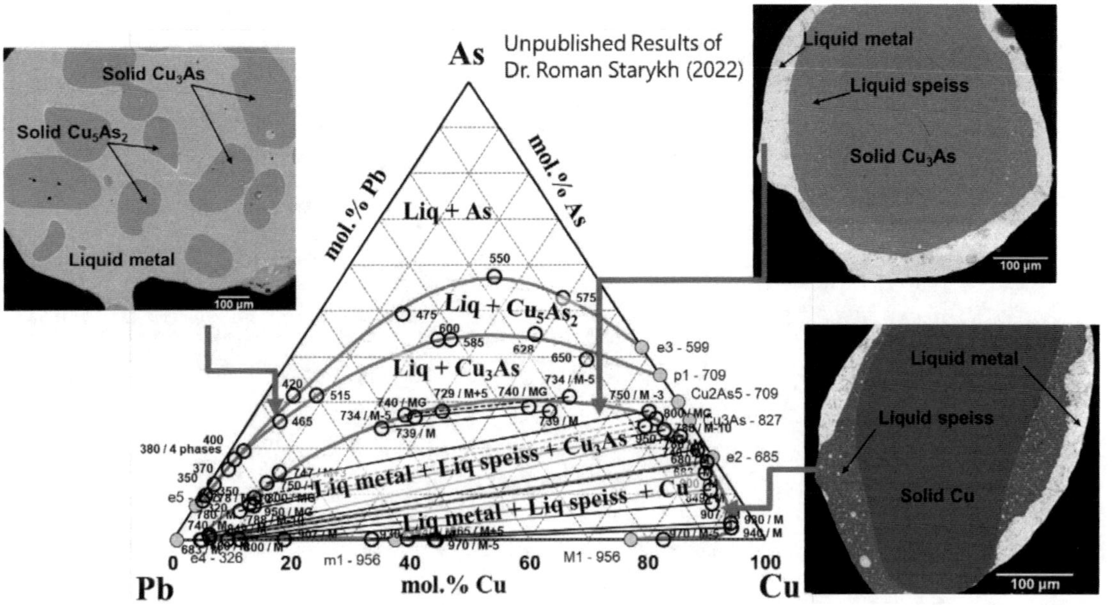

Figure 4. Example of experimental results: backscattered electron images of microstructures of multi-element, multiphase equilibria and compositions of liquid phase for that equilibria plotted on the Pb-Cu-As triangle

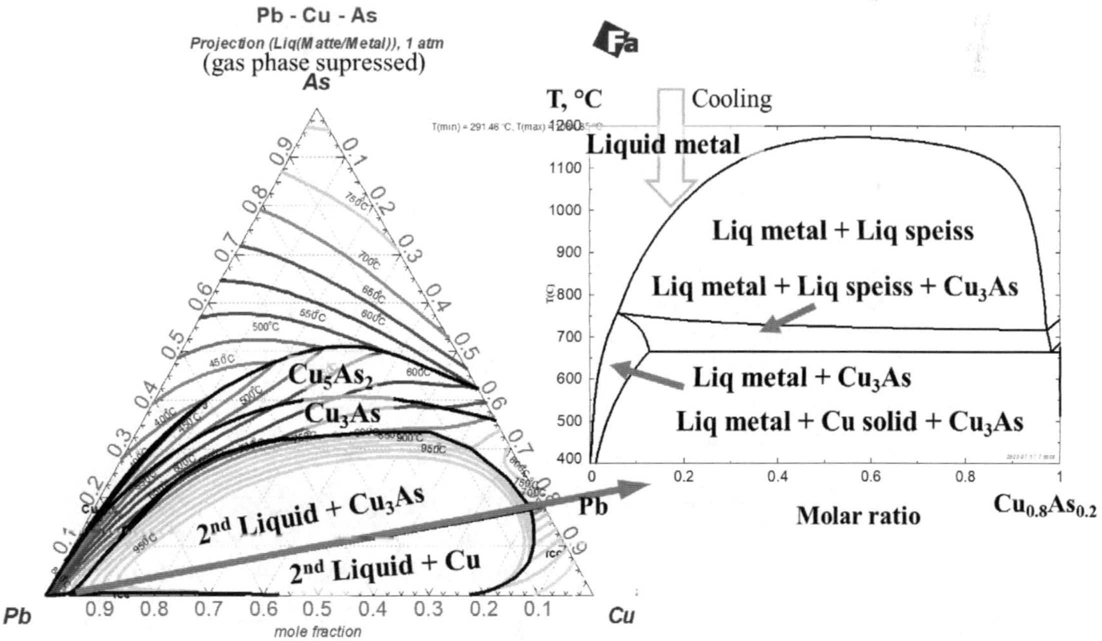

Figure 5. Example of modelling results: equilibria in the Pb-Cu-As system calculated using the thermodynamic model, updated since the earlier publication [42]

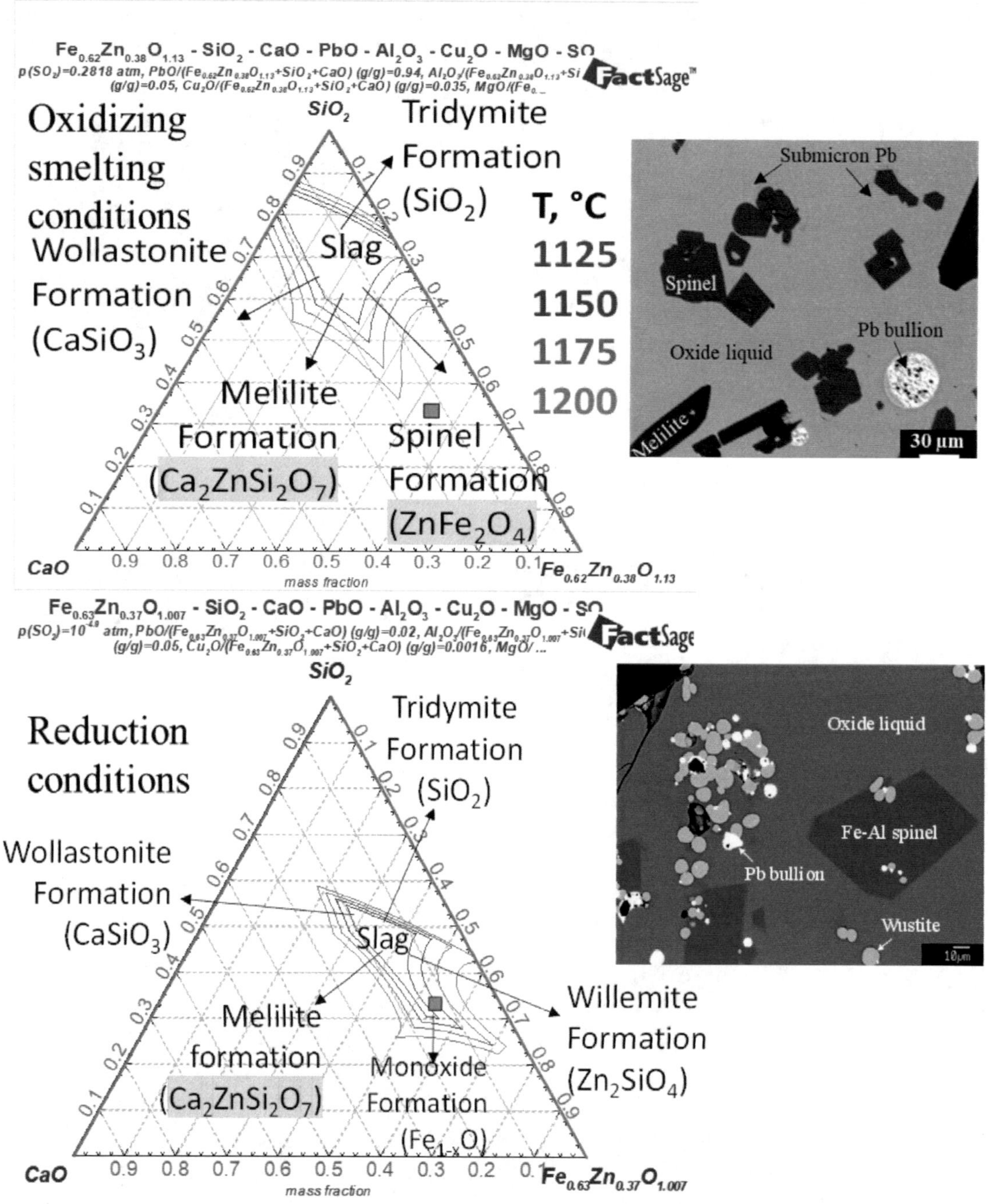

Figure 6. Fluxing diagrams for the oxidizing smelting and reducing stages of Pb/Zn smelting examining the impact of CaO/SiO_2 ratio on slag liquidus and primary phase formation predicted using PYROSEARCH database. Also shown are microstructures of the typical plant samples.

The current 20-component database under construction for the in system with "Cu_2O"-PbO-ZnO-FeO-Fe_2O_3-CaO-SiO_2-Al_2O_3-MgO-S major, Cr and Na slagging and As, Bi, Sn, Sb, Au, Ag, Ni, Co other minor elements contains oxide, matte, and speiss phases. In order to prepare an accurate description for all compositions and process conditions the numbers of subsystems to be characterised is illustrated in Figure 7. To date the PYROSEARCH team has completed ~20,000 experiments in 340

Journal of Physics: Conference Series **2738** (2024) 012007 doi:10.1088/1742-6596/2738/1/012007

subsystems (investigated or in process). Data from 88 systems are available in the literature, 109 systems are impossible to experimentally investigate using the current methodologies. There are no measured equilibrium data on 109 major + 550 minor systems. The current thermodynamic database contains 26 major solutions, 79 small solutions, >380 stoichiometric solid phases, > 100 gaseous species, > 1148 endmembers + stoichiometric compounds, defining S_{298}, ΔH_{298} and c_P vs T for each. The slag solutions include data on 24 metal cations multiplied by [O^{2-}, S^{2-}] anions = 48 endmembers, 276 binary and 2024 ternaries systems, involving the determination of 990 excess parameters.

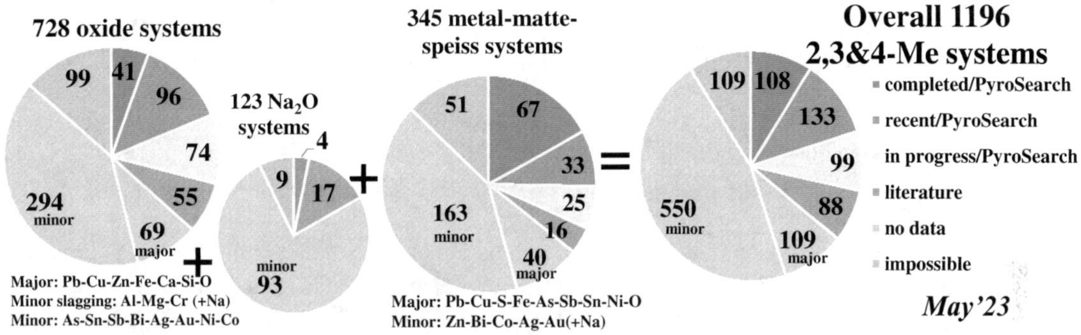

Figure 7. Summary of chemical systems in 20-component system with "Cu_2O"-PbO-ZnO-FeO-Fe_2O_3-CaO-SiO_2-Al_2O_3-MgO-S major, Cr and Na slagging and As, Bi, Sn, Sb, Au, Ag, Ni, Co, other minor elements and current status of database development.

4. Implementation

The collaboration between the industry partners and PYROSEARCH has enabled the development of a set of powerful predictive tools, which can be used to analyse a broad range of industrial systems independent of the technologies used. The collaborative program does not stop there – the important next step is demonstrating the value of these advanced tools and their implementation in industrial practice. The tools can be used by companies to improve process efficiencies and productivities, optimise the utilisation of existing plant, predict changes to plant practice to adjust for changes in process feed compositions, design new processes or operations. It has been recognized and acknowledged by industry sponsors that the successful implementation of the tools into industrial practice requires adequate training in their use. This has been achieved through a variety of actions in collaboration with sponsor companies.

Upskilling of engineering professionals is achieved through providing

- Dedicated courses in chemical thermodynamic theory, pyrometallurgy fundamentals and the application of FactSage tools to industrial problems, delivered by distance and in person.
- Extended visits, placements and secondments of engineers with the research team.
- Undertaking one-on-one projects with the research team on industrial problems.
- Undertaking Research Masters and PhD studies at PYROSEARCH.

The active collaborations between industry and the research team provide important opportunities to demonstrate the capabilities of the predictive tools. Testing the tools against industrial practice provides confidence in their potential for extended use in an industrial context – a critical factor in technology transfer.

5. Education

Universities are viewed traditionally as institutions where knowledge is stored to be passed on through academic debate and learning, in this way they serve society through providing continuity in knowledge and expertise. Through teaching they transfer advanced knowledge and skills to undergraduate and postgraduate students. Much less understood are other important roles played by University academics;

those with specialist knowledge in the metallurgical engineering can actively support the metallurgical industry in a number of ways, by

- Undertaking knowledge generation through fundamental and applied research,
- Developing and maintaining collaborative research links with industry,
- Attracting additional financial support for research from Government sources,
- Developing research infrastructure within institutions, facilities that can be utilised to support industry related research,
- Facilitating communication and support relationships between the metallurgy industry and senior university management,
- Promoting the important role of the metallurgical industry to young people and the broader community.

In addition, by insuring the sustainability of strong research teams with specialist knowledge and expertise, University academics can

- Develop and maintain industry consortia for research and education of common interest to sponsors,
- Actively assist in the optimisation of existing and development of new metallurgical processes,
- Ensure the successful implementation of research outcomes into industrial practice
- Ensure the availability of expert capabilities and advanced technological for application to individual confidential R&D support,
- Facilitate active technological exchange and links between sponsors metallurgists and between sponsor-nominated Universities,
- Minimise the adverse impacts of periodic economic cycles on the retention of knowledge and expertise for the metallurgical industry.

Such is the importance of the discipline that every country should have a strong University academic team with expertise and research capabilities in metallurgical engineering.

The number of students selecting studies in metallurgical engineering and the number of university programs offering educational opportunities in metallurgical engineering have been in decline for many years – this is a worldwide trend. There appear to be a number of contributing factors,

- Unfavourable public perceptions of the industry and lack of information on the role of metallurgy in our society.
- The lack of industry engagement and support in promoting studies and careers in the field.
- The funding formulae used by Government to support Universities, which are in the main based on the number of enrolments rather than industry or societal needs.
- University and Journal Ranking systems do not favour small specialist fields.

The lack of professionals in the field poses significant risk to the successful implementation of process improvements to existing operations and is an impediment to the major changes required for the future economies. The general public and students are now acutely aware of the impacts caused by climate change and the need to transform our industrial operations to meet these challenges. These changes in attitude represent an opportunity to change the narrative and perceptions about the industry from a negative to a positive view. The dramatic electrification of all aspects of our economies from renewable energy, transportation and electrical and computational devices has raised awareness of the need for critical metals to enable these transitions to take place. Added to this is the need for materials conservation and increased recycling of metals to minimize environmental impact and energy usage and achieve circularity in the use of valuable non-renewable resources. These are messages that the industry can use to advantage to explain the role of lead metallurgy in facilitating the recycling of critical, strategic and precious metals.

Increasing public awareness of the key role of metallurgy is necessary but is not necessarily sufficient to attract students and graduates. The industry needs to be proactive in improving workplace environments and practices so that they are attractive to prospective students and engineers. Students are looking for opportunities where they feel they can make a positive contribution to the environment and are able to use the latest technologies to achieve this. Industry can take positive steps in attracting undergraduates in their early years at university when they have an opportunity to redirect their careers into the discipline, through for example, providing on-site experiences in the form of vacation work, plant visits and internships, financial support scholarships, professional development and social activities.

The allocation of university funding is for the most part determined by class size rather than strategic needs of industry or the country. Small classes for specialist courses are more likely to be removed from the curriculum, small programs are closed in the name of efficient use of limited financial resources. This is the pattern around the world – the only way to change the current trend and provide educational resources is for industry to be more proactive. Industry leaders need to inform university management and governments of the importance of the discipline and actively engage with and provide support to those institutions providing programs in metallurgical engineering.

We strongly believe there are opportunities to increase the number of metallurgy graduates if Industry, Universities and Governments are proactive and work collaboratively together to solve this problem to their mutual benefits.

Summary

As a society we are faced with significant technical, societal and environmental challenges that must be addressed in a relatively short period of time. The urgency is such that we should use all means at our disposal to meet these important goals.

To respond to the changes in metals sources and the usage of new combinations of metallic elements changes must be made to metallurgical practice and the technologies employed. Process systems are becoming more chemically complex and process optimization is becoming increasingly more difficult. To address these changes we need to develop both the technologies and the workforce for the future.

Closer collaborations between industry and universities can assist in meeting these goals and can bring benefits to both parties by

- Delivering the new scientific information and advanced predictive tools required by industry
- Providing training on the use and implementation of these tools in industry
- Increasing enrolments in metallurgical engineering at our tertiary institutions

Acknowledgments

The PYROSEARCH team acknowledges the financial and technical support from the following: Australian Research Council (ARC) Linkage program, for the Pb research program by Aurubis AG (Germany), Boliden (Sweden), Kazzinc Ltd, Glencore (Kazakhstan), Nyrstar (Australia), Peñoles (Mexico) and Umicore NV (Belgium); for the Cu research program: Anglo American Platinum (South Africa), Aurubis (Germany), BHP Billiton Olympic Dam (Australia), Boliden (Sweden), Glencore Technology (Australia), Metso-Outotec Oy (Finland), Peñoles (Mexico), RHI Magnesita (Austria), Rio Tinto Kennecott (USA) and Umicore (Belgium); the Fe and slag recycling program by Aurubis AG (Germany), BHP Fe ore (Australia), Rio Tinto ore (Australia), Swerim (Sweden) and Umicore NV (Belgium).

References

[1] Jak E, Hayes PC. Phase chemistry of lead smelting slags. In: Lead-Zinc 2010, Vancouver, Canada. TMS, Warrendale; 2010. 1161-1176.
[2] Hayes PC, Jak E. Exploring compositional space in metallurgical slag systems. In: 9th Intl. Conf. on Molten Slags Fluxes and Salts, MOLTEN12, Beijing. USTB; 2012. paper 076.
[3] Sahu SK, Ganesan R, Gnanasekaran T. Studies on the phase diagram of Pb-Fe-O system and standard molar Gibbs energy of formation of '$PbFe_5O_{8.5}$' and $Pb_2Fe_2O_5$. J. Nucl. Mater. 2012; 426: 214-222, https://doi.org/10.1016/j.jnucmat.2012.03.032.
[4] Perez-Labra M, Romero-Serrano A, Hernandez-Ramirez A, Almaguer-Guzman I, Benavides-Perez R. Effect of CaO/SiO2 and Fe/SiO2 ratios on phase equilibria in PbO-ZnO-CaO-SiO2-"Fe2O3" system in air. Transactions of Nonferrous Metals Society of China. 2012; 22: 665-674, https://doi.org/https://doi.org/10.1016/S1003-6326(11)61230-2.
[5] Sahu SK, Ganesan R, Gnanasekaran T. The standard molar enthalpies of formation of $Pb_2Fe_2O_5$(s) and $PbFe_5O_{8.5}$(s) by acid solution calorimetry. J. Chem. Thermodyn. 2013; 56: 57-59, https://doi.org/10.1016/j.jct.2012.07.004.
[6] Oudich F, David N, Vilasi M. Phase equilibria investigations and thermodynamic modeling of the PbO-Al_2O_3 system. Int. J. Mater. Res. 2015; 106: 832-840, https://doi.org/10.3139/146.111248.
[7] Vinnik DA, Zherebtsov DA, Trofimov EA. Experimental Study and Thermodynamic Modeling of Phase Equilibria in the PbO-Fe_2O_3 System. Materials Science Forum. 2016; 870: 282-285.
[8] Shevchenko M, Jak E. Experimental Phase Equilibria Studies of the PbO-SiO_2 System. J. Am. Ceram. Soc. 2017; 101: 458-471, https://doi.org/10.1111/jace.15208.
[9] Shevchenko M, Jak E. Experimental Liquidus Studies of the Pb-Fe-Ca-O System in air. J. Phase Equilib. Diff. 2019; 40: 128-137, https://doi.org/10.1007/s11669-018-0703-7.
[10] Shevchenko M, Jak E. Experimental Liquidus Studies of the Pb-Fe-Si-O System in air. J. Phase Equilib. Diff. 2019; 40: 319-355, https://doi.org/10.1007/s11669-019-00727-x.
[11] Shevchenko M, Jak E. Experimental Liquidus Studies of the Pb-Fe-Si-O System in Equilibrium with Metallic Pb. Metall. Mater. Trans. B. 2018; 49: 159-180, https://doi.org/10.1007/s11663-017-1136-0.
[12] Shevchenko M, Jak E. Experimental liquidus study of the binary PbO-ZnO and ternary PbO-ZnO-SiO_2 systems. Ceramics International. 2019; 45: 6795-6803, https://doi.org/10.1016/j.ceramint.2018.12.172.
[13] Shevchenko M, Jak E. Experimental liquidus study of the binary PbO-CaO and ternary PbO-CaO-SiO_2 systems. J. Phase Equilib. Diff. 2019; 40: 148-155, https://doi.org/10.1007/s11669-019-00705-3.
[14] Shevchenko M, Jak E. Thermodynamic optimization of the binary systems PbO-SiO_2, ZnO-SiO_2, PbO-ZnO, and ternary PbO-ZnO-SiO_2. Calphad. 2019; 64: 318-326, https://doi.org/10.1016/j.calphad.2019.01.011.
[15] Shevchenko M, Jak E. Thermodynamic optimization of the PbO-FeO-Fe_2O_3-SiO_2 system. Calphad. 2019; 67: 101670, https://doi.org/10.1016/j.calphad.2019.101670.
[16] Shevchenko M, Jak E. Thermodynamic optimization of the binary PbO-"Cu_2O", "Cu_2O"-SiO_2 and ternary PbO-"Cu_2O"-SiO_2 systems. Calphad. 2020; 69: 101774, https://doi.org/10.1016/j.calphad.2020.101774.
[17] Shevchenko M, Jak E. Thermodynamic optimization of the binary PbO-CaO and ternary PbO-CaO-SiO_2 systems. Calphad. 2020; 70: 101807, https://doi.org/10.1016/j.calphad.2020.101807.
[18] Shevchenko M, Ilyushechkin A, Abdeyazdan H, Jak E. Integrated experimental phase equilibria study and thermodynamic modelling of the PbO-SnO-SnO_2-SiO_2 system in air and in equilibrium with Pb-Sn metal. J. Alloys Compd. 2021; 888: 161402, https://doi.org/10.1016/j.jallcom.2021.161402.

[19] Abdeyazdan H, Shevchenko M, Hayes PC, Jak E. Integrated experimental and thermodynamic modelling investigation of phase equilibria in the PbO-MgO-SiO_2 system in air. Metal. Mater. Trans. B. 2021; 53: 954-967, https://doi.org/10.1007/s11663-022-02440-6.
[20] Wen X, Shevchenko M, Nekhoroshev E, Jak E. Phase equilibria and thermodynamic modelling of the PbO–ZnO-"$CuO_{0.5}$"-SiO_2 system. Ceram. Int. 2023, https://doi.org/10.1016/j.ceramint.2023.04.223.
[21] Shevchenko M, Jak E. Experimental Liquidus Studies of the Binary Pb-Cu-O and Ternary Pb-Cu-Si-O Systems in Equilibrium with Metallic Pb-Cu Alloys. J. Phase Equilib. Diff. 2019; 40: 671–685, https://doi.org/10.1007/s11669-019-00754-8.
[22] Shevchenko M, Jak E. Experimental Phase Equilibria Studies of the Pb-Fe-O System in Air, in Equilibrium with Metallic Lead and at Intermediate Oxygen Potentials. Metall. Mater. Trans. B. 2017; 48: 2970-2983, https://doi.org/10.1111/jace.15208.
[23] Shevchenko M, Nicol S, Hayes PC, Jak E. Experimental Liquidus Studies of the Pb-Cu-Si-O System in Equilibrium with Metallic Pb-Cu Alloy. Metall. Mater. Trans. B. 2018; 49: 1690-1698, https://doi.org/10.1007/s11663-018-1249-0.
[24] Shevchenko M, Chen L, Jak E. Experimental phase equilibria study and thermodynamic modelling of the PbO-"FeO"-SiO_2, PbO-"FeO"-CaO and PbO-"FeO"-CaO-SiO_2 systems in equilibrium with metallic Pb and Fe. J. Phase Equilib. Diff. 2021; 42: 452–467, https://doi.org/10.1007/s11669-021-00911-y.
[25] Shevchenko M, Chen L, Jak E. Experimental phase equilibria study and thermodynamic modelling of the PbO-"FeO"-SiO_2-ZnO, PbO-"FeO"-SiO_2-Al_2O_3 and PbO-"FeO"-SiO_2-MgO systems in equilibrium with metallic Pb and Fe. Ceramics International, Available online 8 February 2022. 2021, https://doi.org/10.1016/j.ceramint.2022.02.042.
[26] Shishin D, Fallah-Mehrjardi A, Shevchenko M, Hidayat T, Jak E. Experimental study, thermodynamic calculations and industrial implications of slag-matte-metal equilibria in the Cu-Pb-Fe-O-S-Si system. Journal of Materials Research and Technology. 2022; 19: 899-912, https://doi.org/10.1016/j.jmrt.2022.05.058.
[27] Hidayat T, Fallah-Mehrjardi A, Abdeyazdan H, Shishin D, Shevchenko M, Hayes PC, et al. Integrated Experimental and Thermodynamic Modeling Study of the Pb-Fe-O-S-Si System: Effect of Temperature and $p(SO_2)$ on Slag-Matte-Metal-Tridymite Equilibria. Metall. Mater. Trans. B. 2023; 54: 536-549, https://doi.org/10.1007/s11663-022-02707-y.
[28] Hidayat T, Fallah-Mehrjardi A, Shevchenko M, Hayes PC, Jak E. Development of Experimental Techniques for the Phase Equilibrium Study in the Pb-Fe-O-S-Si System Involving Gas, Slag, Matte, Lead Metal and Tridymite Phases. Processes. 2023; 11: 372.
[29] Wen X, Shevchenko M, Jak E. Experiment and thermodynamic modelling of phase equilibria in PbO–"CuO0.5" and PbO–"CuO0.5"–"FeO1.5" slag systems with metal. Transactions of Nonferrous Metals Society of China. 2023; 33: 2183-2205, https://doi.org/https://doi.org/10.1016/S1003-6326(23)66252-1.
[30] Gierlotka W, Łapsa J, Fitzner K. Thermodynamic Description of the Ag-Pb-Te Ternary System. Journal of Phase Equilibria and Diffusion. 2010; 31: 509-517, https://doi.org/10.1007/s11669-010-9791-8.
[31] Vaajamo I, Johto H, Taskinen P. A thermodynamic assessment of the Cu-Fe-Pb ternary system. European Metallurgical Conference, Duesseldorf, Germany, June 26-29, 2011. 2011; 4: 1283-1301.
[32] Johto H, Taskinen P. Stabilities of phases in the Cu_2S-FeS-PbS system. International Smelting Technology Symposium (Incorporating the 6th Advances in Sulfide Smelting Symposium), Proceedings of a Symposium held during the TMS Annual Meeting. 2012: 205-211, https://doi.org/10.1002/9781118364765.ch25.
[33] Johto H, Taskinen P. Experimental research and thermodynamic model of the FeS-PbS system. European Metallurgical Conference, Duesseldorf, Germany, June 26-29, 2011. 2011; 4: 1127-1136.

[34] Onderka B, Jendrzejczyk-Handzlik D, Fitzner K. Thermodynamic properties and phase equilibria in the ternary Cu-Pb-Fe system. Arch. Metall. Mater. 2013; 58: 541-548, https://doi.org/10.2478/amm-2013-0033.

[35] Johto H, Taskinen P. Phase stabilities and thermodynamic assessment of the system Cu-Pb-S. Miner. Eng. 2013; 42: 68-75, https://doi.org/10.1016/j.mineng.2012.11.009.

[36] Huang S, Liu Y, Chen Y, Kang Z. Thermodynamic descriptions and phase diagrams for Pb-S and Bi-S binary systems. J. Electron. Mater. 2014; 43: 1237-1243, https://doi.org/10.1007/s11664-014-3004-x.

[37] Johto H, Taskinen P. Phase diagram of the system Cu_2S-PbS-ZnS at 873-1273 K. Can. Metall. Q. 2014; 53: 326-332, https://doi.org/10.1179/1879139513y.0000000105.

[38] Johto H, Taskinen P. Phase diagram of system As_2S_3-Cu_2S-PbS at 773-1173 K. Can. Metall. Q. 2014; 53: 362-366, https://doi.org/10.1179/1879139514y.0000000133.

[39] Yang H, Zhang C, Yang B, Xu B, Liu D. Vapor–liquid phase diagrams of Pb–Sn and Pb–Ag alloys in vacuum distillation. Vacuum. 2015; 119: 179-184, https://doi.org/https://doi.org/10.1016/j.vacuum.2015.05.017.

[40] Prostakova V, Shishin D, Jak E. Thermodynamic optimization of the Cu-As-S system. CALPHAD. 2021; 72: 102247, https://doi.org/10.1016/j.calphad.2020.102247.

[41] Shishin D, Chen J, Jak E. Thermodynamic Modeling of the Pb-S and Cu-Pb-S Systems with Focus on Lead Refining Conditions. J. Phase Equilib. Diffus. 2020; 41: 218-233, https://doi.org/10.1007/s11669-020-00811-7.

[42] Shishin D, Chen J, Hidayat T, Jak E. Thermodynamic modelling of the Pb-As and Cu-Pb-As systems supported by experimental study. J. Phase Equilib. Diff. 2019; 40: 758–767, https://doi.org/10.1007/s11669-019-00764-6.

[43] Dong Z, Li L, Xiong H, Liu G, Wang Y, Zhou Z, et al. Application of modified molecular interaction volume model for phase equilibrium of PbS–Sb2S3 system in vacuum distillation. Vacuum. 2022; 201: 111067, https://doi.org/https://doi.org/10.1016/j.vacuum.2022.111067.

[44] Kim S-J, Kim E-J, Sohn H-S. Distribution behavior of Bi and Pb between molten PbO-SiO_2 slag and Bi. Chawon Rissaikuring. 2012; 21: 65-71, https://doi.org/10.7844/kirr.2012.21.5.65.

[45] Shishin D, Hidayat T, Sultana U, Shevchenko M, Jak E. Experimental measurement and thermodynamic model predictions of the distributions of Cu, As, Sb and Sn between liquid lead and PbO-FeO-Fe_2O_3-SiO_2 slag. International Journal of Materials Research. 2020; 111: 733-743, https://doi.org/10.3139/146.111942.

[46] Shishin D, Hidayat T, Sultana U, Shevchenko M, Jak E. Experimental study and thermodynamic calculations of the distribution of Ag, Au, Bi and Zn between Pb metal and Pb-Fe-O-Si slag. J. Sustain. Metall. 2019; 6: 68-77, https://doi.org/10.1007/s40831-019-00257-3.

[47] Hidayat T, Chen J, Hayes PC, Jak E. Distributions of As, Pb, Sn and Zn as minor elements between iron silicate slag and copper in equilibrium with tridymite in the Cu–Fe–O–Si system. International Journal of Materials Research. 2021; 112: 178-188, https://doi.org/doi:10.1515/ijmr-2020-7857.

[48] Jak E, Hayes PC, Lee H-G. Improved methodologies for the determination of high temperature phase equilibria. Korean Journal of Minerals and Materials Institute (Seoul). 1995; 1: 1-8, https://doi.org/10.1007/BF03055319.

[49] Jak E, Hidayat T, Shishin D, Prostakova V, Shevchenko M, Hayes PC. Complex copper pyrometallurgy challenges and opportunities – Integrated Experimental Phase Equilibria and Thermodynamic Modelling Research and Implementation. In: 58th Annual Conference of Metallurgists Copper'2019, Vancouver, Canada. MetSoc; 2019. 595879.

[50] Jak E, Shevchenko M, Shishin D, Hidayat T, Hayes PC. Characterization of phase equilibria and thermodynamics with integrated experimental and modelling approach for complex lead primary and recycling processing. In: PbZn 2020: 9th International Symposium on Lead and Zinc Processing, San Diego, USA. The Minerals, Metals & Materials Series. Springer, Cham; 2020. 337-349,https://doi.org/10.1007/978-3-030-37070-1_30.

[51] Jak E, Shishin D, Shevchenko M, Hayes P. Integrated Experimental Phase Equilibria and Thermodynamic Modelling Research and Implementation in Support of Pyrometallurgical Copper Processing. In: COPPER 2022, Santiago, Chile. 2022.
[52] Bale CW, Belisle E, Chartrand P, Decterov SA, Eriksson G, Gheribi AE, et al. FactSage thermochemical software and databases, 2010-2016. CALPHAD. 2016; 54: 35-53, https://doi.org/10.1016/j.calphad.2016.05.002.
[53] Pelton AD, Decterov SA, Eriksson G, Robelin C, Dessureault Y. The Modified Quasichemical Model. I - Binary Solutions. Metall. Mater. Trans. B. 2000; 31: 651-659, https://doi.org/10.1007/s11663-000-0103-2.
[54] Pelton AD, Chartrand P. The Modified Quasichemical Model. II - Multicomponent Solutions. Metall. Mater. Trans. A. 2001; 32: 1355-1360, https://doi.org/10.1007/s11661-001-0226-3.
[55] Decterov SA. Thermodynamic database for multicomponent oxide systems. Chimica Techno Acta. 2018; 5: 16-48.
[56] Pelton AD, Chartrand P, Eriksson G. The modified Quasichemical Model. IV - Two Sublattice Quadruplet Approximation. Metall. Mater. Trans. A. 2001; 32: 1409-1415.
[57] Shishin D, Hidayat T, Jak E. Thermodynamic assessment of the $CaO-Cu_2O-FeO-Fe_2O_3$ system. CALPHAD. 2019; 68: 101715, https://doi.org/10.1016/j.calphad.2019.101715.
[58] Prostakova V, Shishin D, Shevchenko M, Jak E. Thermodynamic optimization of the $Al_2O_3-FeO-Fe_2O_3-SiO_2$ oxide system. Calphad. 2019; 67: 101680, https://doi.org/10.1016/j.calphad.2019.101680.
[59] Hidayat T, Shishin D, Decterov S, Jak E. Critical assessment and thermodynamic modeling of the Cu-Fe-O-Si system. CALPHAD. 2017; 58: 101-114, https://doi.org/10.1016/j.calphad.2017.06.003.
[60] Shevchenko M, Shishin D, Jak E. Integrated phase equilibria experimental study and thermodynamic modeling of the Cr-Si-O, Fe-Cr-O and Fe-Cr-Si-O systems. Ceram. Int. 2022, https://doi.org/10.1016/j.ceramint.2022.07.286.
[61] Shishin D, Shevchenko M, Jak E. Experimental study and thermodynamic calculations in the $CaO-Cu_2O-FeO-Fe_2O_3-SiO_2$ system for applications in novel copper-based processes. Journal of Sustainable Metallurgy. 2021, https://doi.org/10.1007/s40831-021-00341-7.
[62] Shishin D, Jak E. Critical assessment and thermodynamic modeling of the Cu-As system. CALPHAD. 2018; 60: 134-143, https://doi.org/10.1016/j.calphad.2017.12.005.
[63] Shishin D, Decterov SA, Jak E. Thermodynamic assessment of slag-matte-metal equilibria in the Cu–Fe–O–S–Si system. J. Phase Equilib. Diff. 2018; 39: 456-475, https://doi.org/10.1007/s11669-018-0661-0.
[64] Sineva S, Shishin D, Shevchenko M, Hayes PC, Jak E. Experimental study and thermodynamic modeling of distribution of elements among slag, matte and metal in the Cu-Fe-O-S-Si-(Zn)-(Al, Ca, Mg) system for copper slag cleaning applications. Journal of Materials Research and Technology. 2023, https://doi.org/10.1016/j.jmrt.2023.02.120.
[65] Shishin D, Fallah Mehrjardi A, Shevchenko M, Hidayat T, Jak E. Experimental study, thermodynamic modeling and industrial implications of slag/matte/metal equilibria in the Cu-Pb-Fe-O-S-Si system. J. Mater. Res. Technol., in Press. 2020, https://doi.org/10.1016/j.jmrt.2022.05.058.
[66] Sineva S, Shishin D, Starykh R, Hayes PC, Jak E. Equilibrium Distributions of Pb, Bi, and Ag between Fayalite Slag and Copper-Rich Metal, Calcium Ferrite Slag and Copper-Rich Metal. Thermodynamic Assessment and Experimental Study at 1250 °C. Journal of Sustainable Metallurgy. 2021; 7: 569-582, https://doi.org/10.1007/s40831-021-00360-4.

Hydraulic experiment on macro-instability in the lead smelting side-blown furnace

Shu-xiao Wang [1, 2], Zhi Wen [1], Yi Xing [1], Xue-gang Chen [2], Xun-liang Liu[1]

[1] School of Energy and Environmental Engineering, University of Science and Technology Beijing, Beijing 100083, China
[2] China ENFI Engineering Corporation, Beijing 100038, China
E-mail address: shuxiao90@sina.com
Key words: Lead; Side-blown furnace;Macro-instability; Hydraulic experiment.

Abstract. Macro-instability phenomenon in side-blown furnace was first reported in this paper, and the occurrence conditions, influencing factors and rules, eliminating measures were studied and analyzed by hydraulic experiment. The results show that the occurrence of macro-instability in side-blown furnace needs to meet certain conditions, which is mainly related to the liquid level above the jet nozzle, the number and arrangement of jet nozzles. However, the frequency macro-instability is irrelevant the nozzle height, liquid level, air flow and the nozzle diameter, the number and arrangement of jet nozzles, and the frequency of this model is 1.2Hz.The macro-instability phenomenon in side-blown furnace can be avoided or eliminated by adoption certain measures which can be used as the reference for side-blown furnace design and operation.

1. Introduction

High-speed and high-pressure gas flow is injected into high-temperature molten bath through the jet nozzle, which is widely used in metallurgical industry [1-3]. Metallurgical furnace can be classified into bottom-blown furnace, side-blown furnace, top-blown furnace, and combined-blown furnace according to the installation location of jet nozzle. Side-blown furnace is extensively applied in the smelting of nonferrous metals such as lead and copper [4]. Side-blown gas flow is injected into side-blown furnace through the lance, strongly stirring the molten bath and enhancing the heat transfer, mass transfer and chemical reaction process in side-blown furnace.

Fluid macro-instability is a large scale low frequency transient quasi-periodic phenomenon existing in stirred vessel [5]. Wizard et al. [6] reported macro-instability phenomenon in stirred vessel for the first time. Bruha et al. [7] found that macro-instability in stirred vessel was caused by the alternation of single-cycle flow and double-cycle flow in the system driven by the impeller during its rotation. Liu [8] studied the influence of double rigid-flexible impeller on the frequency of macro-instability with wavelet analysis and flow field visualization method. There are many studies on macro-instability in stirred vessel [9-11], while the macro-instability in side-blown furnace have not been reported before. The present research mainly focuses on the penetration depth [4], the formation and distribution of bubbles [12-14], behaviors of jet flow [15-17] and emulsification phenomena [18]. The side-blown furnace actually is also a typical stirred system, the molten bath in the side-blown furnace is strongly stirred by the high-speed gas jet flow. Macro-instability will affect the mass and energy transfer behavior in side-blown furnace, corrosion of refractory, and scouring effect on the furnace wall. Therefore, it is necessary to study the macro-instability in side-blown furnace.

2. Hydraulic Model Experiment

2.1. Similarity criterion

The experimental setup in the current paper is similar to a side-blown furnace provided by some company. Similarity criterion must be established to ensure that the model flow and prototype flow are similar. The similarity criterion in this study includes geometric similarity and dynamic similarity.

Geometric similarity was the first condition that must be followed in a similarity criterion. This feature indicates that a model and prototype must have the same geometric shape; their corresponding size must have the same ratio, and their corresponding angles must be similar. The value of geometric similarity is computed using equation (1), modified Froude number was selected in dynamic similarity, as shown in equation (2).The modified Froude number of the model and prototype must be the same, as described in equation (3).

$$\lambda = \frac{L_m}{L_p} \quad (1)$$

where, λ is the number of geometric similarity; L_m is the length of the model, m; and L_p is the length of the prototype, m.

$$\mathrm{Fr}' = \frac{v_g^{\ 2}}{gd_0}\frac{\rho_g}{\rho_l} \quad (2)$$

where, v_g is the velocity of gas, m/s; g is the acceleration of gravity, m/s^2; d_0 is the diameter of a jet nozzle, m; ρ_g is the density of gas kg/m^3, and ρ_l is the density of liquid, kg/m^3.

$$(\frac{v_g^{\ 2}}{gd_0})_m(\frac{\rho_g}{\rho_l})_m = (\frac{v_g^{\ 2}}{gd_0})_p(\frac{\rho_g}{\rho_l})_p \quad (3)$$

2.2. Experimental parameters

The geometric similarity number of 1:2 was established based on various factors, including dimensions of the prototype, the size of the laboratory, and the economic cost. Table 1 illustrates the specific structural parameters of both the model and the prototype.

Table 1. Structural parameters of model and prototype.

Parameter	Prototype	Model
Geometric similarity number	1:1	1:2
Diameter of side-blown equipment/mm	1200	600
Height of liquid level/mm	920	460
Height of jet nozzle/mm	500	250
Outlet area of jet nozzle/mm^2	314	78.5

Water was used instead of molten metal, and air was utilized instead of jetting gas. The parameters of prototype and the model are presented in table 2.

Table 2. Parameters of prototype and the model.

Prototype		Model	
Density of molten metal /($kg \bullet m^{-3}$)	8700	Density of water/($kg \bullet m^{-3}$)	998
Density of N_2/($kg \bullet m^{-3}$)	1.25		
Density of O_2/($kg \bullet m^{-3}$)	1.43		
Density of natural gas/($kg \bullet m^{-3}$)	0.71	Density of air/($kg \bullet m^{-3}$)	1.29
Gas flow/($m^3 \bullet h^{-1}$)	80	Air flow/($m^3 \bullet h^{-1}$)	5/10/15

2.3. Experimental device
The experimental facilities in the experiment included a furnace model, an air supply system, a valve station, a digital image processing system. Figure 1 show the schematic diagram .

The working process for this experiment is as follows:

(1) Dry and clean air was provided by an air supply system, which included air compressor, air storage tank, air dryer and QPS air filter.

(2) A valve station, which included Flow regulating valve, flowmeter, pressure gauge, gas pipelines and ball valves can realize functions of air flow regulation and display, pressure display.

(3) The air was jetted into a furnace model through a jet nozzle.

(4) Experimental phenomena were recorded by a high-speed camera and digital images were processed by the computer using graphics processing software.

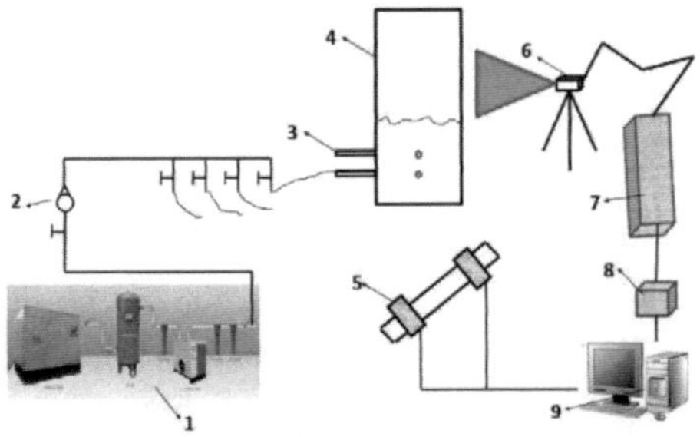

1. Air supply system; 2.Valve station; 3.Jet nozzle; 4. Furnace model;
5.High-speed camera; 6. Light sheet ; 7. Dual-cavity laser; 8.Workstation

Figure 1. Schematic diagram of experimental facility.

3. Experimental Results

3.1. Macro-instability phenomenon
During the normal injection process, the melt pool will experience irregular fluctuation and turbulence-like splashing, as shown in figure 2.

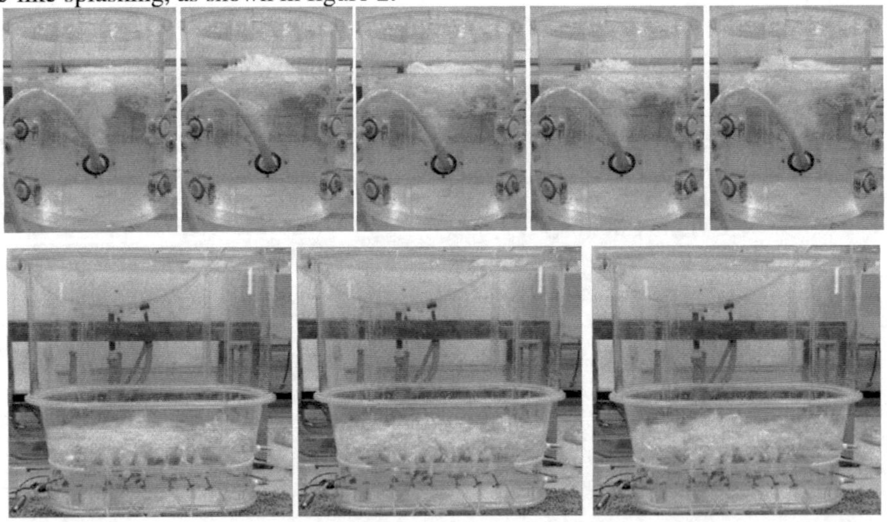

Figure 2. Normal injection process without macro-instability phenomenon.

The fascinating macro-instability phenomenon in the side-blown furnace was serendipitously observed during a hydraulic experiment. The molten pool in the furnace model presented a periodic fluctuation and swaying phenomenon, which was referred to as macro-instability phenomenon due to its large-scale oscillation at a fixed frequency. Figure 3 illustrates the liquid level fluctuation in a half cycle after the occurrence of macro-instability phenomenon. The side-blown air jet flow was ejected from the nozzle, and after a certain penetration depth, it began to float up due to buoyancy and regularly oscillate at a fixed frequency. As depicted, regular fluctuations and swaying phenomenon occur in the melting pool of the side-blown furnace, which can be directly observed behind the side-blown jet nozzle. If serious macroscopic instability occurs in the side-blown furnace, it will lead to increased corrosion of the refractory and scouring effect on the furnace wall. This will result in a shorter service life for the furnace and a higher frequency of maintenance. Therefore, it is necessary to study the factors and rules of macro-instability in the side-blown furnace.

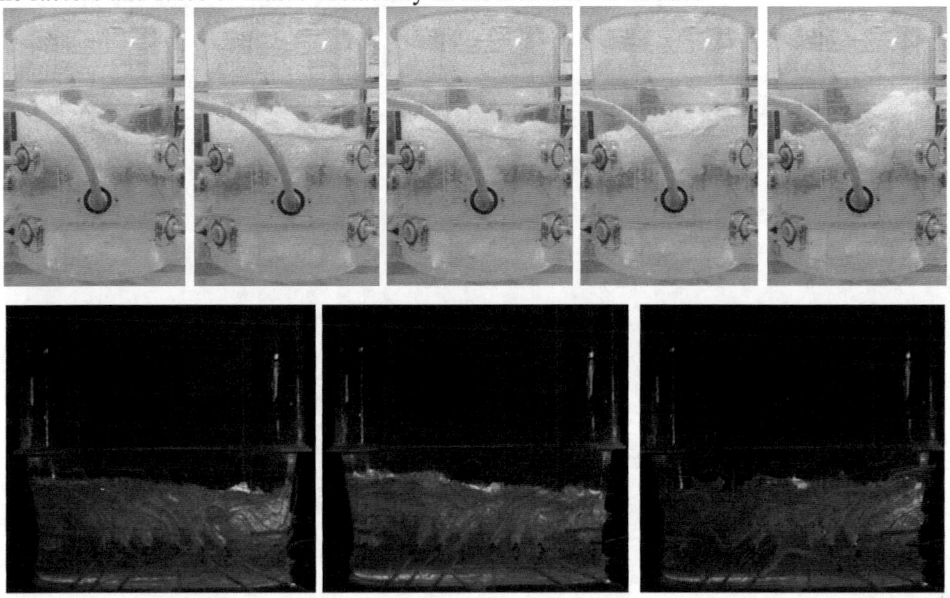

Figure 3. Macro-instability phenomenon in a half cycle.

3.2. Effects of liquid level and jet nozzle height

This paper conducted an experiment to investigate the effect of nozzle height and liquid level on the macro-instability phenomenon in the side-blown furnace. We maintained a steady liquid level of 460mm in the molten pool at 460mm and air flow rate of $10m^3/h$, while adjusting the nozzle height. Macro-instability phenomenon occurred in the side-blown furnace, at nozzle height of 125mm and 250mm, but not at a nozzle height of 375mm. We also kept air flow rate at $10m^3/h$ and maintained a constant the liquid level above the nozzle of 210mm, and observed that the macro-instability phenomenon occurred at nozzle heights of 125mm, 250mm and 460mm in the side-blown furnace, with the same frequency of 1.2Hz. Table 3 shows the experimental data, which indicates that the liquid level above the nozzle has a significant impact on the occurrence of macro-instability. In addition the nozzle height plays a key role in the occurrence of macro-instability.

Table 3. Experimental data in different nozzle height.

Nozzle height H_1/mm	Liquid level H_2/mm	Liquid level above the nozzle $H_3=H_2-H_1$/mm	Macro-instability Yes/No	Frequency/Hz
125	335	210	Yes	1.2
250	460	210	Yes	1.2
375	585	210	Yes	1.2
125	460	335	Yes	1.2
250	460	210	Yes	1.2
375	460	95	No	/

Experimental data in table 4 clearly demonstrates that the occurrence of macro-instability in the side-blown furnace is significantly affected by the liquid level above the nozzle that should be maintained within a specific range.(1) Experiments indicate that when the nozzle height was 250mm and air flow is 10m^3/h, the occurrence of macro-instability in the side-blown furnace was observed within the range of 395mm-555mm liquid level, but would not occur when the liquid level was outside the range.(2) Additionally, once the macro-instability occurred, the frequency of the model was 1.2Hz, independent of nozzle height or liquid level.

Table 4. Experimental data in different liquid level.

Nozzle height H_1/mm	Liquid level H_2/mm	Liquid level above the nozzle $H_3=H_2-H_1$/mm	Macro-instability Yes/No	Macro-instability Frequency/Hz
250	380	130	No	/
250	385	135	No	/
250	390	140	No	/
250	395	145	Yes	1.2
250	400	150	Yes	1.2
250	430	180	Yes	1.2
250	460	210	Yes	1.2
250	490	240	Yes	1.2
250	520	270	Yes	1.2
250	550	300	Yes	1.2
250	555	305	No	/
250	560	310	No	/
250	565	315	No	/

3.3. Effects of air flow and nozzle diameter

The effects of air flow and nozzle diameter on side-blown furnace were studied in this experiment, and table 5 summarizes the experimental data. The "time needed" refers to the duration of time between the initial static state and the occurrence of macro-instability inside the furnace after the injection of side-blown air flow through the nozzle. This parameter can serve as a useful measure to evaluate the difficulty of the occurrence of macro-instability and the difficulty of maintaining stability of the side-blown furnace.

Table 5. Experimental data in different air flow and nozzle diameter.

Nozzle diameter /mm	Air flow $/m^3 \bullet h^{-1}$	Macro-instability		Fluctuation of liquid level		
		Time needed/s	Frequency /HZ	The lowest point H_L/mm	The highest point H_H/mm	Range of fluctuation $H_R=H_H-H_L$/mm
10	5	41	1.2	410	550	140
10	10	28	1.2	395	565	170
10	15	23	1.2	390	575	185
4	5	38	1.2	410	535	125
4	10	19	1.2	400	560	160
4	15	9	1.2	395	575	180

Table 5 shows that the fluctuation range of liquid level increased with the increase of air flow and the time needed for the occurrence of macro-instability decreased correspondingly. When the diameter of the nozzle decreased, the fluctuation range of liquid level decreased and the time needed for the occurrence of macro-instability also decreased. The frequency of macro-instability had no relation with the air flow rate and the nozzle diameter, and the frequency of this model was 1.2Hz.

3.4. Effects of the number and arrangement of jet nozzles

The effects of the number and arrangement of jet nozzles were analyzed when liquid level was 460mm and air flow of single nozzle was $5m^3/h$, the experimental data was represented in table 6.

Table 6. Experimental data in different number and arrangement of jet nozzles.

Case	Number and arrangement of nozzles			Macro-instability	
	Number	Height/mm	angle/°	Yes/No	Frequency/Hz
①	1	250	0	Yes	1.2
②	2	125	45	No	/
		250	0		
③	2	250	0	No	/
		250	90		
④	2	250	0	Yes	1.2
		250	180		
⑤	3	250	0	No	/
		250	90		
		250	180		
⑥	4	250	0	No	/
		250	90		
		250	180		
		250	270		

Table 6 reveals that the number and arrangement of nozzles have a significant impact on the occurrence of macro-instability in the side-blown furnace. Specifically, the macro-instability phenomenon occurred in Case① and case④, while it did not in case②、case ③、case ⑤ and case ⑥. Interestingly, the frequency of macro-instability in the side-blown furnace, regardless of the number and arrangement of jet nozzles, was the same for all cases, with a value of 1.2Hz, indicating the need for further investigation.

3.5. Effects of the type of macro-instability phenomenon

Under different conditions, the side-blown furnace would occur different types of macro-instability phenomenon which were shown in Figures 4-6. The macro-instability frequency is related to the type of macro-instability phenomenon, type A "left and right shaking " is 0.93 Hz, type B "central tearing" is 1.47 Hz and type C "front and rear shaking" is 1.63 Hz。

Figure 4. Type A "left and right shaking" macro-instability phenomenon in a half cycle

Figure 5. Type B "central tearing" macro-instability phenomenon in a half cycle

Figure 6. Type C "front and rear shaking" macro-instability phenomenon in a half cycle

4. Discussions

The side-blown air jet flow was ejected from the jet nozzle, the high-pressure and high-speed air flow formed a gas column, and then penetrated a certain distance into the molten pool. After a certain penetration depth, the airflow started to float up due to buoyancy force and regularly swing with a fixed frequency. The process of "penetration-floatation-swing" was repeated, during which the kinetic energy of airflow was transformed into the kinetic energy of the molten pool in the side-blown furnace continuously. The molten pool in the furnace started to fluctuate and sway because of the floating and swinging airflow, so the molten pool had gravity potential energy and kinetic energy, and the gravity potential energy and kinetic energy were transformable. When the airflow and bubbles escaped from the molten pool, some airflow would blow the surface of the molten pool like wind. The side-blown furnace was a closed vessel with furnace wall made of refractory material, the gas-liquid two flow in the furnace would be limited by the geometrical size and affected by the wall effect. Under the combined action of various factors, the macro-instability phenomenon with fixed frequency occurred in the molten pool above the jet nozzle in the side-blown furnace, while the molten pool below the jet nozzle kept a "macro-static" state, as shown in figure 2.

The continuous injection of side-blown airflow was the main power source of the stirring of the molten pool in the side-blown furnace, and the occurrence of macro-instability was mainly related to the liquid level above the jet nozzle. If the liquid level above the jet nozzle was too high, it was difficult for the molten pool to oscillate regularly and there was no macro-instability phenomenon in the side-blown furnace because of the great inertia force. If the liquid level above the jet nozzle was too low, and the swing frequency of the side-blown air flow was very fast and its amplitude was very small, there was no macro instability because the molten pool was very shallow. Therefore, macro-instability in the side-blown furnaces occurred only within a certain range of liquid level. The number and arrangement of jet nozzles would also affect the occurrence of macro-instability. Under single-nozzle injection and face-to-face double-nozzle injection conditions, it was easy to occur macro-instability in the side-blown furnace. While when arrangement of the jet nozzles was double-nozzle injection with an angle of 45° or three-nozzle injection or four-nozzle injection, the macro-instability phenomenon would not occur in the side-blown furnace, because the regular swinging phenomenon of air jet flow can be broken by multi-airflow.

The macro-instability in the side-blown furnace have not been reported before, this paper provided a preliminary analysis of the macro-instability in the side-blown furnace. In the next step, we will do more theoretical research based on the theory of wave formation and wall effect, hydraulic model experiment and numerical simulation.

5. Conclusions

(1) The macro-instability in the side-blown furnace is a frequently occurring and complex phenomenon that requires specific conditions. Its occurrence is mainly affected by the liquid level above the jet nozzle, several factors including the nozzle height, liquid level, number and arrangement of jet nozzles can affect the occurrence of macro-instability. However, once the macro-instability occurs, the frequency remains constant and independent of the nozzle height, liquid level, air flow and the nozzle diameter.

(2) The frequency of the macro-instability in the side-blown furnace is related to the furnace specifications and macroscopic instability types. In the circular side-blown furnace model, the frequency of the macro-instability phenomenon is 1.2Hz. While in the rectangular side-blown furnace, the frequency of the type A "left and right shaking " macro-instability phenomenon is 0.93 Hz, the type B "central tearing" is 1.47 Hz and type C "front and rear shaking" is 1.63 Hz。

(3) The time needed for the occurrence of macro-instability varies depending on the air flow and diameter of the jet nozzle. An increase in air flow or decrease in jet nozzle diameter reduces the time needed for the occurrence of macro-instability, which means the macro-instability was easier to occur.

(4) Optimizing the number and arrangement of the nozzles, along with increasing the liquid level and reducing the jet nozzle height, can help avoid or eliminate the macro-instability phenomenon in the side-blown furnace, providing guidance for furnace design and operation.

References

[1] Huda N , Naser J , Brooks G A , et al. 2012 Computational Fluid Dynamics (CFD) Investigation of Submerged Combustion Behavior in a Tuyere Blown Slag-fuming Furnace. *Metall. Mater.Trans. B.* 43(5):1054-1068.

[2] Khajavi L T, Barati M.2010 Cold Model Study of Emulsification Behavior in Bottom Blown Metallurgical Baths Covered with Thick Slag. *Transactions of the Iron & Steel Institute of Japan.***50** (5):654-662.

[3] Bjurstrom M, Tilliander, Anders, et al. 2006 Physical-modeling study of fluid flow and gas penetration in a side-blown AOD converter. *ISIJ. Int.* **46**(4):523-529.

[4] Ma J, Zhou P, Cheng W, et al. 2016 Dimensional analysis and experimental study of gas penetration depth model for submerged side-blown equipment. *Exp. Therm. Fluid. Sci.* **75**:220-227

[5] Zhu J, Liu Z H, Zheng X P, et al. 2015 Macro-instability regulated by double rigid-flexible combination impeller. *CIESC Journal.* **66** (3):896-904

[6] Winardi S, Nagase Y. 1991 Unstable phenomenon of flow in a mixing reactor with a marine propeller . *J. Chem. Eng. Jpn.* **24**(2):243-249

[7] Bruha O, Fort I, Smolka P. 1993 Large scale unsteady phenomena in a mixing vessel. *Acta Polotechnica /Journal of Advanced Engineering.* **27**(6):27-33.

[8] Liu Z H, Tang Q, Wang Y D, et al.2014 Enhancement of macro-instability in a mixer-settler with rigid-flexible impeller. *CIESC Journal.***65** (1): 78-86

[9] Hasal P, Montes J L, Boisson H C, Fort I. 2000 Macro-instabilities of velocity field in stirred reactor: detection and analysis. *Chem. Eng. Sci.* **55**(2):391-401

[10] Eng M, Rasmuson A. 2012 Influence of solids on macro-instabilities in a stirred tank. *Chem. Eng. Res. Des.* **90**(8):1052-1062.

[11] Roussinova V T, Kresta S M, Weetman R. 2004 Resonant geometries for circulation pattern macro-instabilities in a stirred tank. *AICHE. J.* **50**(12):2986-3005.

[12] Kulkarni A A , Joshi J B. 2005 Bubble formation and bubble rise velocity in gas–liquid systems: a review. *Ind. Eng. Chem. Res.* **44**(16):5873-5931.

[13] Lau Y M , Sujatha K T , Gaeini M , et al. 2013 Experimental study of the bubble size distribution in a pseudo-2D bubble column. *Chem. Eng. Sci.* **98**:203-211.

[14] Lau Y M, Deen N G, Kuipers J A M. 2013 Development of an image measurement technique for size distribution in dense bubbly flows. *Chem. Eng. Sci.* **94**(5):20–29.

[15] Li T , Pougatch K , Salcudean M , et al. 2008 Numerical simulation of horizontal jet penetration in a three-dimensional fluidized bed. *Powder Technol.* **184**(1):89-99.

[16] Wang F , Yu Z , Marashdeh Q , et al. 2010 Horizontal gas and gas/solid jet penetration in a gas–solid fluidized bed[J]. *Chem. Eng. Sci* .**65**(11):3394-3408.

[17] Harby K, Chiva S, Mu?Oz-Cobo J L . 2014 An experimental study on bubble entrainment and flow characteristics of vertical plunging water jets. *Exp. Therm. Fluid. Sci.* **57**:207-220.

[18] Li X L , Liu Y , Wang D X , et al. 2018 Physical Simulation of Emulsion Phenomena in Copper Side-blown Smelting Process. *J. Northeast. Uni.* **39**(5):649-653.

A decade's Production Practice of JCC Kivcet Technique

Li Yangren[1], Jing Xudong[2,*]

1.Jiangxi copper lead & zinc metal co.,Ltd, Jiujiang, Jiangxi, China, 332500

2.Jiangxi copper lead & zinc metal co.,Ltd, Jiujiang, Jiangxi, China, 332500

Keywords: Kivcet technique; Direct production of matte; Zinc leaching residue; Technique feature; Application practice; Integrated lead and zinc smelting complex

Abstract: This article provides an overview of the Kivcet technique in Jiangxi Copper Corporation, reviews the production status and problems encountered in the past decade, introduces its technological advantages and prospect. It is believed that the Kivcet process has advantages such as strong adaptability to raw materials, stable operation, high metal recovery rate, and low cost of zinc leaching residue treatment, which is the vitality of lead-zinc smelting enterprises in the future.

1 Introduction

At the beginning of the 21st century, large-sized integrated nonferrous metal smelting enterprises named Jiangxi Copper Group Co., Ltd., based on its "Relevant Diversification" development strategy, established an integrated lead and zinc smelting plant in Hukou county, Jiujiang city, Jiangxi province in 2009. The group launched a 200,000 t/a integrated lead and zinc smelting and comprehensive utilization of lead-zinc bearing material project, its annual production includes 100,000 t/a lead and 100,000 t/a zinc. The lead smelter utilizes Kivcet lead material direct smelting process, the zinc smelter utilizes conventional hydrometallurgical zinc production process[1-3] with advanced automation technique, such as automatic zinc stripping machine[4-7]. The project was completed and put into operation in the first quarter of 2011.

The project apply an integrated lead and zinc smelting process: the lead smelter processes the leaching residue of zinc smelter with Kivcet furnace, while the zinc smelter processes the zinc-containing fume produced by the lead smelter[8]. A material circulation system is built within the plant to achieve closed loop of lead and zinc materials. This process has improved the comprehensive recovery efficiency of valuable metals, achieved maximum resource utilization and minimized emissions of "3 wastes", furthermore, met the clean production require of the time. The application of Kivcet process is of great significance for promoting the energy conservation and emission reduction career in China's lead-zinc industry.

2. JCC Kivcet Process

As shown in Figure 1, the Kivcet technique includes severe processes: mixing of metal-bearing materials and fluxes (charge preparation system), drying and ball-milling of charge, mixing of charge, coke and oxygen (feeding system), Kivcet flash smelting process, as well as slag fuming, dust collection and return, waste heat recovery, and continuous copper drossing. The Kivcet process has a wide adaptability to raw materials and can process various materials, such as lead concentrate, complex ore, complex lead-zinc materials, lead-containing residue, etc.

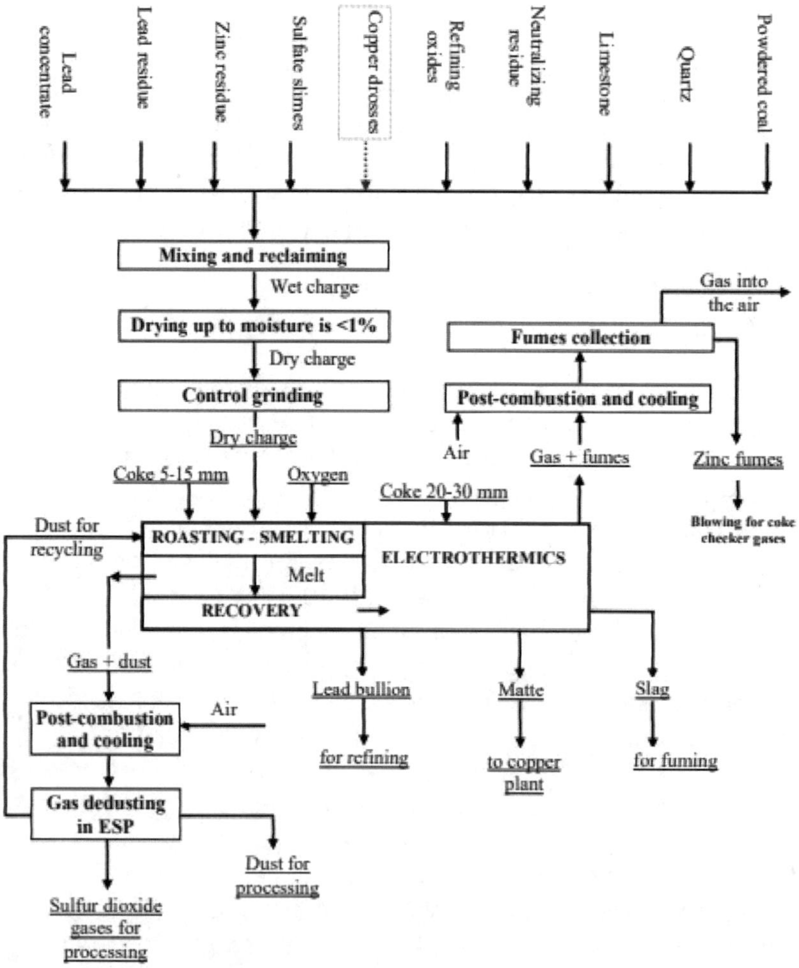

Figure 1. General flow sheet of JCC Kivcet process

2.1 Charge preparation process

In order to achieve fine control of Kivcet charge, the smelter utilizes bin proportioning method together with computer software and manual supervisory to complete the charge preparation process. After preliminary screening, the charge is sent to the dry grinding process through a belt conveyor. As for dried and screened coke, part of it is fed into the reaction section burner by the belt weigher, forming a hot coke layer, while other coke is fed into the electrothermal section of the Kivcet furnace.

2.2 Drying and ball-milling process

After the charge preparation process, a φ3000×23000 mm steam dryer is used for charge drying to ensure that the moisture content of the charge is less than 1%. Before the dried charge is transported by positive-pressure pneumatic conveying system to the top silo or large storage bin of Kivcet furnace, it is ball-milled and screened to ensure that the granularity is about 1-2 mm.

2.3 Kivcet smelting process

The Kivcet equipment layout is shown in Figure 2, which is divided into two parts according to smelting function: the smelting section and the electrothermal section. The smelting section is mainly composed of reaction shaft, setting bath, uptake shaft, etc. The electrothermal section is mainly composed of furnace roof, side wall, setting bath and electrode system. The two setting baths are integrated. During operation, the smelting section is separated from the electrothermal section by means of a partition wall, made of water cooled copper elements, partially immersed into the slag bath.

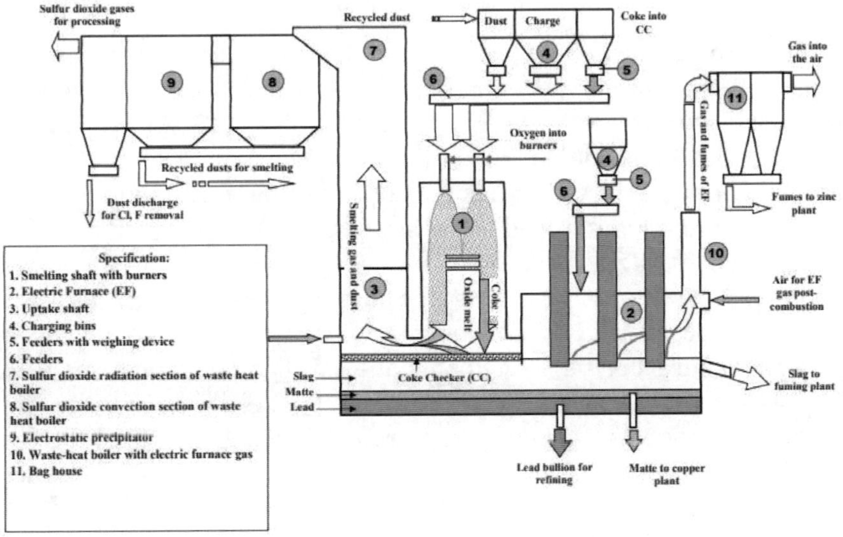

Figure 2. The Kivcet equipment layout[9]

Accurately measured furnace charge and oxygen are fed into the reaction shaft through four venturi charge burner to complete reactions such as smelting, oxidative desulfurization, decomposition, interaction and slagging. Due to its large size, coke falls faster, exothermic oxidation reaction occurs on coke's surface as it falls, then a "coke checker" floats on the surface of setting bath. The formation of coke layer is one of the essential technical features of the Kivcet process, where 80%-85% lead oxide is reduced in the coke layer, and the remaining lead oxide is reduced in the electrothermal section. The flue gas in the reaction shaft will be separated by the partition wall, and the melt will spontaneously floats to the electrothermal section under the effect of the potential energy difference between the liquid level.

In the electrothermal section, a variety of reactions can occur, including: the unreduced lead oxide in the slag is reduced again by coke, the gravitational separation of lead and dross, and the reduction volatilization of zinc oxide in the slag. The upper dross and lower bullion layered in the electrothermal section will be intermittently discharged from the slag taphole and bullion taphole.

Sulfurous gas generated from the reaction shaft goes into the shaft boiler to generate medium-pressure steam for electricity generation through heat recovering. The dedusted gas is directed to the sulfuric acid workshop for acid production, and the collected dust is returned to the reaction shaft of Kivcet furnace.

The zinc vapor generated by reduction process in the electric furnace needs to be re-oxidized into zinc oxide gas by blasted air. The zinc oxide gas will be transported to the zinc smelter after dust collection and purification. The cooled, cleansed gas can measure up the national emission standard, and is finally vented to atmosphere.

2.4 Continuous copper drossing process of bullion

The bullion at the bottom of the Kivcet furnace is discharged from the lead taphole by punching, the molten bullion is conducted to the continuous copper drossing furnace, where the bullion liquid circulates[10]. Sulfur is added to the furnace, so that the copper in the bullion will be converted into matte for copper removal. After this process, the copper content of bullion is generally less than 0.1%. Then the bullion is added to the smelting pot and pumped into the anode casting pot, so that large anode plate (each weigh about 370 kg) can be casted in vertical mold casting machine. The cooled anode plate is sent to the bullion electrolysis process through an automatic conveyor line. The anode residue generated from electrolysis is returned to the smelting pot through the automatic conveyor line for recasting. The 3D effect of the Kivcet furnace and the continuous copper drossing furnace is shown in Figure 3.

Figure 3. 3D effect of the Kivcet furnace and the continuous copper drossing furnace[11]

2.5 Slag fuming process

An 18m^2 slag fuming furnace is built to deplete the metal-rich slag generated from the Kivcet furnace, the slag is intermittently discharged into the slag fuming furnace through the chute for fuming. The essence of Kivcet slag fuming is reduction volatilization of valuble metals, where fine coal acts as reducing agent. The fine coal measured by an annular balance, is conveyed by primary air to the mixing burner to mix with the secondary air. The air pressure for fine coal transportation is generally 0.6-0.8 MPa. The pressure of primary and secondary air, blown into the slag fuming furnace, is about 0.1 MPa, which can effectively disperse the fine coal, stir the slag melt inside the furnace, drive the reduction volatilization process of metal oxides. The high-temperature flue gas generated from fuming process in the slag fuming furnace is treated by heat recovery and purification. The secondary zinc oxide dust collected by the procedure above will be sent to the zinc smelter for treatment. The granulation slag of fuming furnace is sent to the stock yard for storage or sale through a belt conveyor.

3. Production Practice of Kivcet technique

3.1 Operation overview

The initial construction of JCC Kivcet furnace was completed in February

2012, then a trial operation was launched in March for practical production inspection of Kivcet technique. The furnace has been in operation for 11 years since 2012, only three shutdown maintenances were conducted in 2014, 2018 and 2021. Table 1 provides an overview of the previous shutdown maintenances.

Table 1. Overview of the Kivcet shutdown maintenances

Number	Year	Operation time (month)	Maintenance project
1	2014	24	1)Replace lead tapping jacket; 2)Replace a small amount of boiler tubes in the uptake shaft.
2	2018	47	1)Replace chrome bricks on the side wall and roof of the electric furnace; 2)Replace lead tapping jacket; 3)Repair overlay of uptake shaft.
3	2021	42	1)Replace chrome bricks of furnace bottom; 2)Replace setting bath jacket; 3)Repair overlay of uptake shaft.

3.2 Problems encountered in production

During years of operation practice, there are some equipment problems and process issues encountered, for example:

(1) Leakage of the Kivcet boiler. One same boiler tube in the uptake shaft leaked several times, there might be problems in the installation process of the boiler.

(2) Large amount of furnace accretion. This is a disadvantage for melt discharge.

(3) The inefficiency of the coke layer. This leads to a higher lead content in the slag.

(4) Large amount of accretion in furnace and electrostatic precipitator. Therefore, the operation of the furnace had to be interrupted for cleaning.

(5) Excessive crack and consumption of electrode.

3.3 Technical optimization of Kivcet technique

In face of problems arising in production, the company has established a technical team to further study the process. They carried out a large number of experiments, and actively tried to find new solution, so that the production can accord with or superior to the designed standards.

(1) Mastering the control method of Kivcet process. This has significantly reduced the amount of accretion and improved the productivity of the furnace. The daily feeding amount has increased from 1164t (designed) to over 1250t

(actual).

(2) The Kivcet furnace is designed to process slag generated from production of 100,000 t/a zinc, but the processing capacity of zinc slag in lead smelter can match 130,000 tons of zinc in 2022. The significant increase in slag processing capacity has led to a change in the Kivcet charge structure. In particular, the amount of lead in charge decreased from 30% (designed) to 20%.

(3) Improvement of the coke layer efficiency. This results in the lead content decrease to 2%-3%, lower than the planning value 4%.

(4) The prolongation of the Kivcet furnace production life, the corresponding maintenance cycle is also prolonged from the original 2 years to 4 years.

(5) Two 40mm diameter matte tapholes were installed on the side wall of the electric furnace to carry out the practice of directly producing matte (copper content aroud 40%) in the Kivcet furnace.

4 Conclusion

The decade's production practice of the Kivcet furnace amply proves that the Kivcet process has advantages such as high energy utilization rate, strong raw material adaptability, high zinc leaching slag treatment capacity, low operating cost and effective comprehensive recovery.

At present, the domestic and international processing charge of lead ore continue to hover low. In the future, we will further enhance our comprehensive recovery capacity, not only aims at increasing the production of matte in the Kivcet furnace, but also studying the advanced treatment process of matte to further recover metals such as lead, zinc, silver and indium. Moreover, we will try to find detailed solution for arsenic and selenium recovery and treatment. We believe that the feature of Kivcet technique - processing complex polymetallic raw materials is the vitality for lead-zinc smelting enterprises of the future.

Reference:

[1] Wei Chang, Deng Zhigan, Yang Fuxian. Reduced leaching process of hydrometallurgical zinc leaching residue [J]. Nonferrous Metals (Smelting Part), 2022 (5): 1-7

[2] Qiu Wenting. Research on the process of zinc metal recovery from zinc purification residues in hydrometallurgical processes [J]. World Nonferrous Metals, 2021 (16): 175-176181

[3] Liu Yuqin, Chang Yongfeng. Exploration of green metallurgical process for comprehensive utilization of secondary materials in zinc hydrometallurgy [J]. Non ferrous Mining and Metallurgy, 2020,36 (5): 51-53,56

[4] Huang Yuyong. Research situation and development of lead based anodes for zinc electrowinning [J]. Hunan Nonferrous Metals, 2012,28 (5): 37-39,73.

[5] Zhang Yuping. Research on anodes for zinc electrodeposition and its development [J]. Hydrometallurgy, 2001 (4): 169-171

[6] Wu Wenbing. Application practice of zinc electrodeposition large plate and

automated zinc stripping [J]. China Nonferrous Metallurgy, 2013,42 (6): 3
[7] Tong Yizhe. Large plate zinc electrowinning process and its production Automation [J]. Mining and Metallurgy, 2018,27 (3): 93-96
[8] Zhao Hongmei, Zhang Dianbin, Zhang Dewei, Fan Kebiao. Reducing zinc content in fuming furnace slag [J]. World Nonferrous Metals, 2021 (24): 14-17
[9] Jing Xudong. The first application practice of Kivcet direct lead smelting technology in China [J]. Copper Engineering, 2015 (4): 58-61
[10] Zhou Rong. Application of continuous copper drossing furnace in lead pyrometallurgy [J] Shanxi Metallurgy, 2015157 (5): 98-99
[11] Jiang Jianxing, Li Yangren, Guo Haijun. Practice of Kivcet lead pyrometallurgy [J] World Nonferrous Metals, 2018 (18): 7-9

Journal of Physics: Conference Series **2738** (2024) 012010 doi:10.1088/1742-6596/2738/1/012010

Evaporation and its Effects on Physico-Chemical Properties of PbO-containing Slag

Y Kang[1], J X Zhao[1,*], Y R Cui[1], B LI[1], M M Ren[1] and Z M Wang[2]

[1] School of Metallurgical Engineering, Xian University of Architecture and Technology, Xian, Shanxi, China,710055.
[2] Hanzhong Zinc Co. LTD, Hanzhong, Shanxi, China,724200.

zhaojunxue1962@126.com

Abstract. PbO as well as other volatiles will evaporate at high temperature. In the measure of relevant slag property, the slag will be kept at high temperature for a long time (even several hours). The continuous evaporation will result in uncertainty of the measured results. In another point of view, that means the measured result is no longer in consistent with the pre-determined slag composition, but a new one changed obviously. For high or medium PbO content slag, the weight loss can reach to 20~40 % at high temperature. The effect is not negligible. In this work, the evaporation and physicochemical property (melting point and viscosity) measurement of typical PbO-containing slag were jointly analyzed based on calculation, TG-DSC, melting temperature and viscosity measurement. A deviation method based on traditional measurement ways was suggested and practiced. It is proved that to take physicochemical property value with traditional method and the get the in-time composition to match the measured values is feasible. how obtain in-time composition is also discussed and tried. The work will be of extensive reference significance to other slag system with higher volatiles.

1. Preface

Up to now, pyro-metallurgy is still the main route for Pb and part of Zn production. In recent years, a lot of pyro-metallurgical process have been developed for Pb smelting, such as QSL, SKS, Kivcet process, Ausmelt/Isasmelt, Kaldo, and bottom blowing/side submerged blowing and so on [1, 2]. In all these processes, PbO and ZnO containing slag is common. How to control the slag physicochemical properties, such as melting point, viscosity, surface tension, heat and electric conductivity, heat transfer property, component activity. etc., is of great importance to smelting process designing, process control, inclusions removal, product quality, as well as other technical and economic indexes.

It is well known that PbO and ZnO, flouride and chloride are material trend to evaporate at high temperature [3-7]. In the physicochemical property measurement, these volatiles will evaporate continuously at heating or high temperature keeping process, so to the production process. The higher, the temperature is, or the longer the keeping time is, the great the loss of volatile from the slag is. That means the slag composition will be changed continuously. As a result, the measured slag physicochemical properties will change [5-6]. It is referred as uncertainty of slag property for brief.

There have been a lot of research work on the properties of slag containing PbO and ZnO. To focus on slag of Pb smelting, the slag can be devided into three sorts: high PbO content (40~60 % Pb), medium PbO (10~39 % Pb) content, lower PbO content (<10 %). Most of the work are in the field of lower PbO content slag, i.e. the final slag after PbO reduction. The relevant slag system are $FeO-SiO_2$, $FeO-CaO-SiO_2$, $FeO-Fe_2O_3-SiO_2-CaO$, $FeO-SiO_2-CaO-CaF_2$, $FenO-MgO-CaO-SiO_2$, $FeO-Fe_2O_3-SiO2-CaO-MgO$, $FeO-ZnO-CaO-SiO_2$, $FeO-ZnO-CaO-SiO_2-Al_2O_3$, $FeO-ZnO-CaO-SiO_2-MgO$, $FeO-ZnO-CaO-SiO_2-Al_2O_3-MgO$ [8-12].

In oxidation smelting or PbO direct reduction step, the main slag is $PbO-FeO-Fe_2O_3-CaO-SiO_2-ZnO$ system with higher PbO and ZnO content. Only a few of research related to these slag [13-16]. but they focused on phase equilibrium and gave little practical results on slag properties. As what they pointed out that the PbO and ZnO evaporation can affect the accuracy of physicochemical properties measurement. That means that up to now there is no reliable PbO and ZnO containing slag physicochemical property data to support pyro-metallurgical production process development and optimization.

Our group measured the weight loss of PbO slag with TG-DSC. It is detected that the weight loss (i.e., evaporation) to slag with initial PbO content (determined in slag sample preparing) of 21.54 % and 43.09 % can reach to 21.62 % and 38.69 %, respectively, at temperature between 736 °C to 1450 °C [17]. Thus, in the physicochemical property measurement of slag at high temperature, the evaporation is absolutely not neglectable. The traditional slag property measurement method in not fit these slags again [5, 6, 17]. Then how much the evaporation effect is? What is the effect mechanism? How to make it to get the correct property data? How to get the real value? These questions should be replied urgently for PbO and ZnO containing slag control as well as to pollution control.

2. Solution to uncertainty of slag property caused by evaporation

2.1. Weight loss of some PbO and ZnO containing slag at high temperature

The weight loss of some PbO and ZnO containing slag were measured at high temperature as listed in table 1 and figure 1 [17]. To avoid the effect of moisture in the samples, only the weight loss at higher temperature, i.e. above 410 °C were taken. It can be seen that the weight loss is very large. That means the slag composition will change obviously. The higher the PbO content is, the great the weight loss is. To the slag with FeO/SiO_2, CaO/SiO_2 and PbO at 1.54, 0.4 and 43.09 % respectively, the weight loss can reach to 38.69 % when the temperature arises from 736 °C-1450 °C. From figure 1, it is clear that to a sample, the weight loss will increase with the temperature increase. Of course, the weight loss will also increase with the keeping time increase at a high temperature.

To all of the physicochemical property measurement, the slag (or flux) will experience heating, melting and high temperature keeping. If the slag evaporation happens and the weight loss is great like listed in table 1 and figure 1. The effect of evaporation to slag property measurement result is certainly not negligible. Especially at the temperature above melting point.

Table 1. The weight loss of slag sample with PbO at high temperature [17].

Slag type	Composition			Weight loss	Weight loss
	FeO/SiO_2	CaO/SiO_2	PbO	410 °C-736 °C	736 °C-1450 °C
Slag with low PbO content	1.54	0.4	2.69 %	12.88 %	3.95 %
	1.8	0.6	2.69 %	13.3 %	4.66 %
Slag with medium PbO content	1.54	0.4	21.54 %	9.34 %	21.62 %
	1.8	0.6	20 %	9.69 %	24.4 %
Slag with high PbO content	1.54	0.4	43.09 %	7.92 %	38.69 %
	1.8	0.6	40 %	8.09 %	43.3 %

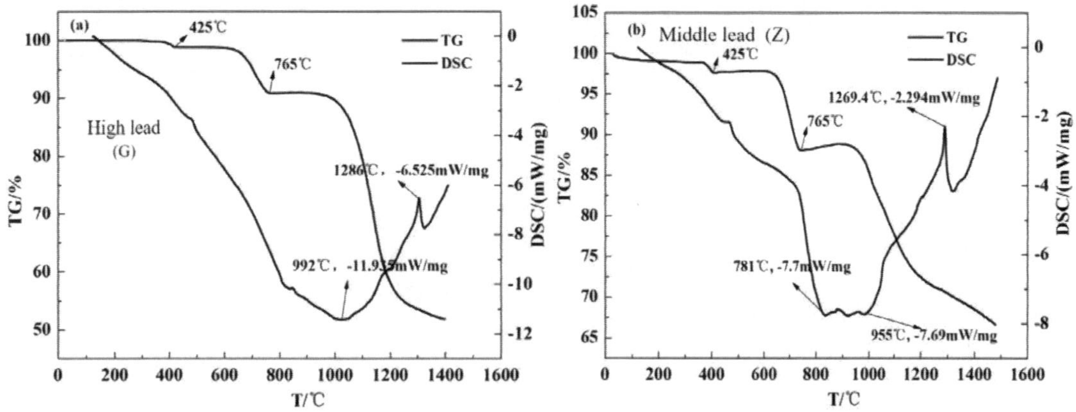

Figure 1. The TG-DSC results of PbO-containing slags [17].

2.2. The consideration to the uncertainty of slag property measurement

In traditional method, to take melting point measuring as an example, the slag composition is determined first, the sample is prepared, and then heating slowly, until the sample smelt at a temperature. The temperature is considered as melting point of the slag with the initial composition. This is no question to the slag sample without strongly evaporation. But to the samples like in table 1, the slag composition will change continuously, because the evaporation. The final temperature corresponding to the melting point with traditional method is not in accordance with the initial composition again, but a composition changed from initial one [5-6, 18, 19]. In another point of view, that means all the physicochemical property results are not accurate without consider the evaporation. The difference in heating route, temperature, and keeping time can deduce to the difference weight loss, as a result, the difference composition changes and different property measurement result. That the main cause of uncertainty of slag property measurement.

Then how to solve this uncertainty? Since the evaporation leads to the chemical composition change, results in the property measurement result not in accordance with the original composition, to take the in-time composition to corresponding with the property measurement result by traditional method will be reasonable theoretically. Here how to acquire the in-time composition become the key problem.

2.3. The new method to get the property of slag with volatile

To take the melting point measurement as an example. The Hemispheric temperature (traditional melting point measuring) method was used. The sample was heated at rate of 15 °C/min, until the height drops down to 50 % of initial height. The temperature is taken as the melting point T_m. The weight loss at the same heating rate was carried out in the same time. Details is shown in figure 2.

To the pre-melted slag sample, the T_m is 1285 °C and the weight loss is less. The evaporation is neglectable. To the mixed slag sample, the measured T_m is 1385 °C, and the weight loss at 1385 °C is about 8.3 %. If the composition at 1385 °C can be obtained or calculated based on weight loss, the measured T_m should be in corresponding to the new composition, not the initial one.

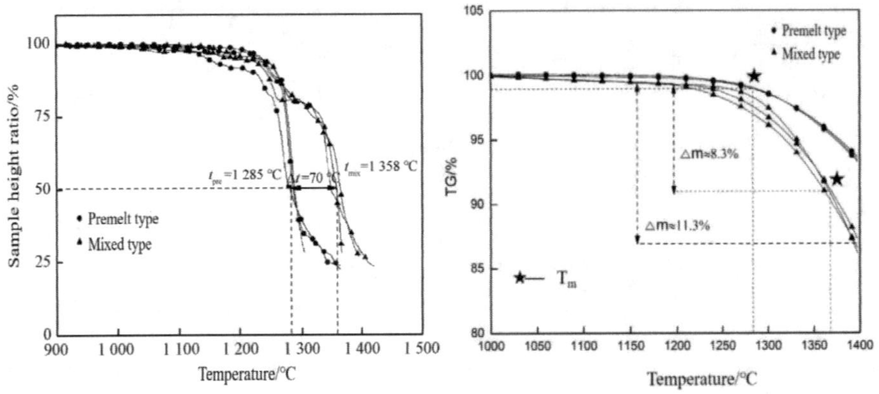

Figure 2. Schematic program of the new melting point measurement [4].

3. The Property measurement to slag with PbO and ZnO

3.1. Determination of evaporation component and in-time composition

The vapor pressures of PbO, ZnO, Pb, Zn at different temperature was calculated as table 2. From the results, it can be seen that if no reduction happens, the evaporation component in slag should be PbO first because of its higher vapor pressure and then ZnO. At reduction atmosphere, PbO, ZnO can be reduced and evaporates at the metal gas state. The test results support this conclusion. The collectives of condensed material are mainly PbO as listed in table 3, especially to higher PbO slag.

Table 2. Calculated vapor pressures of Pb, Zn, PbO and ZnO at different temperature, Pa.

T/°C	P_{Pb}	P_{PbO}	P_{Zn}	P_{ZnO}
900	4.224×10	4.037×102	0.932×105	
1000	1.858×102	2.660×103	2.334×105	
1100	6.552×102	1.307×104	5.078×105	
1200	1.937×103	5.082×104	9.878×105	
1300	4.968×103	1.637×105	1.756×106	2.0×102
1400	1.134×104	4.515×105	2.901×106	4.0×102

Table 3. The vapour condensed material composition by test, %.

Sample No.	PbO	ZnO	FeO	CaO	SiO_2	FeO/SiO_2	CaO/SiO_2	Condensed materials, %	
								PbO	ZnO
1	2.69	13	44.97	11.24	28.10	1.6	0.4	73.3	26.7
2	10	11	42.13	10.53	26.33	1.6	0.4	85.98	14.02
3	20	10	37.33	9.33	23.33	1.6	0.4	93.5	6.59
4	40	6	28.80	7.20	18.00	1.6	0.4	98.8	1.2

So, it is easy to combine the TG results to get the in-time composition at the melting point by traditional way, and then to match the melting point with the in-time composition.

Apart from this way, another method can be used to get the in-time composition by roasting and composition analysis, especially to the complicated evaporation [20]. To roast the slag, sample up to different temperature at the rate of melting point measurement and then cooling down quickly to analyze composition of the rest sample. The device is shown as figure 3. One of the typical test results of the slag sample 3 with 20% PbO is shown in figure 4. It can be seen that with the temperature increase, PbO and ZnO content in slag decreased. At the same time, content of another component (not volatile) increased. If the melting point of the sample is at T2, the matching composition is shown in T2 marked vertical line.

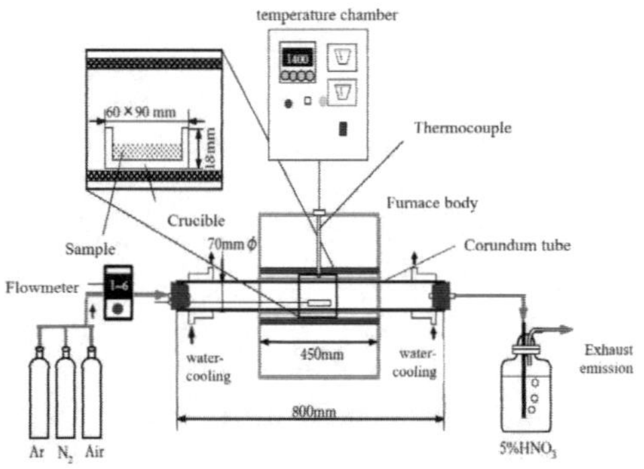

Figure 3. Device diagram for roasting test [5].

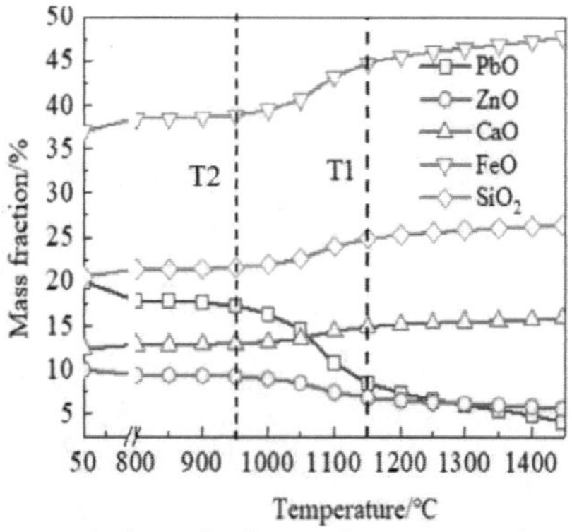

Figure 4. A typical roasting test result, sample 3.

3.2. The practice property measurement to slag with PbO and ZnO

It has been clear that the new melting point T_m measuring method is composed of two steps, measuring T_m in traditional way first, and then getting the in-time composition to match T_m. So, to the viscosity. Generally, slag viscosity is measured with rotating cylinder viscometer at different temperature, i.e., heating the sample up to a temperature above melting point first, viscosity measuring, and then lower temperature to another point, measuring repeatedly. Finally, relation between viscosity and temperature can be obtained as shown in figure 5. According to the new method, the viscosity of sample 3 at 1150 °C is about 0.25 Pa·s. The matching in-time composition is taken as vertical line T1 in figure 4, not the composition of sample 3 in table 3. The in-time composition is about 9%PbO-7%ZnO-25%SiO_2-45%FeO-14%CaO, a great difference from initial composition. In the same way, the viscosity of a serious different slag composition at different temperature can be picked up from the η-T relation of one sample by traditional method.

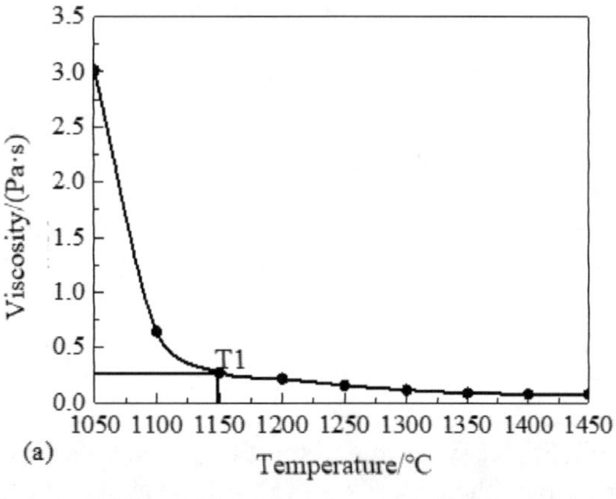

Figure 5. The η-T relation by traditional method, sample 3 with 20 % PbO [21].

4. Evaluation of measurement error with traditional method

Measurement uncertainty in another point reflects the deviation from the real value that is often difficult to get. In melting temperature measuring, the result can be affected by heat transfer, delay of particle reaction, partial melting, sample evaporation and so on. The effect of different factors is shown in figure 6 [20].

To a slag with no volatile, the general result is as line ①. T_m measured will increase with the heating rate. If heating rate is close to zero, we get the approximate real value (line ②, point D). But to the slag with volatile, such as PbO, T_m measured will decrease with the heating rate as line ③. The trend is completely different. If the heating rate is tended to zero, the volatile will evaporate completely and the measured Tm trend to point D, i.e., the same value as no volatile slag. Suppose the effect of all other factors is the same, the only difference is volatile evaporation. The temperature difference between line ① and line ③ is the effect of evaporation. With the increase of heating rate, the evaporation is restrained and lead to T_m measured decreased. Evaporation effect is more than other effect. To sum up, the result of line ③ appeared. Then what is the real value? If at the rate of 30 °C/min, the evaporation can be restrained effectively, the error of T_m measured can be modified only refer line ① to line ②. To move line ① down parallelly, we got line ④ that shows the result of evaporation effectively controlled. The point F is the approximate real value. Gap between line ④ and line ③ as GH reflects the error caused by evaporation.

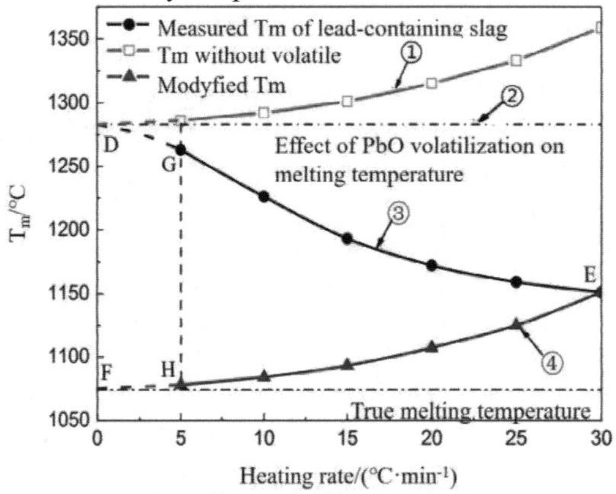

Figure 6. The effect of different factors on melting temperature deviation [20].

5. Conclusion

Focused on slag with PbO, ZnO or other volatiles, the following points can be concluded.

(1) PbO will evaporate obviously at high temperature, lead to composition change and property measuring error that not negligible. Traditional method is not fit PbO containing slag again.

(2) The higher the temperature is, the more the volatile content in slag, and longer the keeping time at high temperature, the greater the weight loss (caused by evaporation), composition change and the effect to property measurement is.

(3) Measuring the physicochemical property by traditional method and then take the in-time composition to match measured result is reasonable and workable. It suits for slag with volatiles, such as PbO-containing slag.

(4) In-time composition is the key problem for the new method. It can be obtained through different ways. If the volatile component is only one, the in-time composition can be gotten with TG. If the evaporation is complicated, the in-time composition can be obtained by slag roasting parallelly. Predicting kinetic model for in-time composition prediction based on evaporation mechanism is the final resolution.

(5) A method for real melting temperature value and measurement error evaluation based on test at different heating time is suggested.

References

[1] Wang C Y, Gao W and Yin F 2012 *Nonferrous Metals (Extractive Metallurgy)* **4** 1–5
[2] Li W F, Zhang X G, Guo X Y and Zhang C F 2010 *Min. Metall.* **19** 54–7
[3] Zhao J X, Chen Y M, Li X M, Cui Y R and Lu X L 2011 *J. Iron Steel Res. Int.* **18** 24–53
[4] Zhao Z Y, Zhao J X, Tan Z X, Qu B Q, Lu L and Cui Y R 2020 *Iron and Steel* **55** 123–8
[5] Zhao Z Y, Zhao J X, Cui Y R, Lu L and Wang G H 2020 *J. Min. Metall., Sect.* B **56** 51–7
[6] Wang Z, Zhao J X, Kang Y, Hu A L, Xin X and Cui Y R 2022 *Steel Res. Int.* **93** 2100556.1–8
[7] Cui Y R, Wang G H, Zhao J X, Guo Z L, Tang H L and Fan H Y 2018 *Chin J Process Eng* **18** 393–8
[8] Shahbazian F. 2001 *Scand. J. Metall.* **30** 302–8
[9] Seki K and Oeters F 1984 *Trans. Iron Steel Inst. Jpn.* **24** 445–54
[10] Ji F Z, Sichen D and Seetharaman S 1997 *J. Min. Metall., Sect.* B **28** 827–34
[11] Dai X, Gan X P and Zhang C F 2003 *Trans. Nonferrous Met. Soc. China* **13** 1451–53
[12] Jak E, Zhao B J and Hayes P C 2002 *J. Min. Metall., Sect.* B **33** 877–90
[13] Jak E and Hayes P C 2002 *J. Min. Metall., Sect.* B **33** 817–25
[14] Jak E and Hayes P C 2003 *J. Min. Metall., Sect.* B **34** 369–82
[15] Jak E, Zhao B J, Hayes P C and Liu N G 1999 *J. Min. Metall., Sect.* B **30** 21–7
[16] Cui Y R, Guo Z L, Chen A L, Li M K, Li X M and Zhao Z Y 2016 *Chin. J. Rare Met.* **40** 928-33
[17] Wang G, Cui Y R, Yang Z, Guo Z L, Zhao L, Li X M, Zhao J X and Tang W D 2020 *J. Min. Metall., Sect.* B **56** 59-68
[18] Ge B L, Dang Y M, Zhao J X, Tang W D, Li X M and Qiu S T 2015 *Proceedings of the 6th International Congress on the Science and Technology of Steelmaking* (Beijing: CHINA MACHINE PRESS) pp 226-9
[19] Cui Y R, Li M K, He J X, Li X M, Zhao J X and Chen A L 2013 *Chin. J. Rare Met.* **37** 473-8
[20] Zhao J X, Wang Z, Wang G H, Hu A L, Xin X, Ren M M, Cui Y R, Li B and Zhang Z K 2021 A method to evaluate the effect to slag melting point of volatile evaporation[P], No. 202111417614.6, China
[21] Zhao J X, Wang Z, Wang G H, Hu A L, Xin X, Ren M M, Cui Y R, Li B and Wang H 2021 A method to evaluate the effect to slag viscosity of volatile evaporation[P], No. 202111415419.X, China

Acknowledgments

The work was supported by NSFC（NO.51674185，51674186).

Analysis of bioleaching characteristics and multi-element dissolution behavior of complex zinc ores

Jiafeng Li [1,*], Xuemin Qiu[2], Xiaobin Qiu[2] and Guili Zhou[3]

[1] School of Mines, China University of Mining and Technology, Xuzhou, China
[2] Zijin Mining Group Co., Ltd, Shanghang, China
3 Xuzhou Construction Machinery Technician College, Xuzhou, China
Jiafeng Li, Email: lijiafengsy@163.com

Abstract. In order to recover low-grade complex zinc ore in a reasonable way, this study adopts bioleaching method to study it. The ore samples contain 1.52%, 2.03% and 14.4% zinc, respectively, which occurs in the form of sphalerite. Other major minerals include pyrite, galena, quartz and mica. The inoculation of the domesticated strain was basically free of adaptation period, and the cell concentration could be rapidly increased after a short decrease. The leaching extent of zinc increased continuously, while the leaching rate decreased gradually. After the bioleaching process, sliver, lead and iron were mainly present in the residue phase. X-ray diffraction spectroscopy analysis showed that sphalerite, galena and pyrite were dissolved, and the latter two further precipitated to produce $PbSO_4$ and jarosite. In addition, the dissolution of calcium compounds can lead to the formation of gypsum precipitation. These precipitates covered the fresh ore surface may hinder the further bioleaching process. The Exponential model was used to simulate the bioleaching process, and it was found that the fit coefficients were all greater than 0.98, and a reasonable leaching cycle was further discussed. The results provide a good basis for the economic and environmentally friendly recovery of low-grade complex zinc ores.

1. Introduction

Zinc is abundant in the earth's crust and has good corrosion resistance. It is widely used in electroplating, alloys, dry cell batteries, rubber, paints, printing, enamelling, fibers, medicine, and other industries. Especially, it has a critial position in the production of batteries (battery surface) [1, 2]。 With the development of the mining industry, rich and easy-to-handle zinc ore are decreasing, but the demand for zinc is increasing day by day. As a major producer and consumer of zinc resources, China produced 6.29 million tons of refined zinc in 2022, while the demand exceeds 7 million tons. With the expansion of zinc production, the conflict between supply and demand becomes more prominent. Therefore, the abundant low-grade complex zinc ores must be exploited [3, 4]. Traditional pyrometallurgical and hydrometallurgical processes, although effective, are energy intensive, expensive, prone to secondary pollution, and unsuitable for treating low-grade ores. Bioleaching technology has the advantages of mild condition, no secondary pollution, low production cost, and environmental friendliness. Thus, it has economic advantages and promoting value in treating low-grade complex ores and refractory ores [5, 6, 7].

Bioleaching is a process that uses the oxidation/reduction properties or metabolites of microorganisms themselves to separate useful components from the original material in a soluble or precipitated form. It has been applied in commercial production for bioleaching of copper and uranium ores, as well as in the bio-oxidation of complex refractory gold ores [8, 9]. As the bioleaching mechanism is continuously researched and explored in depth, the technology has also shown good prospects for the extraction of other metals such as zinc, cobalt, nickel, molybdenum, and manganese [10, 11]. The bioleaching process of zinc ore was proposed by Billiton in 1998, known as the BioZINC process, and a bioleaching pilot plant was established at the Pering zinc mine in South Africa in 2000 [12]. In the case of sphalerite, the bioleaching mechanism is shown in Eqs. 1-4 [13, 14]. Shake flask experiments showed that the extraction of zinc from polymetallic sulfide ores could reach 97.1% after 10 days of bioleaching [15]. Some scholars used bioleaching-sulfide precipitation method to remove metals from lead-zinc ore tailings. The results showed that the zinc bioleaching rate was about 97.38% after 50 days at 10% pulp density, which obeyed the shrinking core model [16]. Bioleaching experiments using mixed acidic *Thiobacillus* found that the redox potential was higher than that of the chemical control method [17]. It can be seen that the bioleaching method is suitable for the treatment of zinc sulfide ores.

$$ZnS + 2H^+ + 0.5O_2 \xrightarrow{Microbe} Zn^{2+} + S^0 + H_2O \tag{1}$$

$$ZnS + 2Fe^{3+} \longrightarrow Zn^{2+} + S^0 + 2Fe^{2+} \tag{2}$$

$$2S^0 + 3O_2 + 2H_2O \xrightarrow{Microbe} 2SO_4^{2-} + 4H^+ \tag{3}$$

$$2Fe^{2+} + 2H^+ + 0.5O_2 \xrightarrow{Microbe} 2Fe^{3+} + H_2O \tag{4}$$

In order to better provide guidance for the rational development of low-grade complex zinc ores, this study explores the feasibility of bioleaching technology with three zinc ores as research targets. At the same time, the evolution of system characteristics and elemental leaching patterns during the bioleaching process were analyzed. The results lay the foundation for the bioleaching application of similar ores.

2. Materials and methods

2.1. Ore sample

The complex zinc ores were taken from Inner Mongolia, China, and defined as a, b and c ores in ascending order of zinc content. The main elements of these ores are shown in Table 1. The a and b low-grade complex zinc ores containing 1.52% and 2.03% zinc, respectively, while the medium-grade complex c ore containing 14.40% zinc. In addition, they contains small amounts of silver, lead and iron. The results of X-ray diffraction spectroscopy (XRD) analysis (see "Results and Discussion" section) show that the main minerals are sphalerite, galena, pyrite, quartz, muscovite, and black zinc-manganese ore. Zinc and lead are mainly in the form of sulfides, and quartz is the main gangue.

Table 1. Main element content in the three zinc ores.

Element	a	b	c
Zn, %	1.52	2.03	14.40
Ag, g/t	16.5	9.4	85.0
Pb, %	0.57	0.79	2.85
S, %	1.94	2.43	13.7
Fe, %	5.12	6.05	7.86

The three zinc ores were first crushed and then ball milled for 20 min to obtain three ore samples with the particle sizes shown in Figure 1. It can be seen that the particle size less than 40 μm accounted for more than 80% in all three ores. Their surface areas were 0.91 m^2/g, 1.25 m^2/g and 1.2 m^2/g, respectively. The particle sizes at 90% of the ore weight were 18.59 μm, 84.04 μm and 67.01 μm, respectively.

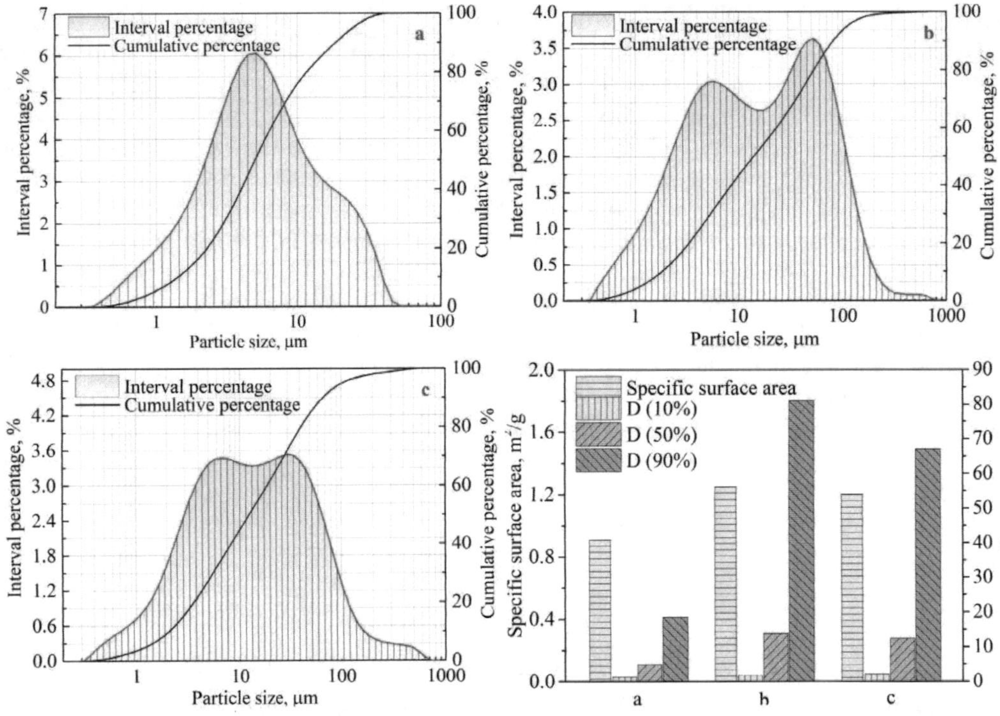

Figure 1. Particle size distribution of finely ground zinc ore samples.

2.2. Microbial strain and culture

The mixed strain used in this study mainly consisted of *Leptospirillum* spp., *Sulfobacillus* spp. and *Acidiplasma* spp. 9 K culture medium was chosen for this work and the component was: $(NH_4)_2SO_4$-3.00 g/L, KCl-0.1 g/L, K_2HPO_4-0.5 g/L, $MgSO_4 \cdot 7H_2O$-0.5 g/L, $Ca(NO_3)_2$-0.01 g/L, $FeSO_4 \cdot 7H_2O$-44.3 g/L.

2.3. Experimental process

This study was carried out in a 500 mL erlenmeyer flask with a total solution volume of 300 mL. The reagents other than $FeSO_4 \cdot 7H_2O$ were weighed into the erlenmeyer flask, 270 mL of distilled water was added, and the pH was adjusted to about 1.8. Then $FeSO_4 \cdot 7H_2O$ was weighed into it and the pH was adjusted to about 1.8 again. Finally, 30 mL of bacterial solution was added with a cell concentration of 20×10^7/mL. The erlenmeyer flasks were put into a thermostatic shaker at a speed of 170 rpm and a temperature of 45°C. When the potential (Eh) of the solution reached 600 mV or more, 30 g of ore sample were added and started bioleaching.

2.4. Analytical method

During the experiment, the supernatant in the erlenmeyer flask was periodically taken for ion concentration detection. At the end of the experiment, the slurry was filtered to obtain the filtrate and residue. The residue was washed and dried for further detection. The sulfur content in the residue was obtained by combustion iodometric method, and the other elemental contents were detected by atomic absorption spectrometry (AAS). A LeiCi PHBJ-260 portable pH meter was used for Eh and pH determination. The Fe^{2+} and total Fe concentrations were determined by potassium dichromate titration. The XRD of the ore samples and residues were obtained by XRD-7000 X-ray diffractometer from Shimadzu, Japan. The cell concentration was obtained by the cell counting plate method.

2.5. Process Fitting

By observing the trend of zinc leaching rate, the exponential model (Eq. 5) was assumed to be satisfied. The bioleaching data were fitted based on this model to evaluate the bioleaching process.

$$\text{Exponential model: } y = y^0 + A \cdot e^{R^0 \cdot x} \tag{5}$$

where y is the dependent variable, denoting the zinc bioleaching rate; x is the independent variable, denoting time; y^0, A, and R^0 are all model parameters.

3. Results and Discussion

3.1. Variation of bioleaching parameters

During the experiment process, the change of cell concentration, Eh, pH, Fe^{2+} concentration, Zn^{2+} concentration and zinc leaching rate are shown in Figure 2. Due to environmental stress, some microorganisms died after inoculation. The remaining microorganisms rapidly adapted to the new environment and repopulated rapidly again. The Eh is related to the Fe^{3+}/Fe^{2+} magnitude, which first decreased and then increased during the bioleaching process. This is due to the rapid oxidation of sulfide ore (Eq. 2) by Fe^{3+} at the initial stage, and the Fe^{3+}/Fe^{2+} value decreases. Under the action of microorganisms, Fe^{2+} is re-oxidized to Fe^{3+}, resulting in an increase in Eh. pH increased rapidly in the initial stage probably due to acid consumption by oxides on the ore surface. With the acid production by bio-oxidation reaction (Eq. 3), the pH is in a state of gradual decrease or dynamic equilibrium. The trend of Fe^{2+} change is consistent with pH, microorganisms quickly oxidize Fe^{2+} to Fe^{3+} after adapting to the environment, resulting in a gradual decrease in Fe^{2+} concentration. Sphalerite is continuously dissolved under microbial catalytic oxidation, and finally all zinc basically enters the solution.

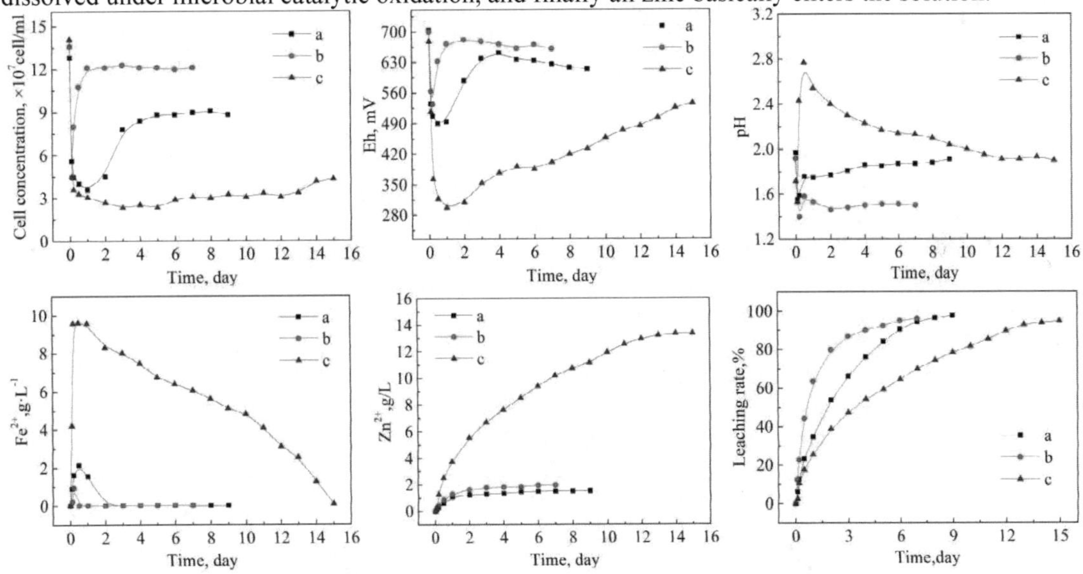

Figure 2. Variation curve of parameters of ore bioleaching process.

According to Figure 2, the various indicators in the bioleaching process of a and b ore are superior to that of c ore. Specifically, the zinc bioleaching rate of b ore was close to 90% after 3 days of leaching. This may be due to the higher amount of lead and iron associated with the high-grade c ore, which entered the solution and inhibited the microbial activity, thus attenuating the bacterial corrosion of sphalerite [18]. The higher sulfur content in b ore serves as a microbial nutrient substrate to provide energy for biochemical processes, thus achieving better microbial activity and leaching effect.

3.2. Distribution rate of major metal elements

After the bioleaching, the slurry was filtered and cleaned to obtain the leaching residue. The elemental content was measured, and the leaching rate of metal elements in the three ore samples was calculated, which are shown in Figure 3. It can be seen that the bioleaching efficiency of a ore is the best, while ore sample c has the worst leaching effect. The particle size of ore samples b and c is relatively large, and the complete dissolution of zinc cannot be achieved within a limited time. When reaching a stable point, it is the initial point for achieving maximum economic benefits. Due to the high zinc content in the sample c, it takes 15 days to achieve a high leaching rate. We cannot disregard the slow rate of bioleaching when compared to typical pyrometallurgical and hydrometallurgical processes. To speed bioleaching, strategies such as bio-domestication, genetic engineering, parameter optimization, process control, and additive enhancement must be further investigated [6, 19].

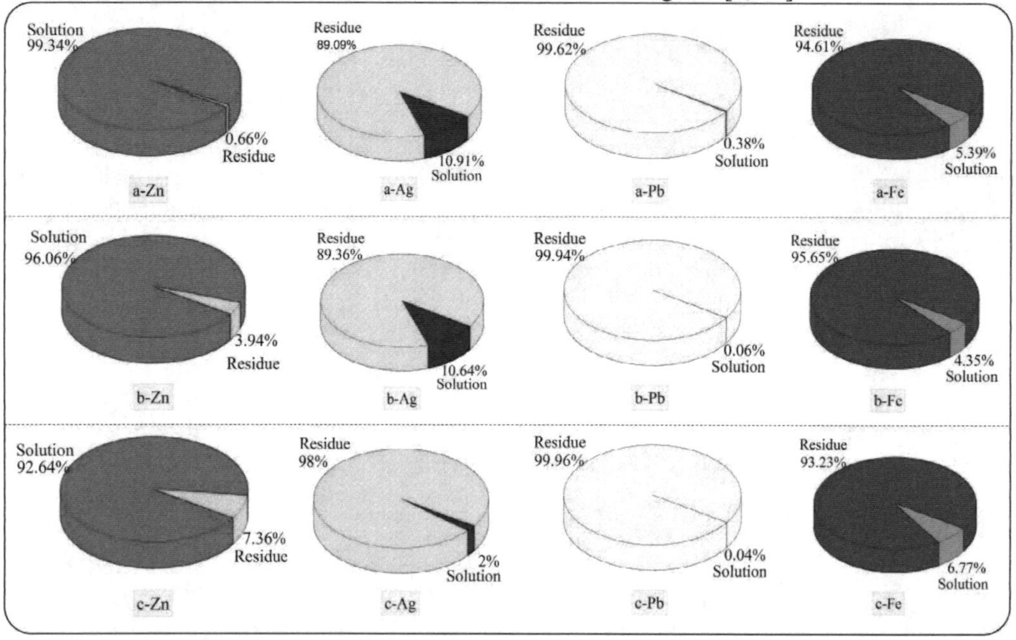

Figure 3. Distribution of metal elements in the ore after bioleaching.

Most of the other metal elements remained in the residue phase, including 89%-98% silver, more than 99% lead, and 93%-96% iron. However, their presence in the residue phase does not mean that they were not leached. Element silver is generally insoluble in acidic solutions and requires cyanidation for recovery. However, some silver in the ore sample entered the liquid phase, which could be the complexation of organic acids secreted by microorganisms (Eq. 6) [20]. Lead exists in the ore in the form of galena (PbS), and bio-oxidation makes it enter the solution as Pb^{2+} (Eq. 7), but it then precipitated with SO_4^{2-} to form $PbSO_4$ and enters the residue phase (Eq. 8). Elemental Fe enters the solution by bio-oxidation (Eq. 9) and is then oxidized to Fe^{3+} and produces jarosite (Eqs. 4 and 10). Therefore, the leaching rate of iron is not high. $PbSO_4$ and jarosite may cover the surface of the sulfide minerals and prevent further bioleaching [21].

$$Microbe \longrightarrow Ac^- + Ag \longrightarrow Ag(Ac)^- \tag{6}$$

$$PbS + 2H^+ + 0.5O_2 \xrightarrow{Microbe} Pb^{2+} + S^0 + H_2O \tag{7}$$

$$Pb^{2+} + SO_4^{2-} \longrightarrow PbSO_4 \downarrow \tag{8}$$

$$FeS_2 + 4Fe^{3+} + O_2 + 6H_2O \xrightarrow{Microbe} 5Fe^{2+} + 12H^+ + 2SO_4^{2-} \tag{9}$$

$$(NH_4^+, K^+, H_3O^+) + 3Fe^{3+} + +2SO_4^{2-} + 6H_2O \xrightarrow{Microbe} (NH_4^+, K^+, H_3O^+)Fe_3(SO_4)_2(OH)_6 \downarrow + 6H^+ \tag{10}$$

3.3. XRD analysis

The ore samples and residues were washed and dried for XRD analysis, and the results are shown in Figure 4. It can be seen that non-metallic minerals are the main components of the ore, including quartz and mica. Metal minerals occupy a relatively low peaks and mainly include sphalerite, pyrite and galena. The peaks of sulfide and quartz in sample c with high metal content basically overlap. In addition, chalcophanite peaks can be detected in the b and c samples. Bioleaching causes the dissolution of metal minerals and the corresponding peaks disappeared. Gypsum may be caused by the dissolution of a small amount of calcium compounds in the ore, and it can also act as a passivation substance to hinder the further bioleaching process. To avoid passivation of gypsum precipitation, a pre-acidification step is performed prior to microbial inoculation to eliminate most of the calcium compounds in practice, such as calcite [22, 23].

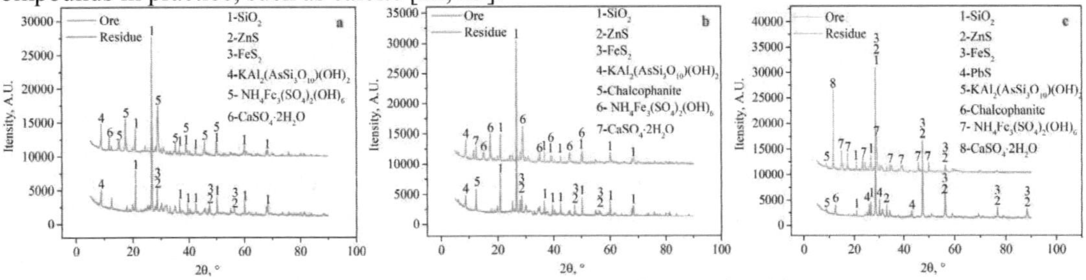

Figure 4. XRD analysis results of ore and residue.

3.4. Process fitting

The inoculated strain underwent a long-term domestication process and started leaching with little adaptation period. Based on the leaching trend of zinc, the Exponential model was selected and the bioleaching process was fitted using the software Origin 9.1, and the results are shown in Figure 5.

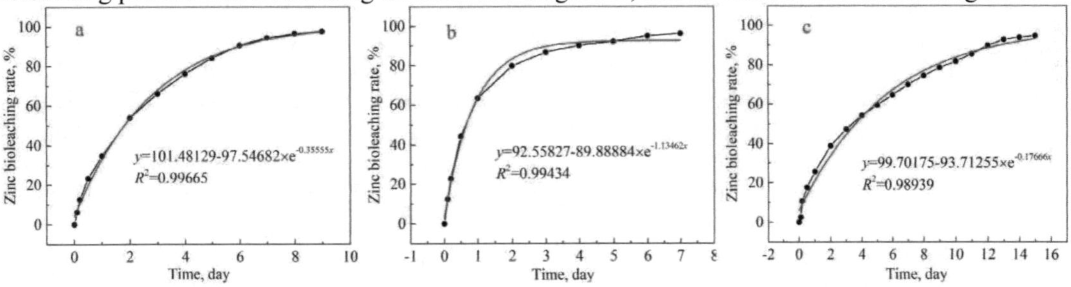

Figure 5. Exponential model fitting results for zinc bioleaching rate.

According to the fitting results, correlation coefficient (R^2) of the three ores are greater than 0.98 during the leaching cycle, which means that the sphalerite bioleaching is consistent with the Exponential model. From the three fitted equations, the leaching time was extended by 5 days for each of the three ores, and the Zn leaching rates were obtained as 100.81%, 92.56% and 96.96%, respectively. The leaching rate of the ore sample a exceeded 100%, indicating that the model were no longer applicable beyond this experimental cycle. The increase in leaching rate was minimal when the leaching time was increased for the ore sample c. Therefore, the leaching cycle was set to 15 days as the most appropriate, which not only ensures the effective recovery of metals, but also controls the time cost.

4. Conclusion

(1) The zinc ores used in this study contain 1.52%, 2.03% and 14.4% zinc, respectively, and are all occurrence in the form of sphalerite. Other major minerals include pyrite, galena, quartz and mica.
(2) Within 7-15 days, the bioleaching rates of zinc reached 99.34%, 96.06%, and 92.64%, respectively. Silver is only partially dissolved by microorganisms, while sphalerite, galena and pyrite are largely dissolved.
(3) Pb^{2+} and Fe^{3+} further precipitated to produce $PbSO_4$ and jarosite. Ca^{2+} generated by the dissolution of calcium compounds combined with SO_4^{2-} to produce gypsum.
(4) The Exponential model can simulate this bioleaching process with coefficients greater than 0.98.
(5) While bioleaching can produce ecologically benign and effective metal resource extraction, there are still issues with long cycle duration and passivation. Further study in strain cultivation, genetic engineering, and process control should be conducted in the future.

Acknowledgments

This work was financially supported by the Fundamental Research Funds for the Central Universities (2022QN1005).

References

[1] Foroutan A, Bazpors Ghaziani S, Abbas Zadeh Haji Abadi M, Shirazian S and Ghadiri M 2021, Intensification of Zinc Bioleaching from a Zinc–Iron Bearing Ore by Condition Optimization and Adding Catalysts. *Transactions of the Indian Institute of Metals*, **74** 1-8.

[2] Ejtemaei M, Gharabaghi M and Irannajad M 2014, A review of zinc oxide mineral beneficiation using flotation method. *Advances in Colloid and Interface Science*, **206** 68-78.

[3] Saririchi T, Azad R R, Arabian D, Molaie A and Nemati F 2012, On the optimization of sphalerite bioleaching; the inspection of intermittent irrigation, type of agglomeration, feed formulation and their interactions on the bioleaching of low-grade zinc sulfide ores. *Chemical Engineering Journal*, **187** 217-221.

[4] Zahiri A, Ahmadi A, Foroutan A and Ghadiri M 2021, Improvement of zinc bioleaching from a zinc flotation concentrate using mechanical activation. *Minerals Engineering*, **163** 106793.

[5] Liao M X and Deng T L 2004, Zinc and lead extraction from complex raw sulfides by sequential bioleaching and acidic brine leach. *Minerals Engineering*, **17** 17-22.

[6] Li J, Yang H, Tong L and Sand W 2022, Some Aspects of Industrial Heap Bioleaching Technology: From Basics to Practice. *Mineral Processing and Extractive Metallurgy Review*, **43** 510-528.

[7] Sajjad W, Zheng G, Zhang G, Ma X, Xu W and Khan S 2018, Bioleaching of copper- and zinc-bearing ore using consortia of indigenous iron-oxidizing bacteria. *Extremophiles*, **22** 851-863.

[8] Li J, Tong L, Xia Y, Yang H, Sand W, Xie H, Lan B, Zhong S and Auwalu A 2020, Microbial synergy and stoichiometry in heap biooxidation of low-grade porphyry arsenic-bearing gold ore. *Extremophiles*, **24** 355-364.

[9] Zhong W, Zhong S, Tang D, Chi X and Weng W 2023, Understanding the mechanism of microcrack-enhanced bioleaching of copper. *Hydrometallurgy*, **218** 106045.

[10] Li S, Zhong H, Hu Y, Zhao J, He Z and Gu G 2014, Bioleaching of a low-grade nickel–copper sulfide by mixture of four thermophiles. *Bioresource Technology*, **153** 300-306.

[11] Vukovic M, Strbac N, Sokic M, Grekulovic V and Cvetkovski V 2014, Bioleaching of pollymetallic sulphide concentrate using thermophilic bacteria. *Hemijska industrija*, **68** 575-583.

[12] Natarajan K A 2018, Bioleaching of zinc, nickel, and cobalt. In: Biotechnology of Metals: Principles, Recovery Methods and Environmental Concerns. Elsevier, 151-177.

[13] Mehrabani J V, Shafaei S Z, Noaparast M, Mousavi S M and Rajaei M M 2013, Bioleaching of sphalerite sample from Kooshk lead–zinc tailing dam. *Transactions of Nonferrous Metals Society of China*, **23** 3763-3769.

[14] Sundramurthy V P, Rajoo B, Srinivasan N R and Kavitha R 2020, Bioleaching of Zn from sphalerite using Leptospirillum ferriphilum isolate: effect of temperature and kinetic aspects. *Applied Biological Chemistry*, **63** 44.

[15] Wang J, Zhao H, Qin W and Qiu G 2014, Bioleaching of complex polymetallic sulfide ores by mixed culture. *Journal of Central South University*, **21** 2633-2637.

[16] Ye M, Li G, Yan P, Ren J, Zheng L, Han D, Sun S, Huang S and Zhong Y 2017, Removal of metals from lead-zinc mine tailings using bioleaching and followed by sulfide precipitation. *Chemosphere*, **185** 1189-1196.

[17] Schippers A, Tanne C, Stummeyer J and Graupner T 2019, Sphalerite bioleaching comparison in shake flasks and percolators. *Minerals Engineering*, **132** 251-257.

[18] Mehrabani J V, Shafaei S Z, Noaparast M and Mousavi S M 2014, Bioleaching of high pyrite carbon-rich sphalerite preflotation tailings. *Environmental Earth Sciences*, **71** 4675-4682.

[19] Yin S, Chen W, Fan X, Liu J and Wu L 2021, Review and prospects of bioleaching in the Chinese mining industry. *International Journal of Minerals, Metallurgy and Materials*, **28** 1397-1412.

[20] Kong M, Huang L, Li L, Zhang Z, Zheng S and Wang M K 2014, Effects of oxalic and citric acids on three clay minerals after incubation. *Applied Clay Science*, **99** 207-214.

[21] Li J, Yang H and Tong L 2021, Microbial solubilization of gangue minerals and their influence on pyrite bio-oxidation. *Geochemistry*, **81** 125790.

[22] Rakhshani Y, Rahpeyma S S, Tabandeh F, Arabnezhad M, Azimi A and Raheb J 2023, Multi-Objective Optimization of Copper Bioleaching: Comparative Study of Pure and Co-Cultured Cultivation. *Iranian journal of biotechnology*, **21** e3278-e3278.

[23] Qin W, Zhen S, Yan Z, Campbell M, Wang J, Liu K and Zhang Y 2009, Heap bioleaching of a low-grade nickel-bearing sulfide ore containing high levels of magnesium as olivine, chlorite and antigorite. *Hydrometallurgy*, **98** 58-65.

Study on bioleaching behavior of sphalerite from lead concentrate in agitator tank

Liu Xue[1,2,3,4], Sun Jianzhi[1,2,3,4], Liu Shuang[1,2,3], Shang He[1,2,3,4], Gao Wencheng[1,2,3], Yang Xinlong[1,2,3,4], Wen Jiankang[1,2,3,4], Chen Bowei[1,2,3]

1.GRINM Resources and Environment Tech. Co., Ltd., Beijing 101407, China;

2.National Engineering Research Center for Environment-friendly Metallurgy in Producing Premium Non-ferrous Metals, Beijing 100088, China;

3.Beijing Engineering Research Center of Strategic Nonferrous Metals Green Manufacturing Technology, Beijing 101407, China

4. GRIMAT Engineering Institute Co., Ltd., Beijing 101407, China

Abstract. Sphalerite which is highly continuous with galena cannot be effectively separated by flotation, the extraction of zinc from sphalerite in bioleaching process has become an alternative method. By testing the initial iron concentration, pulp density, grinding fineness and other parameters as well as monitoring pH, potential and dissolved oxygen, the zinc content of the lead concentrate decreased from 8.41% to less than 1% after bioleaching in 8L agitator tank. The initial concentration of ferric is the most significant factor affecting the zinc extraction efficiency of sphalerite, and when the concentration of ferric ions reaches 9g/L, acceptable leaching yield can be achieved.

1. Introduction

As a nonferrous metal, zinc is an important resource necessary for modern industry. It ranks fourth among all metals in terms of world production after iron, aluminum, and copper. Zinc sulfide ore is the most important zinc-producing mineral in the world and it often coexists with galena.

In many cases the crushed ore particles are not sufficiently small to permit liberation of the desired mineral so that separation can be achieved. The process of utilized lead-zinc minerals gives rise to waste of resources and affects the subsequent smelting process. It also induces tremendous pressure to production efficiency and ecological environment.

Due to the complex intergrow relationship between sphalerite and galena, the particle size of lead-zinc mineral is ground relatively fine. For the purpose of separating lead and zinc, the raw ore often needs to be ground to over 90% of the ore particle size through the 0.074mm sieve. Nowadays the flotation process of lead-zinc is becoming more difficult and costlier than before. As lead and zinc unevenly distributed in gangue with a wide range of particle size, the lead concentrate produced by flotation has a high content of zinc and the yield of lead concentrate is low.

With industrial development promoting, the bioleaching technology, as a green and low-carbon hydrometallurgy technology, meets the current environmental protection policy requirements. Bioleaching of zinc sulfide minerals can significantly increase the recovery of zinc, especially in terms of extracting metals from low-grade sulfide deposits. Thereby it shows advantages on the economic [1].

Through selective adsorption and biological oxidation of minerals, bacterial leaching can effectively solve the problem of encapsulation and make sulfide minerals be selectively dissolved [2,3]. With the valuable metals in the minerals mobilized to solution, microbial can accelerate the oxidation from ferrous to ferric and promote to change S^{2-} in sulfides into sulfur or sulfate [4,5].
Metal sulfides are conductors, semiconductors, or insulators in which both sulfur and metal atoms are bound to the lattice [6]. Both molecular orbital theory and valence band theory indicate that atoms and molecular orbitals form electron bands with different energy levels. The valence band generated by the zinc sulfide orbital is an acid-soluble "polysulfide" pathway. In sphalerite, both of H^+ and Fe^{3+} remove valence band electrons, resulting in the fracture of zinc-sulfur bond [7]. Microorganisms in the oxidation of these intermediate also plays a crucial role in the process, and turn them into sulfate and H^+ [8]. Many literatures indicted that zinc sulfide is acid soluble metal sulfide, and its bioleaching follows polysulfide way: by the ore leaching bacteria, the electron on valence band of metal sulfides was combined with H^+ and Fe^{3+}, releasing the sulfur compounds, which is turned into sulfate by ferric and O_2 oxidation [9,10].
The investigate focuses on selective bioleaching of zinc in agitator tank which contains in lead concentrate produced by traditional flotation. The possibility of selective bioleaching of zinc from zinc-bearing lead concentrate is determined by the investigation of effects. It provides the new approach to solve the beneficial recovery problems of complex lead-zinc ore.

2. Materials and methods

2.1 Ore samples

One sample (S1) is lead concentrate which came from a flotation concentrator in Yunnan province. The particle size of the lead concentrate is 90% under 0.074mm screen. The other sample (S2) is lead concentrate after finely ground to a particle size (over 90%) of 0.038 mm. Prior to use, both of the lead concentrate was washed with dilute sulfuric acid (pH=3) to remove the surface-oxidized deposits and the residue of flotation agent, rinsed repeatedly with distilled water and dried at 70°C for 4 h.
The elemental composition was determined after mineral dissolution in aqueous solution by inductively coupled plasma – optical emission spectroscopy (ICP-OES) measurements.
Mineral liberation analysis system (MLA)combined with scanning electron microscopy (SEM)were employed for the mineralogical identification of the ore samples. residues after bioleaching were cold-inlaid with epoxy resin and observed by SEM after polished.

2.2 Microbial culture

The bacteria were originally grown with 120 mM ferrous sulfate in a mineral salts solution that contained 3.0 g/L $(NH_4)_2SO_4$, 0.5 g/L $MgSO_4{\cdot}7H_2O$, 0.01 g/L $Ca(NO_3)_2$, 5 mol/LH_2SO_4, 0.5 g/L K_2HPO_4 and 0.1 g/L KCl. The pH was adjusted to 1.9 with H_2SO_4. The cultures were acclimated to grow with 5% (w/v) zinc concentrate in the mineral salt solution through several subcultures before the leaching experiments reported in this paper were conducted. The microbial culture was inoculated (initial cell number 10^6 cells/ml) in the pulp. The bacteria were cultured on a stirred fermentation cylinder with air inflation at 30°C. The population of culture was identified and characterized with molecular level technique that metagenomic sequencing based on 16S rDNA analysis.

2.3 Bioleaching experiment

The bioleaching experiments were carried out at 35°C in agitator tanks (Maximal volume 8L) containing 6L pulp. The initial pH of the pulp in all agitator tanks was around 1.9, but no pH control was executed during the experiment. All agitator tanks are with aeration and water bath. Inflating volume was adjusted at 3L/min and rotating speed of mixer was set at 300rpm in the experiment. Bioleaching was sampled to analyze the concentration of Zn^{2+}, total Fe, pH, redox potential (Eh) and solvent oxygen at regular intervals. The factors and correlative value of experiments were shown by table 1.

Table 1. Factors setting in agitator tanks.

Number of tanks	Pulp density (%)	Initial ferric concentration (g/L)	Code of samples
1	5	9	S2
2	10	4.5	S1
3	5	9	S1
4	5	4.5	S1
5	5	4.5	S2

3. Results and discussion

3.1 Characterization of lead concentrate

3.1.1 Elemental analysis by XRF and ICP-OES

The ICP-OES analysis on the composition of lead concentrate used for this work has been shown in table2. The results of the elemental analysis indicate that zinc in lead concentrate affects the quality of lead concentrate and cannot be removed by flotation.

Table 2. The results on ICP-OES analysis.

Elements	% (Wt.)
Pb	63.50
Zn	8.41
S	19.04
Fe	7.32

3.1.2 Minerals analysis by MLA

The mineral ore used in this study mainly consisted of galena and sphalerite. The specific mineral analysis results are shown in table3.

Table 3. Mineral composition list by MLA.

Minerals	Content, %
Galena	68.72
Bournonite	0.91
Cerussite	4.08
Sphalerite	12.03
Marmatite	0.45
Zincite	0.15
Willemite	0.09
Chalcopyrite	0.01
Tennantite	0.01
Pyrite	9.52
Arsenopyrite	0.06
Limonite	0.69
Rhodochrosite	0.01
Feldspar	1.27
Quartz	0.61
Hornblende	0.46
Ankerite	0.35
Calcite	0.09
Barite	0.2
Muscovite	0.2
Biotite	0.01
Garnet	0.05
Rutile	0.02
Total	100

3.2 Microbial investigation

As identified and characterized by metagenomic sequencing based on 16S rRNA analysis, the composition of the population becomes clear. Compared with initial inoculation in the agitator tank (unacclimated), the proportion of iron oxidizing bacteria such as *Acidophilium* and *Leptospirum* in the microbial community structure after leaching(acclimated) decreased, while the proportion of sulfur oxidizing bacteria increased. The results of microbial population analysis are shown in the Figure1.

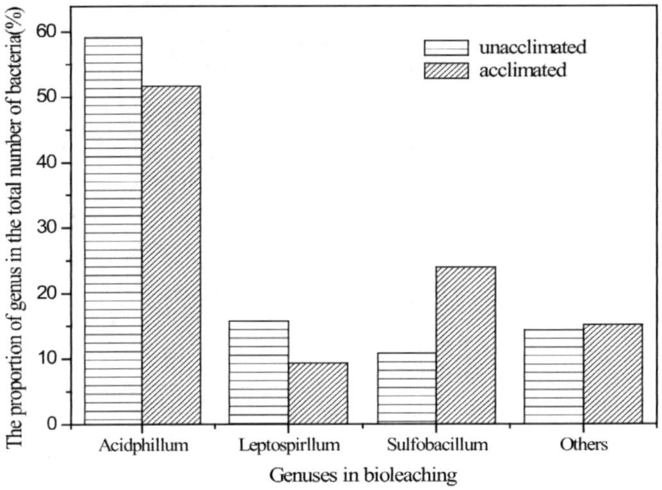

Figure 1. Microbial population in agitator tank.

Microbe adhere on the mineral particles by the extracellular polymeric substance (EPS) to oxidize the sulfides in zinc-bearing lead concentrate [11]. There are many substrates on which *Acidiphilium* strains act, including S^{2-}in sulfides, element S, ferrous iron and other materials [12-14]. The dominant bacteria *Acidiphilium* oxidize pyrite to release Fe^{2+} and produce a large amount of Fe^{3+}. Due to the potential difference between sphalerite and galena, sphalerite is more susceptible to be oxidized. As the production of lead sulfate on the surface of galena, *Sulfobacillus* tends to absorb on the surface of sphalerite and effectively dissolves Zn^{2+}into leachate. The function of *Leptospirllum* is oxidation of ferrous iron. As the main species, they promote the oxidation of sphalerite in the leaching system.

With the dissolution of pyrite in zinc and lead concentrate, the formation of jarosite, the consumption of Fe^{2+} concentration gradually decreased, *Leptospirllum* is not suitable for survival on the lack of energy substrate; However, a variety of reduced sulfur, such as elemental sulfur and polysulfides [15,16], produced in the oxidation leaching process of zinc and lead bearing concentrate can be used by *Sulfobacillus*, so the leaching system is suitable for the growth of *Sulfobacillus.*

Therefore, considering that the extreme acidophilic population, it is the positive factor of the zinc-selective bioleaching from lead concentrate. The microbial population is suitable for redox reactions with sulfide minerals in acidic conditions, which is beneficial to the dissolution of valuable metal ions from sulfide minerals into solution. It is also the internal reason for leaching selectivity of zinc from zinc-bearing lead concentrate.

3.3 Bioleaching of Sphalerite in agitator tank

3.3.1 Changes of parameters in bioleaching

The changes of pH in each stirring tank are different, and the curves of pH are shown in the figure 2.

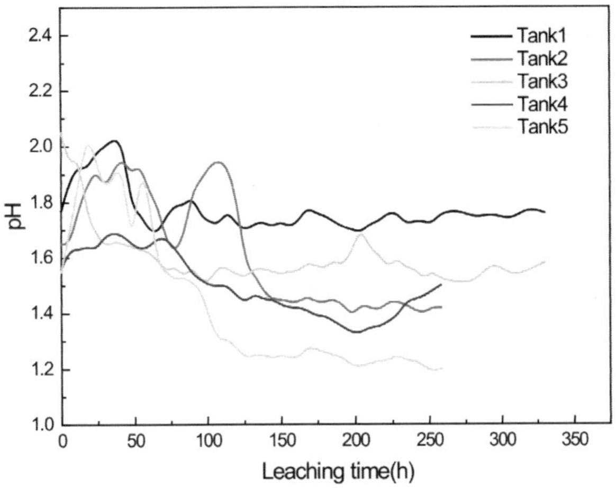

Figure 2. Change of pH in each stirring tank.

The pH value of tank 1# is relatively stable. Except for a small range of fluctuations in the early stage, it has been maintained at about 1.7. The trend of the pH value in 4# is basically decline. The pH value of the solution in tanks 2# and 5# showed a trend of increasing first and then decreasing slowly. The pH value of tank 3# fluctuated greatly, which first increased and then decreased, increased at the later stage of leaching.

Based on properties of solution, the main reason for the fluctuation of pH value is related to the initial Fe^{3+} concentration used in the initial stage of the experiment. It needs to be adjusted to a more suitable pH range by sulfuric acid in time to avoid Fe^{3+} hydrolysis caused by pH increase and affect the zinc efficiency of sphalerite.

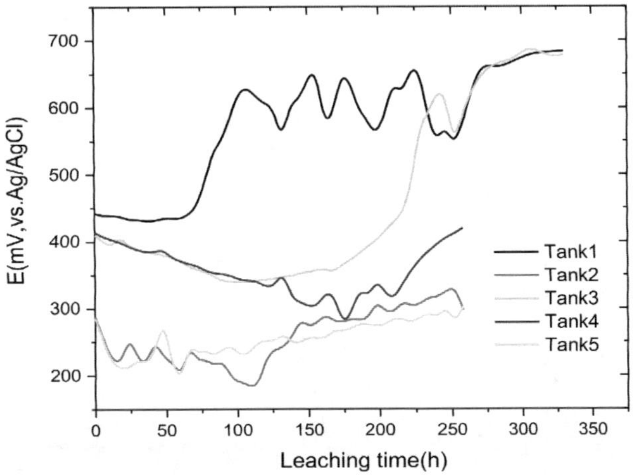

Figure 3. Changes of potential in each stirring tank.

Potential in bioleaching (shown in Figure 3) reflects that the bacteria in the 1# and 3# tanks mainly maintained the concentration ratio of oxidized ions and reduced ions in the early stage of leaching. In the middle of leaching, with the stability of the pH value of the pulp, the adaptability of bacteria to the pulp gradually increases, and the oxidation-reduction potential gradually increases. In the late stage of leaching, the potential began to stabilize above 670 mV (reference electrode Ag/AgCl), indicating that the oxidation of bacteria is extremely significant, and the oxidation of the pulp is suitable for the oxidation of zinc-containing sulfide minerals and the dissolution of zinc.

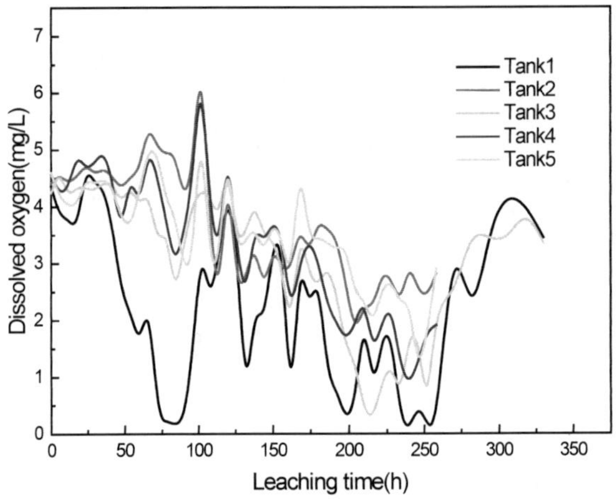

Figure 4. Changes of the dissolved oxygen in each stirring tank.

The dissolved oxygen in solution indirectly reflects the transfer rate of the sulfides oxidized by bacteria with oxygen as electron acceptor. The changes of the dissolved oxygen are shown in figure 4. The dissolved oxygen in 1# and 3# tank change greatly, from 4mg/L to less than 0.3mg/L, indicating that bacteria are rapidly consuming the dissolved oxygen in the pulp. Microorganisms use oxygen as an electron acceptor to convert Fe^{2+} into Fe^{3+} and the rapid dissolution of valuable metals in sulfide minerals is followed by oxygen consuming.

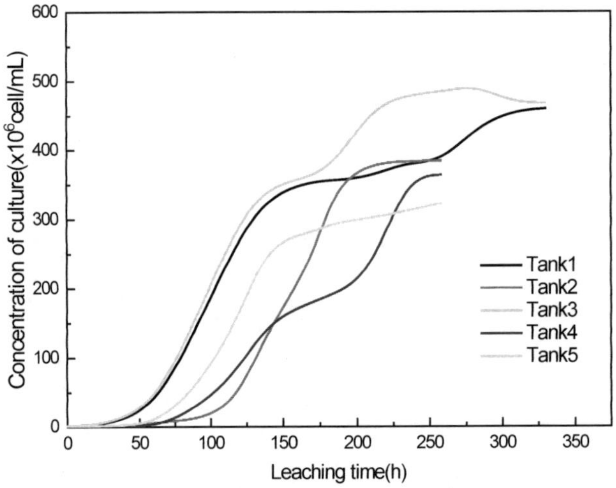

Figure 5. Number of bacteria in bioleaching of agitator tank.

The core problem in the lead concentrate selective bioleaching is the growth and generation of bacteria on leaching zinc. Therefore, it is particularly critical that the concentration of bacteria is monitored in the leaching process. The number of bacteria in experiment was shown in figure 5. In the zinc bioleaching, the change trend of the bacterial concentration in each stirring tank was gradually increasing. It indicates that the bacteria provide energy for their own metabolism pathway by oxidizing S^{2-} in sulfide minerals and the electron transport respiration chain in themself during the leaching process. With rapid growth and reproduction, the bacteria in the 3# and 1# tanks reached 1×10^8 cells/ml within 100 hours. It shows significant adaptability to the zinc leaching of the lead concentrate.

During the leaching process, the zinc ions gradually increased with the extension of the leaching time, especially the Zn^{2+} concentration in the pulp of 1# tank increased rapidly (shown in figure 6). The zinc concentration had reached 2.4g/L within 120h, which is equivalent to the zinc content of about 17g. By this time 57.75% of the zinc in sphalerite has been dissolved in solution. Although the rate of increase in the Zn^{2+} concentration after 120h is slower than that of the previous 120h, the zinc ion can eventually reach 3.7 g/L. The zinc ion concentration in the 3# tank rose slowly within 240h, but after that there was a rapid rise. At the end of the experiment the zinc ion reached 3.6g/L. The zinc ion concentration in the other tanks increased extremely slowly and did not exceed 1g/L until bioleaching finished.

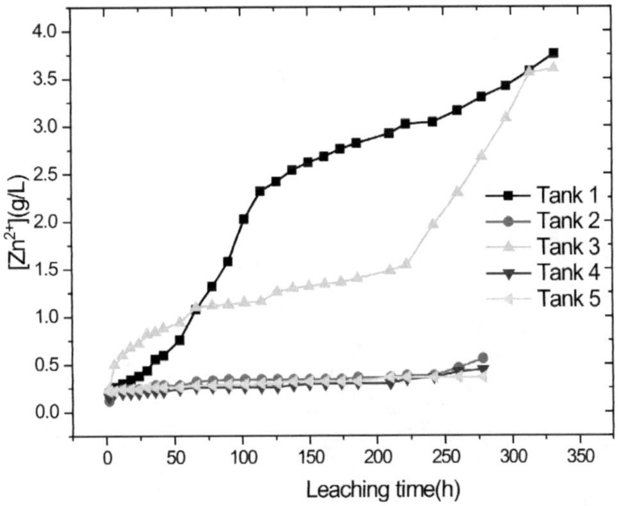

Figure 6. The changes of Zn^{2+} concentration in bioleaching of agitator tank.

3.3.2 Analysis of bioleaching residues

As shown in Figure 7, the grade of zinc in tanks 1# and 3# gradually decreased, the final residues of zinc grade in tank 1# could be reduced to less than 0.9%, and the slag of zinc grade in tank 3# also decreased to about 1.2% at the end of the bioleaching. It is not obvious that the zinc grade decreased in tanks 2#, 4# and 5# and the zinc grade of residues were above 7.6% after the experiment.

Journal of Physics: Conference Series **2738** (2024) 012012 doi:10.1088/1742-6596/2738/1/012012

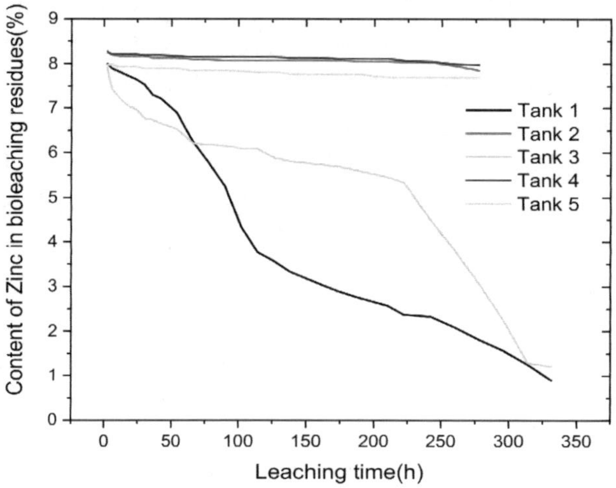

Figure 7. The changes of zinc grade in bioleaching residues.

By the examination of leaching residue by SEM-EDS, it was shown that sphalerite could hardly be found. Except for a small amount of jarosite produced, lead compounds, among which the representative one was that the surface of galena was coated with lead sulfate, as shown in Figure 8.

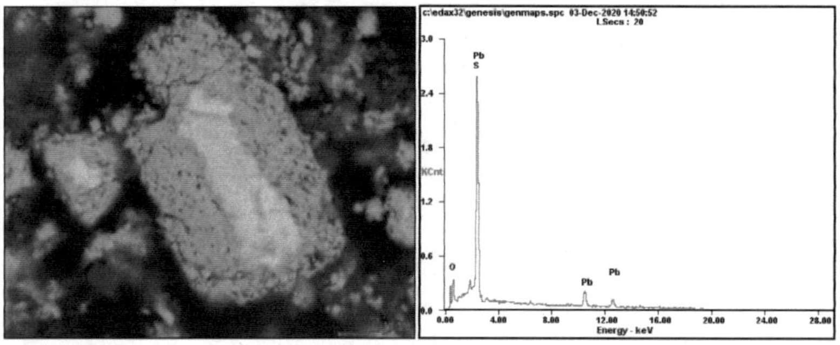

(a1)

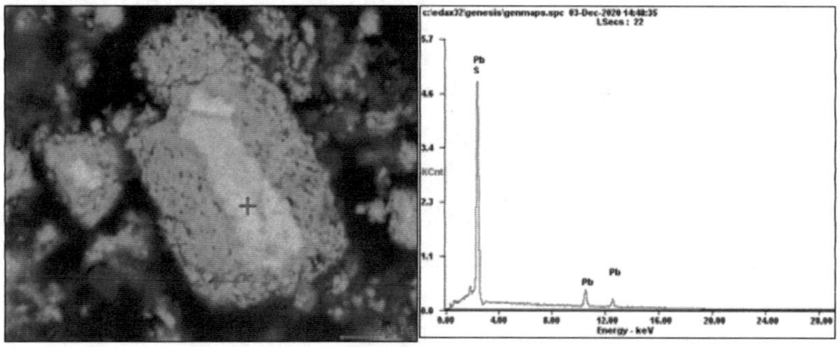

(a2)

Figure 8. SEM-EDS images of leaching residues.

(a1: Lead sulfate on leaching residue surface ; a2: Unreacted galena in the core of leaching residue)

Because the domesticated bacteria of zinc concentrate have better tolerance and selective adsorption, they act on the surface of zinc lead concentrate in the process of bioleaching and give priority to the release of Zn^{2+}. After the release of Zn^{2+}, galena is exposed and then covered by lead sulfate produced by oxidation on the mineral surface to prevent further leaching. However, other domesticated bacteria had poor tolerance and selective adsorption, and the oxidation rate of lead sulfate was faster, and the premature production of lead sulfate prevented the further oxidation of ZnS. Therefore, using domesticated bacteria to leach zinc and lead concentrates has certain advantages in selective zinc leaching.

4. Conclusions

Bacterial leaching of metal values from the substrate is a complex process and is affected by multiple factors. Effects of important parameters including different partial size, different concentration of ferric addition, different pulp density and their interactions on bioleaching of zinc sulfide ores were investigated by regular monitor in 5 agitator tanks. Bioleaching results in agitator tanks were encouraging in that more than 70% of zinc was extracted from in 300h, which is so remarkable to compare with conventional bioleaching. The key factor in zinc recovery during the bioleaching is initial ferric concentration. Higher concentration of ferric ions leads to disintegrate zinc from sphalerite faster than lower ferric concentration. Ferric ion reduces itself to ferrous ion, and promotes the rapid generation and growth of bacteria.

5. Acknowledgment

The research was funded by the National Natural Science Foundation of China (No: 52074243).

6. References

[1] Saririchi T., Azad R. R., Arabian D., et al. 2012 On the optimization of sphalerite bioleaching; the inspection of intermittent irrigation, type of agglomeration, feed formulation and their interactions on the bioleaching of low-grade zinc sulfide ores. *Chem Eng. J.* **187** 217-21

[2] Wang L, Xia JL, Zhu HR, et al. 2017 Progress on research of microbe−mineral interaction and interfacial micro-analysis *Microbiol. China.* **44** (3) 716-21

[3] Brierley C. L. 2010 Biohydrometallurgical prospects *Hydrometallurgy* **104** (3-4) 324-28

[4] Mehrabani J. V., Shafaei S. Z., Noaparast M., et al. 2013 Bioleaching of sphalerite sample from Kooshk lead –zinc tailing dam *Trans Nonferrous Met. Soc. China* **23** (12) 3763-69

[5] Garcia O. Jr., Bigham J. M., Tuovinen O. H. 1995 Sphalerite oxidation by Thiobacillus ferrooxidans and Thiobacillus thiooxidans *Can. J. Microbial.* **41** 578-84

[6] Baba A. A., Adekola F. A., Atata R. F., Ahmed R. N., Panda S. 2011 Bioleaching of Zn (II) and Pb (II) from Nigerian sphalerite and galena ores by mixed culture of acidophilic bacteria *Trans. Nonferrous Met. Soc. China.* **21** 2535-41

[7] Olubambi P. A., Ndlovu S., Potgieter J. H., Borode J. O. 2007 Effects of ore mineralogy on the microbial leaching of low grade complex sulphide ores *Hydrometallurgy* **86** (1-2) 96-104.

[8] Hao XD, Zeng WM, Peng TJ, Hu Q, Liu XD 2015 Analysis of microbial diversity during bioleaching of low grade copper ore from Zambia using high-throughput sequencing technology at different temperatures *The Chinese Journal of Nonferrous Metals* **25** (9) 2558-64

[9] Yu RL, Shi LJ, Zhou D, et al. 2013 Research development of microorganism synergy mechanisms during bioleaching *The Chinese Journal of Nonferrous Metals* **10** 3006-15

[10] Zhang GJ, Yuan QH, Fang ZH, Yang C 2007 Effects of external addition of Fe^{2+} and inoculum on bioleaching of marmatite flotation concentrate *The Chinese Journal of Process Engineering* **7** (3) 536-40

[11] Schippers A., Tanne C., Stummeyer J., Graupner T. 2019 Sphalerite bioleaching comparison in shake flasks and percolators, *Minerals Engineering* **132** 251-57

[12] Lu Z, Yao W, Liang J, et al. 2013 Decomposition of hydrogen sulfide in non-thermal plasm aided by supported CdS and ZnS semiconductors *Green Chemistry* **15** (6) 1509-13

[13] Schippers A. 2004 Biogeochemistry of metal sulfide oxidation in mining environments, sediments, and soils. *Sulfur Biogeochemistry- Past and Present* (Geological Society of America) pp49-62

[14] Rohwerder T., Gehrke T., Kinzler K., et al. 2003 Bioleaching review Part A: Progress in bioleaching: fundamentals and mechanisms of bacterial metal sulfide oxidation *Applied Microbiology &Biotechnology* **63** (3) 239-248

[15] Schippers A., Sand W. 1999 Bacterial leaching of metal sulfides proceeds by two indirect mechanisms via thiosulfate or via polysulfides and sulfur *Applied and Environmental Microbiology* **65** (1) 319-21

[16] Schippers A., Hedrich S.,Vasters J., et al. 2014 Biomining: metal recovery from ores with microorganisms *Advances in Biochemical Engineering/biotechnology* **141** 1-47

Study on Optimization of Deep Purification Process Design of Zinc Oxygen Pressure Acid Leaching Solution

Ziliang Liu, Yizhe Tong, Xingmin He, Longyi Chen

CINF Engineering Co., Ltd., Changsha 410011, Hunan, People's Republic of China

lzl09137@163.com

Abstract: According to the characteristics of high impurity content and complex composition of oxygen pressure leaching (OPL) solution of zinc sulfide concentrate, combined with the quality requirements of large plate electrowinning and automatic plate stripping, the advanced purification process of solution was designed. Through the deep purification process consisting of neutralization indium precipitation and low iron purification, high temperature antimony salt nickel and cobalt removal, copper slag chlorine removal, low temperature zinc powder replacement copper and cadmium removal, fine purification, calcium and magnesium removal, the contents of harmful impurities such as iron, indium, copper, cadmium, nickel, cobalt, fluorine and chlorine in the solution are strictly controlled to ensure that the quality of the deeply purified zinc sulfate solution is qualified and can meet the needs of continuous and stable production. The energy-saving effect of advanced purification process is obvious, and the average DC power consumption per ton of electric zinc is 2880kWh, reaching the advanced level of the same industry in China. The good effect of advanced purification process ensures the successful application of OPL and large plate zinc electrowinning, and improves the technology and equipment level of zinc smelting industry, which has important popularization and application value.

Key words: Deep purification; Hydrolysis precipitation; Displacement precipitation; Oxygen pressure leaching; Comprehensive recovery

1. Introduction

Oxygen pressure leaching (OPL) is a hydrometallurgical process with short flow and enhanced leaching, which has the characteristics of strong adaptability of raw materials, environmental friendliness and good comprehensive recovery effect[1-2]. OPL of sulfuric acid system can be used to leach and recover zinc, copper, nickel, cobalt, molybdenum and other metals from concentrates and secondary materials[3-5] . In the domestic zinc smelting industry, OPL zinc smelting technology has been applied in lead and zinc smelting enterprises such as Zhongjin Lingnan Danxia Smelter, Hulunbeier Chihong Mine Limited Company and Western Mining Industry[6-7].

2. Deep Purification

As a zinc hydrometallurgy technology, the main processes of OPL also include leaching, purification and electrowinning. Compared with traditional technology, deep purification of

OPL solution has become an important guarantee for continuous and stable production. Mainly because:

(1) Compared with the traditional "roasting-leaching" zinc smelting technology, OPL eliminates roasting and adopts direct leaching of zinc sulfide concentrate. The cancellation of roasting shortens the technological process and thoroughly solves the problem of sulfur dioxide emission in traditional zinc smelting technology, but at the same time, it also cancels the open circuit path of fluorine, chlorine and other impurity elements in smelting flue gas, which is easy to cause the accumulation and over-standard of fluorine and chlorine impurities in zinc hydrometallurgy production.

(2) The leaching rate of copper, cadmium, cobalt, nickel, indium and other valuable metals in the concentrate is improved by means of enhanced leaching in OPL. For example, the leaching rate of copper is increased from about 40% to over 92% in conventional method, and the leaching rate of indium which is difficult to be leached in conventional method is also over 90%[8] . More associated valuable metals enter the leaching solution, and the impurity content of the leaching solution increases accordingly.

(3) The application of large plate electrodeposition and automatic zinc stripping technology puts forward higher requirements for the quality of zinc sulfate solution[9] . The application and development process of large plate electrowinning and automatic zinc stripping technology in China are almost synchronized with OPL zinc smelting technology, and the two technologies represent the new progress of electrowinning and leaching respectively. As an intermediate process of zinc hydrometallurgy, advanced purification combines OPL with large plate electrowinning to form a complete technical process of clean zinc smelting. The traditional electrodeposition period is 24 h, but the electrodeposition period of automatic zinc stripping technology for large plate zinc electrodeposition needs to be extended to more than 36 h, so that the zinc sheet can reach a certain thickness and meet the requirements of automatic zinc stripping; However, with the prolongation of electrowinning period, the adverse effects of impurity elements such as nickel, cobalt, copper, cadmium, iron, indium, magnesium, fluorine and chlorine in OPL solution on zinc electrowinning process are further enlarged. Therefore, it is of great significance to design a reasonable deep purification process of zinc sulfate solution, so as to reduce the impurity content to the allowable range, and effectively enrich valuable metals such as nickel, cobalt, copper, cadmium and indium in the removal process, thus realizing the comprehensive recovery of secondary resources.

See Table 1 for the composition requirements of zinc sulfate solution for large plate electrowinning and automatic zinc stripping.

Table 1 Composition requirements of zinc sulfate solution for large plate electrowinning and automatic zinc stripping, g/L

Zn^{2+}	Cu^{2+}	Cd^{2+}	Co^{2+}	Ni^{2+}
155	< 0.0002	< 0.0002	< 0.0002	< 0.00001
In^{3+}	Cl^-	F^-	Fe^{2+}	Fe
< 0.005	< 0.2	<0.02	< 0.005	<0.02

3. Study on process scheme

3.1. Typical leaching solution

A zinc smelting project adopts two-stage OPL process. The leaching solution produced by it contains impurity elements such as nickel, cobalt, copper, cadmium, indium, fluorine, chlorine and magnesium. The composition of the leaching solution is shown in Table 2. The electrowinning process of this process is 3.2 m^2 In the process of large plate electrowinning and automatic zinc stripping, the leaching solution needs deep purification.

Table 2 Main components of OPL solution of a zinc project, g/L

Zn^{2+}	Cu^{2+}	Cd^{2+}	Co^{2+}	Ni^{2+}
155	0.60	0.62	0.016	0.005
In^{3+}	Cl^-	F^-	Fe^{2+}	Fe
0.026	0.347	0.091	0.60	1.83

The components in Table 2 are typical components of two-stage countercurrent low acid OPL solution. The advantages of two-stage countercurrent low acid OPL are low acidity and iron content of leaching solution, and only less neutralizer is needed. During the leaching process, most of the iron dissolved in the leaching process precipitates into the slag in the autoclave with the final acid reduced to about 10 g/L, and the impurities such as arsenic which can co-precipitate with iron are basically removed from the solution. However, residual iron remains in the solution, while impurity elements such as nickel, cobalt, copper, cadmium, indium, fluorine, chlorine and magnesium basically remain in the leaching solution.

3.2. Harm of impurity elements

3.2.1. Ni, Co, Cu, Cd. Nickel and cobalt cause corrosion of zinc sheet in electrowinning, resulting in burning plate, and black spots or horn-shaped round holes are corroded on zinc sheet.

Copper is a positively charged metal, which is easy to precipitate on the cathode, and does not redissolve after precipitation, thus reducing the quality of precipitated zinc. At the same time, copper can form a micro-battery with zinc, which makes zinc redissolve and forms a round "burning plate" with irregular periphery.

Because the precipitation voltage of cadmium is relatively positive than that of zinc, it is easy to precipitate at the cathode, which reduces the quality of zinc sheet.

3.2.2. Fe, In. Iron and indium generally do not precipitate at the cathode, which will not affect the quality of precipitated zinc, but will reduce the current efficiency and increase the power consumption.

3.2.3. Ca, Mg. Calcium and magnesium are more negatively charged than zinc, and will not precipitate at the cathode or oxidize at the anode during electrodeposition. However, when the content of calcium and magnesium ions in the solution is too high, it will bring other adverse effects:

1) Increase the density, viscosity and resistance of the solution, reduce the current efficiency of zinc electrowinning and increase the power consumption.

2) The sulfate solubility of calcium and magnesium is relatively small. In the solution circulation system, when the temperature of electrolyte decreases, it will precipitate from the solution, which will cause scaling of equipment and pipelines in the system. The crystallization of calcium magnesium sulfate is difficult to remove, which causes equipment damage and pipeline blockage in severe cases, and brings great harm to normal production.

3.2.4. F, Cl. Fluorine and chloride ions corrode the electrode plate, which causes trouble to zinc electrowinning production.

Fluoride ion corrodes cathode aluminum plate, which makes it difficult for zinc sheet to peel off from cathode plate. Aluminum is a more active metal than zinc. It can be used as a electrode in acidic electrolyte mainly because of the existence of a corrosion-resistant oxide film on the surface. The existence of oxide film makes the zinc sheet easily peeled off from the electrode plate. However, when fluorine ion corrosion destroys the oxide film of cathode aluminum, the precipitated metal zinc will form a solid solution with aluminum and tightly combine together, which makes it difficult to strip zinc. Modern zinc electrowinning production adopts large plate and mechanical zinc stripping technology. In order to ensure uninterrupted zinc stripping, the concentration of fluoride should not exceed 10mg/L.

Chloride ion corrodes anode lead plate, shortens the service life of anode, increases the lead content of precipitated zinc, and reduces the product quality. After coating, a layer of PbO was formed on the surface of anode plate$_2$ Protective film. Chloride ion has a small radius, which easily penetrates into the anode to react with lead, corrodes the anode plate and increases the lead content in the electrolyte.

Therefore, the impurities such as nickel, cobalt, copper, cadmium, iron, indium, calcium, magnesium, fluorine and chlorine in the OPL solution will have adverse effects on the product quality, current efficiency and automatic zinc stripping in the subsequent zinc electrowinning process, and must be removed from the solution. At the same time, valuable metals such as nickel, cobalt, copper, cadmium and indium are effectively enriched in the removal process to realize comprehensive recovery.

3.3. Purification method

3.3.1. Fe removal and In precipitation. The removal methods of iron and indium are similar, but iron can only be removed as impurities, while indium needs to be recovered. Therefore, iron removal and indium precipitation should be carried out step by step to avoid the mixing of iron and indium, so as to improve the enrichment rate of indium. At present, the methods of removing iron from solution mainly include precipitation, ion exchange or extraction. Among them, precipitation method is more suitable for iron removal in zinc hydrometallurgy.

The iron content in OPL solution can be controlled below 2 g/L, and most of the iron content in OPL solution is Fe^{3+} Exist in form, a small amount with Fe^{2+} Morphological existence. Fe^{3+} And Fe^{2+} The precipitation methods of Fe are different, and Fe^{3+} The hydrolysis precipitation method is directly adopted, and the reaction formula is:

$$Fe_2(SO_4)_3+6H_2O=2Fe\ (OH))_3\downarrow+3H_2SO_4$$

Fe^{2+}Can be oxidized to Fe^{3+} Re-hydrolysis precipitation, or goethite precipitation, the reaction formula is:

$$FeSO_4+1/2O_2+H_2\ O = FeOOH\downarrow + H_2SO_4$$

The precipitation of indium also adopts hydrolysis precipitation, and the reaction formula is

$$In_2(SO_4)_3+6H_2O=2In\ (OH))_3\downarrow+3H_2SO_4$$

Whether in dilute solution or concentrated solution, Fe^{3+}, Fe^{2+}, In^{3+} The precipitation conditions are different. In the process of decreasing acidity and increasing pH of solution, when acidity is 2~3 g/L, Fe^{3+} firstly begin to precipitate. When the pH of the solution rises to about 3.0, In^{3+} Hydrolysis precipitation began to occur; When the acidity of the solution continues to decrease and pH rises to about 5.5, Fe^{2+} It is possible to precipitate. Therefore, the process of iron removal and indium precipitation can be divided into three steps: high valence iron precipitation, indium precipitation and low valence iron precipitation, so as to obtain indium enriched slag with less iron and higher indium grade.

In the process of iron removal and indium precipitation, the residual impurities such as arsenic and antimony in the leaching solution are also precipitated and removed.

3.3.2. Removal of Ni and Co. Cobalt and nickel are the most difficult impurities to remove from zinc leaching solution. There are two commonly used purification methods: one is electrochemical method, that is, displacement precipitation method; The second is chemical precipitation, such as cobalt precipitation with xanthate and cobalt precipitation with naphthol. Because xanthate and naphthol will bring organic contamination to the solution, which is not conducive to the subsequent electrodeposition, displacement precipitation is a reasonable method for removing nickel and cobalt[10]。

Displacement is an oxidation-reduction reaction in which the added metal loses electrons and is oxidized, while the precipitated material gains these electrons and is reduced. It is often used to separate other elements from the main metal solution[11]. The displacement reaction can be divided into two steps:

1) Anodic reaction (metal zinc is oxidized to Zn^{2+} Transfer to solution);

2) Cathode reaction (impurity metal ions are reduced to metal and precipitated).

Zn is more active than most metal impurities in leaching solution, and Zn^{2+} Entering the solution will not cause secondary pollution, so zinc powder is selected as displacement agent. In order to make the replacement process have enough thermodynamic driving force, it is necessary to add activator to realize the replacement precipitation of nickel and cobalt with zinc powder. Commonly used methods include arsenic salt purification method, antimony salt purification method. Compared with the arsenic salt purification method, SbH_3 is easy to decompose than AsH_3, and the toxic gas is relatively small, so the antimony salt purification method is preferred based on safety considerations. The reaction formula is:

$$2Co^{2+}+2Sb^{3+}+5Zn = 2CoSb\downarrow + 5Zn^{2+}$$

$$2Ni^{2+}+2Sb^{3+}+5Zn = 2NiSb\downarrow + 5Zn^{2+}$$

The antimonide of nickel and cobalt is more stable than nickel and cobalt themselves. In the presence of antimony salt, the replacement of nickel and cobalt by zinc powder has greater electromotive force and equilibrium constant in thermodynamics, and the reaction is more efficient and rapid

3.3.3. Removal of Cu and Cd. When copper and cadmium are removed by zinc powder replacement, copper precipitates preferentially. However, excessive zinc powder is needed in the process of cadmium removal to meet the purification requirements. The reaction formula is as follows:

$$Zn + Cu^{2+}=Zn^{2+} + Cu\downarrow$$

$$Zn + Cd^{2+}=Zn^{2+} + Cd\downarrow$$

Journal of Physics: Conference Series **2738** (2024) 012013 doi:10.1088/1742-6596/2738/1/012013

3.3.4. Removal of F and Cl. Chlorine mainly comes from concentrate, neutralizer zinc calcine or zinc oxide dust. The methods used to remove chlorine from solution are silver sulfate precipitation, copper slag removal of chlorine and ion exchange[12] . Among them, copper slag dechlorination method is that copper and copper ions react with chloride ions in solution to transform copper into insoluble cuprous chloride precipitation and remove it. In practical production, sponge copper (Cu 30%) slag produced in the process of recovering cadmium from copper and cadmium slag treated by comprehensive recovery process of zinc hydrometallurgy can be used as chlorine precipitator. Chlorine removal by copper slag dechlorination method, the reaction formula is as follows:

$$Cu_{(Copper\ sponge)} + 2Cl^- + Cu^{2+} = Cu_2Cl_2\downarrow$$

Because the solution will be contaminated by copper ions after chlorine removal, in order to avoid repeated copper removal, chlorine removal should be carried out before purifying copper and cadmium.

Fluorine is removed by adding lime to convert it into calcium fluoride. Lime is also a kind of neutralizer, which is needed in the process of indium precipitation and iron removal. Therefore, a small amount of lime is mixed into the neutralizer zinc calcine to achieve the purpose of fluorine removal.

3.3.5. Remove Ca and Mg. Calcium and magnesium are removed by cooling method. According to the different solubility of calcium and magnesium ions at different temperatures, when calcium and magnesium are close to saturation, the temperature of solution is reduced from 50 °C to 35 °C by forced cooling, so that calcium and magnesium ions are removed by crystallization of calcium sulfate and magnesium sulfate.

4. Process flow design

Impurities elements such as Fe^{3+}, Fe^{2+}, Cu^{2+}, Cd^{2+}, Co^{2+}, Ni^{2+}, Mg^{2+}, In^{3+}, Cl^-, F^- in the leaching solution, must be removed according to the requirements of deep purification. According to the principle of impurity element removal, the deep purification process of OPL solution is designed, as shown in Figure 1. The main processes include neutralization precipitation indium and low iron purification, nickel and cobalt removal by high temperature antimony salt, chlorine removal by copper slag, copper and cadmium removal by zinc powder replacement at low temperature, deep purification and calcium and magnesium removal by cooling method.

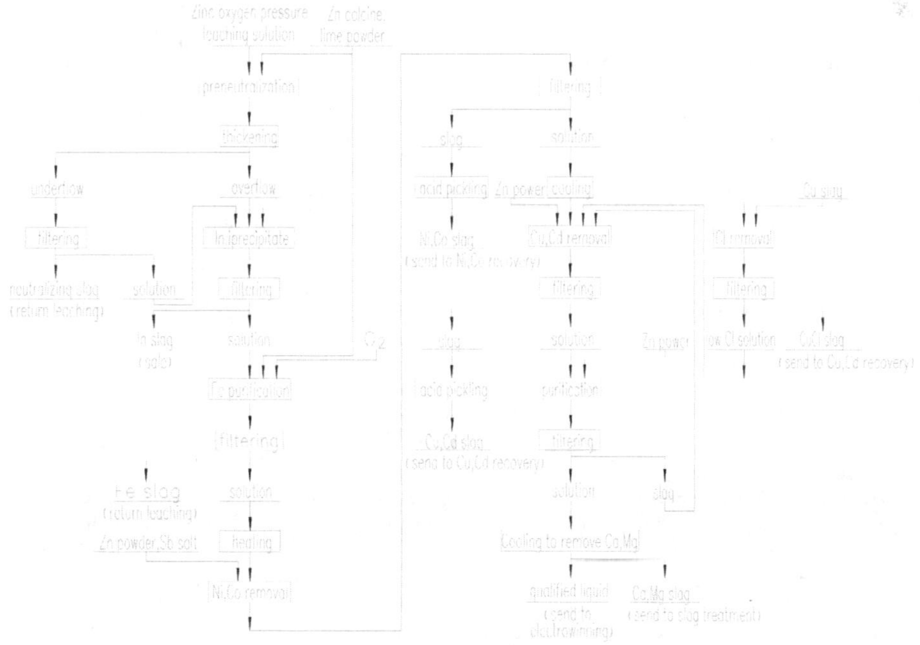

Fig. 1 Deep Purification Process

4.1. Removal of Fe and In

The supernatant containing about 10 g/L of acid produced in the OPL process is pre-neutralized with zinc calcine as neutralizer. The neutralization temperature is 80 ~ 85 °C, the neutralization reaction time is 1.5 ~ 2 h. Fe^{3+} will hydrolyze and precipitate when the solution is pre-neutralized to the acidity is 2 ~ 3 g/L, and the Fe^{3+} form neutralizing slag with excessive calcine and insoluble substances in calcine, but indium will not form hydroxide precipitate and still exist in solution in the form of ions.

The neutralized pulp is pumped into a neutralization thickener for preliminary liquid-solid separation, and the underflow of the thickener is further separated by a filter press. The filtrate and thickened supernatant are pumped to the indium precipitation tank for neutralization and indium precipitation. After slurry ball milling, the filter residue is pumped back to the second stage leaching of OPL to recover the residual zinc in the residue.

Indium precipitation adopts neutralization indium precipitation process with calcine as neutralizing agent. When calcine slurry is added into the indium deposition tank, the neutralization temperature is controlled at 70 ~ 75 °C, the neutralization reaction time is controlled at 1.5 ~ 2h, and the pH value of the solution after indium deposition is controlled at about 3. Indium in the solution will precipitate into the slag in the form of hydroxide. Indium slag can be obtained by liquid-solid separation of the slurry. Liquid-solid separation adopts box filter press, and the filter cake of the filter press is indium slag, which is transported to indium recovery system to produce indium ingots. The solution enters the low iron purification process.

In the process of neutralization and indium precipitation, iron in solution, especially Fe^{3+} Most of them were also hydrolyzed and precipitated, and the iron content of the solution decreased to 0.1 g/L, which was mainly Fe^{2+} However, the iron content in the solution required by electrowinning process is less than or equal to 10mg/L, so the solution after indium precipitation still needs to be further purified with low iron. For low iron purification, a mechanical stirred tank was used. By adding calcine into the iron removal tank, the pH value of the solution was adjusted to about 5.5, and oxygen was blown into the solution to remove Fe^{2+} Oxidation to Fe^{3+}. The reaction temperature was controlled at about 80 °C, and the continuous reaction was carried out for 2 ~ 3 hours. After iron removal, the content of Fe in the solution decreased from 0.1 g/L at the initial stage to less than 10mg/L at the end stage. Meanwhile, impurities such as As, Sb and F in the solution also co-precipitated with Fe into the solid phase.

After iron removal, the slurry is separated from liquid and solid by a box filter press, and then sent to a ball mill for ball milling together with neutralized slag, and then sent to a second-stage leaching feeding tank for OPL to recover zinc. The filtrate is pumped to the high temperature nickel and cobalt removal process.

In the conventional zinc hydrometallurgy production system, the process flow of removing Cu and Cd at low temperature, Ni and Co at high temperature, and Cu and Cd at high temperature is generally adopted. According to the characteristics of the technological process: after iron removal, the liquid temperature reaches about 80°C. If Cu and Cd are removed first, the solution temperature should be reduced to about 55 °C . However, when Ni and Co are removed by replacement in the second stage, the solution needs to be heated to 80~90°C, and then cooled down in the third stage. This is not conducive to the use of energy. Therefore, the technological process of removing nickel and cobalt with high temperature antimony salt, copper and cadmium with low temperature zinc powder and copper and cadmium with zinc powder is adopted.

4.2. Removal of nickel and cobalt

The impurities such as Fe, As and Sb in the solution have been removed by neutralization. Before electrodeposition, the impurities such as Cu, Cd, Ni and Co in zinc sulfate solution need to be further removed.

The liquid after iron removal at about 80°C is heated by a spiral plate heat exchanger, and the heated solution at 90 °C flows into a continuous displacement purification stirred tank, and the slurry zinc powder and antimony salt are added to remove nickel and cobalt. The reaction temperature in the control tank for removing nickel and cobalt is 85~90 °C, and the reaction time is about 2.5~3h. The purified solution is continuously fed into a nickel and cobalt removal filter press for filtration, and the filtered filtrate flows into an outdoor filtrate storage tank, and then is pumped to remove copper and cadmium for purification and chlorine for purification. The reaction temperature of removing Cu and Cd by zinc powder replacement is 55~ 60 °C, and that of removing Cl from copper slag is 40 °C, so the filtrate after removing Ni and Co should be cooled by blast air cooling tower first. The filter residue enters the slurring tank under the filter press respectively, and the washing water is added into the slurring tank for slurring, and the slurring residue slurry is sent to pickling.

The pickling of nickel-cobalt slag adopts intermittent operation, adds waste electrolyte in the pickling tank, adjusts the acidity of the solution to a pH value of 2~ 3, and directly heats it with a steam pipe in the tank. The temperature is controlled to be about 70 ° C, so that the zinc in the slag is dissolved, and the metal zinc in the slag is recovered. The filter residue after pickling is used as the raw material for the comprehensive recovery of nickel and cobalt.

4.3. Removal of Cl

The main task of the dechlorination step is to remove chloride ions from neutral solutions. Continuous dechlorination operations are based on the open circuit of part of the solution for dechlorination, and the chlorine content in the decontaminated solution determines the amount of open circuit to the dechlorination section. According to the chlorine content in the solution, after removing nickel and cobalt, 1/3 of the solution needs to enter the dechlorination process.

In a continuously stirred reaction tank, copper slag precipitated chloride ions from the neutral stock solution at a temperature of 40℃. After filtration, the filter slag is sent to the comprehensive recovery system of copper and cadmium slag, and the filtrate enters the low temperature zinc powder to replace and remove copper and cadmium.

4.4. Removal of Cd

After mixing the solution from nickel and cobalt removal and copper slag dechlorination, the solution temperature is about 55 ℃, and it first flows into multiple series of copper and cadmium removal stirring tanks to form continuous purification. The addition of zinc powder is also in the form of pulp, and the residence time of the material in the tank is about 1.5~2h. The purified slurry is sent to a chamber filter press for filtration. The filtrate flows into the filtrate storage tank located outdoors, and then pumps the fine purification process. After the filter slag is slurried, it pumps the copper-cadmium slag pickling tank.

The pickling of copper-cadmium slag is to reduce the zinc content in the slag, and it also adopts intermittent operation. The process flow is similar to that of nickel-cobalt slag pickling. After pickling and filtration, the filter residue is sent to the copper-cadmium residue recovery system to recover copper and cadmium.

4.5. Deep purification

In order to meet the requirements of mechanized zinc stripping on the impurity content of the solution, the solution purified by removing copper and cadmium is then purified by adding zinc powder. The purification step is a process step to ensure safety and quality. Mainly, the cadmium leaked in the first two purification steps will be removed in this purification stage.

Fine purification is also low temperature purification, the purification temperature is~ 50 °C, the reaction time is about 1.5~ 2h, and multiple mechanical stirring tanks are used for continuous operation. Zinc powder was added in the fine purification tank to further remove copper and cadmium. The copper-cadmium slag obtained by fine purification contains more zinc powder, and after slurring, it is returned to remove copper and cadmium for purification. The filtrate is pumped into the cooling tower to cool down and remove calcium and magnesium.

4.6. Removal of Ca and Mg

The removal of calcium and magnesium is realized by using the solubility of calcium sulfate to decrease with the decrease of temperature. The temperature of refined purified filtrate is about 50 °C. After cooling by cooling tower, the temperature drops to about 35 °C, calcium and magnesium ions form crystals and precipitate, and liquid-solid separation is carried out by thickener. The underflow discharged by thickener is pumped into calcium and magnesium slag slurry tank, and waste electrolyte is added into the tank to dissolve zinc sulfite crystals, and the pulp is filtered by filter press. The clear liquid of thickener contains about 155 g/L of zinc, which is sent to zinc electrowinning process as qualified zinc sulfate solution.

4.7. Removal Efficiency

After deep purification, the removal efficiency of harmful impurities is shown in Table 3.

Table 3 Removal efficiency of harmful impurities, %

Zn^{2+}	Cu^{2+}	Cd^{2+}	Co^{2+}	Ni^{2+}
/	>99.97%	>99.97%	>98.75%	>99.80%
In^{3+}	Cl^{-}	F^{-}	Fe^{2+}	∑Fe
>80.77%	>42.36%	>78.02%	>99.17%	>98.91%

5. Conclusion

The main results are as follows: 1) The content of impurities in OPL solution is higher than that in conventional zinc hydrometallurgy because of its complex composition. Through reasonable process design, the advanced purification process consisting of neutralization indium precipitation and low iron purification, high temperature antimony salt removal of nickel and cobalt, copper slag removal of chlorine, low temperature zinc powder replacement removal of copper and cadmium, fine purification, calcium and magnesium removal can ensure the production of qualified zinc sulfate solution, which can meet the actual production requirements of large plate zinc electrowinning. In solutionCl^{-}, F^{-} The content is stably controlled below the limit (Cl^{-} < 200mg/LF^{-} < 20mg/L), continuous production can be realized by mechanical zinc stripping operation. The average DC power consumption per ton of electric zinc is 2880kWh, reaching the advanced level of the same industry in China.

2) The deep purification of OPL solution is not only a process of impurity removal, but also a process of enrichment of associated valuable metals. The enrichment rate of valuable metals in indium slag, copper-cadmium slag, nickel-cobalt slag and other purified slag is high, which realizes the high value of purified slag.

3) The advanced purification process has achieved good application effect, ensured the successful application of zinc OPL and large plate zinc electrowinning, and improved the technology and equipment level of zinc smelting industry, which has important application value.

Acknowledgments

The work was financially supported by the Guangxi Science and Technology Major Project (No. AA22067076).

References

[1] TONG Yizhe, YAN Hao, YUAN Qiqi. Scheme Comparison and Selection for Zinc Hydrometallurgy Projects [J]. Nonferrous Metals Engineering & Research, 2018, 39 (02): 16-18 +28.

[2] TONG Yizhe. Application and Popularization of Zinc Pressure Leach Process [J]. Mining and Metallurgy, 2011, 20 (04): 94-97.

[3] Luo Honglin, Feng Zeping. Experiential Study on Oxygen Learning Slag Flotation of a Copper Sulfide [J]. Multipurpose Utilization of Mineral Resources, 2020 (05): 125-129.

[4] Chen Qinggen. Recovery of Molybdenum from a Molybdenum Ore Concentrates by High-press Oxygen Digestion & Ion-exchange [J]. Multipurpose Utilization of Mineral Resources, 2014 (04): 40-44.

[5] LI Yun-gang. Ammoniacal Pressure Learning of A Low-grade and Refraction Copper Ore [J]. Multipurpose Utilization of Mineral Resources, 2001 (01): 1-4.

[6] M.J.Collins,R.J. Kalanchey,I.M. Masters,N.E., et al. Sherritt zinc pressure leach study for China Western Mining Company: high zinc extraction at high altitude[J]. China Nonferrous Metallurgy, 2009(03):16-21.

[7] HU Dong-Feng, XY Liu. Analysis on Production Practice of Oxygen Pressure Learning of Zinc Concentrate [J]. Nonferrous Metals (Extractive Metallurgy), 2018 (03): 19-21 +29.

[8] W Luo, J Wang, LU Guohong. Test Study on Pressure Learning of Zinc, Iron and Indian from Indian-iron containing Sphalerite Using Sulfuric Acid [J]. Hydrometallurgy of

China, 2016, 35 (01): 25-29.
[9] TONG Yizhe. Large Cathode Zinc Electric Process and its Automatic Production [J]. Mining and Metallurgy, 2018, 27 (03): 93-96.
[10] XIE Ting-fang, LUO Yong-guang, LI Guo-jiang, et al. Cobalt removal process of high cobalt sulfate zinc solution with two stages of high temperature purification [J]. The Chinese Journal of Nonferrous Metals, 2020, 30 (10): 2410-2417.
[11] QS Zhong, HE Xu-Zhen, YQ Liu, et al. Separation of Silver, Selenium and Tellurium from Copper Anode Slime Pressure Learning Solution with Low Concentration [J]. Nonferrous Metals (Extractive Metallurgy), 2014 (09): 51-54 +58.
[12] WEN Jian. Study on the Choice of Dechlorination in the Production of Electric Zinc of Jinshi Metallurgy Chemical Plant [J]. Hunan Nonferrous Metals, 2008, 24 (06): 34-36.

Deep purification of copper from zinc sulfate solution using controllable aging FeS nanoparticles through slow-release sulfide precipitation

Xingfei Zhang, Jia Tian, Liqiang Zeng, Yufeng Wang, Wei Sun, Haisheng Han*, Yue Yang

School of Minerals Processing and Bioengineering, Central South University, Changsha 410083, China

E-mail: hanhai5086@csu.edu.cn

Abstract. The purification of impurity ions within zinc hydrometallurgical solutions significantly impacts the subsequent quality of zinc electrodeposition. Traditional methods for copper removal from zinc solutions suffer from issues like limited reactivity, high consumption, and elevated costs. To address these challenges, controllable aging of FeS nanoparticles was designed, employing temperature variations to facilitate comprehensive copper removal from neutral zinc leachates. The aging process imparts a crystalline structure to the FeS nanoparticles, enabling the slow release of sulfide ions. A temperature of 60°C is regarded as the optimal aging temperature, at which the resultant FeS nanoparticles exhibit an optimal sustained-release performance during the aging process. Temperature has been found to be a crucial factor influencing the sustained-release performance of FeS. Once the reaction temperature surpasses 60°C, the utilization efficiency of S ions in the sustained-release agent can reach over 95%, with a maximum of 98.5%. Under optimal conditions featuring a reaction temperature of 60°C, a reaction duration of 30 minutes, and a solution pH maintained at 1, the dosage aged FeS at the S/Cu (II) ratio of 1.2 times, nearly 100% copper removal rate can be achieved. The maximum removal rate for cadmium does not exceed 2.5%, and the residual copper concentration in the zinc leachate is below 0.2 mg/L. The precipitate obtained is high-purity copper sulfide residue, which contains virtually no zinc. Slow-release sulfidation can maximize the utilization efficiency of sulfide ions in the solution, reduce the occurrence of competing reactions, and minimize the intermixing of products. It offers advantages such as high selectivity and deep purification.

Keywords: FeS nanoparticles; aging; sulfide precipitation; copper removal; zinc sulfate solution.

1. Introduction

More than 80% of global zinc production originates from hydrometallurgical procedures, in which zinc concentrates or other zinc-containing raw materials undergo leaching to yield neutral leachate. Throughout this leaching process, compounds of impurity metals, including Cu, Cd, Co, Ni, and other ions, can dissolve into the zinc sulfate solution [1, 2]. These impurity ions wield considerable influence on ensuing zinc electrowinning processes, culminating in diminished current efficiency, heightened energy consumption, and compromised quality of zinc cathodes [3, 4]. Consequently, the imperative emerges to purify the solution, with the impurity metal-laden filter cake retrieved post-purification potentially serving as a recyclable source for recovering valuable metals like copper. At present, the elimination of copper and cadmium from zinc leachate involves displacement reactions with zinc powder [5-7]. Nonetheless, this approach is not without its limitations, as the displacement reaction transpires solely at the surface of zinc powder, resulting in suboptimal efficiency and vulnerability to encapsulation and passivation phenomena. Furthermore, this method necessitates a substantial surplus of zinc powder (exceeding 5 times the stoichiometric requirement), leading to escalated expenditures [8]. Furthermore, owing to incomplete reactions, the concentration of valuable elements within the zinc residue remains notably low, rendering the retrieval of these metals from the zinc residue a challenging and costly endeavor. Thus, the imperative of diminishing zinc powder consumption, reducing smelting expenses, and achieving optimal utilization of key constituents (including copper and cobalt) has emerged as a critical quandary within the realm of hydrometallurgical zinc refining processes.

Sulfide precipitation finds wide application in metal extraction owing to its manifold benefits. These encompass the capability for selectively extracting metals, rapid reaction kinetics, achieving higher product quality, and the potential for recycling precipitates through smelting [9]. Still, conventional ionic sulfidizing agents exhibit rapid spontaneous hydrolysis and sulfur ion release in aqueous solutions, resulting in low sulfur utilization efficiency, severe product intermixing, and fine particle size that hinders settling [10]. The key to addressing these issues is the slow-release rate of S^{2-} ions, as it can provide a lower sulfur ion saturation in the solution, allowing sufficient time for the binding of other metal ions with sulfur ions [11]. Iron sulfide (FeS), characterized by a considerable solubility product ($Ksp= 1.59 \times 10^{-19}$), proves to be an exceptional material for the controlled release of S^{2-}[12]. Synthesized through chemical precipitation methods, amorphous FeS nanoparticles display remarkable reactivity. However, appropriate modifications are necessary to hinder their rapid breakdown in acidic conditions [13]. Aging is a traditional method for nanoparticle modification, which affects their structure and reactivity [14, 15]. Theoretically, different degrees of aging can yield FeS nanoparticles with ideal sustained-release performance.

In this paper, the design of temperature-controlled aging of FeS nanoparticles was explored for achieving deep copper removal and effective resource recovery from neutral zinc leachate in hydrometallurgical zinc refining processes. Aging enabled the development of a crystalline structure in FeS nanoparticles, facilitating the slow release of sulfide ions. The improvement in sulfide ion sustained-release performance and subsequent enhancement of removal efficiency were achieved by modifying the influencing conditions. The study also investigated the impact of reaction condition factors on the sustained-release performance and explored suitable process parameters for the sustained-release sulfidizing agent. Finally, the mechanism of FeS as a sustained-release agent for deep copper removal from zinc sulfate solutions was summarized.

2. Materials and methods

2.1. Materials

The synthesis of amorphous FeS particles involved the use of $FeSO_4 \cdot 7H_2O$ and $Na_2S \cdot 9H_2O$, both obtained from Sinopharm Chemical Reagent Co., Ltd. To create a zinc leaching solution containing impurity ions, $ZnSO_4 \cdot 7H2O$, $CuSO_4 \cdot 5H_2O$, and $CdSO_4 \cdot 8/3H_2O$ were employed. All reagents adhered to analytical grade standards, and deionized water was utilized for solution preparation.

2.2. Experimental procedures

2.2.1. Methods for synthesis and aging of FeS nanoparticles. The FeS synthesis was carried out through a chemical precipitation method within a 100 mL volumetric flask. The initial phase involved preparing a FeS suspension through the combination of 0.05 L of 1 mol/L S (-II) solution with an equivalent volume of 1.2 mol/L Fe (II) solution. The synthesis duration was rigorously confined to 10 minutes. In particular, the $FeSO_4$ solution underwent vigorously stirring as the Na_2S solution was gradually added, resulting in the formation of a concentrated FeS suspension. The representation of the synthesis procedure is given by Eq. (1). Subsequently, the FeS suspension was moved to a water bath with precise temperature control and subjected to different temperatures (25, 40, 60, and 80°C) for aging periods of 3 hours. The aging process of the nanoparticles was significantly influenced by the varying temperatures. The samples were labelled as FeS-25, FeS-40, FeS-60, and FeS-80, corresponding to the respective aging temperatures. After centrifugation, the solid FeS samples were freeze-dried and used for characterization purposes. On the other hand, the removal experiments utilized the FeS suspension directly, without prior filtration.

$$Na_2S+FeSO_4 \rightarrow FeS\downarrow +Na_2SO_4 \quad (1)$$

2.2.2. Copper slow-release separation experiment. The preparation of the simulated zinc leaching solution followed a standard procedure, with the Zn (II) concentration fixed at 130 g/L and the Cu (II) concentration held at 0.64 g/L, unless stated otherwise. A solution of sulfuric acid (1%) and sodium hydroxide (1%) was adjusted to obtain the desired pH value. The reaction liquid, amounting to 100 mL, was employed in a 200 mL conical flask, and the reaction was initiated by adding the precalculated FeS suspension. After the reaction, the filtrate underwent filtration with a 0.22 μm filter membrane. The calculation of removal efficiencies was performed using Eq. (2). The filter residue underwent three rinses and was subsequently dried at 40°C for 12 hours. The experimental procedures were replicated at least twice, and the results were averaged.

The calculation of ion removal efficiency in the simulated zinc leaching solution was performed using the following equation:

$$\chi = 1-\frac{C_1 \times V_1}{C_0 \times V_0}\times 100\% \quad (2)$$

where C_0 (mg/L), C_1 (mg/L) represent the initial and residual concentrations of the elements in the solution, respectively. Additionally, V_0 (L), V_1 (L) denote the initial and residual volumes, respectively.

3. Results and discussion

3.1. Characterization of FeS nanoparticles

The comprehensive exploration of the aging time impact on FeS nanoparticles was undertaken through the implementation of various characterization techniques. As depicted in Fig. 1(a), the extended aging time reveals the conversion of FeS from amorphous to crystalline, evident in the increasing intensity of the four peaks (2θ=17.62°, 30.06°, 38.96°, 50.43°) associated with mackinawite. Furthermore, the incremental decrease in the half-maximum width of FeS points to an enlargement in the size parameters of FeS nanoparticles. Notably, a novel greigite phase becomes apparent upon aging of FeS, with distinct peaks at 2θ=29.96°, 36.34°, and 52.35°. Several studies have reported the production of unstable intermediates of greigite under mild oxygen reduction, and the process of greigite formation can be depicted through Eqs. (3) and (4).

$$FeS+\frac{1}{2}O_{2(aq)}+2H^{+}=S_{(s)}+Fe^{2+}_{(aq)}+H_2O \quad (3)$$

$$3FeS+S^0 \rightarrow Fe_3S_4 \quad (4)$$

As depicted in Fig. 1(b), the surface the surface morphology of FeS undergoes gradual transformations as the aging time increases. Initially, FeS exhibits a fine sheet-like structure with a notable specific surface area. Yet, as aging progresses, it undergoes a transformation into structures characterized by growth and collapse. The diminishing of surface voids and their transformation into thicker plates over time may pose challenges in the removal of subsequent ions. The information in Table 1 reveals that a continuous increase in the lattice parameter along the a-axis as the aging time extends while the c-axis (interlayer spacing) decreases from 6.36 Å to 5.05 Å, closely resembling the spacing between layers seen in crystalline FeS [16]. The ongoing shortening of the Fe-S bond length indicates an incremental stabilization of the structure. The lattice energy, which quantifies the energy generated when a stable crystal breaks down to produce ions, plays a critical role in determining solubility and phase transition behavior. Over an extended aging period, the lattice energy of FeS decreases from -3071 to -3653 kJ/mol, highlighting an increased stability in its crystal structure [17, 18]. Based on the aforementioned calculations and analyses, an illustration depicting the structural modifications of FeS is presented in Fig. 1(c). The shift from a disordered state to an ordered state within the FeS structure is associated with an augmentation in stability.

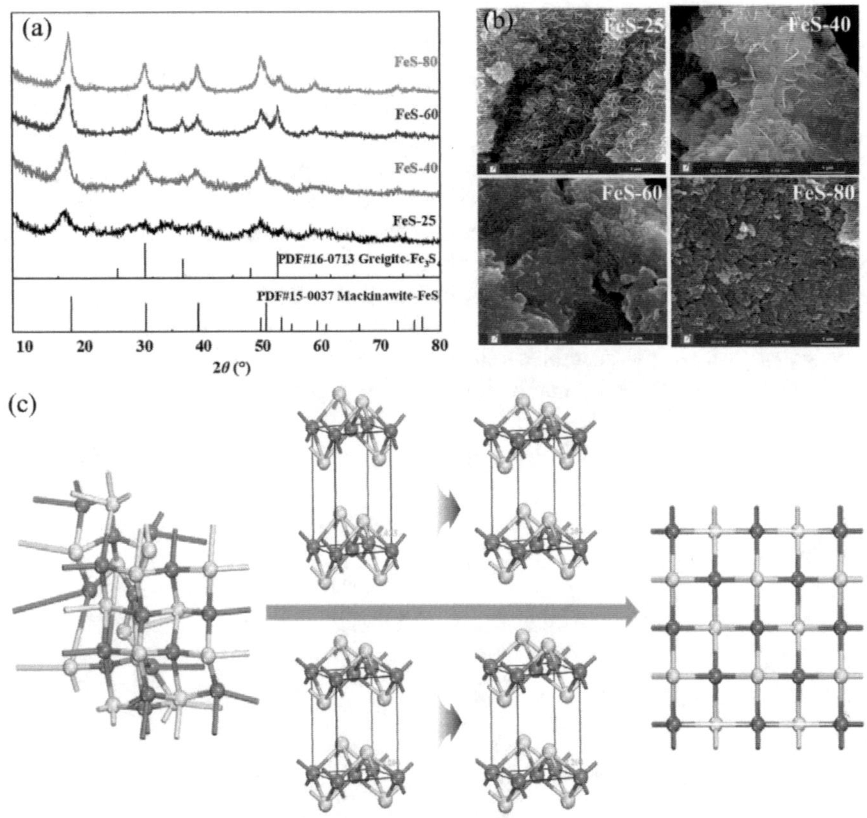

Fig. 1. The XRD (a), PSD (b) and structural evolution (c) of FeS with varying aging times.

Table 1 The lattice parameters, bond length and lattice energy of FeS with varying aging temperatures

Type	lattice parameter (Å)			Fe -S bond distance(Å)	Lattice energy(kJ/mol)
	a	b	c		
FeS-25	3.540	3.540	6.361	2.423	-3071
FeS-40	3.644	3.644	5.241	2.288	-3599
FeS-60	3.673	3.673	5.154	2.280	-3615
FeS-80	3.681	3.681	5.058	2.268	-3653

3.2. Screening of FeS nanoparticles at different aging temperatures

Unlike ionic sulfidizing agents such as Na_2S, the release of sulfur by sustained-release sulfidizing agents occurs slowly. This requires the sustained-release agents to possess a certain level of stability to maintain the release of sulfur, while also avoiding excessive stability that hinders the release of sulfur. This delicate balance is worthy of further in-depth research and exploration. Na_2S was used for control experiments, while FeS nanoparticles obtained at four different aging temperatures were utilized for copper ion removal experiments at different pH levels to select the optimal sustained-release sulfidizing agent. The reaction conditions included a reagent dosage with a S/Cu (II) molar ratio of 1:1, a reaction temperature of 25°C, and a reaction time of 30 minutes. As shown in the Fig. 2, it can be observed that at pH 1, Na_2S exhibits rapid reaction during the copper removal process, reaching equilibrium within 2 minutes with a removal rate of 86.9%. FeS-25, on the other hand, reaches equilibrium after 5 minutes of reaction, achieving a removal rate similar to that of sodium sulfide for copper. This indicates that the addition of FeS has a certain effect in slowing down the reaction rate, demonstrating its sustained-release performance. As the aging temperature increases, the sustained-release performance of FeS continues to improve. FeS-40 and FeS-60 reach equilibrium at 10 minutes and 20 minutes, respectively. However, it is important to note that this comes at the expense of a lower removal rate. FeS-80, on the other hand, has not reached equilibrium within 30 minutes. Under the pH 4 conditions (Fig. 2(b)), all reactions exhibit a slow behavior. The observed behavior is probably a result of the diminished concentration of hydrogen ions, which diminishes the influence of competing reactions. As a result, observable kinetic phenomena can occur between sulfide ions and metal ions.

Therefore, taking into consideration the overall findings, FeS with optimal sustained-release performance should possess a certain degree of stability. The FeS product aged at 60°C can be selected for further investigations on sustained-release copper removal experiments. In the subsequent experiments, the removal efficiency can be improved by adjusting the influencing conditions to obtain adjustable sulfide ion sustained-release performance.

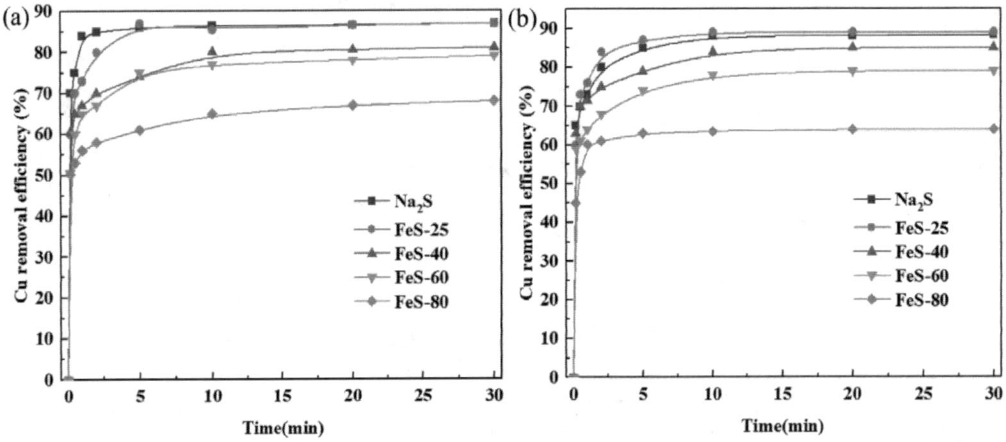

Fig. 2. The removal efficiency of Cu (II) by Na_2S and FeS from simulated zinc sulfate solution at pH 1 (a) and pH 4 (b).

3.3. Removal behavior

3.3.1 Effect of pH values on Cu (II) removal. The influence of pH on copper ions removal is shown in Fig. 3 (a). However, it was found that pH had minimal impact on the copper removal under the investigated pH conditions, as the removal efficiency remained relatively stable at around 80% across all conditions. This may be attributed to the fact that the decomposition of aged FeS requires specific acidity, which was not achieved under the pH conditions studied in the paper. Interestingly, a slight increase in the copper removal efficiency was observed at pH 5. The residual concentration of copper ions in the solution is reduced from 0.14g/L at pH=1 to 0.12g/L, which could be attributed to the higher utilization rate of sulfide ions at elevated pH levels. From a thermodynamic perspective, a higher solution pH is more favourable for the removal of copper ions.

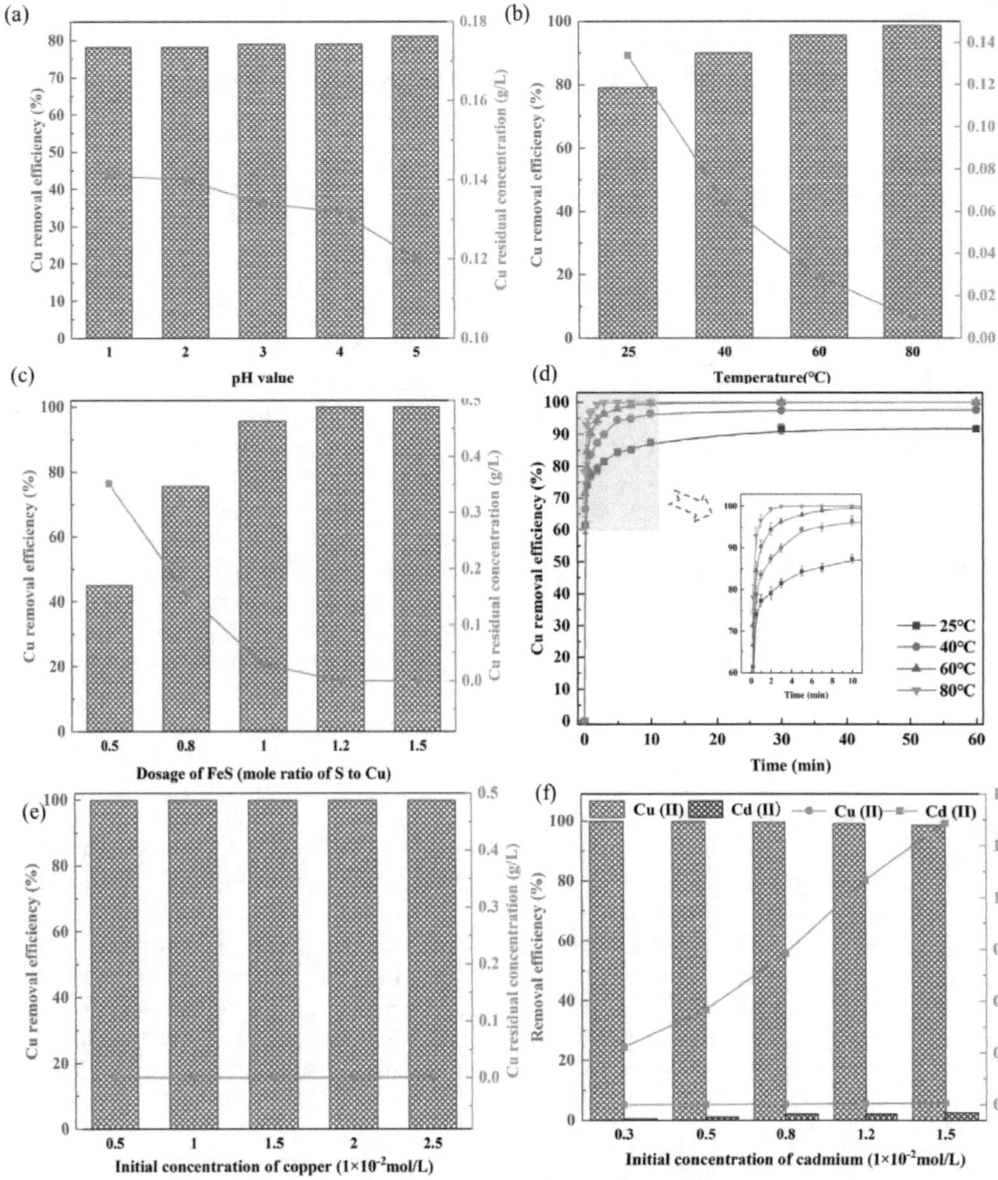

Fig. 3. Cu (II) removal from simulated zinc sulfate solution by FeS as the function of: (a) pH value, (b) reaction temperature, (c) FeS to Cu (II) mole ratio, (d) reaction time, (e) initial concentration of Cu (II), and (f) initial concentration of Cd (II).

3.3.2 Effect of reaction temperatures on Cu (II) removal. Temperature is an important factor influencing the sustained release efficiency and decomposition activity of FeS, as shown in the Fig. 3 (b). As the reaction temperature increases, the removal efficiency of copper ions is significantly enhanced. As the reaction temperature reaches 80°C, the copper ions removal efficiency increases from 79.5% at 25°C to 98.4%. This can be attributed to the endothermic nature of the FeS decomposition reaction [19], where higher temperatures facilitate the release of S ions from FeS, thereby improving the removal efficiency of copper ions. This also indicated that the maximum utilization efficiency of sulfur ions in the sustained-release agent was 98.4%. Considering that the temperature of zinc leachate in industrial production is maintained between 50-70°C, a reaction temperature of 60°C will be used in the subsequent experiments. At this temperature, the removal efficiency of copper ions exceeds 95%.

3.3.3 Effect of FeS dosage on Cu (II) removal. As depicted in the Fig. 3 (c), with an increase of FeS/Cu (II) molar ratio from 0.5 to 1.2, the Cu (II) removal efficiency increases from 45.4% to 99.9%. The residual concentration of copper ions decreases from 352 mg/L to 0.15 mg/L. Hence, to achieve the efficient removal and recovery of Cu(II), it is concluded that the optimal molar ratio of FeS to Cu(II) is 1.2:1.

3.3.4 Effect of reaction time on Cu (II) removal. As shown in the Fig. 3 (d), the copper removal gradually increases with varying time, reaching a plateau around 30 mins, indicating the effective sustained-release performance of aged FeS. Moreover, higher temperatures notably enhance the rate of copper removal by FeS. At a reaction temperature of 25°C, the copper removal rate rises gradually with time, reaching equilibrium at 30 mins. However, with a temperature increase to 80°C, the reaction nearly attains equilibrium within just 3 mins.

3.3.5 Effect of copper concentration on Cu (II) removal. Considering the fluctuation in copper content in zinc leachate in different smelters (0.2-1.2 g/L), the impact of various concentrations of initial copper ions on the removal efficiency was explored. As indicated in the Fig. 3 (e), Under specified conditions with FeS dosage 1.2 times the theoretical amount, 60°C reaction temperature, and 30 minutes reaction time, the removal efficiency of copper ions exceeded 99.9% within the investigated concentration range. The copper content in the solution was consistently below 0.2 mg/L, achieving the objective of deep copper ion removal. This indicates that the sustained release of copper removal by FeS exhibits excellent adaptability, making it suitable for selectively removing copper within a wide concentration range.

3.3.6 Effect of Cadmium concentration on Cu (II) removal. In addition to copper, zinc leachate also contains a significant amount of other impurity ions, with cadmium ions (0.6-1.2 g/L) being a crucial interfering factor to consider during the copper removal. The impact of interference by different cadmium ion concentrations on the FeS copper removal is illustrated in the Fig. 3 (f). The copper level in the solution was fixed at 0.64 g/L, with FeS dosage set at 1.2 times the theoretical amount, a reaction temperature of 60°C, and 30 minutes. Within the range of cadmium concentrations from 0.3 to 1.5×10-2 mol/L, the copper removal efficiency by FeS remained above 98.5%, while the cadmium removal rate in the solution did not exceed 2.5%. This indicates that FeS, as a sustained-release sulfidizing agent, exhibits excellent selectivity by avoiding interference from impurity ions like Cd during the removal process. It allows for the highly precise separation of copper from the solution and enables the efficient recovery of copper resources.

3.3.7 Characterization of removal products. The XRD analysis of arsenic removal products from Na_2S and FeS at different aging temperatures revealed insights into the phase transformation during the reaction. Fig. 4 illustrated that the products of Na_2S and Fe-25 were identified as copper sulfide (CuS), indicating that Na_2S and FeS-25 possess high activity and transformation efficiency. Additionally, the intensity of the copper sulfide diffraction peak in FeS-25 was found to be enhanced compared to Na_2S, pointing an increase in the crystallinity of the copper removal product in FeS-25. As the aging temperature increased (exceeding 40°C), the enhanced intensity of the diffraction peak at 2θ=29.365° and the presence of a novel diffraction peak at 2θ=49.039° were confirmed as characteristic peaks of $CuFeS_2$. The gradual increase in the corresponding peak intensity signifies a diminishing reactivity of FeS, which is consistent with its enhanced sustained release performance. The strengthening of the FeS structure and the weakening of its decomposition were observed as the aging temperature increased, suggesting the formation of $CuFeS_2$ phase during the sustained release of S ions in FeS. Therefore, it is necessary to select a suitable temperature for aging modification of amorphous FeS, which will determine its slow release performance in zinc leaching solution.

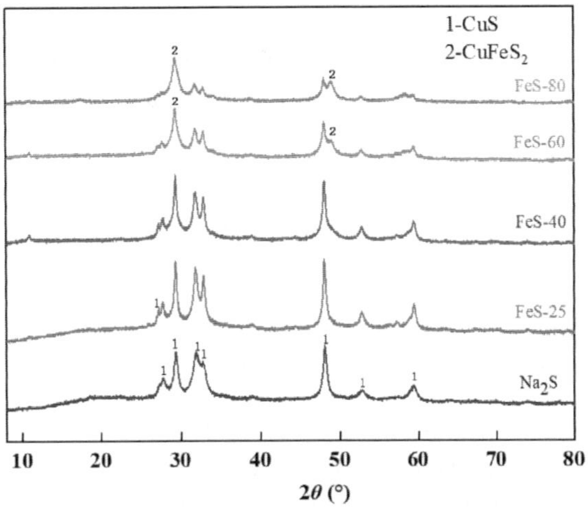

Fig. 4. XRD pattern of the copper removal precipitates of Na_2S and FeS with different aging temperatures.

The analysis of copper-containing precipitates from Na2S and different types of aged FeS involved SEM and elemental mapping to examine their characteristic morphology and elemental composition, as shown in the Fig. 5. The copper removal product in Na_2S exhibited numerous loose and irregular-shaped particles (Fig. 5(a)), which can be attributed to the local oversaturation of S ions in the solution caused by the high concentration of Na_2S, resulting in nucleation bursts and the generation of numerous amorphous fine particles. After copper removal using FeS, the point-like particles gradually disappeared, and the copper removal product mainly appeared as block-shaped particles (Fig. 5(b)) with an increased crystallinity, consistent with the XRD analysis. With the increase in aging temperature, the characteristics of the block-shaped particles became more prominent (Fig. 5(c-e)). This structure indicates that S is initially fixed in FeS and slowly released during the reaction process. Elemental analysis results (Fig. 5(f)) revealed the copper grade in the copper removal product of FeS-60 exceeded 60%, indicating a relatively pure copper sulfide, and its purity was close to 85%. The grade of Fe was 5.34%, suggesting that some Fe did not completely decompose, demonstrating the slow release of S ions in FeS.

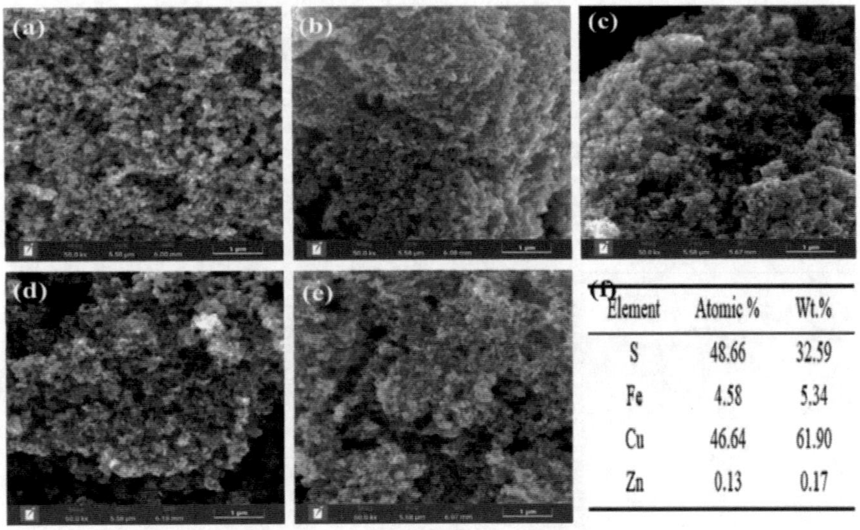

Element	Atomic %	Wt.%
S	48.66	32.59
Fe	4.58	5.34
Cu	46.64	61.90
Zn	0.13	0.17

Fig. 5. SEM image and mapping results of the copper removal products of Na_2S (a) and FeS with different aging temperatures. (b) FeS-25, (c) FeS-40, (d) FeS-60, (e) FeS-80, and (f) FeS-60.

3.4. Proposed removal mechanism

The solubility product of a metal sulfide governs its dissolution characteristics, with precipitates having higher solubility products transforming into those with lower solubility products. Sulfides with larger solubility products transform into those with smaller solubility products. A smaller solubility product for a given metal ion implies a higher probability of preferential precipitation. Therefore, theoretically, the order of precipitation for the four metal ions is as follows: Cu (pK=35.85) > Cd (27.19) > Zn (23.10) > Fe (16.47) [20]. In addition to the solubility product principle, inherent characteristics such as metal ion radius and ion charge also have important implications for Cu(II) removal using FeS in high-concentration zinc sulfate solutions. Studies have shown that metal ions with smaller ionic radii are more likely to form strong coordination bonds with sulfur-containing groups [21, 22]. Copper, having an ionic radius of 0.72 Å, exhibits a smaller size compared to zinc (0.74 Å), cadmium (0.97 Å), and ferrous iron (0.78 Å). The difference in size may result in the generation of new bonds between copper ions and sulfur-containing groups on the surface of aged FeS, enabling selective recovery. Metal ions with higher ion charges have an increased ability to attract electrons and tend to form complexes with sulfur-containing groups [23]. The order of ion charges for the four metals is as follows: Cu (2.78) > Zn (2.70) > Fe (2.56) > Cd (2.06).

However, the separation between similar elements is often difficult to achieve in an ideal manner as predicted by theoretical calculations, mainly due to external conditions, particularly the type of sulfidizing agent used [20]. To approach the ideal separation state, certain measures need to be taken. As shown in the Fig. 6, sulfur ions gradually bind and become fixed to iron, transitioning from sodium sulfide to amorphous FeS and finally to aged FeS. The strengthening of the Fe-S bond enhances the fixation of sulfur, leading to an improved sustained-release performance of FeS. It has been found that the sustained-release rate can be selectively controlled through heating. The slow release of sulfide ions minimizes the supersaturation of sulfur ions, thereby increasing the utilization efficiency of sulfide ions. This approach brings the separation order of elements closer to the theoretical calculations based on solubility product or other principles, enabling high selectivity and deep purification of copper.

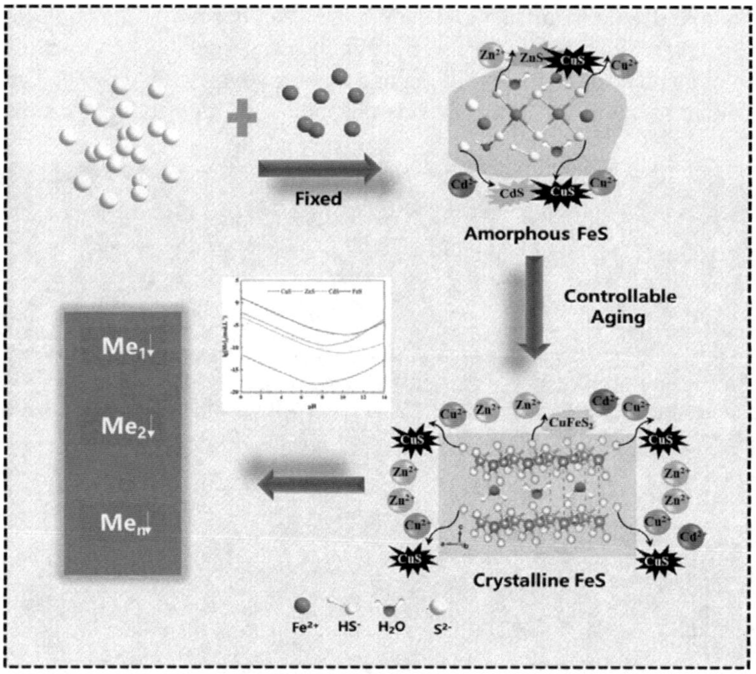

Fig. 6. The mechanism and schematic diagram of deep copper removal from zinc leachate using controllable aging FeS.

Journal of Physics: Conference Series **2738** (2024) 012014 doi:10.1088/1742-6596/2738/1/012014

4. Conclusions

In this study, controllable aging of FeS nanoparticles was implemented by varying the temperature to achieve deep removal of copper and effective recovery of resources from neutral zinc leachate in hydrometallurgical processes. The aging process resulted in the development of a crystalline structure within the FeS nanoparticles, facilitating the gradual release of sulfide ions. Through the manipulation of influencing conditions, it was possible to attain adjustable sulfide ion sustained-release performance and enhance the copper removal efficiency.

Temperature has been found to be a crucial factor influencing the sustained-release performance of FeS. When the temperature exceeds 60°C, the utilization efficiency of S ions in the sustained-release agent can reach over 95%, with a maximum of 98.5%. By optimizing the conditions, the copper removal can approach 100% under the optimal conditions. The residual copper content in the zinc leachate is reduced to 0.2 mg/L, as the removal rate of cadmium does not exceed 2.5%. Slow-release sulfidation can maximize the utilization efficiency of sulfide ions in the solution, reduce the occurrence of competing reactions, and minimize the intermixing of products. It offers advantages such as high selectivity and deep purification. By employing sustained-release sulfidation for copper precipitation, not only can the deep purification and resource recovery of copper from leachate be achieved with high selectivity, but it can also greatly decrease the usage of zinc powder.

References

[1] S. Song, W. Sun, L. Wang, R. Liu, H. Han, Y. Hu, Y. Yang, Recovery of cobalt and zinc from the leaching solution of zinc smelting slag, Journal of Environmental Chemical Engineering, 7 (2019) 102777.
[2] L. Tang, C. Tang, J. Xiao, P. Zeng, M. Tang, A cleaner process for valuable metals recovery from hydrometallurgical zinc residue, Journal of Cleaner Production, 201 (2018) 764-773.
[3] J. Lu, D. Dreisinger, K. Rees, Simultaneous removal of Co, Cu, Cd and Ni from zinc sulfate solution by zinc dust cementation, Hydrometallurgy, 197 (2020) 105479.
[4] B.S. Boyanov, V.V. Konareva, N.K. Kolev, Purification of zinc sulfate solutions from cobalt and nickel through activated cementation, Hydrometallurgy, 73 (2004) 163-168.
[5] B. Zhang, C. Yang, H. Zhu, Y. Li, W. Gui, Kinetic Modeling and Parameter Estimation for Competing Reactions in Copper Removal Process from Zinc Sulfate Solution, Industrial & Engineering Chemistry Research, 52 (2013) 17074-17086.
[6] M.D. Rao, A. Meshram, H.R. Verma, K.K. Singh, T.R. Mankhand, Study to enhance cementation of impurities from zinc leach liquor by modifying the shape and size of zinc dust, Hydrometallurgy, 195 (2020) 105352.
[7] X. Liu, S. Wang, Z. Peng, G. Zhang, Q. Gui, L. Zhang, Removal of toxic cadmium (II) from zinc sulfate solution with zinc powder enhanced by ultrasound: Kinetics and mechanism, Separation and Purification Technology, 308 (2023) 122995.
[8] R. Raghavan, P.K. Mohanan, S.K. Verma, Modified zinc sulphate solution purification technique to obtain low levels of cobalt for the zinc electrowinning process, Hydrometallurgy, 51 (1999) 187-206.
[9] A.H.M. Veeken, L. Akoto, L.W. Hulshoff Pol, J. Weijma, Control of the sulfide (S2−) concentration for optimal zinc removal by sulfide precipitation in a continuously stirred tank reactor, Water Research, 37 (2003) 3709-3717.
[10] L. Kong, X. Hu, X. Peng, X. Wang, Specific H2S Release from Thiosulfate Promoted by UV Irradiation for Removal of Arsenic and Heavy Metals from Strongly Acidic Wastewater, Environmental Science & Technology, 54 (2020) 14076-14084.
[11] X. Zhang, J. Tian, Y. Hu, H. Han, X. Luo, W. Sun, T. Yue, L. Wang, X. Cao, H. Zhou, Selective sulfide precipitation of copper ions from arsenic wastewater using monoclinic pyrrhotite, Science of The Total Environment, 705 (2020) 135816.

[12] R. Liu, Z. Yang, Z. He, L. Wu, C. Hu, W. Wu, J. Qu, Treatment of strongly acidic wastewater with high arsenic concentrations by ferrous sulfide (FeS): Inhibitive effects of S(0)-enriched surfaces, Chemical Engineering Journal, 304 (2016) 986-992.
[13] X. Zhang, H. Fan, J. Yuan, J. Tian, Y. Wang, C. Lu, H. Han, W. Sun, The application and mechanism of iron sulfides in arsenic removal from water and wastewater: A critical review, Journal of Environmental Chemical Engineering, 10 (2022) 108856.
[14] V. Sarathy, P.G. Tratnyek, J.T. Nurmi, D.R. Baer, J.E. Amonette, C.L. Chun, R.L. Penn, E.J. Reardon, Aging of Iron Nanoparticles in Aqueous Solution: Effects on Structure and Reactivity, The Journal of Physical Chemistry C, 112 (2008) 2286-2293.
[15] X. Zhang, C. Lu, J. Tian, L. Zeng, Y. Wang, W. Sun, H. Han, J. Kang, Artificial intelligence optimization and controllable slow-release iron sulfide realizes efficient separation of copper and arsenic in strongly acidic wastewater, Journal of Environmental Sciences, 139 (2024) 293-307.
[16] X. Zhang, J. Tian, H. Han, W. Sun, Y. Yang, X. Jiang, Y. Cao, Mechanistic insights into the structure-activity relationship of FeS for arsenic removal in strongly acidic wastewater, Journal of Water Process Engineering, 53 (2023) 103904.
[17] A.J. Devey, R. Grau-Crespo, N.H. de Leeuw, Combined Density Functional Theory and Interatomic Potential Study of the Bulk and Surface Structures and Properties of the Iron Sulfide Mackinawite (FeS), The Journal of Physical Chemistry C, 112 (2008) 10960-10967.
[18] S. Kaya, Relationships between lattice energies of inorganic ionic solids, Physica B: Condensed Matter, 538 (2018) 25-28.
[19] X. Zhang, J. Yuan, J. Tian, H. Han, W. Sun, T. Yue, Y. Yang, L. Wang, X. Cao, C. Lu, Ultrasonic-enhanced selective sulfide precipitation of copper ions from copper smelting dust using monoclinic pyrrhotite, Transactions of Nonferrous Metals Society of China, 32 (2022) 682-695.
[20] X. Zhang, L. Zeng, Y. Wang, J. Tian, J. Wang, W. Sun, H. Han, Y. Yang, Selective separation of metals from wastewater using sulfide precipitation: A critical review in agents, operational factors and particle aggregation, Journal of Environmental Management, 344 (2023) 118462.
[21] X. Yang, X. Hu, L. Kong, X. Peng, Selective recovery of Cu(II) from strongly acidic wastewater by zinc dimethyldithiocarbamate: Affecting factors, efficiency and mechanism, Journal of Environmental Sciences, 129 (2023) 115-127.
[22] J. Jiang, Z. Wang, Y. Chen, A. He, J. Li, G.D. Sheng, Metal inhibition on the reactivity of manganese dioxide toward organic contaminant oxidation in relation to metal adsorption and ionic potential, Chemosphere, 170 (2017) 95-103.
[23] F.S. Hoseinian, B. Rezai, M. Safari, D. Deglon, E. Kowsari, Separation of nickel and zinc from aqueous solution using triethylenetetramine, Hydrometallurgy, 202 (2021) 105609.

Journal of Physics: Conference Series **2738** (2024) 012015 doi:10.1088/1742-6596/2738/1/012015

A permanent insulator system for hydrometallurgy with a significant increase in electric efficiency and without short circuit

Robert Dufresne[1], Hamid Arabzadeh[2, *]

[1]. President at Pultrusion Technique, St-Bruno, QC, Canada, J3V 6B9.
Email: rdufresne@pultrusiontech.com

[2]. PhD, R&D Director at Pultrusion Technique, St-Bruno, QC, Canada, J3V 6B9.
Email: harabzadeh@pultrusiontech.com

Abstract. A permanent insulator system for hydrometallurgy has been developed with a significant increase in electric efficiency and without short circuits. The system consists of a double electric contact point for each cathode and anode and a new design of the insulating and contact system of the electrodes. The combination of these technologies provides a more uniform electric density inside each cathode and anode, in each cell, and also for the whole electric circuit in the cellhouse. Keeping a consistent maximum performance necessitates using parts with extreme durability against chemical attacks and short circuits. Hence, a comprehensive research into chemical stoichiometry and the behaviour of different molecules resulted in the development of an outstanding chemical resistance level to sulfuric acid and extremely high temperatures throughout the entire hydrometallurgical process. Field observations show that this formulation has been able to resist up to 19 years of continuous operation without maintenance. The goal of these developments combined with a new contact system was to reduce any type of short circuit by 70 to 80%, which in turn increases the electric current effectiveness by 8 to 10%. As a result, higher contact efficiencies have doubled the lifetime of the cathode and anode with this system. Consequently, the overall electric efficiency in a refinery cellhouse can be improved by 10-12%. Observations from real-world applications over the past decades have proven the efficiency of the earlier versions of this new concept. The system was found to be applicable to any type of contact system and has been used in various hydrometallurgical processes with great success. With the increasing price of energy and CO2 production, it has become more important to operate ecologically and efficiently to maximize profitability and this permanent insulator system can address many challenges which lead to a more efficient and cost-effective process.

1. Introduction

The widespread use of copper stems from its exceptional conductivity, corrosion resistance, and versatility. As the backbone of modern infrastructure, copper has witnessed an exponential surge in demand. In 2022 alone, global copper consumption surpassed 24 million metric tons, reinforcing its position as a critical resource for industrial development. With global demand for copper reaching unprecedented levels, major producers like Chile, Peru, and China face significant challenges in meeting the market needs, leading to price fluctuations. To enhance efficiency and reduce production costs in copper refineries, continual exploration of innovative solutions becomes imperative. Such efforts aim to boost the competitiveness of copper producers and cater to the ever-growing global demand [1-3].

A crucial factor impacting refining efficiency lies in the maintenance of cellhouse equipment and the associated production losses. Within the refinery cellhouse, traditional capping boards, serving as insulators, undergo significant wear and tear during the refining process, resulting in frequent maintenance and replacement needs [4,5]. The labor-intensive and time-consuming nature of insulator replacement poses challenges, and can also result in damages to the cells.

Fig. 1. Damages observed in the electrolytic cells in refineries

Failure to adequately maintain or replace insulators when necessary can lead to operational issues, including uneven current distribution and difficulty in maintaining optimal operating conditions within the cells. These issues ultimately reduce efficiency and production rates. Moreover, the maintenance and replacement demands of traditional insulators can result in production losses and potential shutdowns of the cellhouse, exacerbating the impact on efficiency [6].

In light of these challenges, a comprehensive examination of maintenance practices and potential solutions is essential for optimizing hydrometallurgy refining operations. This paper explores innovative approaches to improve the lifespan and performance of insulators, aiming to minimize maintenance requirements, enhance production efficiency, and reduce associated costs.

2. Design and build of equipotential insulators

Traditional insulator technologies often employ substandard or non-premium materials, and due to their multiple-piece design concepts, installation procedures are unnecessarily complex and time consuming, leading to operational challenges for the cellhouse. These issues include uneven current distribution and difficulties in maintaining optimal operating conditions. Consequently, efficiency is compromised, resulting in lower production rates.

Recognizing the critical role of insulators in establishing electrical connectivity and achieving consistent potential across all cells, a new generation of insulators, termed Permanent Equipotential Insulators, has been developed. These insulators exhibit exceptional durability and chemical resistance, making them suitable for the demanding conditions found in cell houses. The implementation of high-quality equipotential insulators can profoundly impact the efficiency of hydrometallurgy cell houses.

These advanced insulators ensure that cells operate at optimal voltage and current levels, thereby improving the overall process efficiency. Furthermore, utilizing pioneer technologies of manufacturing and material engineering, their superior resistance to wear and tear is extremely higher than other products in the industry. Such an extraordinary performance extends the lifespan of the cells, minimizing the need for maintenance and repairs for the cells. As a result, the efficiency and safety of the cell house are maximized, leading to reduced operating costs.

Remarkably, early versions of these insulators have demonstrated exceptional performance, with over 22 years of operation in copper refineries and over 19 years in zinc refineries, in both cases, totally without maintenance. They have exhibited unmatched protection against electrochemical reactions and accelerated corrosion, even when subjected to highly concentrated acids up to 100% (Fig. 2). The aim for design and utilization of Permanent Equipotential Insulators has been to provide a significant advancement in cell house technology, providing extended lifespan, improved efficiency, and enhanced corrosion resistance.

Fig. 2. Early versions of Permanent Equipotential Insulator in operation since 1994

3. Parametric Study

In order to evaluate the effectiveness and benefits of Equipotential insulators, a comprehensive parametric study was conducted, comparing them to traditional insulators. The study was based on a multi-criteria decision analysis (MCDA) with weighted benefits and costs and was focused on seven key aspects: enhanced cell protection, cost efficiency enhancement, reduced level or elimination of copper contamination, superior metal refining quality and purity, heightened safety measures, ecologic advantages and carbon footprint, and return on investment. The findings of this study provide insights into the significant advantages offered by Equipotential insulators and their potential to revolutionize the industry.

3.1. Enhanced Cell Protection

Equipotential insulators exhibited superior capabilities in safeguarding electrolytic cells. The study showcased their ability to mitigate some common issues that used to cause severe damage to the cells, reducing the risk of cell damage and facilitating optimal cell performance and longevity. These insulators are available in various shapes and designs, offering diverse functionalities. One such design incorporates side flanges, commonly referred to as the umbrella shape, which allows for a secure fit over two adjacent cells. The design features a dedicated slot for positioning the contact bar and serves as a protective shield against thermal chocks resulting from short circuits and chemical corrosion (Fig. 3). In the event of a short circuit, the insulator acts as a barrier, preventing the heat flow created by the electric current from reaching the electrolytic cells. This safeguards the cells from the detrimental effects of excessive heat, thermal impacts and mechanical stress associated with short circuits.

During a short circuit, the normal path of electric current is diverted, resulting in the generation of excessive heat that quickly raises the temperature of the insulator. This rapid temperature rise induces mechanical stress and other forms of damage. Additionally, the excessive heat generated by a short circuit can adversely affect the electrolytic cells positioned beneath the insulator. These cells are designed to operate within specific temperature limits, and exposure to excessive heat can diminish their efficiency or lead to complete failure.

(a) (b) (c)

Fig. 3. Different types of Permanent Equipotential Insulators: a) Standard Equipotential Insulator, b) Coronary Insulator, c) Equipotential Insulator with Umbrella-Shape Design

Furthermore, certain equipotential insulators are specifically designed to combat chemical corrosion. They feature a protective coating that resists the corrosive effects of acid spills commonly encountered in electro-refineries. In the event of an acid spill, the permanent insulator acts as a shield, redirecting the acid away from the electrolytic cells, thus preserving their integrity and preventing damage.

Fig. 4. Coronary insulator combined with permanent Equipotential insulator protecting two adjacent cell walls from thermal, mechanical and chemical stresses.

3.2. Cost Efficiency Enhancement

The study revealed a notable reduction in electrical costs with the implementation of Permanent Equipotential insulators. Compared to traditional insulators, these advanced counterparts demonstrated improved electrical efficiency, resulting in substantial cost savings for the refining process.

(a) Without Coronary Insulator (b) With Coronary Insulator

Fig. 5. Thermal test on the performance of coronary equipotential insulator when subjected to intense thermal load due to a short circuit

This can be attributed to significant influence of the alignment and tolerances of electrode seats on the performance of a refinery cell house. However, achieving precise alignment poses challenges when using traditional multi-piece insulators.

The choice of insulator design traditionally depends on factors such as cell size, shape, available space, and other specific electrorefining requirements. In contrast, permanent equipotential insulators are manufactured in a single piece form, reinforced with pultruded fiberglass bars, with a highly adaptive design. These insulators are produced using a unique process that ensures a homogeneous, continuous-length structure with exceptional dimensional stability.

With precision reaching up to 0.5 mm centre-to-centre and 2.8 mm along an 11-meter length from the first to the last anode, they offer a secure fit and accurate positioning for electrodes, minimizing vulnerability to mechanical stress and other damage. The permanent insulator, consisting of a single monolithic piece, incorporates various components such as contact bars, anodes, and cathode seats insulation. The insulating seats are meticulously arranged and are perfectly synchronized with the contact bar positioned at the centre of the permanent insulator. This synchronization is achieved by utilizing the embedded longitudinal pultruded fiberglass rebars, which seamlessly integrate with the equipotential insulator. The contact system is securely anchored, rendering any relative movement impossible. The precision of this contact system is maximized, ensuring its longevity throughout the entire lifespan of the structure. As a result, the lifespan of these insulators is extended, reducing the likelihood of electrical failures caused by misaligned electrodes, such as short circuits and uneven electroplating.

3.3. Reduced Level or Elimination of Copper Contamination into Pure Cathodic Zinc

During the daily zinc harvesting process, an abrasion occurs between the contact bar and the aluminium cathode copper contact, leading to the generation of fine powder. This powder subsequently falls into the cell and plates onto the cathode, causing a significant impurity issue.

To address this challenge, certain types of equipotential insulators are designed with a unique built-in solution. Specifically, equipotential permanent insulators are designed with an integrated copper powder dust collector that features two slopes, facilitating the evacuation of the copper powder when the insulator is washed. Consequently, the expelled copper powder can easily be recovered and eliminated using an appropriate system.

3.4. *Superior Metal Refining Quality and Purity*

The implementation of Permanent Equipotential insulators resulted in elevated standards of metal refining. The study demonstrated that these insulators contribute to higher quality and purity of the refined metal products, enabling producers to meet stringent industry requirements and enhance their market competitiveness.

Inconsistent electrical distribution in a cellhouse can negatively impact the quality and purity of extracted copper. Uneven copper deposition on cathode plates, imbalanced dissolution rates of anodes, and disrupted electrolyte boundary layers are some factors contributing to this issue. Maintaining a consistent electrical distribution is crucial to avoid variations in copper coating thickness, impurity incorporation, and inefficient refining conditions. By addressing these challenges, refineries can optimize their processes and achieve high-quality copper extraction.

Permanent equipotential insulators act as a protection against electrical disturbances, such as short circuits, by providing superior electrical insulation. This maintains a consistent electro-refining process, minimizing the occurrence of irregularities and impurities in the copper plating. The result is plating with a higher-quality end products that achieves improved market value.

3.5. Heightened Safety Measures

One of the significant findings of the study was the enhanced safety provided by Equipotential insulators. These insulators effectively minimize the potential for accidents and operational hazards, ensuring a safer working environment within the cellhouse.

The improved safety measures for the working environment in hazardous settings like zinc electro-refineries can be attributed to the fact that the continuous and consistent level of insulation that permanent equipotential insulators provide plays a vital role, specially at 200,000 Ampere with 900 Ampere/m^2. These insulators effectively mitigate the risk of electrical accidents and injuries that can occur in such environments (Fig. 6). In the absence of proper insulation, electrical leaks, operators are susceptible to electrical shocks arising from potential differences between different parts of the electrical system. However, by functioning as intended, insulators prevent electric current from flowing through a person's body if accidental contact with the system occurs, effectively eliminating the chances of serious injury. The reliable insulation provided by permanent equipotential insulators was found to enhance the overall safety of the refinery, safeguarding workers from potential electrical hazards and ensuring a secure operational environment.

Fig. 6. High separating walls help with eliminating short circuits and electrical leakage between adjacent electrodes, and cell to cell.

3.6. Ecologic Advantages and Carbon Footprint

Permanent equipotential insulators offer not only numerous operational advantages but also contribute to ecological benefits by reducing CO2 emissions and minimizing carbon footprint. The implementation of these insulators in copper electro-refineries results in a more efficient electro-refining process, which in turn reduces energy usage and increases productivity. By improving electrical efficiency and preventing irregularities in the refining process, permanent equipotential insulators help optimize the overall performance of the refinery.

This enhanced efficiency leads to reduced energy consumption, thereby lowering CO2 emissions associated with energy production. Additionally, the improved quality and purity of the extracted copper achieved with equipotential insulators reduce the need for additional refining steps, minimizing carbon-intensive processes and reducing the carbon footprint of the entire refining operation. Therefore, the adoption of permanent equipotential insulators not only offers operational benefits but also contributes to a more sustainable and environmentally friendly approach to copper electro-refining.

3.7. Return on Investment

Permanent equipotential insulators offer a multitude of advantages that make them a valuable investment for refineries. It is true that these insulators are made of premium materials, and hence, there is a premium upfront cost for them. However, despite their initial cost compared to traditional insulators, their exceptional longevity and durability result in a reduced need for frequent replacements, leading to substantial cost savings over the long term.

By minimizing the frequency of insulator replacements, refineries can significantly reduce maintenance expenses and operational downtime, optimizing their overall cost-effectiveness. Most traditional insulators need replacement over a life of two years or less, compared to field observations that have shown equipotential insulators are able to operate continuously for up to 20 years without requiring maintenance. The cost of replacing traditional insulators includes not only the material cost but also labour, downtime, and potential operational disruptions. This can result in a potential cost saving of up to 70% on the insulators over a 20-year period.

Moreover, these insulators play a pivotal role in enhancing the efficiency and effectiveness of the electrorefining process. Their consistent maintenance of electrical potential ensures a stable and reliable electrical distribution, minimizing the occurrence of voltage fluctuations, short circuits, and uneven electroplating. This improved operational efficiency translates into increased productivity, reduced wastage, and improved product quality, contributing to higher profits for the refinery.

Considering a reduction of 70 to 80% in the number of short circuits by using the equipotential insulators, an 8 to 10% increase in electric current efficiency will be achieved, thanks to their small dimensional tolerances of less than 2.5mm in insulators with up to 10.5m and being single-piece instead of multi-piece. This will be combined with better protection and extended lifespan of the electrodes themselves up to 3 times (12 years instead of 4 years), and the lifespan of the cell up to two times. Moreover, the higher mechanical and thermal resistance of these insulators lets the operation to go under higher current intensity. Hence, it was estimated to result in up to 12% less electricity consumption per ton of copper in the cost of operation.

The initial investment in premium raw materials for equipotential insulators can be offset by the long-term cost savings from reduced replacements and the enhanced productivity achieved through their consistent electrical performance. As a result, refineries can experience major favourable financial outcome and a higher return on investment (ROI) by choosing permanent equipotential insulators. This combination of long-term cost savings, improved operational efficiency, and higher productivity makes equipotential insulators a compelling choice for refineries seeking to maximize profitability and achieve sustainable growth in the electrorefining industry.

4. Conclusion

In conclusion, the comprehensive research conducted on permanent equipotential insulators and their chemical formulation has yielded remarkable results. These insulators demonstrate exceptional mechanical, chemical and thermal resistance with permanent physical consistency, withstanding the corrosive effects of sulfuric acid and extreme temperatures throughout the hydrometallurgical process.

The primary objective of these developments, along with the implementation of a new contact system, was to significantly reduce the occurrence of short circuits. As a result, the extended lifespan of the cathode and anode has effectively doubled, leading to improved overall electric efficiency in refinery cellhouses.

Real-world applications have demonstrated the efficiency of earlier versions of this concept, proving its applicability in various hydrometallurgical processes. Given the rising energy costs and concerns about CO2 production, the utilization of these permanent insulators offers an ecological and efficient approach to maximize profitability.

Lastly, while they may initially involve a higher investment due to their premium materials, their durability and longevity translate into fewer replacements and long-term cost savings. These advantages, coupled with the improved efficiency and consistency they provide to the electrorefining process, contribute to a higher return on investment. Therefore, permanent equipotential insulators represent a wise investment choice for refineries, offering both financial and operational benefits.

References

[1] Bastos, J. P. A., & Sadowski, N. (2003). Electromagnetic modeling by finite element methods. CRC press.

[2] Dufresne, R., Arabzadeh, H., (2023). Permanent Equipotential Insulators: Put an end to the frequent replacement of capping boards in cell house refineries. Canadian Mining Journal.

[3] Wiechmann, E. P., Morales, A. A., Aqueveque, P. E., & Burgos, R. P. (2007, September). Measurement of cathodic currents in equipotential inter-cell bars for copper electrowinning and electrorefining plants. In 2007 IEEE Industry Applications Annual Meeting (pp. 2074-2079). IEEE.

[4] Krishna, P. G., & Das, S. C. (1992). Enhancement of operating current density in a copper electrowinning cell. Hydrometallurgy, 31(3), 243-255.

[5] Davenport, W. G., King, M., Schlesinger, M. E., & Biswas, A. K. (2002). Extractive metallurgy of copper. Elsevier.

[6] Wiechmann, E. P., Morales, A. S., & Aqueveque, P. (2010). Improving productivity and energy efficiency in copper electrowinning plants. IEEE transactions on industry applications, 46(4), 1264-1270.

Design and production practice of large fluidized roasting furnace

Yuan Liang[1, *]

1.China ENFI Engineering Corporation, Beijing, Beijing, China, 100038.

Email: liangyuan2023@126.com

Abstract. With the expansion of the scale of zinc smelting enterprises and the progress of new equipment and technology, 152 m^2 fluidized roasting furnace has been developed and applied to industrial production, with a single unit capacity of 150,000 to 170,000 tons of electrolytic zinc per year. This paper analyzes and introduces an engineering example of zinc concentrate treatment by a dual-series fluidized roasting furnace. The project was designed by China ENFI Engineering Corporation and put into operation in Hunan province of China in 2018. This paper firstly describes the process flow and main equipment of the preparation system, roasting furnace system, waste heat utilization system, dust collection system from a design perspective. Then, based on production practice, it analyzes and compares aspects such as furnace operation, actual production data, major technical and economic indicators, and the characteristics and optimization of the roasting system equipment. Finally, it is concluded that the dual-series large-scale fluidized roasting furnace has stable production, good product quality, less investment, and fewer labor requirements, which brings good economic benefits to the enterprise. Its operational experience has accumulated parameters and experience for the development of larger roasting furnaces in the future.

1. Introduction

Modern zinc smelting production processes are mainly divided into two categories: pyrometallurgical zinc smelting and hydrometallurgical zinc smelting. The production of hydrometallurgical zinc smelting has accounted for more than 85% of the world's zinc production. Hydrometallurgical zinc smelting is mainly composed of roasting, leaching, purification, electrowinning, and casting processes. Roasting, as the first process of hydrometallurgical zinc smelting, has a significant impact on the hydrometallurgical process due to its roasting capacity and the quality of roasted ore. In recent years, China, as a major zinc-producing country, has developed a 152m^2 fluidized roasting furnace through continuous exploration of new equipment and technologies, with a single unit capacity reaching 150,000 to 170,000 tons of electrolytic zinc per year.

Hunan Zhuye Nonferrous Metals Co., Ltd. has constructed a zinc and zinc alloy factory with an annual production capacity of 300,000 to 340,000 tons in Songbai Town, Changning City. The roasting system adopts two 152m^2 fluidized roasting furnaces. The project was completed and put into operation in December 2018 and has been running for 5 years .

2. Roasting process flow

The roasting system mainly consists of two parts: the material preparation section and the roasting section.

The material preparation section includes: storage of zinc concentrate, drying, crushing and screening processes.

The roasting section includes: roasting, calcine cooling, calcine grinding processes.

In addition, there are waste heat boiler system, dust collection system and other auxiliary production systems.

The process flow chart of the roasting system is shown in Figure 1.

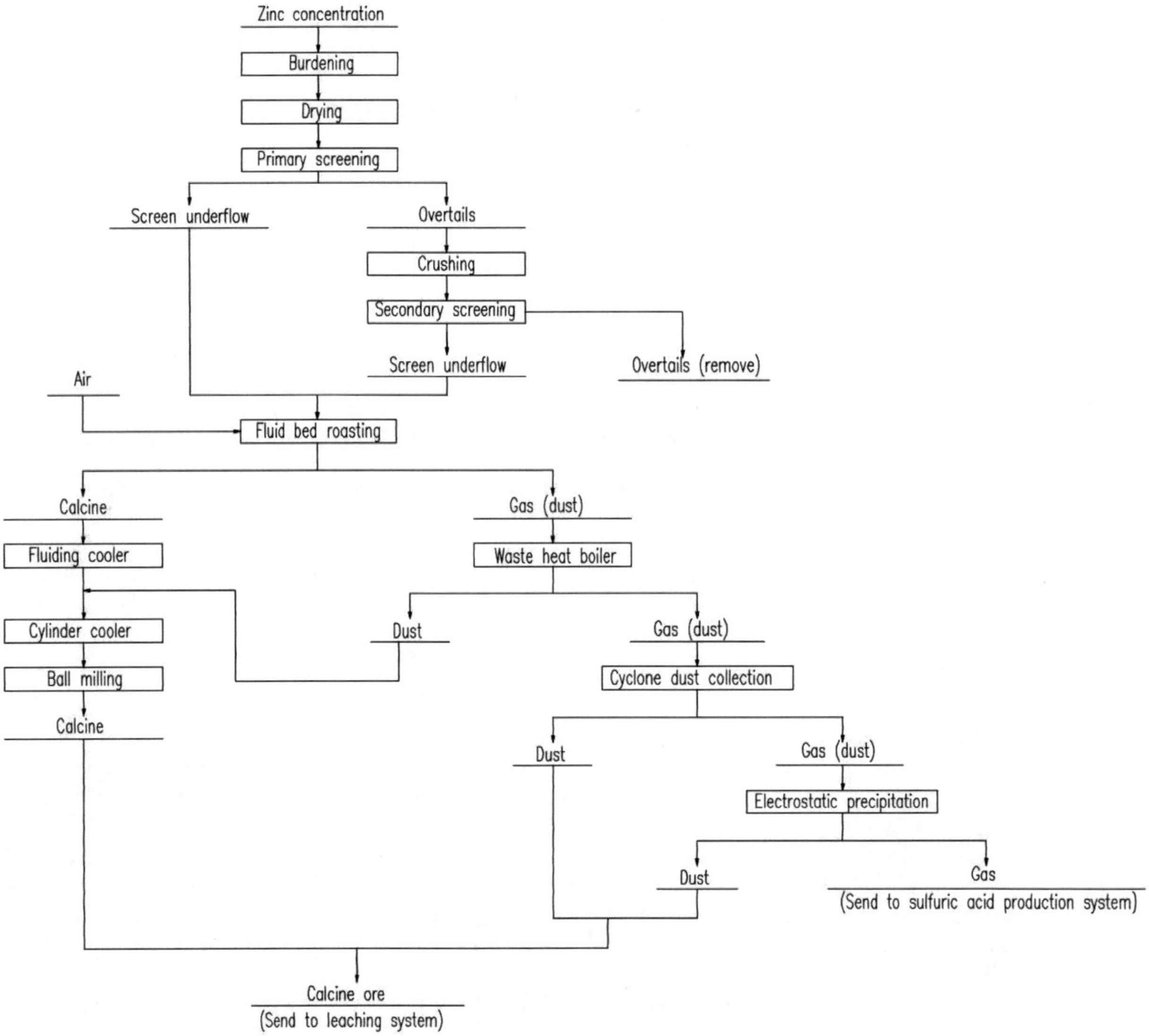

Figure 1. The process flow chart of the roasting system.

2.1 Material preparation system

The concentrate warehouse is a semi-underground warehouse. Zinc concentrate is transported to the warehouse by trucks or trains. The concentrate is fed into the proportioning bins through a bridge-type grab crane and a variable speed disc feeder. After the concentrate is weighed by the electronic belt scales, it is conveyed to the concentrate drying section by a conveyor belt. Depending on the incoming material in the concentrate warehouse, the drying section is divided into two routes. One route is for the concentrate that needs to be dried, which is sent to the drying kiln through a conveyor belt for drying and then to the screening and crushing section. The other route is for the concentrate that does not need to be dried, which is directly sent to the screening and crushing section by the conveyor belt without going through the drying kiln.

The screening and crushing system consists of a crusher and two high-efficiency vibrating screens. The material is first fed into the vibrating screen. The undersized material is sent to the feeding conveyor belt, while the oversized material enters the crusher for crushing. After the second screening, the oversized material is manually sorted to remove impurities, the undersized material is transported back to the conveyor belt by a bucket elevator. The screened material from both screenings is collected and sent to the pre-furnace storage bin in the roasting plant.

The entire plant is equipped with one set of drying system, one set of screening and crushing system, and one set of feeding system to provide raw materials for the two roasting furnaces.

2.2 Roasting furnace system

The two roasting workshops are arranged in a north-south mirror relationship on the general drawing. Each workshop is equipped with a $152m^2$ fluidized roasting furnace. A bin is set in front of each roasting furnace to store concentrate for four hours. the zinc concentrate is measured by the double belt quantitative scales at the bottom of the bin and then thrown to the roasting furnace at a speed of 15 ~ 18 m/s by the throwing machine.

Under the action of the distribution plate airflow, the zinc concentrate is in a boiling state. Eight sets of cooling coils are set up in the boiling layer to take away the remaining heat and maintain a reaction temperature of about 930℃ in the furnace. Under this condition, a series of physical and chemical reactions such as roasting desulfurization occur in the zinc concentrate. The produced zinc calcine flows out through the overflow port on the furnace side and is cooled to around 150 ℃ after two stages of cooling, then the calcine is transported by a buried scraper conveyor to the calcine ball milling workshop for grinding treatment. After grinding, the calcine is relayed to the leaching workshop by the buried scraper conveyors and bucket elevators.

The gas produced by the fluidized roasting furnace first enters the waste heat boiler to recover the waste heat, and then goes through the two-stage cyclone dust collector and the electrostatic precipitator to collect the dust, and finally is sent to the acid production system by the exhaust fan. The collected dust is sent to the leaching workshop.

Figure 2 shows the external view of a $152m^2$ fluidized roasting furnace. The diameter of the fluidized bed is 13.9m, the diameter of the expansion section is 17.58m, and the height of the furnace chamber is 18.5m. The height of the fluidized bed is approximately 1m. The furnace wall materials, from inside to outside, consist of refractory bricks, insulation bricks, steel shell, and external insulation coating. The furnace structure, from bottom to top, includes the air box, distribution plate, fluidized bed, conical furnace chamber, expanded furnace chamber and furnace top.

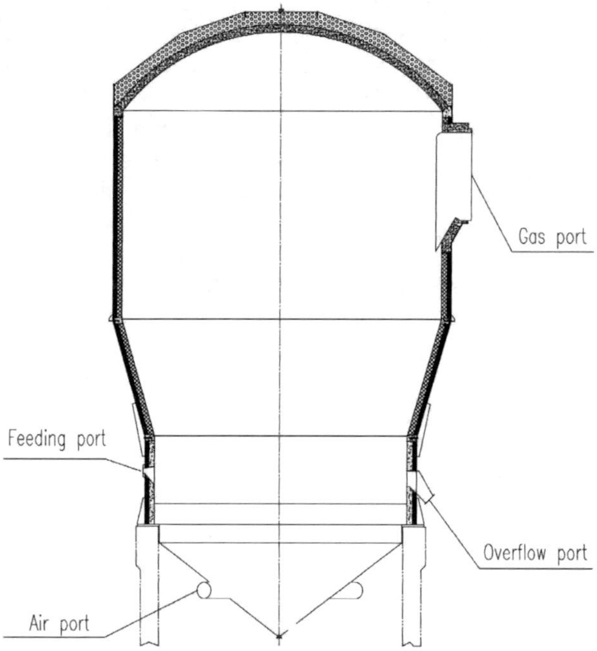

Figure 2. Schematic diagram of the roasting furnace.

2.3 Waste heat utilization system

The gas flow rate of a single roasting furnace is 79185 Nm^3/h, with a temperature of 950°C, a dust concentration of 195 mg/Nm^3, and a SO_2 content of 9.97% by volume. The steam production of a single boiler is approximately 43t/h, which is used for power generation and supplying the wet process production.

Each roasting furnace is equipped with a waste heat boiler, and the inlet of the waste heat boiler is connected to the gas outlet of the roasting furnace through a flexible fabric compensator. The waste heat boiler consists of a horizontal straight-through flue, a radiation chamber, and a convection section, and operates in a forced circulation mode with an outdoor layout. The furnace wall adopts a membrane water-cooled wall structure, with the boiler shell covered by an external insulation structure. The high-temperature gas passes through the radiation chamber horizontally and is cooled to 700°C. After the radiation cooling chamber is the convection section, which is equipped with condensing and convection tube bundles. Both the condensing and convection tube bundles are made of $\Phi 38\times 5$ boiler steel tubes and arranged in series. The gas exits the waste heat boiler at a temperature of around 360°C and enters the dust collection system.

The waste heat boiler is equipped with 77 sets of elastic hammer-type ash cleaning devices and 24 explosion ash cleaning points, which can timely and effectively remove the accumulated ash on the heating surface, ensuring the normal operation of the waste heat boiler. The lower part of the ash hopper of the waste heat boiler is equipped with a buried scraper conveyor. The settled dust in the waste heat boiler and the ash residue shaken off by the ash cleaning device are discharged from the boiler and transported to the calcine cooling cylinder by the buried scraper conveyor.

2.4 Dust collection system

The gas generated by the roasting furnace has the characteristics of high SO_2 content, high dust content, corrosiveness, high viscosity of dust, fine particle size, and high specific resistance, making it difficult to recover. The design adopts a multi-stage dust collection system, with the following process: roasting furnace → waste heat boiler → primary cyclone dust collector → secondary cyclone dust collector → electrostatic precipitator → fan → acid production.

The gas enters the dust collection system at a temperature of 360°C. The gas is first passed through two stages of cyclone dust collectors to recover larger particles of dust, and then purified by an electrostatic precipitator to recover fine particles of dust that are difficult to recover. The purified gas is then sent to the acid production system by a high-temperature fan. The dust collected by the cyclone dust collectors and electrostatic precipitator is sent to the leaching workshop without ball milling.

The high-temperature fan is located at the outlet of the electrostatic precipitator, which reduces the wear of the fan impeller by the dust in the gas. It also allows the electrostatic precipitator to operate under negative pressure, improving the operating and working conditions of the electrostatic precipitator. The high-temperature fan adopts variable frequency speed regulation to adjust the air volume and pressure of the system.

A bypass duct is set at the outlet of the second-stage cyclone dust collector, which allows the flue gas generated during drying and firing to be directly sent to the desulfurization system for firing gas treatment.

Each dust collection system consists of two φ2600×4 cyclone dust collectors and one 130m^2 five-field electrostatic precipitator. Two sets of dust collection systems are configured for the two roasting furnaces.

3. Production practice

3.1 Roasting furnace preheating and start-up

After the construction of the roasting furnace, it needs to undergo a drying process. The main purpose of furnace drying is to remove free water and bound water from the refractory materials. During the drying process, a series of physical and chemical changes occur in the refractory materials of the roasting furnace, so that the furnace can obtain high temperature stability, high temperature strength, and longer service life.

The furnace drying process follows a temperature ramp-up curve. Initially, wood is used for drying, followed by diesel burners. The drying process follows the principles of multi-point heating, even heating, and slow drying. The initial drying cycle takes about 20-30 days, with a diesel consumption of approximately 150 tons.

3.2 Roasting furnace production indicators

The zinc concentrate used in the production consists of 50% imported ore and 50% domestic ore, with a particle size of less than 260um and a moisture content of 14%. The main components of the zinc concentrate are shown in Table 1.

The design includes one φ3x22m drying kiln, with a processing capacity of 120t/h. The natural gas consumption is 3.5Nm3 for per ton of ore, and the moisture content of the dried concentrate is reduced to 7%. The drying process helps reduce the gas volume and increase the processing capacity of the roasting furnace.

Table 1. Composition of zinc concentrate fed into the furnace (dry basis)(wt %).

Material Name	Zn	Pb	Fe	S	SiO_2
Zinc concentrate (design value)	48.49	1.37	10.22	31.73	2.22
Zinc concentrate (production value)	46.71	1.64	10.70	30.97	2.38

The design feed rate of the roasting furnace is 42t/h, while the actual feed rate on-site is 47-50t/h. The design and actual specifications for the sulfur content of dust and calcine are detailed in Table 2 and Table 3.

Table 2. Comparison of design and actual value for sulfur content of calcine (wt %).

Material name	S_{total}	$S_{suloble}$	$S_{insoluble}$
Calcine(design value)	1.4	1.1	0.3
Zinc concentrate (production value)	0.98	0.71	0.27

Table 3. Comparison of design and actual value for sulfur content of dust (wt %).

Material name	S_{total}	$S_{suloble}$	$S_{insoluble}$
Dust(design value)	2.64	2.14	0.5
Dust (production value)	4.05	3.4	0.66

The mixture of calcine and dust is collectively referred to as roasted ore, which contains approximately 1.8% soluble sulfur, 0.5-1.0% insoluble sulfur, and the zinc content varies depending on the zinc content of the concentrate, generally around 55%. The roasted ore is sent to the wet process for leaching, and the resulting acid leaching residue contains about 16% zinc, which is better than the design value of 18%, indicating a higher zinc leaching rate.

The main production technical indicators are shown in Table 4. According to the table, the bed capacity of the roasting furnace can reach a maximum of 7.8t/m^2·d, and the operating rate can reach a maximum of 345 days, both of which are better than the design value. Therefore, it has the ability to increase production by 20%.

The main production technical indicators are shown in Table 4.

Table 4. Main production technical indicators.

Numbers	Technical indicator	Unit	Design value	Production value
1	Roasting furnace area	m^2	152	152
2	Number of units	unit	2	2
2	Annual operation time	days	330	330~345
3	Roaster specific capacity	t/m^2·d	6.5	6.5~7.8
4	Blower volume	Nm^3/h	78900	75000~96000
5	Roasting temperature	°C	930	900~1000
6	Superficial air velocity	m/s	0.6	0.6~0.7
7	Annual Zinc content in roasted ore	t/a	~300000	300000~360000

3.3 Operation control of the roasting furnace

During normal production, the furnace temperature is relatively stable. Typically, the temperature of the boiling layer is controlled at around 930°C and can reach a maximum of 1000°C. There are a total of 8 sets of cooling coils, but currently only 3 sets are needed to maintain thermal balance. The temperature of the boiling layer is adjusted by adjusting the feeding amount and the air volume on-site.

Based on on-site experience, the blower of the roasting furnace is relatively large and not easy to adjust. Therefore, once the air volume is adjusted properly during the initial production stage, it is generally not further adjusted.

When there is a high concentration of $SO4^{2-}$ in the wet system, the feeding amount is increased to raise the furnace temperature, leading to the decomposition of sulfates and a reduction in residual sulfur in the roasted ore.

3.4 Roasting furnace system equipment

In previous projects, two outlets from the front bin of the furnace were connected to two short belt conveyors, one belt conveyor, one metering belt conveyor and one distribution disc, totaling five conveying devices to transport the concentrate to the roasting furnace. The multiple transfer points resulted in multiple dust points, and the distribution disc hanging below the platform was inconvenient for maintenance. In this project, after preliminary analysis and investigation, the distribution disc was eliminated, and two double-belt quantitative scales were used instead of the five devices to achieve the functions of feeding, metering, and distribution. Practice has proven that this approach is effective, saving investment, reducing equipment footprint, and reducing the height of the plant, which can be promoted in future projects.

For this project, one non-frequency-controlled domestic fan was used. Pipe elbows are fewer ,silencers were installed at both the inlet and outlet to keep the on-site noise below 85 decibels. The original design of the furnace bottom air box inlet was a vertical inlet, but in this project, three tangential inlets were used, and the wind caps on the distribution plate were designed as straight-through type to ensure uniform and stable air distribution inside the furnace.

The roasting furnace is equipped with two bottom discharge ports, which are connected to the bottom discharge bin. The outer wall of the bin is made of steel, and the interior is lined with refractory castable to ensure that the temperature of the outer wall is below 80°C. Two water-cooled buried scraper conveyors are connected to the outlet of the discharge bin to handle the bottom discharge. The bottom discharge system is not used during normal production. When planned shutdown is required, the feed inlet and overflow outlet are closed, the feeding machine stops adding material, and the fan increases the air volume to discharge the materials from the bottom discharge port into the bottom discharge bin. The design of the bottom discharge device greatly reduces the labor intensity of manual furnace scraping in high-temperature and high-dust environments.

3.5 Site retrofit and optimization measures

The following improvements have been made on-site:

(1) The buried scraper conveyor transporting from the cooling cylinder to the ball mill has been improved by increasing the height of the baffles and upgrading the motor, after the modification, the occurrence of motor overload protection and shutdown due to sudden large material volume was avoided. Therefore, when selecting equipment, the instantaneous maximun material flow rate should be fully considered.

(2) An additional outdoor feeding device was installed in both roasting workshops. In the event of a sudden failure in the feeding system that cannot be repaired in time, the outdoor feeding device can be used to supply the roasting furnace, ensuring uninterrupted production.

(3) During the initial operation of the roasting furnace, the temperature of the furnace shell was found to be excessively high, reaching around 350°C. It was discovered that the external insulation layer was too thick, preventing the dissipation of heat from the furnace shell. The insulation layer was subsequently thinned and the furnace shell temperature is controlled at around 200°C now.

4. Conclusions

(1) In the feasibility study and design phase, the project initially planned to use three sets of 109m² fluidized roasting furnace. However, after many times negotiation, it was decided to use two sets of 152m² fluidized roasting furnace, which proved to be the correct decision. The two roasters share one

set of raw material preparation system, resulting in a simple configuration, smaller overall footprint, fewer equipment units, lower construction costs, and a shorter design and construction period. The two roasting systems are arranged in a completely mirrored layout, making them easy to manage. The labor force required is reduced by 30% to 40% compared to the three sets of 109m^2 roasters. The overall economic benefits of the dual-series 152m^2 fluidized roasting furnace scheme are much higher than the three sets of 109m^2 roasters scheme.

(2) Currently, there are a total of four 152m^2 fluidized roasting furnaces in China. In chronological order of production, they are: one at the Baiyin Nonferrous Group Co., Ltd. Northwest Lead-Zinc Smelter (2016), two at the Hunan Zhuye Nonferrous Metals Co., Ltd (2018), and one at the Henan Jinli Gold-Zinc Co., Ltd (2022). The two 152m^2 roasters in this project have made some improvements based on the first roasting furnace. The project has achieved production and exceeded production in a short period of time. The high annual operating rate and low-sulfur calcine ore indicating that the design technology of the 152m^2 fluidized roasting furnace is advanced and the on-site operational level is high. The whole operation production experience of the 152m^2 roasters have accumulated rich practical data for the development of larger roasters.

(3) With technological advancements, zinc smelting will have the following development trends in the future: weakly oxidizing roasting, intelligent control, large-scale production, and cleaner processes. Considering the significant potential for increasing capacity in roasting furnaces, future design of roasting furnace systems should take into account the oxygen consumption, and the supporting equipment should be matched with the increased production capacity.

5. References

[1] Li Ruo-gui. Development and application of large-scale fluidized bed roasting furnace. Chinese Nonferrous Metallurgy, 2021(5): 40-43.
[2] Chen Yong-jin. Analysis of large scale fluidized calciner operation stability. World Nonferrous Metal, 2022(7 down): 38-40.
[3] Xu Liang. Analysis of production capacity of large-scale fluidized roaster. Chinese Nonferrous Metallurgy, 2020.6(3): 40-43.

Optimization and Upgrading of Key Technologies for Lead Electrolysis Equipment Sets

H D Peng [1,2,]*, A M Deng[1,2], Y J Zong[3] and J Z Xiong [1,2]

[1]China Nerin Engineering Co., Ltd., Nanchang, Jiangxi , China, 330032,
[2]Jiangxi Nerin Equipment Co.,Ltd., Nanchang, Jiangxi , China, 330032,
[3] Yuguang Gold&Lead Co., Ltd., Jiyuan, Henan, China, 459000

Abstract. In response to the current key issues in the production of existing lead electrolysis equipment, such as poor verticality of lead cathode, inferior welding quality, short service life of conductive rod and high contact resistance, we have developed several technologies based on the principles of elastoplastic mechanics, bending and shaping, resistance spot welding and neural network control.These technologies include the stiffness-enhanced plate surface embossing, the dynamic multi-eigenvalue fusion of BP neural network resistance welding control, and the efficient rod-plate separation for cathode lead plates and conductive rod. They have been applied in a newly constructed 480kt/a project at a large domestic lead smelting enterprise. The results have demonstrated significant improvements in overall equipment production capacity, reducing the production time from 15s/plate to 11.5s/plate, decreasing the short-circuit rate caused by cathode verticality to 0.43%, and extending the service life of conductive rod by 12.8%.

Keywords: lead electrolysis; equipment sets; electrolysis short-circuit; service life of conductive rod

1. Introduction

Lead is an essential foundational metal in the industry and is generally purified through electrolytic refining. There are two production processes for lead electrolysis: small plate and large plate. Due to the fact that the comprehensive energy consumption of lead electrolytic refining with large plates is only 70 kgce/t, which is 50 kgce/t less than the 120 kgce/t of small plate electrolysis [1], in the past decade, most newly established lead electrolytic refining projects have adopted the large plate process along with corresponding equipment. Since the introduction of complete sets of large plate production equipment from Japan by a Yunnan-based company in China in 2005 [2], domestic enterprises have been continuously assimilating, absorbing and improving based on production experience. They have made remarkable contributions to the localization of equipment, essentially meeting the equipment requirements of the electrolytic process. The equipment is with high automation, low investment, and convenient maintenance.

With the continuous expansion of single lead electrolysis workshop production capacity, ranging from 100kt/a to gradually reach 150kt/a (even 180kt/a), the demands for performance and stability of a single set of lead electrolysis equipment have become increasingly stringent. Currently, domestically produced equipment struggles to meet the production requirements. Simultaneously, in order to reduce energy consumption during electrolysis and decrease maintenance and operation costs of the equipment, smelting enterprises have diverse and complex demands for equipment performance. This paper focuses on the research of key issues in traditional equipment such as poor verticality of lead starting sheet, low welding quality, short service life of conductive rod and large contact resistance. Corresponding solutions and technologies have been proposed and developed through this research. The equipment has been applied in real projects, yielding certain positive outcomes.

2. Current status of lead electrolysis equipment technology

2.1. Verticality of starting sheet

The verticality of lead starting sheet is a crucial parameter for the product and also one of the significant reasons for potential short-circuit during later-stage electrolysis. In general, the verticality of lead starting sheet before it is placed into the tank needs to be controlled within 10mm to 15mm. Figure 1 illustrates the situation of improper verticality between anode and cathode. Currently, key factors affect the verticality of lead starting sheet during its manufacturing process include:

1) Unreasonable design of the embossed pattern on the anode, inadequate embossing depth, and poor board rigidity.

2) After embossing and leveling the anode, secondary deformation occurs before it enters the electrolytic cell.

Figure 1. Unqualified verticality of distance between cathode and anode

Through production practice, while ensuring that the welding seams of the embossing machine frame do not detach and the design of the embossing die pattern is reasonable, the embossing depth can be maintained stably. However, the main factors affect the deterioration of the verticality are primarily concentrated on factor 2. After completion of lead starting sheet , it needs to be alternately arranged with the anode, facilitating the lifting of both anodes and cathodes into the electrolytic cell. Currently, the technical solutions for spacing between anodes and cathodes involve inserting the cathode from the top down into the space between the anodes. The thickness of starting sheet of cathode itself is about 0.9mm, with a thin profile and large dimensions, making it prone to deformation. Before descending, the motion of starting sheet is with vibration, combined with the effects of air resistance or guide baffle, leading to potential deformation. In severe cases, during the descent, the bottom of starting sheet may collide with the anode plate, resulting in bending and damage to the cathode and necessitating equipment shutdown. Figure 2 and figure 3 show the on-site use and spatial situation of the downward insertion spacing scheme.

Figure 2. Downward insertion spacing arrangement

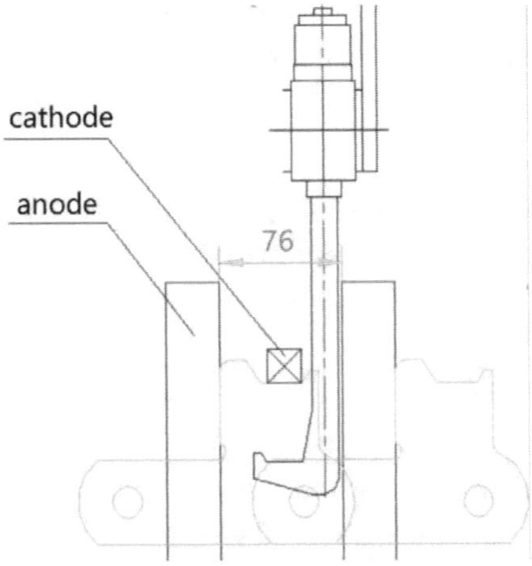

Figure 3. Collision protection space for cathode during downward insertion

2.2. Connection Quality of Lug of Starting Sheet

During the processing of starting sheet, it's necessary to wrap the conductive rod with lead foil and assemble it, followed by securing it to the folded side. The originally introduced technique involved uniformly pressing three circular holes in the length direction of conductive rod on the first floor of the workshop, and then using an inclined lifting mechanism to elevate the starting sheet to the second floor. At a welding station, 14 to 16 points are welded in the length direction of the folded area of the lead foil, enhancing the starting sheet conductivity and verticality strength [3]. Figure 4 depicts the locations of the pressed hole and the spot welding on the cathode plate.

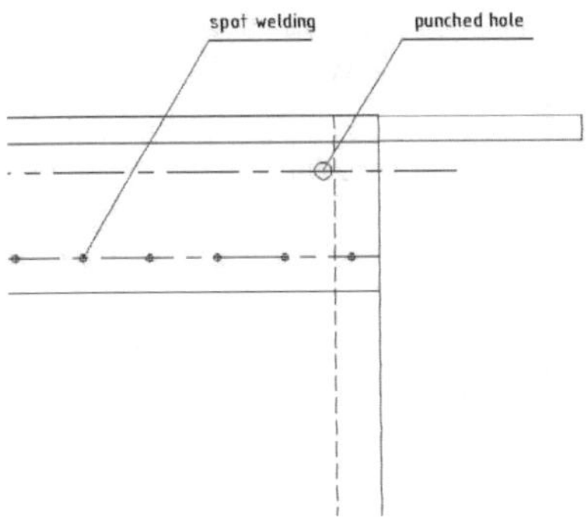

Figure 4. Pressed hole and spot welding on the cathode plate

Figure 5. Starting sheet with loosened pressed hole

The quality of the connection at starting sheet lug is closely tied to the quality of pressed holes and spot welding. During hole pressing, if the punch-matrix or assembly dimensions are poor, there is a tendency for loosening or opening of the connection. During the inclined lifting process, the lead foil might strip or shift, as shown in Figure 5. In the case of spot welding, if there is incomplete welding, during the transport or electrolysis after spacing the anode and cathode, the lead foil may strip and slide off the conductive rod, resulting in scrap pieces.

2.3. Lifespan of Conductive Rod

The conductive rod functions as conducting electricity and bearing the cathode lead plate during the electrolysis process. It is supplied in the cathode machine and withdrawn in the cathode lead cleaning machine. The conductive rod is circulated internally between the two machines in the workshop and the electrolytic cell. Made from copper-clad steel with layer thickness of less than 3mm, the conductive rod carries a high replacement cost. Its service life is primarily connected to damages, which predominantly occur during the rod withdrawal and cleaning processes.

Currently, there are two methods to withdraw the conductive rod, both aimed at removing it from the cathode lead . One method is manually pulling the conductive rod from one side of the deposited lead and then collecting it in a storage frame. The other adopts a mechanical arm to grip the conductive rod from the side of the deposited lead, and during this process, due to the dry friction between the lead foil and the rod, the copper layer on the surface of the rod gets damaged. Meanwhile, the scraped-off copper results in secondary contamination of the lead product.

There are currently two cleaning methods in use: one is used with introduced equipment, employing steel brushes to polish and remove impurities and oxide layers from the conductive rod. This method causes significant wear on the conductive rod and generates a considerable amount of dust, leading to poor working conditions. The other method involves using an acid cleaning tank equipped with an ultrasonic generator, where high-frequency ultrasonic vibrations are employed to remove oxides during the cleaning process. This solution has low efficiency, high operating costs, and the acid fumes negatively impact the workshop environment.

3. Key Equipment Technologies

3.1. Enhanced Rigidity Plate Embossing Technology

In order to enhance the flatness and rigidity of the starting sheet surface, optimization of the embossing technology is required. In this project, measures have been taken including:

1) Embossing Die Design: Based on the trace direction of subsequent transfer of lead starting sheet, the embossing die pattern has been rationally optimized. This results in the formation of a regular embossed pattern on the surface of the lead starting sheet, ensuring even pressure distribution during the embossing process to increases rigidity while preventing excessive stress concentration. The traditional pattern from the imported technology is shown in Figure 6, while the pattern used in this project is as depicted in Figure 7. This adapted pattern can accommodate the deformation effects caused by rapid lateral transfer of starting sheet by the robot.

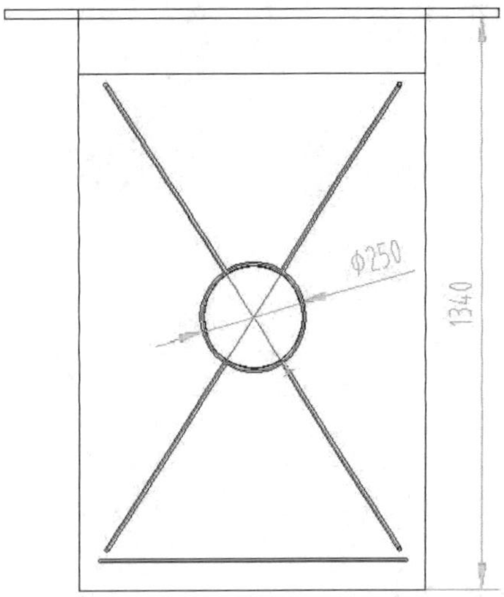

Figure 6. Cathode pattern in the traditional imported technology

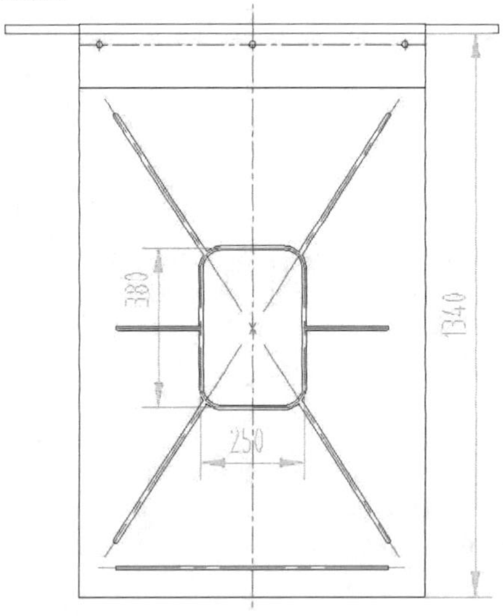

Figure 7. Optimized cathode pattern

Journal of Physics: Conference Series **2738** (2024) 012017 doi:10.1088/1742-6596/2738/1/012017

2) Embossing Parameter Control: the control of parameters during the embossing process, such as pressure, speed and time, is essential to achieve the optimal rigidity-enhancing effect. In this project, the oil cylinder pressure used is within the range of 170 to 182 kN, the embossing speed ranges from 20 to 30mm/s, and the time varies from 3.0 to 3.5s.

3) Die Material Selection: materials that possess good rigidity and wear resistance for the embossing die. Implement suitable heat treatment or surface treatment methods to enhance their mechanical properties. Additionally, ensure that the flatness of the embossing die surface is controlled within 2mm, which contributes to the uniform and consistent alignment of embossed patterns. This reduces the risks of bending and deformation after embossing.

3.2. Neural Network-Controlled Resistance Welding Technology

Resistance welding technology is widely applied, particularly suited for connecting metals with low melting points and poor thermal conductivity. It has been widely adopted in the production of lead staring sheet. Due to the thin thickness of lead foil, multiple spot welding is necessary to ensure load-bearing capacity. Typically, 14 to 16 spots are used in the industry. To simplify the welding equipment, each set of welding equipment is shared by every 7 to 8 welding heads, two sets in total. Several parameters, such as welding current, welding time, electrode spacing and pressure influence the quality of resistance welding during the process. In order to enhance the uniformity of connection quality across multiple welding points, control technology for resistance welding is crucial.
In this project, a BP neural network-controlled resistance welding technology with dynamic multi-feature fusion has been adopted. It makes use of multiple characteristic values in combination with the back propagation (BP) neural network to control the resistance welding process. The topology structure of this technology is depicted in Figure 8.

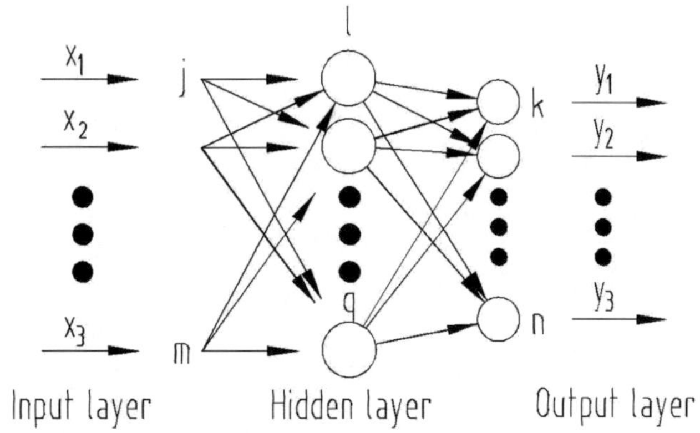

Figure 8. Topology structure of BP neural network

The BP neural network establishes a model between the characteristic values and welding quality by learning from historical data and utilizing the back propagation algorithm. The transformation from the input layer to the hidden layer is nonlinear, while from the hidden layer to the output layer is linear. In this project, the network structure and parameters are obtained by extracting a subset of experimental data for network training, and the remaining data is used as samples for network prediction[5]. The network structure used consists of an input layer with 5 nodes, a hidden layer with 15 nodes and an output layer with 1 node. The iteration count is set at 120, the learning rate is 0.1. The sigmoid function is used as the activation function. The performance of the neural network is evaluated using the root mean square error of shear strength for welding points[6], as follows:

$$f(x) = \frac{1}{1+e^{-x}} \quad (1)$$

3.3. Robot Material Transfer Technology by Using S-Curve Algorithm

After the initial production of starting sheet, it is flipped to a vertical position using a flipper machine. The sheet then undergo embossing and straightening processes before being transferred to a spacing machine for arrangement. To minimize space occupation and equipment investment, an industrial robot is used in this project to transfer the starting sheet between the flipper , embossing, straightening and spacing machines. Managing the overall transfer time and preventing deformation of starting sheet during the process is critical.

In this project, the S-curve algorithm is adopted to plan the robot's motion trajectory. The S-curve exhibits smooth and continuous characteristics. By considering the physical properties of the robot and materials, adjusting acceleration and deceleration at turning points, optimizing the trajectory path and controlling the radius of curvature, the robot can achieve smooth acceleration and deceleration during material transfer process. It avoids sudden acceleration changes that could impact or damage the material, thereby enhancing transfer stability and precision. Ultimately, it achieves optimal material transfer quality and efficiency. Figure 8 illustrates the optimized path during robot material transfer.

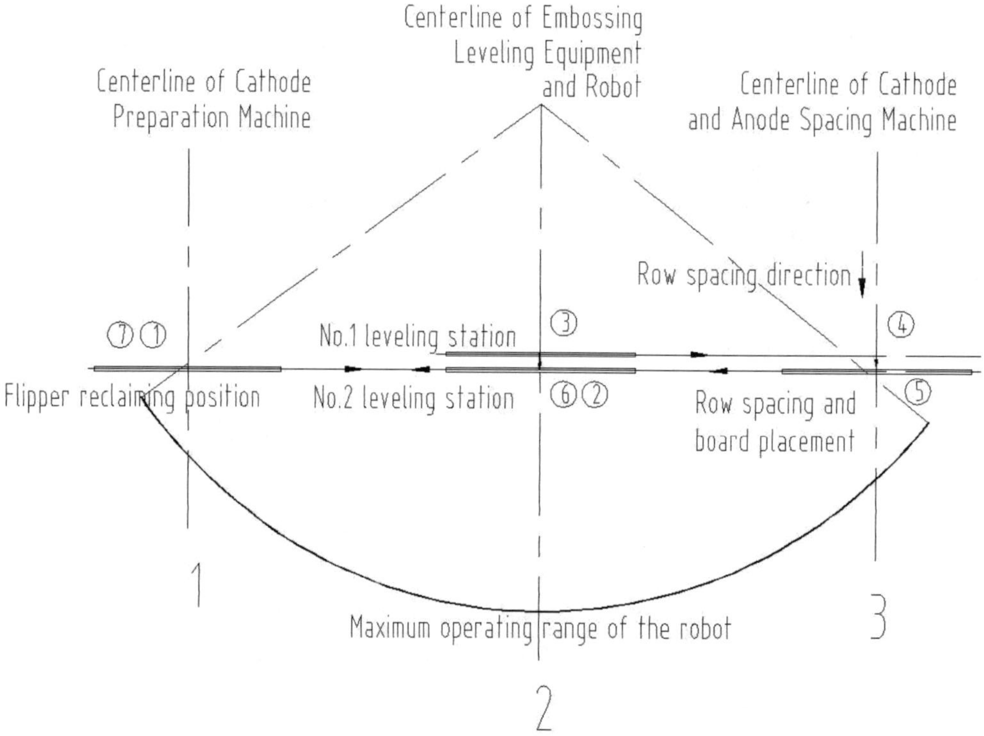

Figure 9. Optimized path for material transfer by robot

In the project implementation, the robot follows steps ① to ⑦ after retrieving the plates from the flipper machine. This sequence facilitates the movement of the cathode between the flipper, straightening and spacing workstations. The S-curve algorithm is applied at each turning point, ensuring smooth material transfer. It has achieved a favorable outcome with a single cycle transfer time controlled within 11.5s.

3.4. High-Efficiency Rod-Plate Separation Technology

In order to reduce the wear on the conductive rod caused by traditional withdrawal, a high-efficiency rod-to-plate separation technology is adopted. Before the separation, the conductive rod and lead plate are securely fastened. A cutting tool is then used to start cutting from the V-shaped notch on one side of the cathode lead plate, moving towards the other side. After separation, the lead plate slides into a shredder from below. Figure 10 illustrates the mechanism of rod-to-plate separation.

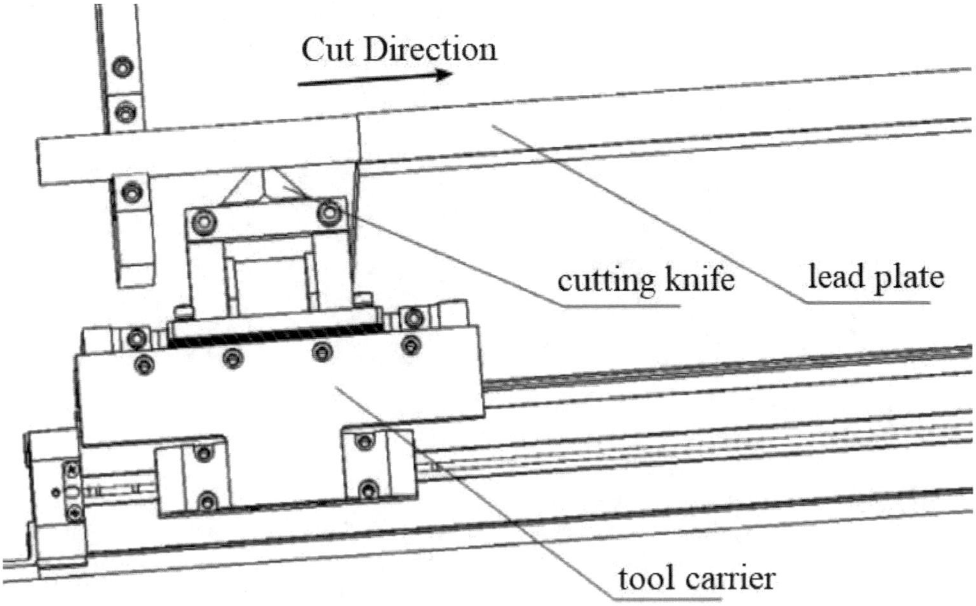

Figure 10. Mechanism of rod-plate separation.

To achieve efficient and reliable rod-plate separation, the following measures have been implemented in the project:

1) Optimization of Separation Mechanism Design: this includes the design of the cutting tool holder, positioning device and auxiliary force control device. The shape, size and material of the cutting tool, especially the optimization of its design, have been addressed to enhance separation efficiency and life service. Various cutting tool shapes like double-edged, internal and external figure-eight, sickle-like tools have been designed. Materials with sufficient hardness and wear resistance have been chosen to ensure efficient cutting of the lead foil.

2) Detection and Feedback Control: by using sensors and feedback control systems, the force, position and effectiveness during the separation process are monitored in real time. Adjustments and optimizations are made based on the feedback information to ensure an efficient separation operation.

3) Cutting Speed Control: the speed of cutting tool is controlled. Excessive cutting speed might lead to wear and damage of the tool, while a too low speed could decrease separation efficiency and impact the cutting process. Generally, the cutting speed is controlled within 1 to 1.2 m/s.

4. Application Effects

4.1. Efficiency Improvement of Machine

By implementing the enhanced rigidity plate embossing technology, the robot material transfer technology by S-curve algorithm and the high-efficiency rod-plate separation technology, the production yield of starting sheet has been improved. The quality of cathode-anode spacing has also been enhanced, and the efficiency of material transfer and separation has been ensured. As a result, the overall operational efficiency of workshop machine has been elevated to 11.5s/sheet, and operation rate of machine has reached over 96% (excluding the impact of manual bending). Figure 11 shows the cathode-anode spacing status.

Figure 11. Cathode-anode spacing status

Figure 12. Verticality condition of starting sheet

4.2. Electrolytic Energy Consumption Reduction

By adopting the enhanced rigidity plate embossing technology and the robot material transfer technology by S-curve algorithm, the surface tolerance of the starting sheet has been effectively controlled. Additionally, the material transfer process has been executed smoothly and steadily, while the subsequent deformation of the material has been maintained within an acceptable range. The verticality of starting sheet after spacing is depicted in Figure 12, showing that the verticality is less than 8mm.

Based on the BP neural network-controlled resistance welding technology, its strong nonlinear fitting ability comprehensively considers the mutual influence among multiple characteristic values. This enhances the precision and stability of the welding process control. The technology ensures a more uniform distribution of strength and resistance values between multiple welding points, leading to a reduction in the contact resistance of the starting sheet itself.

The combination and optimization of these measures effectively reinforce the surface rigidity, improve the starting sheet verticality, and reduce the short-circuit rate during electrolysis. The average short-circuit rate has decreased from 1.12% (in the direct smelting plant electrolytic workshop) to 0.43%. Additionally, the technology lowers the material's intrinsic contact resistance. This multifaceted improvement in the lead electrolysis workshop enhances energy efficiency, resulting in a 15kWh reduction in lead electrolytic energy consumption per ton. The short-circuit rate data over a three-month period is presented in Table 1.

Table 1. Short-circuit rate statistics in the electrolytic workshop.

No.	Item	Value		Unit
		Smelter	Refinery	
1	Number of Electrolytic Cells (In Operation)	344	320	cell
2	Number of Single-Cell Cathodes	39	51	pcs/cell
3	Number of Cathode Plates	13416	16320	pcs
4	Electrolytic Cycle	7	8	day
5	Calculating Production Time	90	90	day
6	Number of Electrolytic Batches	12.86	11.25	batch
7	Total Number of Electrolytic Cathodes	172530	183600	pcs
8	Total Number of Short-Circuited Cathodes	1932	789	pcs
9	Average Short-Circuit Rate	1.12%	0.43%	

4.3. Extension of Conductive Rod Life Service

The high-efficiency rod-plate separation technology enhances the efficiency, accuracy and stability of separation, safeguarding the conductive rod from dry wear caused by the lead foil during the withdrawal process, which effectively protects the copper layer. Additionally, during the cleaning of the conductive rod, high-pressure water cleaning technology can be used. By using high-pressure cleaning liquid without abrasive particles, the surface of conductive rod is impacted from a specific angle with a certain pressure and flow rate. This process removes the adhered oxides and impurities from the surface, avoiding damage to the copper layer matrix caused by other physical cleaning methods. Consequently, the life service of the conductive rod is further prolonged. After experimental verification, when compared to the conventional steel brush cleaning, the copper layer's thickness change is slow, resulting in a 12.8% extension in the life service of the conductive rod copper layer.

5. Conclusion

Addressing various critical challenges in traditional lead electrolytic equipment and recognizing the increasing demands on equipment processing capacity, stability and user-friendliness with capacity enhancements, this study has delved into technologies such as enhanced rigidity plate embossing, dynamic multi-feature BP neural network resistance welding control, high-efficiency rod-plate separation for cathode lead plate and conductive rod. These technologies are then implemented in practical production equipment, confirming their feasibility and reliability. The practical application of these technologies resulted in an overall production capacity of 11.5s/plate, a reduced electrolytic short-circuit rate of 0.43%, and an extended conductive rod life service of 12.8%. These positive outcomes have brought considerable economic benefits to enterprises and played a proactive role in advancing the upgrade of equipment in the lead electrolysis industry.

For a certain large domestic smelting enterprise constrained by project investment limits, manual cleaning for conductive rod is temporarily adopted. In the future, considering the physical and chemical properties of metallic copper and its surface oxide layer, options such as high-pressure water cleaning or laser cleaning could be tried. These approaches aim to achieve effective cleaning while ensuring the copper matrix remains undamaged, which could contribute positively to enhancing the life service and electrolytic efficiency of the conductive rod. It is anticipated that practical implementation will be realized in future production.

6. References

[1] QIAN Q and ZHANG Y P 2008 Introduction to the New Process of Lead Electrolytic Refining Large Plate Equipment *Proceedings of the 10th National Symposium on Lead & Zinc Metallurgy Technology and Production Equipment* p 252-255.

[2] LI T and ZHANG J L 2015 Technology and Application of Large Plate Lead Electrolysis Process *World Nonferrous Metals*. **02** p 31-32.

[3] CHEN Z H and ZHANG G S 2007 cheed Lead Smelting Process and Equipment *Engineering Design and Research* **122** p 7-8.

[4] LU X K and GAO Z L 2021 Intelligent Control of Electric Resistance Welding Based on BP Neural Network and PID Self-adaptive Tuning *Hot Machining Process* **50** p 141-142.

[5] LIN W Y, REN X Y and ZHOU T T 2018 A novel robust algorithm for position and orientation detection based on cascaded deep neural network *Neurocomputing* **308** p 138 – 146.

[6] LUO L X and XING Y F 2019 CMT Spot Welding Deformation of Sheet Metal Based on BP Neural Network and Genetic Algorithm *Transactions of the China Welding Institution* **40** p 80-81.

Development and Innovative Application of Main Equipment for Zinc Electrowinning

Yizhe Tong, Ziliang Liu

CINF Engineering Co., Ltd., Changsha 410011, Hunan, People's Republic of China

tyz@cinf.com.cn

Abstract. The latest development and application of zinc electrowinning equipment in China are introduced. The characteristics and innovative applications of main equipment for zinc electrowinning, such as cathode and anode, electrolytic cell, cathode and anode multifunctional crane, zinc stripping and cathode washing machine, zinc sheet conveyor, anode temper mill and solution cooling tower, are analyzed. It is pointed out that the application of advanced equipment characterized by large-scale and automation will dominate the upgrading and transformation of existing zinc electrowinning capacity.

Key words: Zinc electrowinning; Multifunctional crane; Electrolytic cell; Zinc stripping

1. Introduction

Zinc hydrometallurgy was put into industrial production in 1916, and developed rapidly, soon surpassing pyrometallurgy. Since 1980s, zinc hydrometallurgy has reached more than 80% of the world's zinc production[1]. Zinc electrowinning is an important link in zinc hydrometallurgy process, which is to extract metal zinc from zinc sulfate solution by electrowinning, including electrowinning, electrode plate loading and trough, cathode stripping and washing, anode leveling, solution cooling and circulation, etc. Because of the complexity of zinc electrowinning operation, the equipment level of zinc electrowinning represents the equipment level of zinc hydrometallurgy to a great extent. In 2022, Chinese zinc output was 6.802 million tons, accounting for more than 40% of the world's zinc output, ranking first in the world for more than 20 consecutive years. With the improvement of zinc smelting capacity and quality, in recent 10 years, some newly-built zinc hydrometallurgy projects in China have successively adopted advanced equipment such as large plate, large electrolyzer and automatic zinc stripping, and achieved good results, which has positive reference significance for promoting the transformation and upgrading of traditional zinc hydrometallurgy production system and improving the overall equipment level of zinc smelting industry.

2. Cathode and anode

2.1. Cathode

Cathode is usually composed of electrode plate, conductive rod, conductive head and insulating strip. The electrode plate is pure aluminum plate, and its hardness and flatness have great influence on the service life and production operation of the electrode plate. The conductive rod is made of aluminum, the conductive head is made of pure copper or copper-aluminum composite material, and the insulating strip is generally made of PP and other insulating materials. In addition, in order to improve the rate of mechanical zinc stripping, some cathodes with mechanical zinc stripping are provided with zinc stripping opening and insulating tape at the liquid level line on the electrode plate[2]。

The development trend of cathode is the continuous enlargement of the effective area of cathode, the reduction of the number of cathode sheets and the operation of stripping zinc. Cathode effective area refers to the area where zinc metal can be deposited on a single cathode, ranging from 1.13 m^2 to 1.6 m^2 , 2.6 m^2 , 3.2 m^2 , 3.6 m^2. Due to the weight of the plate, below to 1.6 m^2 cathodes can be manually stripped of zinc, and above 2.6 m^2 cathodes generally need to be mechanically stripped of zinc. See Table 1 for typical cathode specification series and production operation parameters.

Table 1 Typical specifications and parameters of the cathodes

Project	Effective area, m^2/Piece				
	1.13	1.6	2.6	3.2	3.6
Length, mm	1000	1190	1460	1760	1960
Width, mm	666	800	1000	1000	1000
Same polar distance, mm	62	75	90	90	90
Zinc stripping form	Zinc can be stripped manually	Zinc can be stripped manually	Mechanical stripping of zinc	Mechanical stripping of zinc	Mechanical stripping of zinc
Number of sheets required for 100,000 t/a zinc electrowinning scale, Sheet/(100,000 t/a electric zinc)	26680	18240	11880	9120	7992

2.2. Anode

Anodes are generally composed of plates, conductive rods and insulators. The electrode plate is usually lead-silver alloy plate containing 0.75% ~ 1% silver, and ternary alloy plate of lead, silver and calcium or multi-element alloy plate can also be used to reduce the amount of silver[3] . The plate is produced by calendering or casting, and the plate surface is mesh or flat. The mesh helps the anode slime to precipitate on the plate and take it out of the cell, thus reducing the frequency of digging anode slime from the electrolytic cell. The conductive rod is pure copper, and the insulator is PP and other insulating materials.

Anode and cathode are used together, and they are also developing towards large scale. The number of anodes per tank is generally one more than that of cathodes to ensure the maximum deposition area. Typical anode specification parameters are shown in Table 2.

Table 2 Typical specifications and parameters of the anodes

Project	Effective area,m^2/ Piece				
	1.13	1.6	2.6	3.2	3.6
Length, mm	975	1170	1410	1710	1910
Width, mm	620	780	945	945	945
Same polar distance, mm	62	75	90	90	90
Number of sheets required for 100,000 t/a zinc electrowinning scale, Sheet/(100,000 t/a electric zinc)	27140	18620	12060	9200	8806

3. Electrolytic cell

The development of electrolyzer is mainly reflected in two aspects: large-scale specifications and diversified materials.

With the application of large plates and mechanical zinc stripping, the specifications of electrolytic cells gradually increase, the length develops from 4m to about 10m, and the number of plates accommodated in each cell increases from dozens to more than 100 groups, which effectively reduces the number of electrolytic cells and the occupied area of electrolytic cells under the same production scale. The concentration polarization phenomenon in large electrolyzer can be avoided by increasing the circulating amount of solution and other measures. See Table 3 for typical electrolyzer specifications.

Table 3 Typical specifications and parameters of the cells

Project	Electrolytic cell 1	Electrolytic cell 2	Electrolytic cell 3	Electrolytic cell 4	Electrolytic cell 5
Internal length, mm	4190	4030	6060	10050	10050
Internal width, mm	810	940	1230	1330	1330
Internal depth, mm	1500	1580	1960	2260	2460
Cathode effective area,m^2/ Piece	1.13	1.6	2.6	3.2	3.6
Cathodes per tank, sheet/tank	58	48	72	114	108
Number of anodes per tank, sheet/tank	59	49	73	115	109
Number of electrolytic cells required for 100,000 t/a electric zinc, Tank/(100,000 t/a electric zinc)	460	380	180	80	74

The area of electrolytic cell required for 100,000 t/a electric zinc, m^2/(100,000 t/a galvanic zinc)	3600	3700	2500	1900	1750

Electrolytic cell materials mainly include concrete lining anticorrosive materials, steel frame lining plastic, glass fiber reinforced plastic, resin concrete and so on. Resin concrete monolithic electrolyzer has developed rapidly in recent years. Because of its long service life, maintenance-free and high dimensional accuracy, it has been applied in many large-scale zinc hydrometallurgy projects such as lead-zinc smelting project of Jiangxi Copper Group and zinc-oxygen pressure leaching project of Western mining industry.

4. Cathode and anode multifunctional crane

In traditional zinc electrowinning production in China, bridge insulated cranes operated by cab are mostly used to discharge and load cathode and anode into the tank. Because of the acid mist released from the electrolyzer in production, the working environment of crane drivers is very bad. In recent years, some newly-built zinc hydrometallurgy projects have used cathode and anode multifunctional cranes, which can remotely operate the control room, and realize accurate positioning and automatic loading and unloading trough. A zinc electrowinning workshop with a production scale of 100,000 t/a, if 3.2 m^2 is used Large plate process and 48-hour cathode deposition cycle, equipped with two cathode and anode multifunctional cranes can meet the needs of production operation. The cathode and anode multifunctional crane has two sets of gripping and lifting systems, which are used to lift cathode and anode respectively, lifting 1/2 groove cathode or 1/3 groove anode at a time[4]。

5. Cathode and anode processing equipment

5.1. Zinc stripping and cathode washing equipment

Because of the large number of electrode plates and short electrodeposition period, zinc stripping and cathode washing have become the main production operations in zinc electrode workshop. The electrowinning period of manual stripping zinc is generally 24h, and the small plate process is mostly used, so the labor intensity of stripping zinc is very high. Taking a 100,000 t/a zinc electrowinning workshop as an example, use 1.13m^2 cathode and deposition cycle of 24 hours, need to strip 26,000 cathodes and 300t metal zinc every day on average.

Mechanical zinc stripping greatly improves labor efficiency and working environment. In 2005, China successfully realized mechanical zinc stripping for the first time, and successively adopted mechanical zinc stripping equipment in zinc hydrometallurgy projects such as Zhuzhou Smelter Zinc System Reconstruction Project, Danxia Smelter Zinc Oxygen Pressure Leaching Project, Jiangxi Copper Group Lead Zinc Smelting Project and Western Mining Zinc Oxygen Pressure Leaching Project. Mechanical zinc stripping is mostly 3.2 m^2 Large plate, 36h~48h long period electrowinning process, the capacity of 100 kt/a needs to be equipped with two sets of zinc stripping units with a capacity of 200 cathodes/h and one set of cathode washing units with a capacity of 400 cathodes/h. CINF has creatively developed large, medium and small zinc electrolysis plates (effective area 3.2m^2, 2.6m^2 , 1.6m^2 , 1.4m^2 or 1.13m^2) and a full range of intelligent complete sets of equipment (including a full range of zinc stripping units, intelligent driving units, cathode washing units and anode treatment units), and put into use in new projects and renovation projects to provide high-quality technical services for enterprises.

5.2. Zinc sheet conveying equipment

The zinc sheets stripped from the cathode need to be stacked neatly, and then sent to the zinc melting and casting workshop to be cast into zinc ingots. CINF has developed intelligent forward and reverse flat stacking technology, which is higher, more uniform and faster than the prior art (Figure 1), and brings great convenience to the transportation of zinc stacks[5]. Forklift trucks are traditionally used to transport zinc sheets in China. The lead-zinc smelting project of Jiangxi Copper Group completed by CINF has realized the "non-landing" transportation of zinc sheets for the first time, that is, the zinc sheets stripped by the zinc stripping machine are directly sent from the zinc electrowinning workshop to the zinc melting and casting workshop through the automatic zinc sheet conveying line and hoist, and pushed into the zinc melting furnace through the automatic feeder, thus avoiding secondary transportation and reducing manual operation.

Fig.1 The zinc stack (left: use intelligent forward and reverse flat stacking technology, right: not use)

5.3. Anode leveling equipment

Anode mechanical flattening is an operation corresponding to cathode mechanical stripping of zinc, which is an improvement on manual flattening. Manganese dioxide and other solid deposits are produced on the surface of anode plate during electrodeposition, which must be cleaned regularly, and the cleaning period is about 20 days. The cleaning and leveling of anode has great influence on the quality and current efficiency of zinc sheet.

The mechanical leveling process is as follows: the anode is lifted out of the electrolytic cell by the cathode and anode multifunctional crane and put on the anode trolley, and the anode plate is transported to the automatic flattening machine by the anode trolley, and the high-pressure water washing and flattening treatment are carried out on the anode flattening machine. When the anode trolley returns to the electrolysis area, the cathode and anode multifunctional crane hoists the patted and cleaned anode plate back to the electrolysis cell[6] . In the application of many projects in China, the improvement and optimization of anode flattening equipment are mainly reflected in reducing water consumption, reducing water consumption and installed power of high-pressure water pumps by adding accumulators in high-pressure water systems[7].

6. Solution cooling tower

The electrothermal effect of zinc electrowinning process leads to the increasing temperature of electrolyte. In order to ensure the stability of electrolysis temperature, it is necessary to circulate the electrolyte with a cooling range of 6~8 °C to maintain the heat balance of electrolyzer. In industrial production, the solution cooling methods such as heat transfer in tank, vacuum evaporation and air cooling tower have been used. Heat transfer in tank will reduce the utilization of space in tank. The main disadvantages of vacuum evaporative cooling are high energy consumption and difficult maintenance. Practice has proved that air cooling tower is the appropriate way to cool zinc electrowinning solution. At present, air cooling tower is basically used in zinc electrowinning production in China.

The cooling area of air cooling tower has been developed from small to large, and different specifications have been developed: 12.5m^2, 24m^2, 30m^2, 40m^2, 50m^2, 60m^2 , 72m^2 etc. With the expansion of zinc smelting scale, 50m^2 or larger cooling towers is widely used at present, the large-scale equipment reduces the number of equipment used. The capacity of electrolyte cooling equipment required by zinc smelting project is closely related to the climatic conditions in the area where it is located. In general, 100,000 t/a zinc electrowinning workshop needs to be equipped with 10 50 m^2 cooling towers or 7 72 m^2 cooling towers in china. Cooling towers need more in hot and humid areas, whereas less cooling towers can be selected.

Using cooling tower to ventilate and remove acid mist in zinc electrowinning workshop is an ideal measure to improve the operating environment of zinc electrowinning workshop. Traditional zinc electrowinning workshop ventilation is mostly natural ventilation, and the ventilation effect is not good. The fan of electrolyte cooling tower is used to ventilate and remove acid mist, which can cool the solution and realize ventilation in the workshop. The acid mist is also brought into the cooling tower and purified by the droplet catching layer, thus reducing the unorganized discharge of acid mist and achieving the effect of achieving multiple goals at one stroke。

7. Conclusion

With the vigorous development of domestic zinc smelting industry, advanced zinc electrowinning equipment has been applied to many new projects, which promotes the large-scale zinc electrowinning equipment, mechanization and automation of production, reduces labor intensity, and improves the level of safety, hygiene, energy conservation and environmental protection. However, there is still a long way to go to improve the overall level of zinc electrowinning equipment in China, and the penetration rate of automation equipment such as zinc stripping machine and multifunctional crane is less than 1/4 of the domestic zinc smelting capacity. With a series of key equipment gradually completing the process of introduction, digestion, absorption, improvement and re-innovation, and gradually realizing localization, in the future, the improvement of equipment level will dominate the upgrading of existing zinc electrowinning capacity.

Acknowledgments

The work was financially supported by the Guangxi Science and Technology Major Project (No. AA22067076).

References

[1]TAN Zidong, ZHEN Yong, JIN Xin, et al. Study on intensive leaching of indium and

germanium from complex materials containing indium and germanium in zinc smelting[J]. Journal of Central South University (Science and Technology), 2023,54(07):2552-2562.

[2]Tong Yizhe. Design of easy peeling cathode plate [J]. Coloured Equipment, 2012 (02): 25-27 +33.

[3]MA Fei-fei， ZHAN Zongbo. Comprehensive utilization of electrolytic anode slime from zinc hydrometallurgy and Mn^{2+} control in electrolyte[J]. China Nonferrous Metallurgy, 2022,51(03):112-116.

[4]Yang Ping. Application of multifunctional automatic crane in zinc electrowinning workshop [J]. Hunan Nonferrous Metals, 2017, 33 (03): 34-36.

[5]He Jun. The application practice of a new type of small plate zinc stripping machine and automatic palletizing technology [J]. Science and Technology Innovation, 2020(16): 14-15.

[6]Tang Xiaomao. PLC design and application of zinc smelting flattening machine [J]. Colored Equipment, 2017 (06): 38-40.

[7]GAO Lan. Application of on Automation Stripping Zinc Unit in Zinc Electrowining [J]. HUNAN NONFERROUS METALS, 2018,34(03):37-39+46..

A Review on Recycling of Waste Lead-Acid Batteries

Tianyu Zhao[1,*], Sujin Chae[1], and Yeonuk Choi[1,*]

1. The Robert M. Buchan Department of Mining, Queen's University, 25 Union Street, Kingston, Ontario, Canada, K7L3N6

Corresponding author: tianyu.zhao@queensu.ca; choi.y@queensu.ca

Abstract: Lead-acid batteries (LABs) have become an integral part of modern society due to their advantages of low cost, simple production, excellent stability, and high safety performance, which have found widespread application in various fields, including the automotive industry, power storage systems, uninterruptible power supply, electric bicycles, and backup power supplies. Hence, the use of LABs has greatly benefited human society and contributed to advancements in science and technology. However, the extensive use of LABs unavoidably leads to the generation of a significant amount of LABs waste. On one hand, if these waste LABs are not handled properly, any leakage can cause devastating damage to the natural environment and human health. On the other hand, waste LABs represent an important secondary resource for lead, with approximately 64.57% of global lead resources derived from recycled lead, making them a major source of lead worldwide. Moreover, approximately 85% of global lead resources are currently utilized for manufacturing LABs, and the recycling of waste LABs brings favourable prospects for the sustainable development of the energy storage industry. Therefore, the recycling of waste LABs is necessary and inevitable. In this paper, we have comprehensively reviewed the methods of recycling waste LABs. Particularly, we focused on the valuable component of waste lead paste and critically evaluated the pyrometallurgical and hydrometallurgical techniques associated with it. By categorizing and summarizing the characteristics of different methods, we have conducted a detailed comparison of these technologies, aiming to provide a comprehensive assessment of the advantages, disadvantages, status, and trends in LABs recycling technology. Additionally, the paper explores the necessity and impacts of recycling waste LABs from the perspectives of resource, energy, economy, environment, and society. It discusses the challenges faced by waste LABs recycling and presents the development prospects from both technical and non-technical point of views.

Keywords: Waste Lead-acid Batteries, Recycling, Resources, Energy, Environment.

1.Introduction

The extensive consumption and exploitation of traditional fossil fuels may led to a global energy crisis **[1, 2]**. Furthermore, the environmental consequences of using fossil fuels, such as air pollution, the greenhouse effect, and acid rain, pose significant threats to Earth's fragile ecosystem **[3-5]**. To address these challenges, the adoption of renewable energy sources has emerged as an ideal solution. Renewable energy, including solar power, wind energy, geothermal energy, and tidal energy, can be harnessed to generate electricity for human consumption. In this context, energy storage devices play a critical role in facilitating the application of renewable energy **[6-9]**.

Among the various energy storage technologies, lead-acid batteries (LABs) have emerged as a prominent and widely utilized option due to their low cost, well-established technology, high availability, safety features, adaptability to different environments, and efficient recycling processes **[9-14]**, Invented by Gaston Planté in 1859 **[15, 16]**, LABs have played a crucial role in numerous domains, including the automotive industry, power storage systems, uninterruptible power supply (UPS), electric bicycles, backup power supplies, and small household appliances **[10, 11]**. Notably, LABs remain indispensable in hybrid and electric vehicles as they serve as independent 12 V power sources, ensuring uninterrupted operation even in adverse weather conditions or in the event of high-voltage battery failure **[17, 18]**. Consequently, LABs have significantly contributed to the progress and development of human society, facilitating convenient and reliable lifestyles. In 2018, LABs account for approximately 70% of the global energy market share **[16, 19]**.

Particularly noteworthy is the substantial role of LABs in the electric bicycle industry. As shown in **Fig. 1**, despite the gradual replacement by lithium batteries, LABs-powered electric bicycles still dominate the market, capturing 77% of China's electric bicycle market as of 2020 **[20]**. Additionally, LABs maintain a significant market share of 20.6% in grid energy storage, second only to lithium-ion batteries (53.8%) in 2021 **[21]**. The grid energy storage sector presents promising prospects for LABs, with a projected market size reaching trillions of dollars in the future **[16]**.

While LABs may have lower energy density, which is 50 watt hours per kilogram, compared to lithium-sulphur batteries (505 watt hours per kilogram), lithium-ion batteries (265 watt hours per kilogram), and sodium-ion batteries (150 watt hours per kilogram) **[22]**, their competitive advantage lies in their cost-effectiveness. Considering the cost factor, LABs exhibit theoretical capacities comparable to many cathode materials used in lithium-ion batteries **[16]**. Therefore, in applications where product volume is not a constraint, LABs still present a rational choice. Moreover, their safety features and low cost make LABs an attractive option. As a result, LABs have garnered significant attention and widespread application, signifying a promising future.

However, the limited lifespan of LABs and their widespread usage result in the generation of substantial waste. Those waste LABs contain hazardous substances, such as sulfuric acid and lead element, which, if not properly handled, can contaminate the environment, including air, soil, and water bodies, posing risks to ecosystems and human health **[23-25]**. Consequently, appropriate disposal of waste LABs is paramount. Moreover, waste LABs represent a valuable secondary resource **[26]**. Recycling waste LABs reduces the demand for primary lead ore, mitigates environmental pollution, and conserves energy and water resources consumed during mining and processing. Lead recycling, in particular, can save approximately 55% to 65% of energy comparing with lead mining **[25]**.

In conclusion, the active recycling and utilization of waste LABs serve as effective strategies for resource reutilization, enhancing material using efficiency, and reducing environmental pollution. These measures are crucial for promoting sustainable and eco-friendly development **[27]**. This paper provides an in-depth exploration of waste LABs recycling technologies, analyses current challenges, and proposes solutions. Furthermore, this paper also examines the impact of waste LABs recycling from multiple perspectives, including resources, energy, economy, environment, and society, discussing the present state and future development trends of waste LABs recycling, both technically and non-technically.

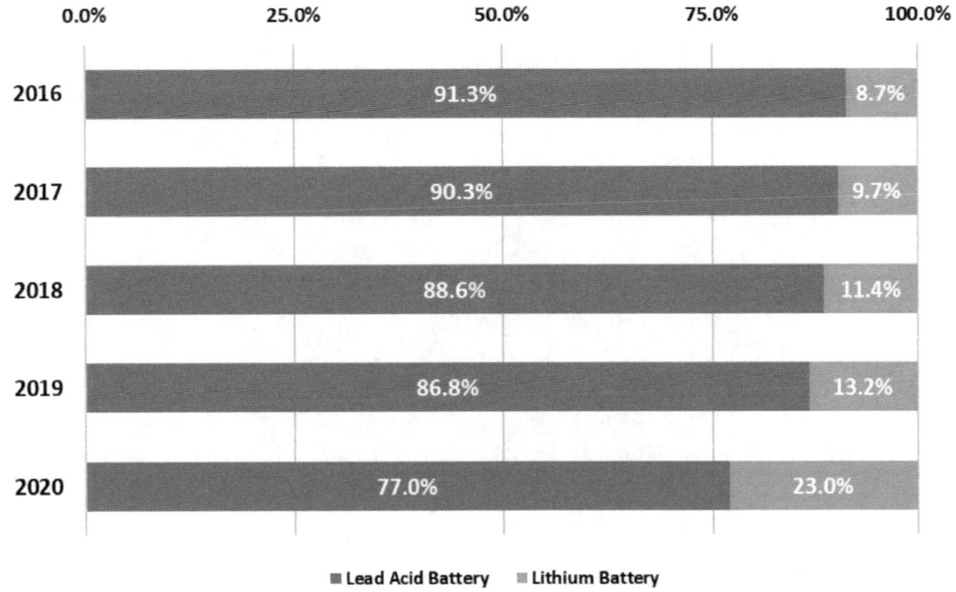

Fig. 1 Market share of battery materials for electric two-wheelers in China from 2016 to 2020

2. Working Principle of Lead-Acid Battery

LABs, which is a rechargeable battery, is a crucial electrochemical energy storage device mainly composed of cathodes, anodes, porous medium with electrolytes, battery shells, vent valves, and current collectors. The cathode primarily consists of the active material lead dioxide (PbO_2), while the anode is composed of pure lead (Pb). The electrolyte used is a concentrated sulfuric acid (H_2SO_4) solution, which facilitates ion transfer between the cathode and anode **[28, 29]**.

To increase the energy storage capacity of LABs during manufacturing, multiple cathode plates and anode plates are connected in parallel. This arrangement effectively increases the storage capacity, which is one of the main reasons for its large energy storage and stable electrical output. For instance, a typical 12 V vehicle LAB is composed of six individual LABs connected in series, resulting in a total voltage that is six times the voltage of each individual LAB **[17]**.

In the discharging process, chemical energy stored in lead compounds converts into electrical energy through anodic oxidation of lead (Pb) and cathodic reduction of lead dioxide (PbO_2) to lead sulphate ($PbSO_4$) **(Eq. (1) - (3)) [9, 17, 29].**

Cathode: $PbO_2 + SO_4^{2-} + 4H^+ + 2e^- \rightarrow PbSO_4 + 2H_2O$ (1)

Anode: $Pb + SO_4^{2-} - 2e^- \rightarrow PbSO_4$ (2)

Overall-discharging: $PbO_2 + Pb + 2H_2SO_4 \rightarrow 2PbSO_4 + 2H_2O$ (3)

While during the charging process of a LAB, lead sulphate ($PbSO_4$) from the electrolyte converts back to lead dioxide (PbO_2) on the anode, and to lead (Pb) on the cathode **(Eq. (4) - (6)) [17, 29].**

Cathode: $PbSO_4 + 2e^- \rightarrow Pb + SO_4^{2-}$ (4)

Anode: $PbSO_4 + 2H_2O - 2e^- \rightarrow PbO_2 + 4H^+ + SO_4^{2-}$ (5)

Overall-charging: $2PbSO_4 + 2H_2O \rightarrow PbO_2 + Pb + 2H_2SO_4$ (6)

3. Necessity of Recycling

Fig. 2 Five impact aspects of recycling waste lead-acid batteries

LABs recycling is crucial, which affects many aspects related to human activities. Here we group them into five categories, which are shown in **Fig. 2**, and they are respectively resources, energy, environment, economy, and society.

- In terms of resources, waste LABs are critical valuable secondary resources rich in lead. Recycling them enables the effective utilization of the lead resources, thereby reducing the demand for lead ore extraction and extending the extractable lifespan of lead mines **[30, 31]**. According to statistical data, the global refined lead production in 2021 was 12.28 million tons, with approximately 64.57% derived from recycled lead **[32]**. This indicates that the recycling of secondary lead resources is a crucial source for maintaining the existing supply-demand balance. Notably, the United States has achieved 100% sourcing of lead smelting from secondary lead. Moreover, in China, over 300 million waste LABs are generated annually, containing nearly 40 million tons of lead **[33]**. This substantial volume of waste highlights the immense potential for recovering lead from waste LABs and realizing significant resource conservation on a large scale.
- For energy aspect, LABs play a significant role as energy storage materials in various fields. The recycling and utilization of waste LABs undoubtedly contribute to the development of the energy industry, particularly in promoting the use of renewable energy sources. For instance, establishing energy storage systems using recycled LABs allows for the storage of electricity

during unstable renewable energy supply and subsequent release of stored energy during peak periods, thereby achieving grid load balancing. Statistics show that approximately 85% of global lead is used for manufacturing LABs **[34]**, with recycled lead being reutilized in LABs manufacturing, thereby facilitating the circular and sustainable development of the energy sector.

- In terms of the environment, LABs contain harmful components such as lead and sulfuric acid, which pose risks to the environment and living organisms. Improper handling of waste LABs can result in the leakage of these hazardous substances into soil and water sources **[23, 30]**, adversely affecting farmland and aquatic life. However, through the recycling of waste LABs, it is possible to isolate these harmful substances from the ecological environment, thereby achieving green and sustainable development **[31]**. Furthermore, the use of secondary lead contributes to a significant reduction in carbon dioxide emissions (up to 99%) **[34]**.
- From an economic perspective, the recycling of waste LABs leads to multiple economic benefits. Firstly, it reduces the costs associated with mining operations and post-mining environmental remediation. Secondly, due to the high lead content and relatively simple composition of spent batteries, the smelting and separation costs of lead are relatively low, facilitating resource reuse and generating economic value. It is estimated that lead scrap changing in value is from $0.20 to $0.40 per pound **[35]**. Furthermore, according to statistical data, in 2022, the global market for the recycling of waste batteries was valued at $1.38 billion, and it is projected to reach $17.08 billion by 2030. Among these, the recycling of waste LABs accounts for 85.41% (in 2022) of the total market value **[36]**. This significant market value highlights the immense economic potential and recycling value of waste LABs. Therefore, through effective recycling and reuse, we can transform waste into wealth.
- Lastly, from a societal perspective, the recycling of waste LABs has positive impacts on community harmony and public health. Proper recycling management can improve environmental quality of communities, reduce the risks and diseases associated with waste LABs, and prevent the disruption of human social activities caused by waste LABs. To achieve effective recycling, it requires the collective cooperation of various social groups, including governments, companies, and communities. Through strong collaborative relationships and effective dialogue, a healthier, greener, and more harmonious society can be established. Moreover, the waste LABs recycling industry promotes employment growth, and increases employment rates, which is also beneficial for social security and reducing crime rates. For example, in India, the waste LABs recycling sector provides job opportunities for approximately 600,000 to 700,000 individuals **[34]**.

In conclusion, the impacts and challenges of waste LABs recycling need to be comprehensively considered in terms of the resources, energy, environment, economy, and society. These aspects are interconnected and mutually influential. By actively promoting the recycling of waste LABs, we can achieve resource circularity, promote sustainable energy development, protect the environment, create economic benefits, and improve social well-being. Therefore, waste LABs recycling is a task of significant scientific nature, requiring the collective efforts and cooperation of society as a whole.

4. Recycling Technologies

The waste LABs mainly consist of the following components: grid plates (pure Pb), organic casing (polypropylene, polyethylene, polyvinyl chloride, etc.), lead paste (PbO, PbO_2, and $PbSO_4$), electrolyte (H_2SO_4), and separators. The specific proportions of these material components are outlined in **Table 1 [10, 37, 38].**

Table 1 Typical phase composition of a waste lead-acid battery

Category	Content, %						
	Pure Pb	PbO	PbO_2	$PbSO_4$	Pb in total	H_2SO_4	Organics
Grid plates	92-95	N/A	N/A	Few	92-95	N/A	N/A
Organic casing	N/A	N/A	N/A	N/A	N/A	N/A	99
Lead paste	Few	10-15	15-20	25-30	67-76	N/A	N/A
Electrolyte	Few	Few	Few	Few	Few	99	N/A
Separators	N/A	N/A	N/A	N/A	N/A	N/A	99

Due to the diverse composition and varying properties of different parts of waste LABs, direct processing is difficult to target and often results in low lead recovery rates and severe environmental pollution. Therefore, it is more reasonable to dismantle waste LABs and sequentially separate their different components for targeted recycling processes. The recycling process of waste LABs includes pretreatment, smelting, and electrolytic refining processes. The pretreatment process involves steps such as acid drainage, crushing, and powder separation **[10, 39, 40]**.

Overall, a typical process of waste LABs recycling is shown as **Fig. 3**.

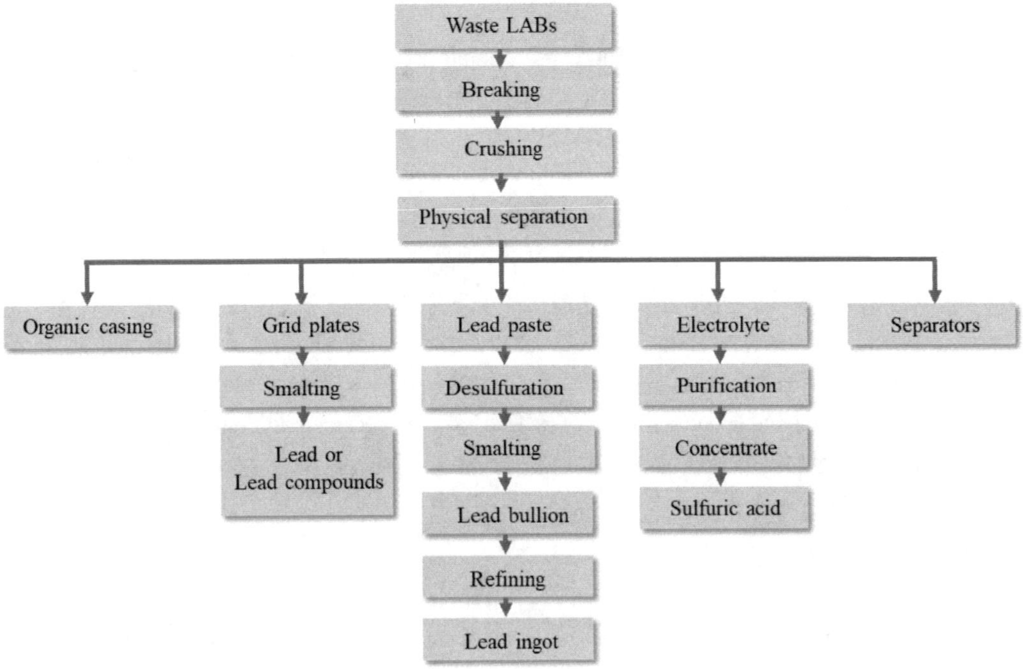

Fig. 3 The process of waste lead-acid batteries recycling

Generally, after the breaking, crushing processes, waste LABs require the physical separation of their different components for effective recycling **[41, 42]**. The plastic component, after being washed, can be separated and recovered using flotation methods **[32]**. The main component of the electrolyte is sulfuric acid, which can be purified and concentrated through techniques such as ion exchange, membrane separation, and distillation for recycling and reuse **[32, 43]**. Moreover, sulfuric acid can also be neutralized and then sent to wastewater treatment plants **[44]**. As indicated in **Table 1**, lead is primarily distributed in the grid plates and lead paste, making these two components highly valuable for recycling and crucial for the circular utilization of lead resources. Moreover, the separation between lead paste and grid is achieved through the process of air classification **[45]**.

The grid plates mainly consist of lead alloys and can be recycled separately. Typically, the grid plates are produced to pure lead or lead alloys through the smelting process **[32, 41, 46]**. Due to the relatively low melting point of lead and its alloys, grid plates recycling can be conducted at relatively lower smelting temperatures. In contrast, the lead paste, which contains a complex composition with a significant amount of lead sulphate and various lead oxide species, presents the most challenging part of the waste LABs recycling process and remains a key focus of current research **[32, 38]**.

Currently, two methods, namely pyrometallurgical and hydrometallurgical processes, are used for lead paste recovery. Nevertheless, regardless of the method chosen, the purpose of this process is to remove impurities and reduce various lead compounds in the lead paste, ultimately obtaining purified lead or lead oxide products **[10, 47]**.

In the following sections, this article provides separate discussions on these two methods.

4.1 Pyrometallurgical Process

The pyrometallurgical method is widely used for the recycling of waste lead paste, and it is currently a mature technology **[47]**. Generally, this method involves high-temperature treatment of the spent lead paste and the addition of a reducing agent such as coke to promote the reduction reaction of the lead paste in a molten state **[37, 47]**. The pyrometallurgical method can be further classified into one-step and two-step processes.

Specifically, the one-step process refers to the simultaneous reduction of lead and desulfurization of lead sulphate during the high-temperature smelting and reduction process (above 1000 °C) **[40, 42, 45]**. In this process, the organic matter and impurities in the waste lead paste are decomposed, resulting in the formation of liquid metallic lead. The relevant reaction equations are as follows (**Eq. (7)-(9)**):

$$2PbO + C \rightarrow 2Pb + CO_2\uparrow \tag{7}$$

$$PbO_2 + C \rightarrow 2Pb + CO_2\uparrow \tag{8}$$

$$2PbSO_4 + 2C \rightarrow 2Pb + 2SO_2\uparrow + 2CO_2\uparrow \tag{9}$$

For example, Hu et al. **[48]** mixed desulfurizing agents (zinc oxide and sodium carbonate) with a reducing agent (carbon powder) and achieved simultaneous desulfurization and reduction. Under optimized experimental conditions, the lead recovery rate was 98.14 wt.%, the obtained crude lead metal grade was 98.85 wt.%, and the sulphur fixation rate was 94.52%. The chemical equations involved in this method are shown as **Eq. (10)-(15).**

$$PbSO_4 + ZnO + 2.5C \rightarrow Pb + ZnS + 2.5CO_2 \tag{10}$$

$$PbSO_4 + Na_2CO_3 + 2.5C \rightarrow Pb + Na_2S + 3.5CO_2 \tag{11}$$

$$PbSO_4 + 2C \rightarrow 2Pb + CO_2 \tag{12}$$

$$2PbO + C \rightarrow 2Na + 3CO_2 \tag{13}$$

$$PbO_2 + C \rightarrow Pb + CO_2 \tag{14}$$

$$Na_2S + ZnO + CO_2 \rightarrow Na_2CO_3 + ZnS \tag{15}$$

While the one-step method for recovering lead paste is relatively simple and has a shorter process flow, it operates at very high temperatures (The reduction of $PbSO_4$ requires a temperature of 1499°C or higher.) and generates a significant amount of SO_2 **[49]**, resulting in environmental pollution and posing a threat to human health **[10]**. Therefore, considering the aspects of economy, energy consumption, and environment, this method is not competitive and has been gradually phased out. As a result, a two-step process, known as pre-desulfurization smelting, has been developed.

The two-step process in the recovery of waste lead paste involves a pre-desulfurization step followed by smelting. The purpose of pre-desulfurization is to reduce the sulphur content in the lead paste before the smelting process, thereby minimizing the amount of SO_2 generated during subsequent smelting. This is achieved by adding desulfurizing agents such as sodium carbonate, sodium hydroxide, ammonium carbonate, ammonium bicarbonate, and potassium carbonate, which react with the lead paste at room temperature. The reaction equations are as follows (**Eq. (16)-(20)**) **[37, 49-51]**:

$$2PbSO_{4\ (paste)} + 4NaOH_{\ (aq)} \rightarrow 2Na_2SO_{4\ (aq)} + Pb(OH)_{2\ (paste)} \tag{16}$$

$$PbSO_{4\ (paste)} + Na_2CO_{3\ (aq)} \rightarrow PbCO_{3\ (paste)} + Na_2SO_{4\ (aq)} \tag{17}$$

$$PbSO_{4\ (paste)} + (NH_4)_2CO_{3\ (aq)} \rightarrow PbCO_{3\ (paste)} + (NH_4)_2SO_{4\ (aq)} \tag{18}$$

$$PbSO_{4\ (paste)} + NH_4HCO_{3\ (aq)} \rightarrow PbCO_{3\ (paste)} + NH_4HSO_{4\ (aq)} \tag{19}$$

$$PbSO_{4\ (paste)} + K_2CO_{3\ (aq)} \rightarrow PbCO_{3\ (paste)} + K_2SO_{4\ (aq)} \tag{20}$$

Through the pre-desulfurization step, the lead sulphate in the waste lead paste reacts with the desulfurizing agents, resulting in the formation of insoluble lead carbonate or lead hydroxide (referred to as new lead paste) and a byproduct solution of sodium sulphate. The obtained new lead paste is then subjected to reduction smelting, which reduces the sulphur content and the generation of SO_2 during the smelting process. Additionally, the smelting temperature can be lowered, thereby reducing environmental pollution and health risks **[45, 52]**. The smelting temperature of desulfurized lead paste is typically at least 100-150 °C lower than that of untreated lead paste **[40]**. Specifically, lead carbonate can decompose at 288 °C **[49]**.

Other than carbon, certain metals can also act as reducing agents. Liu et al. **[53]** treated the waste lead paste with sodium carbonate for desulfurization, and then used lead-antimony alloy as the reducing agent to reduce the waste lead paste under vacuum, achieving both the separation of lead-antimony alloy and the reduction of waste lead paste. Ultimately, the direct recovery rate of Pb from the desulfurized lead paste reached 99.29%. The antimony content in the lead-antimony alloy was reduced from 46% to 0.98%. This method not only enables the efficient and economical recovery of lead paste but also achieves the efficient separation of lead-antimony alloy without generating any waste.

Furthermore, a reverse approach is also one of the methods for treating waste lead paste, such as converting all the components of waste lead paste into high-value lead sulphate. Liu et al. **[54]** used potassium bisulphate as a roasting agent and achieved the conversion of waste lead paste into lead sulphate under the conditions of 300°C, $KHSO_4$/waste lead paste mass ratio of 1:1, and a holding time of 10 minutes. This method significantly reduces the emissions of carbon dioxide and smoke. The results of the life cycle assessment show that recycling 1.0 ton can reduce carbon emissions by 2.45 tons of carbon dioxide and 0.13 tons of smoke.

Overall, the two-step process is widely adopted in the recovery of waste lead paste as it effectively reduces environmental pollution, improves workplace safety, and decreases energy consumption. It is a more sustainable and environmentally friendly approach. However, it should be noted that despite the use of this method, complete removal of sulphur from the lead paste is not achieved, leading to the production of SO_2 **[55, 56]**.

The pyrometallurgical smelting method offers several advantages, including its simple operation, fast reaction rate, high adaptability to various raw materials, and suitability for large-scale production. However, it has certain limitations that contradict the principles of sustainability. The pyrometallurgical process is associated with low recovery rates, high energy consumption, and the emission of significant amounts of sulphur dioxide and lead dust. These emissions contribute to environmental pollution and pose risks to the health of workers and nearby residents. The environmental impact and health hazards associated with the pyrometallurgical method contradict the concept of sustainable development. Therefore, it is crucial to explore alternative approaches that minimize these adverse effects and align with sustainable practices **[10, 37, 44]**.

4.2 Hydrometallurgical Process

Hydrometallurgy has emerged as a viable alternative to pyrometallurgy over the past decades, experiencing rapid development and widespread adoption while gradually replacing pyrometallurgical processes **[47]**. This shift can mitigate the significant drawbacks of pyrometallurgy, including low recovery rates, high energy consumption, and severe environmental pollution. In contrast, hydrometallurgy is regarded as a more environmentally friendly approach, characterized by simpler equipment and processes, as well as high metal recovery rates and wide adaptability. By utilizing aqueous solutions as the medium, hydrometallurgy efficiently and precisely reduces various components of lead paste to pure lead or lead oxide. Furthermore, in order to mitigate environmental pollution, hydrometallurgy incorporates desulfurization pretreatment of lead paste.

Hydrometallurgy can be further classified into two primary methods: conventional hydrometallurgy and electrochemical methods.

Conventional hydrometallurgy typically involves leaching the lead paste, followed by the addition of reducing agents (such as lead powder) to induce the reduction reaction. After undergoing purification processes, the desired lead oxide product is obtained through techniques like crystallization. This method is relatively straightforward and commonly employed in small-scale production environments. As a feature, this method does not require electrochemical assistance **[44]**.

For example, Pan et al. **[50]** used a method where waste lead paste and lead powder are dissolved in NaOH solution to form a mixed solution of Na_2SO_4 and $NaHPbO_2$. Impurities such as iron and barium in the lead paste are removed because they do not dissolve. The solution is then cooled for crystallization to obtain lead oxide. Finally, the lead oxide is dissolved again in a sodium hydroxide solution and subjected to another round of crystallization. The resulting lead oxide has a purity level of over 99.99%, and the lead recovery rate reaches 99.2%.

Wu et al. **[57]** initially sulfurized the waste lead paste to $PbSO_4$ (with a conversion rate of 99.99%) using reagents such as H_2SO_4, H_2O_2, and NaCl. Subsequently, they carried out a simultaneous process of lead leaching and desulfurization by reacting the sulfurized paste with ammonium acetate, achieving a lead leaching rate of up to 99.94%. In the resulting solution, glycerol was added, and after distillation, the glycerol solution containing lead acetate was obtained. This solution was then reacted with ammonium carbonate as the decarburizer, resulting in the precipitation of over 99.9% of the lead.

Electrochemical methods rely on electrochemical reactions to ultimately reduce the lead paste to pure lead or lead oxide. Depending on the form of the lead paste during electrolysis, electrochemical methods can be further divided into liquid-phase electrodeposition and solid-phase electrolysis.

Liquid-phase electrodeposition begins with dissolving the lead paste, and the resulting solution serves as the electrolyte. By applying an external power source, the lead ions in the solution are reduced at the cathode, depositing as elemental lead on its surface, thereby achieving the recovery and utilization of lead paste. However, it should be noted that prior to leaching, generally, the insoluble components in the lead paste need to be converted into soluble PbO, a process that often involves desulfurization as well **[10]**.

For example, Andrews et al. **[58]** utilized a mixed solution of sodium chloride and hydrochloric acid to leach lead paste and obtain lead chloride solution. After solid-liquid separation to remove impurities, lead powder was added to further purify the lead chloride solution. Subsequently, the purified solution underwent electrodeposition, resulting in lead with a purity level as high as 99.995% and a recovery rate of 99.5%.

Chang et al. **[59]** proposed a process for the recovery of waste lead paste using an electro-deposition method. Firstly, $PbSO_4$ was converted in the waste lead paste into soluble $Pb_3(CO_3)_2(OH)_2$ or $NaPb_2(CO_3)_2OH$ in Na_2CO_3 solution. and this process also converted insoluble PbO_2 into PbO simultaneously. Then, the lead paste after pre-treated was leached in methane sulfonic acid solution. The leachate was used as the electrolyte, with the addition of 0.5 g/L calcium lignosulfonate and 0.05 g/L bone gum. Finally, lead was deposited on the cathode through electrodeposition, resulting in a purity of 99.996% for the obtained product. Moreover, the lead recovery rate reached 95.28%, and the cathodic current efficiency and specific energy consumption were 99.31% and 598.91 kWh/t Pb, respectively.

Xing et al. **[60]** utilized an acidic solution (pH=1) consisting of calcium chloride and ferrous chloride tetrahydrate as the leaching agent to dissolve the lead paste at 90°C, achieving a leaching rate of 99%. Subsequently, an electrodeposition experiment was conducted at a current density of 200 A/m^2, resulting in lead product with a purity of 99.6%. The cathodic current efficiency was 96.3%, and the specific energy consumption was 85.9 kWh/t Pb. Importantly, this method eliminates the need for a desulfurization step, and the iron ions and waste electrolyte are recyclable.

In contrast, solid-phase electrolysis involves transforming the lead paste into cathode plates or slurry, which are then subjected to reduction through an external circuit, resulting in the production of sponge lead or lead oxide. The choice of electrolyte in this method can include acids, alkalis, or ionic liquids **[10, 38, 61]**.

Fan et al. **[62]** developed a method to directly electrochemically reduce waste lead paste in an NH_3-NH_4Cl solution, resulting in the production of high-purity metallic lead with a purity of 98.3%. The process exhibited an electrical current efficiency of 86.3% and an energy consumption of 689.4 kWh/t Pb, respectively.

Ultimately, the core idea of electrochemical methods is to use the reduction reaction through electrochemistry to convert the lead compounds in the lead paste into pure lead. Regardless of the exist states of the lead paste (solution, slurry, or electrode plate), regardless of variations in the electrolytic cell structure or the use of different additives, the fundamental principles and production objectives remain the same.

With the increasing emphasis on environmental protection and sustainable development, hydrometallurgy has gained widespread application as a more environmentally friendly alternative. It not only effectively enhances the recovery rate of lead paste and the utilization of metal resources but also reduces energy consumption and environmental pollution. Therefore, hydrometallurgy plays a crucial role in promoting the sustainable development of secondary lead resources **[34, 40, 44, 63]**.

5. Challenges, Solutions and Prospects

5.1 Challenges and Solutions

The current waste LABs recycling industry still faces significant challenges, including both technological (technology and environment) and non-technological (legislation and society) aspects. Here, we discussed these challenges and their corresponding solutions in both aspects.

- Challenges of technology and environment

The technological challenges primarily revolve around the impact on the environment and human life and health.

Although lead recovery has made significant progress, the environmental issues including the lead-containing particulates, SO_2 gas and lead-containing slag require special precautions in the process of waste LABs processing. Nowadays, most of the secondary lead recovery plants are based on pyrometallurgical methods **[64]**. The emission of sulphur oxides, nitrogen oxides and lead-containing particulate matters from the pyrometallurgical route constitute a serious consequence for the local environment and human lives. Despite the fact that the paste smelting process has made important progress in energy efficiency, the issue of dust measured as PM 2.5, containing lead particulates constitute a major lead poisoning threat to humans, animals, birds, soil and drinking water **[65]**. Atmospheric dispersion and deposition are the major transport pathways. An investigation has shown that the lead intakes were 50.1 and 6.2 μg/kg/d for the children and adults which are much higher than the Provisional Tolerable Weekly Intake (PTWI) values in these areas in China **[65]**.

Moreover, in low to middle-income countries (LMICs), the hazards associated with waste LABs recycling are amplified substantially. Notably, LABs recycling is the most environmentally damaging industrial process in the world, resulting in an estimated loss of between 2 million and 4.8 million disability-adjusted life years (DALYs) **[66]**. While developed nations implement stringent control measures to regulate lead exposure and release, LMICs exhibit significantly higher rates. An exhaustive investigation conducted in Kenya revealed that all contamination measurements, including those from office areas, exceeded the Occupational Health and Safety Administration's permissible levels **[67]**. Blood lead levels in Kenyan lead smelter employees ranged from 43.4±6.6 to 62.2±12.7 μg/dL, exceeding levels observed when petrol additives were utilised. The elevated lead contamination levels in Kenya were attributed to deficiencies in engineering controls, work practises, respirator use, and personal hygiene. In addition, the exposure hazards associated with less formal facilities are even greater.

Haefliger et al. also investigated a community in Dakar, Senegal, where informal LABs recycling was taking place for a separate study **[68]**. Within five months, 18 infants succumbed to a rapidly progressive disease of the central nervous system. Blood lead levels ranged from 39.8 to 613.9 μg/dL with an average of 129.5 μg/dL among adults and children residing in the same area. The contamination had a considerable impact not only on the immediate area, but also on the surrounding regions **[69]**. Ericson et al. conducted a comprehensive analysis of informal lead recycling practises in 90 LMICs and estimated the presence of up to 29,241 sites engaged in such activities, exposing approximately 16.8 million people to elevated lead levels **[70]**.

In the face of technological and environmental challenges, the development of more efficient and environmentally friendly new technologies is inevitable. These new technologies should aim for high energy efficiency, high recovery rates, short processing cycles, and environmental friendliness. This would require significant investment and research and development of new production equipment.

In addition, strict adherence to production standards and regulations is necessary during the manufacturing process. Proper disposal of waste LABs should be ensured to prevent any leakage into the environment and avoid serious consequences. Furthermore, it is important to establish long-term production plans and emergency response plans to address various unforeseen circumstances, including ecological restoration and medical support.

- Challenges of legislation and society

The efficient collection of waste LABs is also a crucial challenge. The inadequate collection networks, limited collection infrastructure, lack of recycling awareness and low community participation pose limitations to the recycling of spent LABs. These challenges are not of a technical nature, but they have a significant impact on the overall sustainability of the industry **[34, 71, 72]**.

In many areas, the lack of well-established collection networks and systems is a major barrier. This means that there are insufficient collection points and recycling centres, making it inconvenient for people to deliver their spent LABs to professional recycling facilities **[73]**. Furthermore, due to the lack of a formalized recycling system in some regions, a substantial amount of waste LABs enter the hands of tiny secondary lead enterprises illegally. For example, small illicit secondary lead enterprises are responsible for the disposal of more than forty percent of the nation's scrap lead resources in China, a disposal capacity comparable to that of large-scale enterprises. However, industry concentration is low, and small businesses lack of environmental protection equipment has led to a disregard for zoning and pollution restrictions **[70]**.

Another challenge is the lack of sufficient recycling awareness and participation in recycling. Community residents may not fully understand the importance and impact of recycling waste LABs. They may lack proper knowledge on how to dispose of waste LABs correctly and may lack the motivation to participate in recycling efforts. The lack of effective publicity and educational activities also hinders public engagement **[34, 41, 71-73]**.

Moreover, the limited collection infrastructure is also a problem. In certain areas, there is a lack of adequate collection points and containers, making it difficult for residents to conveniently dispose of their spent LABs at designated collection points. This inconvenience and lack of proper infrastructure pose challenges to the recycling of spent LABs **[71, 72]**.

In addition, Insufficient and inadequate implementation of relevant laws and regulations in certain regions also pose obstacles to the waste LABs recycling industry **[74-76]**.

Overall, to establish an effective and sustainable waste LABs collection system, it is crucial to address these non-technical challenges. This requires improving the collection network, strengthening recycling awareness and education, and investing in appropriate collection infrastructure to facilitate the recycling process. Additionally, enhancing the legal framework and strengthening regulatory measures are of utmost importance.

Suitable subsidies and penalties can help curb illegal practices and enhance people's willingness to participate in it. Ultimately, this will lead to the efficient and environmentally friendly recycling and reuse of secondary lead resources.

5.2 Prospect

In the context of achieving carbon peak and carbon neutrality, as well as the growing prominence of sustainable development principles, technological advancement has become a critical factor in the waste LABs recycling industry. Within this framework, there is an urgent need to develop cleaner, more efficient, cost-effective, streamlined, and user-friendly new technologies. This necessitates increased attention, research interest, and financial investment to drive technological innovation in the field of waste LABs recycling. Moreover, existing technologies are bound to be replaced by new ones in pursuit of economic benefits.

The establishment of a robust recycling system and network is also of utmost importance. This includes the establishment of dedicated responsible recycling institutions, as well as the development of recycling plans and emergency protocols. Such a recycling system ensures the smooth and uninterrupted lifecycle circulation of LABs, mitigating the risk of fatal issues due to any missing links or failures in the recycling process.

Furthermore, it is imperative to enhance relevant policies, regulations, and oversight, which may involve implementing reward and penalty mechanisms. Additionally, strengthening public awareness through educational campaigns is crucial to foster a deeper understanding of the concept of a circular economy. By encouraging proactive public participation in waste LABs recycling and promoting active classification of secondary resources, the difficulty and cost associated with recycling can be significantly reduced.

Enhancing collaboration and cooperation is also of paramount importance. This encompasses not only communication and collaboration among governments, businesses, research institutions, and consumers but also the sharing of technological advancements among different enterprises, regions, and even nations. Particularly, developed countries possessing more advanced recycling technologies should assist developing nations in achieving efficient circularity and recycling within the LABs industry. Moreover, interdisciplinary academic exchanges will further facilitate the development of the waste LABs recycling industry, as interdisciplinary integration is one of the future trends.

6. Conclusion

The recycling of waste lead-acid batteries (LABs) is crucial for sustainable development. The increased usage of LABs has resulted in significant waste generation, posing risks to the environment and human health. Recycling waste LABs offers multiple benefits. Firstly, it aids in reducing the release of hazardous substances such as lead and sulfuric acid, thereby minimizing damage to the natural environment and ecosystems. Secondly, waste LABs are a valuable secondary resource, with a substantial portion of global refined lead production derived from the recycling of LABs, highlighting the resource importance of LABs recycling. Thirdly, considering the substantial consumption of lead by LABs globally, recycling the lead from waste LABs holds positive implications for the energy storage sector. Finally, the recycling of LABs not only mitigates environmental risks but also generates significant economic benefits by converting waste into valuable resources.

In this paper, we provided an overview and classified the main technologies currently used in LABs recycling, discussed the challenges faced, and presented relevant solutions. We also analysed the impact of LABs recycling from various perspectives, including resources, energy, economy, environment, and society. Through a comprehensive examination of both technical and non-technical factors, we have evaluated the status of LABs recycling and identified future trends.

Overall, this study emphasizes the importance of recycling waste LABs and highlights the need for ongoing research and development in this field. By implementing effective recycling practices, we can minimize environmental impacts, preserve precious resources, support the energy storage sector, and contribute to sustainable development.

References

1. Manieniyan, V., M. Thambidurai, and R. Selvakumar, Study on energy crisis and the future of fossil fuels. Proceedings of SHEE, 2009. **10**: p. 2234-3689.
2. Kreps, B.H., The rising costs of fossil-fuel extraction: an energy crisis that will not go away. American journal of economics and sociology, 2020. **79**(3): p. 695-717.
3. Muradov, N.Z. and T.N. Veziroğlu, “Green” path from fossil-based to hydrogen economy: an overview of carbon-neutral technologies. International journal of hydrogen energy, 2008. **33**(23): p. 6804-6839.
4. Din, M.I., et al., Microbial fuel cells—A preferred technology to prevail energy crisis. International Journal of Energy Research, 2021. **45**(6): p. 8370-8388.
5. Awan, A.B. and Z.A. Khan, Recent progress in renewable energy–Remedy of energy crisis in Pakistan. Renewable and Sustainable Energy Reviews, 2014. **33**: p. 236-253.
6. Vanitha, M. and N. Balasubramanian, Waste minimization and recovery of valuable metals from spent lithium-ion batteries–a review. Environmental Technology Reviews, 2013. **2**(1): p. 101-115.
7. Maddukuri, S., et al., On the challenge of large energy storage by electrochemical devices. Electrochimica Acta, 2020. **354**: p. 136771.
8. Liu, C., et al., Advanced materials for energy storage. Advanced materials, 2010. **22**(8): p. E28-E62.
9. May, G.J., A. Davidson, and B. Monahov, Lead batteries for utility energy storage: A review. Journal of energy storage, 2018. **15**: p. 145-157.
10. Li, M., J. Liu, and W. Han, Recycling and management of waste lead-acid batteries: A mini-review. Waste Management & Research, 2016. **34**(4): p. 298-306.
11. Bashir, N., et al. Lifetime maximization of lead-acid batteries in small scale UPS and distributed generation systems. in 2017 IEEE Manchester PowerTech. 2017. IEEE.
12. Murariu, T. and C. Morari, Time-dependent analysis of the state-of-health for lead-acid batteries: An EIS study. Journal of Energy Storage, 2019. **21**: p. 87-93.
13. Lavety, S., R.K. Keshri, and M.A. Chaudhari, A dynamic battery model and parameter extraction for discharge behavior of a valve regulated lead-acid battery. Journal of Energy Storage, 2021. **33**: p. 102031.
14. Yu, Y., J. Mao, and X. Chen, Comparative analysis of internal and external characteristics of lead-acid battery and lithium-ion battery systems based on composite flow analysis. Science of The Total Environment, 2020. **746**: p. 140763.
15. Kandeeban, R., et al., Battery economy: Past, present and future. Materials Today: Proceedings, 2022. **48**: p. 143-147.

16. Lopes, P.P. and V.R. Stamenkovic, Past, present, and future of lead–acid batteries. Science, 2020. **369**(6506): p. 923-924.
17. Treptow, R.S., The lead-acid battery: Its voltage in theory and in practice. Journal of chemical education, 2002. **79**(3): p. 334.
18. Kumar, R.V., A low-cost green technology for recovering lead paste and lead-alloy grid materials for spent lead acid batteries. Mineral Processing and Extractive Metallurgy, 2017. **126**(1-2): p. 89-93.
19. Machado Santos, S., J. Cabral Neto, and M. Mendonça Silva, Forecasting model to assess the potential of secondary lead production from lead acid battery scrap. Environmental Science and Pollution Research, 2019. **26**: p. 5782-5793.
20. Statista. Market Share of Battery materials for electric two-wheelers in China from 2016 to 2020. 2023; Available from: https://www.statista.com/statistics/1300248/china-market-share-electric-two-wheeler-battery/.
21. Statista. Gravimetric energy density of different types of batteries in 2020. 2023; Available from: https://www.statista.com/statistics/1249539/gravimetric-energy-density-of-batteries/.
22. Statista. Distribution of the global market for grid-scale batteries in 2021, by type. 2023; Available from: https://www.statista.com/statistics/1339858/share-of-the-global-grid-scale-battery-market-by-type/.
23. Kumar, S., et al., Lead and other elements-based pollution in soil, crops and water near a lead-acid battery recycling factory in Bangladesh. Chemosphere, 2022. **290**: p. 133288.
24. Zeng, Z., et al., Lead exposure is associated with risk of impaired coagulation in preschool children from an e-waste recycling area. Environmental Science and Pollution Research, 2018. **25**: p. 20670-20679.
25. Organization, W.H., Recycling used lead-acid batteries: health considerations. 2017.
26. Schriner, D., P. Taylor, and J. Grogan. A review of slag chemistry in lead recycling. in Advances in Molten Slags, Fluxes, and Salts: Proceedings of the 10th International Conference on Molten Slags, Fluxes and Salts 2016. 2016. Springer.
27. Yu, J., et al., Emergy based sustainability evaluation of spent lead acid batteries recycling. Journal of Cleaner Production, 2020. **250**: p. 119467.
28. Zhang, Y., et al., Advances and challenges in improvement of the electrochemical performance for lead-acid batteries: A comprehensive review. Journal of Power Sources, 2022. **520**: p. 230800.
29. Guo, S., The application of genetic algorithms to parameter estimation in lead-acid battery equivalent circuit models. 2010, University of Birmingham.
30. Wei, M., J. Ma, and T. Gao. Analysis on pollution prevention and control of waste lead battery recycling process. in IOP Conference Series: Earth and Environmental Science. 2021. IOP Publishing.
31. Tian, X., et al., Design and simulation of a secondary resource recycling system: A case study of lead-acid batteries. Waste Management, 2021. **126**: p. 78-88.
32. Li, Z., Main recycling process and development status of waste lead acid batteries. Nonferrous Metals (Smelting Section). 2022. **9**: p. 9-27.
33. Weizhao, Z., et al., Kinetic analysis of pre-desulfurization process with ammonium bicarbonate method from lead paste in waste lead-acid batteries. Chinese Journal of Environmental Engineering, 2020. **14**(3): p. 772-779.
34. Varshney, K., et al., Current trends and future perspectives in the recycling of spent lead acid batteries in India. Materials Today: Proceedings, 2020. **26**: p. 592-602.

35. BUFFALO, M. Lead Acid Battery Recycling in Buffalo, NY. 2023; Available from: https://www.metalicobuffalo.com/electrical-scrap-recycling/lead-battery-recycling.

36. Research, G.V. Battery Recycling Market Size, Share & Trends Analysis Report By Chemistry (Lithium-ion, Lead Acid, Nickel), By Application (Transportation, Industrial), By Region (Europe, Asia Pacific, North America), And Segment Forecasts, 2023 - 2030. 2023; Available from: https://www.grandviewresearch.com/industry-analysis/battery-recycling-market?utm_source=prnewswire&utm_medium=referral&utm_campaign=CMFE_10-April-23&utm_term=battery_recycling_market&utm_content=rd1.

37. Zakiyya, H., Y. Distya, and R. Ellen. A review of spent lead-acid battery recycling technology in indonesia: comparison and recommendation of environment-friendly process. in IOP Conference Series: Materials Science and Engineering. 2018. IOP Publishing.

38. Hu, B., F. Yang, and L. Chen, Research progress of technology for recycling lead paste from spent lead-acid batteries. Appl. Chem. Ind, 2019. **48**: p. 2742-2748.

39. Yanamandra, K., et al., Recycling of Li-Ion and Lead Acid Batteries: A Review. Journal of the Indian Institute of Science, 2022. **102**(1): p. 281-295.

40. Zhu, X., Study on Leaching Process of Spent Lead Acid Battery Paste with Organic Acid and Preparation of Ultrafine Lead Oxide by Calcination at Low Temperature. 2012, Huazhong University of Science and Technology

41. Sun, Z., et al., Spent lead-acid battery recycling in China–A review and sustainable analyses on mass flow of lead. Waste Management, 2017. **64**: p. 190-201.

42. Pan, H., et al., Sustainability evaluation of secondary lead production from spent lead acid batteries recycling. Resources, Conservation and Recycling, 2019. **140**: p. 13-22.

43. Zhang, X.X., et al., Toward sustainable and systematic recycling of spent rechargeable batteries. Chemical Society Reviews, 2018. **47**(19).

44. Li, M., et al., Review on clean recovery of discarded/spent lead-acid battery and trends of recycled products. Journal of Power Sources, 2019. **436**: p. 226853.

45. Gao, L., Behavior of the Impurities in Recovery Process of Spent Lead-Battery Paste Leaching with the Citric Acid and Its Impacts on the Battery Performance. 2014, Huazhong University of Science and Technology

46. Xie, X., Research progress in lead-acid battery grid material in China. BATTERY BIMONTHLY, 2010. **40**(3): p. 173-176.

47. Liu, W., et al., Lead recovery from spent lead acid battery paste by hydrometallurgical conversion and thermal degradation. Waste Management & Research, 2020. **38**(3): p. 263-270.

48. Hu, Y.J., et al., Reductive smelting of spent lead-acid battery colloid sludge in a molten Na2CO3 salt. International Journal of Minerals Metallurgy and Materials, 2015. **22**(8): p. 798-803.

49. Li, W., et al., Recycling lead from waste lead-acid batteries by the combination of low temperature alkaline and bath smelting. Separation and Purification Technology, 2023: p. 123156.

50. Pan, J., et al., Preparation of high purity lead oxide from spent lead acid batteries via desulfurization and recrystallization in sodium hydroxide. Industrial & Engineering Chemistry Research, 2016. **55**(7): p. 2059-2068.

51. Hua, C., et al., A new desulfation process of spent lead paste via cyclic utilization of CO2–NH3· H2O. Journal of Cleaner Production, 2022. **349**: p. 131307.

52. Li, Y., et al., Novel recycling process for lead-acid battery paste without SO2 generation-Reaction mechanism and industrial pilot campaign. Journal of Cleaner Production, 2019. **217**: p. 162-171.
53. Liu, T.T., Z.Q. Bao, and K.Q. Qiu, Recycling of lead from spent lead-acid battery by vacuum reduction-separation of Pb-Sb alloy coupling technology. Waste Management, 2020. **103**: p. 45-51.
54. Liu, K., et al., Batch Production of Lead Sulfate from Spent Lead-Acid Batteries via an Oxygen-Free Roasting Route: A Negative-Carbon Strategy. Acs Sustainable Chemistry & Engineering, 2023. **11**(18): p. 7244-7252.
55. Ma, Y., et al., Analysis of a more sustainable method for recycling waste lead batteries: Surface renewal promotes desulfurization agent regeneration. Waste Management, 2022. **137**: p. 319-328.
56. Duan, X.J., et al., Preparation of polypyrrole nanocomposites for supercapacitor using spent battery powder as raw materials. Electrochimica Acta, 2016. **210**: p. 646-654.
57. Wu, Y.Z., et al., Preparation of high-purity lead carbonate and lead oxide from spent lead paste. Journal of Cleaner Production, 2022. **372**.
58. Andrews, D., A. Raychaudhuri, and C. Frias, Environmentally sound technologies for recycling secondary lead. Journal of Power Sources, 2000. **88**(1): p. 124-129.
59. Chang, C., et al., Green hydrometallurgical extraction of metallic lead from spent lead paste in the methanesulfonic acid system. Separation and Purification Technology, 2023. **306**.
60. Xing, P., et al., Hydrometallurgical recovery of lead from spent lead-acid battery paste via leaching and electrowinning in chloride solution. Hydrometallurgy, 2019. **189**.
61. Soundarrajan, C., et al., Improved lead recovery and sulphate removal from used lead acid battery through Electrokinetic technique. Journal of hazardous materials, 2012. **217**: p. 452-456.
62. Fan, Y.Y., et al., Preparation of Metal Lead from Waste Lead Paste by Direct Electrochemical Reduction in NH3-NH4Cl Solution. Jom, 2019. **71**(12): p. 4518-4527.
63. Tan, S.-y., et al., Developments in electrochemical processes for recycling lead–acid batteries. Current Opinion in Electrochemistry, 2019. **16**: p. 83-89.
64. Encyclopedia, M. The development of the lead recycling industry. 2023; Available from: http://baike.asianmetal.cn/metal/pb/recycling.shtml.
65. Chen, L., et al., Lead exposure assessment from study near a lead-acid battery factory in China. Science of the total environment, 2012. **429**: p. 191-198.
66. Problems, W.s.W.P. Top Ten Polluting Industries 2016. 2023; Available from: http://www.worstpolluted.org.
67. Were, F.H., et al., Air and blood lead levels in lead acid battery recycling and manufacturing plants in Kenya. Journal of Occupational and Environmental Hygiene, 2012. **9**(5): p. 340-344.
68. Haefliger, P., et al., Mass lead intoxication from informal used lead-acid battery recycling in Dakar, Senegal. Environmental health perspectives, 2009. **117**(10): p. 1535-1540.
69. Poll, C.G., et al., Electrochemical recycling of lead from hybrid organic–inorganic perovskites using deep eutectic solvents. Green Chemistry, 2016. **18**(10): p. 2946-2955.
70. Ericson, B., et al., The global burden of lead toxicity attributable to informal used lead-acid battery sites. Annals of global health, 2016. **82**(5): p. 686-699.
71. Sun, M.X., et al., Consumer behavior and perspectives concerning spent household battery collection and recycling in China: a case study. Journal of Cleaner Production, 2015. **107**: p. 775-785.
72. Ahmed, M., et al., Impacts of waste in Africa. 2018, United Nations Environment Programme.

73. Mmereki, D., et al., The management of hazardous waste in developing countries. Management of hazardous wastes, 2016. **6**(8): p. 39-50.
74. Tian, X., et al., Management of used lead acid battery in China: Secondary lead industry progress, policies and problems. Resources, conservation and recycling, 2014. **93**: p. 75-84.
75. Alam, M. and K. Bahauddin, Electronic waste in Bangladesh: evaluating the situation, legislation and policy and way forward with strategy and approach. Present Environment and Sustainable Development, 2015(1): p. 81-102.
76. Perkins, D.N., et al., E-waste: a global hazard. Annals of global health, 2014. **80**(4): p. 286-295.

Journal of Physics: Conference Series **2738** (2024) 012020 doi:10.1088/1742-6596/2738/1/012020

Technology Selection for Slag Zinc Fuming Process

Elmira Moosavi-Khoonsari[1,2*], Sina Mostaghel[3], Andreas Siegmund[4]

1. Department of Mechanical Engineering, École de technologie supérieure (ÉTS), Montreal, QC H3C 1K3, Canada
2. Department of Materials Sceince and Engineering, University of Toronto, Toronto, ON M5S 1A1, Canada
3. SNC-Lavalin, 191 The West Mall, Toronto, ON M9C 5K1, Canada
4. LanMetCon LLC, 780 Lark Street, Lantana, TX 76266, USA

*elmira.moosavi@etsmtl.ca

Abstract. Circular economy and multi-metal extraction philosophies more and more encourage smelters to reprocess their by-products most significant of which is slag for recovery of valuable metals. Zinc, for instance, can be introduced to lead and copper smelting operations through different sources, including the recycling of waste electric and electronic equipment. During the conventional smelting processes from primary resources, or even those that are particularly developed for recycling purposes, e.g., Black Copper route, the zinc is typically deported to the slag phase as zinc oxide. Recovery of zinc from slag is typically carried out via a slag zinc fuming operation, where a reductant is used to reduce zinc oxide and volatilize zinc metal. In most cases, volatilized zinc is re-oxidized to produce zinc oxide, which can then be sent to hydrometallurgical unit processes for refining. Several technologies and reactors have been developed for efficient and cost-effective fuming processes, none of which can be considered the "best" and most suitable technology for all applications and smelter conditions/slag compositions. This paper reviews the advantages and disadvantages of each of the available technologies and recommends the most suitable process for common conditions.

1. Introduction

Zinc is the fourth most produced metal after Fe, Al, and Cu with a wide range of applications in metallic products, rubber, and medicine, and is primarily produced from the smelting of Zn sulfide concentrates. The main primary Zn smelting processes were the electrolytic process (i.e., roasting, leaching, and electrowinning), imperial smelting process or ISF (i.e., the blast furnace), and retort processes including vertical, electrothermic, and horizontal. The retorting processes have disappeared, and the main production route, today, is the electrolytic process. The ISF process is used for the separation and concentration of Zn values from complex, mainly higher lead-containing zinc sulfide concentrates, the production rate of Zn to Pb is around 2:1. The use of low-cost secondary feeds makes an ISF smelter, consisting of 3 steps of concentration, final reduction, and recovery of Zn metal, competent with the electrolytic process.

Secondary resources, e.g., industrial residues and end-of-life (EoL) products, are of increasing interest because of the depletion of primary sources, ever-increasing demands for materials and especially metals, the importance of sustainable and intelligent resource utilization, and the negative impact of waste on the environment. The main secondary feeds for Zn production are galvanizing ashes and dross and dust from scrap-based electric arc furnace (EAF) steelmaking. Galvanizing ashes are generated from the oxidation of Zn hot dip galvanizing bath, and contain 65-75% of Zn in the form of mainly oxide and some metal and Zn chloride. Galvanizing dross is a Zn-Fe alloy collected from the bath bottom, and contains 3.5% Fe and 96% Zn. The EAF dust consists of mainly oxides of Fe (25-35%) and Zn (10-35%), some volatile chlorides, and slag formers such as lime (5-20%). Other industrial residues such as diecasting scraps, Zn-rich slag from the Pb blast furnace, leach residues containing Zn ferrite, Sb slag, and slag from dirty Cu concentrates and secondary Cu recycling are also treated for recovery of Zn. Oxidized Zn ores such as silicates or carbonates, difficult to separate by flotation, can be treated by fuming too [1,2].

Zinc fuming processes are highly flexible for fuel use and the application of in-situ combustion, and can be used for recovery of Zn metal from ZnO found in low-grade ores or secondary complex materials of low or negative value. The low-value feeds can be treated to cover the operational costs of fuming and reduction. Zinc fuming is based on carbothermic reduction and selective volatilization of ZnO to Zn metal vapor (the overall reaction: $ZnO + C = Zn(v) + CO(g)$), reoxidation of vapor to fine ZnO particulate matters or fume ($Zn(v) + 0.5O_2 = ZnO$), and condensation of fume from the gas phase. Zinc oxide in the feed can be reduced to Zn(v) via the addition of C, or by the use of CO or natural gas (the reaction: $Zn + CO(g) = Zn(v) + CO_2(g)$). Direct condensation of Zn metal from vapor is favorable in high-grade materials processing, eliminating the need for subsequent reduction of ZnO to Zn metal. For the contaminated low-grade feeds and high levels of dross-forming compounds such as dust and halides, high-grade ZnO fume containing submicron-size particles will be produced followed by a reduction stage. The fume often contains Pb, Ag, and Cd and minor impurities such as As and Ge, and an additional refining stage is required [1]. The fume is either cooled and filtered or scrubbed from the gas phase for recovery of ZnO. The ZnO fume can be also used as feed to an ISF or an electrolytic Zn plant [3]. Oxidation of Zn(v) is an exothermic reaction and the generated heat can be used for the reduction of ZnO, which is an endothermic reaction, or recovered as steam.

2. Zinc fuming technologies

Fuming processes can be categorized into three major classes: (a) direct fuming from solid/gas reactions (Walez kiln and shaft furnace processes), (b) fuming from a molten slag (the conventional slag fuming furnace, the submerged lance slag fuming, O_2-enriched air side blown smelting, and submerged plasma fuming), and (c) fuming from particles suspended in a gas phase (Horsehead Flame Reactor and Contop Cyclone Reactor). The most important operational features and advantages and disadvantages of these technologies are summarized below.

2.1. Waelz kiln

The Waelz Kiln is the most often used process for Zn recovery from secondary materials and industrial residues. It is a long rotary kiln similar to cement or lime kiln, patented in Canada in 1909 but developed in Germany with the first real-scale production in 1925 [1].

The charge (ZnO-containing residues, a low volatile anthracitic coal or coke fines, and flux) is dried and preheated to the reaction temperature of 1150 °C in the first section of the kiln. Flux like lime or limestone is added to increase the melting temperature of the generated slag. The process can operate well at the $(CaO+MgO)/(SiO_2+FeO)$ ratios of 0.1-0.2 or >1.3, either the acidic or basic slags, respectively [4]. The feed moves slowly in a countercurrent flow against the heated air. In addition to $ZnO \rightarrow Zn(v)$ reduction, $FeO \rightarrow Fe(v)$ reduction also occurs, where Fe helps in reducing ZnO and other base metal oxides (e.g., $Fe + ZnO = Zn(v) + FeO$). Zinc vapor is directed toward the gas space of the freeboard where it is oxidized into ZnO fume by the excess oxygen in the gas stream above the tumbling bed. Other metals with high vapor pressures such as Pb, In (80%), Tl (80%), and Cd (97%) are also volatilized and reported to the fume. Iron and Cu form the $FeS\text{-}Cu_2S$ matte phase containing Ag, Sb, and As. Germanium partly (33%) reports to the fume and partly stays in the residual slag [1].

2.1.1. Advantages and disadvantages. Waelz process requires a simple installation, and a wide range of feed materials can be treated via process modification (e.g., temperature, residence time, reductant quantity). However, the formation of a molten phase should be minimized for the charge could maintain an open structure, a dry and tumbling condition, and good kinetics of solid-gas reactions. The formation of eutectic compositions at 1180 °C, especially in the presence of Fe, can lead to accretion on the kiln walls avoiding gas and solid proper movements which requires shutdown and accretion removal. The feed Pb content higher than 5% should be also avoided. Lead-containing phases accumulate in the bottom of the kiln and attack the refractory. Particularly in the presence of Cu, they form low melting temperature matte phase eutectics. Excess coke (as large as 40% of the charge in comparison to the normal addition of 25%) can be used as a conditioning agent to absorb molten phases, decreasing the strengths of accretions, or reducing the accretions and enabling their removal from the furnace wall.

The particle size distribution of the feed must be also carefully evaluated to avoid large amounts of feed carryover to the off-gas system and excessive dusting. This consequently lowers ZnO concentration in fume which at the same time may negatively affect the oxidation reactions of the fumes and requires a large volume of the furnace. It is needed to loosely agglomerate the feed in a drum conditioner [1, 5]. It was reported that pellets allow fast reduction and high metallization rates due to high and uniform porous structure [5]. The particle size distribution of coke is also very important (2-6 mm with at least 60% above 2 mm). Too coarse particles remain unreactive while too fine coke will burn excessively and increase the extent of melting.

The oxidation reaction of Zn(v)→ZnO is very exothermic [5], and heat energy is partly recovered to the bed for compensation for heat consumed by the endothermic reaction of ZnO→Zn(v) reduction. The overall thermal efficiency of the process is relatively low per tonne of Zn recovered due to various heat losses [1]. The energy consumption is relatively low but the burner needs to run. Modern furnaces require <200 kg of coke per 1 ton of EAF dust [5].

The refractory lifetime was reported to be an issue, which is about 12 months. About 67% replacement is necessary requiring a shut-down time of about 4 weeks [1]. For Waelz kilns designed for the treatment of Zn-residues, it has been reported that the refractory lining in the feed end area (subject to the abrasive effect of the tumbling material) of the kiln may last between 2 and 2½ years, while other types of refractory material used in the most aggressive zones (e.g., main reduction zone) within the kiln may need replacement between 2½ and 3 months [6]. The most important causes behind the wear and failure of the refractory lining are the deformation of the kiln shell, abrasive properties of the load, chemical aggressiveness of the load, and thermal shocks due to sudden changes in process operating conditions. Failure of the refractory lining may also be caused by the penetration of metallic fumes within the bricks [7]. However, it is not completely clear to what extent is this problem avoidable via the substitution of bricks for a castable refractory.

Waelz slags often conform to leaching requirements from the environmental point of view. Calcium oxide reduces the leachability of heavy metals in slag, reducing the environmental impact.

Overheating could be identified as a potential problem although it may be controlled by allowing excess air ingress [8].

2.2. Conventional slag fuming furnace

The original test work behind the conventional slag fuming furnace technology was developed by the Sulphide Corporation at Cockle Creek (Australia) between 1906 and 1920 [9]. The furnace was developed for Zn recovery from Pb blast furnace slags containing 10-18 wt% Zn and other Zn-rich secondary materials. The first large-scale conventional slag fuming furnace was installed by the Anaconda smelter in Montana (USA) in 1925 [1, 9, 10]. The furnace has been used since then at major lead smelters. Other operations that use this furnace are Boliden's Ronnskar smelter (Sweden), the Filn Flon smelter of Hudson Bay Mining and Smelting Company (Canada), Teck's Pb and Zn metallurgical complex in Trail (Canada), Nyrstar's Port Pirie smelter (Australia), and the Non-ferrous Metals Smelter at Plovdiv (Bulgaria). Since 2018, a fuming furnace has been installed in Guangxi raw antimony-rich Polytron Technologies Inc. (China) to smelt and fume the slag from the Sb blast furnace [2].

The Zn recovery is maximum at the residual Zn level in slag between 2.5 and 3 wt%, below this level, there will be a large drop in furnace capacity. In the case of charging hot slag to the furnace, e.g., from an electric furnace connected to Kivcet Pb processing, the Zn residue in slag can be as low as 2 wt% [11]. Common to the majority of fuming technologies, the furnace should not exceed a limit of Zn recovery (often recommended at 1 wt% Zn in the final slag) to prevent the generation of metallic Fe. The heat extracted from cooling the furnace (40 GWh/year) was recovered at the Boliden plant and used for central heating in the neighbourhood [12].

2.2.1. Advantages and disadvantages. This process is only economic for high concentrations of Zn in slag (a minimum of 16% of Zn in feed is needed). Agglomeration or screening of charged slag should be done to have the proper feed size distribution and prevent it from going to the off-gas system. Too coarse particle size distribution reduces the production capacity due to a slow melting rate [2]. Low ash content coals are preferred due to the dilution effect of ash on the slag Zn content, reducing Zn recovery [1]. If pulverized coal is selected as the source of heat and reduction, then considerable operating and safety issues must be taken into account. Industrial facilities often pulverize coal for their own consumption, and although different pulverizing technologies may be available, all of them require precise controls of free oxygen in the milling and off-gas systems to avoid hazardous conditions that could result in severe explosions. Furthermore, coal pulverization plants are designed with standard inert-gas (usually nitrogen) blanketing systems for additional explosion risk mitigation, which may significantly add to the overall operating cost. Despite a proper design, the risk of explosions is not completely mitigated in this type of facility.

The Zn recovery rates of 85%-90% and 88%-95% were reported at Teck's Trail operation (Canada), and in Guangxi (China), respectively [1, 2].

The slag freezing line on the furnace's inner surface provides extra protection against the aggressiveness, turbulence, and abrasive nature of the slag bath, increasing the lifetime of water-cooled jackets over 10 years and even up to 20 years [1, 12]. However, additional fuel and costs are required for the associated heat loss for forming the slag freeze line.

The best capacity has been reported for batch operation. Operation in continuous mode is not common and often lower capacities were experienced however this can be the subject of further study case by case from one smelter to another. Overall plant availability has been reported at 88% [1].

Conventional fuming furnaces are generally regarded as having relatively high maintenance requirements for tapping blocks, cooling jackets, and waste heat boilers. Tapping blocks are subject to high-temperature stress and the corrosive nature of the slag, which is plant-specific. The failure of cooling jackets has been reported as a common problem in conventional slag fuming furnaces which has been attributed in part to sudden temperature changes associated with the loss of the protective slag layer [13]. Regular inspection and maintenance of water jackets are mandatory to guarantee the furnace's integrity. Continuous cleaning of the boiler tubes and walls is needed because of sticky accretions on the ducts which reduce the boiler's heat extraction efficiency. Soot is often extensively blown away using saturated steam [1].

2.3. Submerged lance slag fuming

The submerged lance slag fuming technology can be used for treating the leach residues from an electrolytic Zn plant, low-grade complex ores or concentrates, and any difficult-to-process feeds [14]. The process principles were developed by Commonwealth Scientific and Industrial Research Organization (Australia). The process was originally applied to Sn fuming in 1978. Later on, it was further developed to Zn fuming by Ausmelt (Whyalla, Australia); the first commercialization was performed by Onsan Smelter of Korea Zinc (South Korea). Mitsui at the Hachinohe ISF smelter (Japan) and Young Poong Corp. (Sukpo, South Korea) also use this technology for Zn fuming. The process was later applied to Pb and Cu smelting by Mount Isa Mines (i.e., Isasmelt process) [1].

Solid feed (Zn-containing materials, coal, and fluxes) is added via a feed port in the roof of the furnace. Fuel (gas, oil, or pulverized coal) is injected with O_2-enriched air through a vertical submerged lance into a molten slag bath in a cylindrical vessel. Bath is highly turbulent with high reaction rates and mass transfer rates. The chemical reactions are similar to the conventional slag fuming furnace and fuming rates are in a similar order. The final slag contains < 3% Zn and <1% Pb [1].

A slag freeze line protects the refractory outer layer of the tube, forming under the cooling effect of air injection [1].

The gas handling and fume collection systems are very similar to the conventional slag fuming furnace. The cooling system consists of a waste heat boiler, a water tube section, and humidification using water sprays. For small plants, only cooling by humidification can be used due to the high cost of a waste heat boiler. Using electrostatic precipitators allow for fume collection at temperatures as high as 400-500 °C, higher than the bag filter, minimizing the required cooling degree [1]. The heat from the secondary post-combustion can be recovered to generate electricity [15, 16].

2.3.1. Advantages and disadvantages. It is a flexible and rather simple technology and can be easily integrated into existing flowsheets. It can also incorporate both solid and liquid feeds and can be applied to a wide range of feed materials from primary resources to secondary Zn leach residues, Zn slags, EAF dust, and complex EoL products. The lowest-cost available fuel sources can be used and difficult-to-source and high-cost cokes are not required. High feed flexibility leads to optimum productivity and economic performance [15, 16]. It was reported that the partial pressure of O_2 can be precisely controlled to achieve the desired bath chemistry and products [15, 17]. Volatile species can be recovered in the fume, valuable non-volatiles in metal, and low-value non-volatiles in slag [1].

Early recoveries at Korea Zinc QSL's slag fumers in series, which at the time were treating granulated slag combined with other Zn residues, were reported at 61% and 55% for the first and second fuming stages, respectively, which corresponds to a global recovery of 83% [18]. Korea Zinc also reported a zinc recovery of 49% for the first stage fuming furnace, which was increased to 64% by using 100% powdered coal [19]. In general, a typical zinc recovery range is given from 85% to 90% for the Korea Zinc fuming plant [20]. A more recent test developed at Onsan with EAF dust resulted in a zinc recovery of 90% [21].

The refractory lifetime is the primary concern in the application of this process and often relining is required. For the submerged lance vessel, the high turbulent bath reduces the life of refractory to below 3 months in some cases, leading to plant availability of less than 80% [1, 18]. The refractory lining of the furnace wall is not possible and water-cooling jackets are mandatory. Using water jackets alone limits the operating temperature, which can barely exceed the slag liquidus temperature. This limitation can be overcome if the refractory lining is used and the temperature is not above the bricks' degradation temperature. Using chrome magnesite fused grain re-bonded bricks, the resistance against iron silica slags, and the lifetime has increased to over 2 years. If the refractory lining is backed up with water-cooling jackets, the lifetime would further increase. Using water jackets, the formation of metallic Fe in fume should be avoided because Fe fuming leads to accretion formation and furnace blockage. Consequently, the residual slag's Zn content will be more than 2%. For refractory lined vessels, Zn content as low as 1% and below can be obtained in the final slag but even with having a refractory lining, Fe fuming should be avoided over a long period of time. Due to the exothermic reaction between Zn and CO(g) in the upper part of the reactor (secondary combustion), the gas reaches a high temperature affecting negatively the refractory lining. Therefore, a water-cooled refractory lining with a slag freeze line has been proposed as a solution, also assisting in cooling the gas and separation of fume [17, 19]. Hachinohe claimed to have achieved a 2-year lifetime for one single campaign [22].

The lance consumption rate is high, and lance design to reach stable operation and long lifetime plays a key role in the process [18, 23].

It was reported that the capital and operational costs are low. Low capital cost is due to simple furnace construction and peripheral system arrangement. Operational cost is low due to low energy consumption and low manpower needs [1, 15, 18]. Having a very turbulent bath, the slag can splash into the gas outlet causing accretion formation and blockages [1, 18]. Therefore, the reactor needs to be tall (height/diameter ~ 3.0). The lance needs to be completely withdrawn from the bath, thus, a high plant structure is needed. Hughes et al. [15] reported that in comparison to conventional slag fuming furnaces, the furnace gas volume can be significantly reduced due to high O_2 enrichment and having a more gas-tight system limiting the air ingress. They did not explain why submerged lance slag fuming is tighter than any other stationary slag fuming furnace.

The process can be operated in batch or continuous mode depending on the plant preference and furnace availability [15, 16]. If multiple vessels are used, the process can be operated in a continuous mode where the first reactor is used for sulfur removal (if the feed contains sulfur) and Zn and Pb fuming, followed by the second step where Zn fuming is completed and low Zn in residual slag is obtained. It is favorable to operate in a continuous mode rather than batch mode. The major plant cost can be reduced because the need for charge slag holding facilities and larger gas handling volume capacity is decreased. In batch mode operation, the fuming rate for the submerged lance slag bath furnace is similar to the average rate in the conventional fuming furnace in batch mode. This leads to a decrease in Zn content of 6 %/h between 17 and 3 % Zn content. For slag containing 2% Zn, the continuous fuming rate leads to 4 %/h decline in Zn content [1].

The environmental and workplace impact is minimal. Fugitive emissions are low because of well-sealed and stationary reactors. The slag is benign and can be used in the construction industry [15, 16]. However, this can be different depending on each country's policy (*private communication*).

An additional design problem, that affected the operation at Hachinohe, was the generation of sulfuric acid (from SO_2 and H_2O) in the off-gas system which resulted in the corrosion of the equipment parts. An adequate off-gas temperature must be maintained in order to avoid the formation of acid [22].

2.4. *Oxygen-enriched air side blown technology*

The O_2-enriched air side blown smelting was developed based on the Vanyukov process by the former Soviet Union, and has been used for Cu and Ni smelting since 1976. The re-innovation and improvement were performed by CINF Engineering Co. (Changsha, China), and in 2002, it was used for the reduction of hot high Pb slag. The commissioning of the furnace for treating the Pb-bearing polymetallic materials was performed in 2014 [24, 25].

The hearth and body of the furnace are made of refractory materials and Cu and steel water cooling jackets, respectively. Several tuyeres embedded in the copper water jacket provide O_2-enriched air into the slag bath. The feed enters the furnace from the upper port and reacts with the O_2-enriched air.

The process is performed continuously in three O_2-rich side-blown furnaces linked by chutes: oxidation-melting-desulfurization (to produce Pb-rich slag), reduction melting (to produce crude Pb and reduction slag), and slag fuming (to produce ZnO). Fuming is carried out in the last furnace of the three furnaces (i.e., only one furnace is required for fuming). During slag fuming, pulverized coal is added through the nozzle to reduce ZnO into Zn(v), which is further oxidized in the tertiary tuyere of the fuming furnace to form ZnO fume. The heat energy of the high-temperature fume is recovered by the waste heat boiler followed by the fume desulfurization and dust collection. The off-gas is emitted after desulfurization. Figure 1 depicts a schematic of an O_2-rich side-blown furnace [24, 25].

Lead slag with 15-17% Zn and 1.5-2.5% Pb was treated together with pulverized coal in the fuming smelting furnace. The residue slag contained <1% and <0.1% Zn and Pb, respectively, and was water-quenched after tapping (to be dumped or sold). The Zn recovery rate was reported to be more than 92%.

The furnace's lifetime was reported to be up to 5 years. The main equipment, process parameters, and main techno-economic indicators of the process are provided by Zhang et al., 2020 [25].

2.5. *High temperature submerged plasma fuming*

High temperature submerged plasma fuming is an intensified version of conventional slag fuming processes. The feed contains Zn-containing residues (e.g., Pb blast furnace slag, secondary Cu smelting slag, EAF dust), petcoke, and fluxes. For enhancing the fuming rate, a bath temperature as high as 1350 °C is required, and the thermal energy is provided via thermal plasma and chemical reactions. Inside electric plasma torches, the cold blast air is transformed into a high enthalpy plasma gas which together with fuel is injected into the slag bath to provide heat and slag agitation for mixing. Consequently, the

ZnO reduction and volatilization occur followed by the afterburning of Zn(v) to ZnO fume, and part of the heat is recycled back to slag increasing the energy efficiency. Slag phase, matte and ZnO fume are the products. The off-gas is treated in an electrostatic precipitator. Figure 2 illustrates a schematic of the submerged plasma fuming process [26, 27].

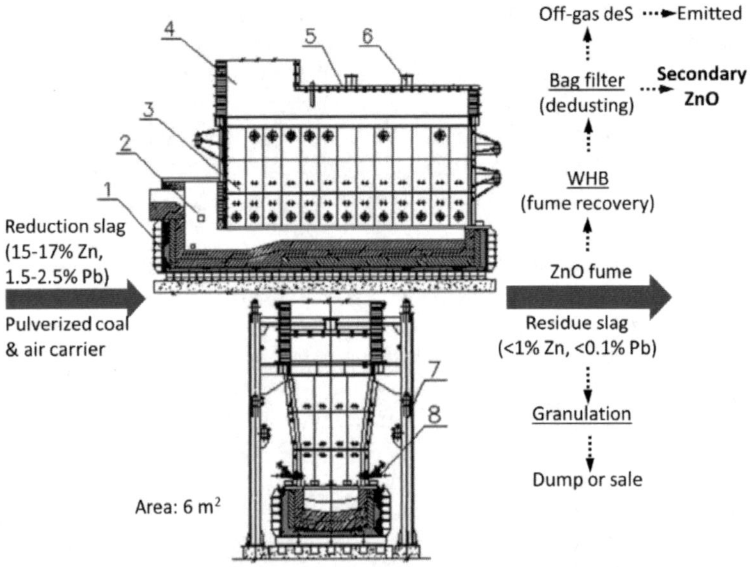

Figure 1. Oxygen-enriched air side blown furnace (adapted from [24, 25]). 1: hearth, 2: siphon chamber, 3: furnace body, 4: uptake, 5: furnace top, 6: charging opening, 7: steel frame, and 8: tuyere.

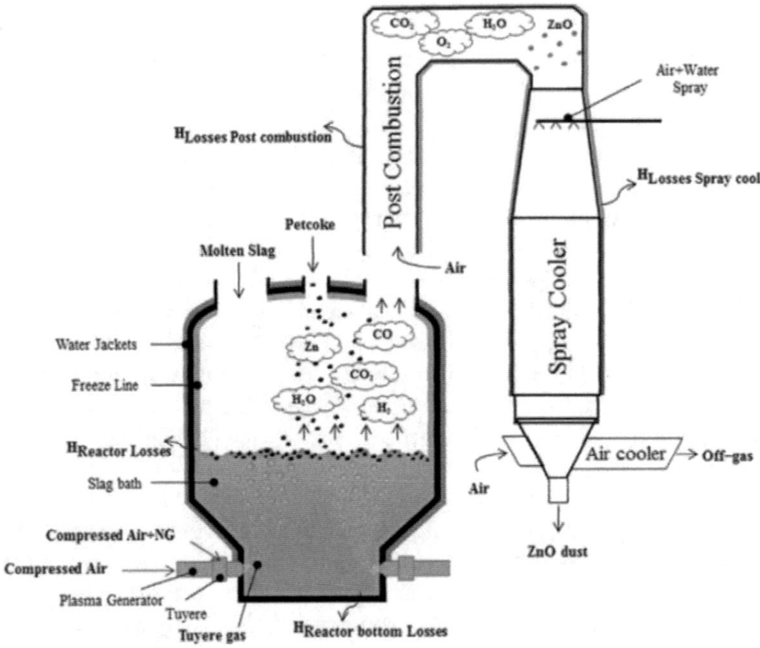

Figure 2. Schematic of high temperature submerged plasma fuming process (from [26]).

High temperature does not favor the furnace lining which could be overcome by forming water-cooled freeze linings on the smelter reactors, the stability of which can be controlled via the bath composition. Slag liquidus temperature should be controlled carefully to form a stable freeze lining to decrease the heat flux via the reactor walls and increase energy efficiency [26, 27].

Plasma-driven fuming processes are emerging as environmentally friendly processes in comparison to conventional top-blown or side-blown slag fuming furnaces. The process is especially convenient in conditions where non-hydrocarbon source electricity such as hydrogen electricity is accessible, and petcoke replaces combustion gases, reducing the net C footprint of the process [26].

The first industrial plant was built for ERAS by ScanArc [28] in Hoyanger (Norway) in 2005. Metallo Belgium also operates this furnace for recycling ZnO-containing residues [26].

2.6. *High intensity fuming processes*

In high intensity fuming processes, Zn-containing charges are introduced into a high-temperature reducing flame. Zinc oxide is reduced to Zn(v) reporting to the gas phase and slag suspended in the gas phase is separated in a settler or a cyclone. Zinc vapor is oxidized into ZnO fume via injection of secondary air, and the fume is filtered. The principles of carbothermic reduction and oxidation-fuming are applied to these processes like the others. The difference is that the reduction is carried out at higher temperatures, having high reaction rates, and a small equipment volume is required. High O_2 enrichment is used to reduce the gas volume and size of the gas handling system.

2.6.1. *Horsehead flame reactor*

The Horsehead flame reactor was developed by Horsehead Resource Development Company Inc. (USA) for processing dust containing 10-35% Zn from scrap-based EAF. The goal was to develop technology at a small scale that can be easily installed at steel plant premises. The slag can be recycled back to the steel process and ZnO fume sold to the Zn smelter. The reductant was initially coal, and later, natural gas was also introduced leading to higher combustion rates and higher temperatures enabling the generation of molten slags without the need for fluxing. There is a possibility of sending gas from the slag separation vessel to a Zn condenser directly without forming ZnO fume. However, conditions for Zn condensation are difficult. The operational costs are largely higher than a conventional slag fuming furnace (e.g., higher energy consumption, 42 GJ of natural gas, and a need for O_2 addition, 2.5 tonne per tonne of Zn recovered) but the capital costs are lower because the plant is smaller (it was not clear with respect to footprint, equipment, or both) [1].

2.6.2. *Cyclone flame reactor*

The cyclone flame reactor was developed in Russia in the 1950s for flash smelting of Cu concentrates to disperse fine concentrates into suspension at the top of a smelting shaft. With the transition from air to O_2, the reactions could occur in the cyclone and there was no need for the shaft. It can treat Zn concentrates, Zn-containing residues, and secondary materials [29]. A pilot furnace was applied to fume Pb blast furnace slag at the Port Pirie smelter (Australia) [30]. It has been commercially applied as a retrofit to Cu reverberatory furnaces. The vessel walls reach very high temperatures and there is a need for an evaporative cooling system and steam rather than circulatory water cooling as used in standard water jackets. The generality of the process is similar to the flame reactor and other fuming processes but with a different initial combustion device. The use of O_2 for combustion reduces the size of the gas handling system and the large capital cost of the plant [1].

2.6.3. *Shaft furnace*

A shaft furnace or blast furnace has the same principles as the ISF furnace except that air or O_2 is injected near the top of the furnace to complete combustion and form ZnO fumes. It is practical for lower-grade residues and especially Zn ferrites. The charge contains the residues mixed with fluxes and coal or coke fines pelletized or briquetted with good strength and size to maintain the open structure in the shaft and good gas permeability similar to Fe blast furnace. Air is blown in the bath through tuyeres like ISF but some amount of fuel like oil or natural gas is also injected as a coke-saving measure. Secondary air is injected at the furnace top and the generated heat of the secondary combustion is transferred to the descending charge increasing the fuel efficiency. However, ZnO sticks to the charge and is not fully recovered, leading to Zn circulation in the furnace needing extra C and leading to low fuel efficiency [1].

This is not a very favorable process. The ZnO accretion in the shaft walls requires regular shutdown for cleaning. Therefore, the shaft furnaces are small in size in a series and on a campaign basis. A variation of the process was developed by the Mitsui Mining and Smelting Company at the Miike Smelter (Japan) for treating Zn leach residues but then changed to processing EAF dust. Westinghouse has a similar process at Hamilton, Ontario (Canada) but uses a plasma torch in the base of the shaft to supply heat, similar to the SKF plasma dust processing, producing ZnO rather than condensing Zn metal [31].

3. Technology selection trade-off

The conventional and rather emerging processes for Zn fuming from secondary resources are listed in Table 1 along with important technological features. Among the above-listed technologies, Cyclone flame and Horsehead flame reactors are not included in the table because they were not commercialized. As was explained in Section 2, several technologies and reactors have been developed for Zn fuming process, each one of which shows certain advantages and disadvantages, which are compared below. It should be indicated that the volume of information available for some of the processes such as plasma fuming slag and O_2-enriched air side blown reactor is rather little making a comparison with the conventional slag fuming processes somehow challenging.

Waelz kiln and shaft furnace are not flexible in terms of feed composition, size, and morphology. On the contrary, the conventional slag fuming furnace, submerged lance slag fuming, and O_2-enriched air side blown can process a wide variety of charged materials.

In terms of the Zn recovery rate, almost all the processes show a Zn recovery rate between 85 and 95%. The reported Zn recovery for the O_2-enriched air side blown technology is the highest among all the processes. The Zn recovery rate of plasma slag fuming was not explicitly reported in the literature.

Regarding maintenance and repair requirements, the conventional slag fuming furnace has the highest furnace campaign life of 10-20 years. The O_2-enriched air-side blown fuming comes to second place with a lining lifetime of around 5 years. Submerged lance slag fuming and Walez kiln have on average 2 years of lining lifetime. So far, based on the available information, it seems that the submerged lance slag fuming has the lowest plant availability in comparison to Waelz kiln, conventional slag fuming furnace, and O_2-enriched air side blown reactor. The lining lifetime of the plasma fuming process controlled by the slag freeze lining was not reported however failure of water-cooling jackets could be a concern because of the loss of slag freeze lining and rapid variations in temperature.

It was reported that the capital cost of recent technologies (top submerged lance and side blown) is low in comparison to more conventional slag fuming processes. However, this can be subject to further investigation. They also have a better energy efficiency which together with their eco-friendly nature play an important role these days in technology selection because they show lower C and environmental footprint.

Considerable numbers of Waelz kiln and conventional slag fuming furnaces have been installed worldwide and are currently in operation (the exact number is not known). For EAF dust, for example, about 80% of the global capacity is treated in Waelz kilns. Also, China has numerous conventional slag fumers. However, it is clear that the top submerged lance furnace with 4 precedents has made a good place among them.

Table 1. Comparison of commercialized technologies for zinc fuming from secondary resources

	Waelz kiln	Conventional slag fuming furnace	Submerged lance slag fuming	O_2-enriched air side blown fuming	Shaft furnace	Submerged plasma fuming
Feed requirements	Pb<5%, Fe control, dry, Coke (2-6 mm, >60% > 2mm), agglomerate	Solid or liquid, high Zn in slag (≥16%), low ash coal, size screening needed for solids, agglomeration	Solid, liquid, high- or low-grade, EoL products, lowest-cost available fuel (gas, oil, pulverized coal)	No need for drying, powder, lump, pulverized coal	Low-grade; coal or coke fine (+ oil or natural gas), pellets, briquettes	Zn residues, petcoke, pellets
Zn recovery	Up to 92%, 88-95%, low for low Zn-slag	85-90%, 88-95%	85-90%, 90%	>92%	80-85%	Up to 98%
Maintenance & repair	Refractory-life time: 12 months, feed end area: 2-2 1/2 yrs, most aggressive zone: 2 ½ - 3 months	Water-cooled jackets + slag freeze line: 10-20 yrs	Low lance lifetime, Refractory lining < 3 months, With chrome magnesite refractory~ 2yrs, Refractory+cooling jackets>2 yrs, refractory+water cooling jackets+slag freeze line~2yrs	Refractory+ Cu & steel water jackets: lifetime up to 5 years	Regular shutdown due to ZnO accretion on walls	Water cooling + slag freeze line (crucial and to be controlled via slag chemistry)
Availability	92%	88%	<80%			
Capital cost			To be investigated	Low		
Operational cost		High	To be investigated			
Energy consumption	?	High	Low	Low		
Energy efficiency	?	Low			Low	High
Batch mode		the best capacity	x			
Continuous mode	x	Low capacity (not common)	x (preferable)	x		
Environment impact	Low		Low	Low		Low
Precedent	HorseHead Corp., Nexa Resources; Global Steel Dust Gulf; Befesa Zinc Aser SA; Befesa Zinc Duisburg GmbH; Befesa Zinc Freiberg GmbH; Befesa Zinc Korea, Ltd.; Befesa Silvermet SL; Recytech SA; Taiwan Steel Union Co., and many others globally	Anaconda, Boliden's Ronnskar, Flin Flon, Teck, Nyrstar, Plovdiv, Polytron Technol. Inc.	Ausmelt, Korea Zinc, Mitsui, Young Poong Corp.	Chenzou, China	Mitsui Min. & Smel., Westinghouse	ERAS, Umicore, Metallo

4. Conclusions

There are six zinc fuming technologies available on an industrial scale for the recovery of zinc from secondary resources, Walez kiln, conventional slag fuming furnace, top submerged lance slag bath, oxygen-enriched air side blown reactor, shaft furnace, and high-temperature plasma slag fuming. These technologies can treat complex zinc secondary resources such as lead blast furnace slag, secondary copper slag, electric arc furnace dust, and zinc leach residues. In the current paper, we reviewed the advantages and disadvantages of each of these technologies and provided an alternative solution to the technological challenges proposed in the literature. This study can help in the selection of an appropriate zinc fuming process based on smelter conditions and feed availability. For common conditions, considering feed flexibility, capital cost, energy efficiency, environmental and workplace impact, and the number of previous installments, the submerged lance slag fuming can be considered a promising technology for zinc recovery from complex to process secondary residues and EoL products. The main inconvenience of this process is the high maintenance and repair and plant availability. The oxygen-enriched air-side blown fuming reactor possesses all of the advantages of top submerged lance plus a higher Zn recovery rate, as well as low maintenance and repair. However, the extent of available detailed information on this process and other rather new processes such as plasma fuming is comparatively scarce, and further information could help in better evaluating the technological pros and cons of the emerging processes.

Reference Lists

[1] Sinclair R J (2005) *The Extractive Metallurgy of Zinc* (Victoria: The Australasian Institute of Mining and Metallurgy)

[2] Xin P, Wei J, Xu L, Wu W, Lan D (2020) The recovery of Pb and Zn in antimony smelting slag *Proc. 9th Int. Symp. on Lead and Zinc Proc.* (February 23-27, 2020, San Diego)

[3] Assis G (1998) Emerging pyrometallurgical processes for zinc and lead recovery from zinc bearing waste materials *Proc. Zinc Lead Proc.* (August 16-19, Calgary)

[4] Mattich C, Hasselwander K, Lemmert H and Bezzar A H (1998) Electrolytic zinc manufacture and Waelz treatment of neutral leach residues *Proc. Zinc Lead Proc* (August 16-19, Calgary)

[5] Morcali M H, Yucel O, Aydin A, and Derin B (2012) Carbothermic reduction of electric arc furnace dust and calcination of Waelz oxide by semi-pilot scale rotary furnace *J Min Metall Sec B-Metall* 48(2) B 173-184

[6] Kazanbaev L A, Kozlov P A, Kolesnikov A V, and Ivakin D A (2004) Wear of basic refractories in the lining of Waelz kiln in the production of zinc (refractories in heat units) Refract. Ind. Ceram. 45(4) 225-227

[7] Clay J E and Schoonraad G P (1976) Treatment of zinc silicates by the Waelz process *J South Af Inst Min Metall* 11-14

[8] Kusuda Y and Imai T (1993) Treatment of zinc leaching residue by Waelz kiln *1st Int Conf Proc Mater Prop* (November 7-10, Hawaii)

[9] Richards G G, Brimacombe J K and Toop G W (1985) Kinetics of the zinc slag-fuming process: Part I. Industrial measurements *Metall Trans* 16B 513-527

[10] Bell R C, Turner G H and Peters E (1955) Fuming of zinc from lead blast furnace slag *JOM* 7 472-477

[11] Ashman D W, Goosen D W, Reynolds D G, and Webb D J (2000) Cominco's new lead smelter at Trail operations *4th Int. Symp. on Lead Zinc* (October 22-25, Pittsburgh) 171-185

[12] Hansson R, Holmgren H, and Lehner T (2010) Recovery of recycled zinc by slag fuming at the Ronnskar Smelter Lead-Zinc 2010 (October 3-6, Vancouver) 1235-1244

[13] Scholey K E, Samarasekera I V, and Richards G G (1994) Thermal stress failure of water-cooled zinc fuming furnace jackets *Can Metall Q* 33(4) 327-343

[14] Buckett G A and Sinclair R J (1998) The BUKA zinc process – a 21st century standard Proc. *Int Symp. Zinc Lead Proc* (August 16-19, Calgary) 579-596

[15] Hughes S, Reuter M A, Baxter R, and Kaye A (2008) Ausmelt Technology for lead and Zinc Processing *Proc Lead Zinc* (February 25-29, South Africa) 147-162

[16] Creedy S, Glinin A, Matusewicz R, Hughes S, and Reuter M (2013) Outotec Ausmelt technology for treating zinc residues *World Metall-ERZMETAL* 66(4) 230-235

[17] Mounsey E N and and Piret N L (2000) A review of Ausmelt technology for lead smelting *Proc TMS Fall Extr Proc Conf* (March 12-16, Nashville) 149-169

[18] Choi C Y and Lee Y H (1999) Treatment of zinc residues by Ausmelt technology at Onsan zinc refinery *Global Symp Recycl Waste Treat Clean Technol* (September 5-9, San Sebastian, Spain) 1613-1622

[19] Kim M-B, Lee W-S, and Lee Y-H (2000) The QSL lead slag fuming process using an Ausmelt furnace *Lead-Zinc 2000* (October 22-25, Pittsburgh)

[20] Choi C-Y (2003) The role of lead smelting at Korea zinc *Yazawa Int Symp Metall Mater Proc: Princ Technol Vol II: High-Temp Metal Prod* (March 2-6, San Diego)

[21] Lee Y H, Moon N I, Kim B M, Choi C Y (2006) Recycling of EAF dust by using top submerged lance technology *Sohn Int Symp, Adv Proc Metals Mater Vol 9 – Legal Manage Environ Issues* (2006)

[22] Sekiguchi T and Azuma S (1998) Slag fuming at the Hachinohe smelter *Zinc Lead Proc* (October 22-25, Pittsburgh)

[23] Floyd J M and Swayn G P (1998) An update of Ausmelt technology for zinc and lead processing *Proc. Zinc Lead Process* (August 16-19, Calgary) 861-874

[24] Wu X (2020) Application of CSC technology in nonferrous metallurgy *9th Int Symp Lead Zinc Process* (February 23-27, San Diego) 201-217

[25] Zhang L, Zhang L, and He Y (2020) The process and application of oxygen-enriched air side blown smelting of lead–zinc materials *9th Int Symp Lead Zinc Process* (February 23-27, San Diego) 291-299

[26] Nagraj S, Chintinne M, Guo M, and Blanpain B (2020) A dynamic model of a submerged plasma slag fuming process *9th Int Symp Lead Zinc Process* (February 23-27, San Diego) 291-299

[27] Verscheure K, Van Camp M, Blanpain B, Wollants P, Hayes P C, and Jak E (2005) *John Floyd Int Symp on Dev Met Process* (July 3-6, Melbourne)

[28] ERAS, www.erasmetal.com

[29] Henrich G et al (1985) *German Patent No 3514717 A1 24*

[30] Blanks R F and Ward D H (1968) Development of a cyclone furnace for slag fuming *Adv Extr Metall* (April 17-20, London) 234-244

[31] Chalfant R V (1996) Recovering zinc and iron from electric arc furnace dusts *New Steel,* September

Copper-cobalt cake as a potential source of germanium

Michał Drzazga*, Mateusz Ciszewski, Sylwia Kozłowicz, Izabela Maj, Sonia Kasierot, Szymon Orda, Adrian Radoń

Łukasiewicz Research Network – Institute of Non-Ferrous Metals, ul. Sowinskiego 5, 44-100 Gliwice, Poland

michal.drzazga@imn.lukasiewicz.gov.pl

Abstract. Germanium is an important element used in crucial industry sectors like optical fibers for telecommunication or IR optics for night vision systems. However, its global output is limited. It is estimated that its annual global production is ca. 140 tons. Therefore, many world economies, including USA and EU, consider it as a critical raw material. One of the potential sources of germanium is zinc metallurgy. It is assumed that only 3% of germanium present in processed zinc ores is recovered. It was found that during the technological processes of Polish zinc smelters some by-products containing elevated germanium concentrations are produced. In the paper, potential germanium sources, including copper-cobalt cake obtained during the purification of zinc electrolyte and dross from the feeding furnace, are presented. The main components of the cake are cadmium, zinc, copper, nickel, cobalt, and lead, while the dross contains mainly zinc (>65%). Results of leaching tests for both materials using aqueous sulfuric acid solutions are shown. In the case of the cake, wet and dried material was investigated. It was found that the germanium leaching yield for the dried material reached 99%, while for the wet one was 46-86%, depending on leaching conditions (without or with oxidant). On the other hand, the germanium leaching yield for the dross reached 55%. Further processing of the solutions obtained after germanium leaching from the copper-cobalt cake was also analyzed. Two ways were proposed, including precipitation with tannic acid and solvent extraction with addition of complexing agents using a trioctylamine extractant.

Germanium is considered by many world economies, including the EU, USA, Australia, and Japan, as a critical raw material [1]. It is associated with its limited output and application in important industry sectors. It is estimated that its annual global production is ca. 140 tons. Moreover, germanium production is not uniformly distributed worldwide – ca. 2/3 of global output comes from China. Other important germanium producers are Russia, USA, Belgium, Canada, Germany, Japan, and Ukraine. [2]. Germanium has several important applications. Almost 1/3 of its supply is used to produce optical fibers, which are the core of fast telecommunication systems. Other critical applications include IR optics (night vision system lenses, contactless thermometers), PET polymerization catalysts, electronics, and solar cells [3].

The typical germanium recovery process may be divided into three steps. Firstly, by-products containing germanium are generated during technological processes associated with zinc metallurgy or coal combustion. They include fly ash from the combustion of Ge-rich lignite, dusts, cakes, or drosses from zinc metallurgy. Some examples of Ge-based sources are presented in Table 1. Secondly, germanium concentrate containing at least 5% Ge is obtained by applying pyro- and/or hydrometallurgical methods. Finally, concentrates are chlorinated and distilled to obtain pure germanium tetrachloride, which may be later processed by water hydrolysis to germanium dioxide and then by reduction to metallic germanium [3-4].

Table 1. Composition of selected germanium sources.

Material	Origin	*Ge content*	Ref.
Coal fly ash	Yunnan, China	4.658%	[5]
	Inner Mongolia, China	1.109-3.517%	[5]
	Primorsky Krai, Russia	0.123-2.908%	[5]
	Spain	194-420 ppm	[6-7]
Zinc refinery residue	Guangdong, China	0.36%	[8]
Roasted residue from neutral leaching	Yunnan, China	0.2205%	[9]
Stockpiled copper cake	Turkey	700 ppm	[10]
Lead blast furnace slag	Namibia	0.194-1.111%	[11]
Flue dust from flash copper smelting	Chile	417 ppm	[12]
Waste optical fiber	-	860 ppm	[13]

Natural sources of germanium in Poland are limited. Especially, Polish coals and zinc ores are not rich in germanium. However, several potential sources, containing 231 ppm to 8.9%, were identified in Polish zinc smelters. They are presented in Table 2.

Table 2. Composition of selected Polish germanium sources.

Material	*Ge content*	Ref.
Polymetallic alloy	6.9%	[14]
Germanium dross	8.9%	[15]
Liquation feeding furnace dross	0.68%	[-]
Copper-cobalt cake from zinc electrolyte purification	231-287 ppm	[-]

It is estimated that only 3% of germanium present in processed zinc ores is recovered [2]. Therefore, it is still huge room for germanium recovery enhancement. This paper is focused on the experimental results of germanium recovery from Polish-based materials. Potential materials, including copper-cobalt cake and liquation-feeding furnace dross, were examined.

1. Materials

1.1. Copper-cobalt cake

One of the steps of electrolytic zinc production is the purification of crude zinc electrolyte. The process is performed to remove impurities that may negatively impact zinc electrolysis. One such impurities is germanium. The typical purification process is composed of three steps: cold, hot, and final. In the cold purification step first portion of zinc powder is added to remove the first part of the impurities, mostly cadmium. In the hot step additional zinc powder is added in the presence of activators. The residue from this step is copper-cobalt cake. The final zinc portion is added to remove any remaining impurities in the final step. The composition of investigated copper-cobalt cake is presented in Table 3. Two different cake forms were investigated – wet unconditioned and dried at 70°C for 8 hours.

Table 3. Composition of the copper-cobalt cake

Element	*Cu*	Cd	Zn	Pb	Ni	Co	Ge	Dry loss
Content [%wt.]	24.7	21.2	16.3	4.50	3.96	3.22	231 ppm	38.7

1.2. Liquation feeding furnace dross

The pyrometallurgical process of zinc production includes zinc refining using New Jersey process. In this process melted GOB (Good Ordinary Brand) zinc is fed to the rectification column. Zinc vapors volatilize to the top of the column, where they condense and liquid SHG (Super High Grade) zinc is collected. On the other hand, liquid zinc with impurities flows to the bottom of the column and is then directed to the liquation furnace. Due to oxidation, dross is formed on the liquid metal surface. The dross is periodically removed and routinely recycled for the zinc production process. However, it was examined as a potential source for germanium recovery due to elevated germanium content. Due to the presence of a large metallic fraction, the dross was sieved at 0,3 mm mesh prior to investigation. The composition of the dross is presented in Table 4.

Table 4. Composition of the liquation feeding furnace dross

Element	*Zn*	O	Sn	Pb	Al	Cu	Ge	Si
Content [%wt.]	68.0	20.1	3.35	2.52	2.22	1.83	0.67	0.25

2. Methods

Leaching tests of copper-cobalt cake in wet and dried form as well as liquation feeding furnace dross were performed. All three analyzed materials were leached in aqueous solutions of sulfuric acid (98%, Avantor, Poland). In some tests, the addition of hydrogen peroxide (30%, Avantor, Poland) as an oxidant was also analyzed. All tests were performed using a 500 cm^3 acid solution. A sample of aqueous solution was obtained dissolving concentrated acid in deionized water. Then, the solution was heated to set temperature, and the appropriate amount of investigated material was added depending on the tested solid-to-liquid ratio (S:L). The influence of several parameters like temperature and initial acid concentration was analyzed.

The calculation of leaching yields for selected elements was based on the chemical composition of the initial solid material and final solution. Concentrations of elements in solutions were analyzed using ICP-OES (Inductively Coupled Plasma—Optical Emission Spectrometry; Agilent 5110 SVDV, Agilent Technologies). Solid samples were melted with Na_2O_2 (p.a. Avantor, Poland) at 600°C and chemically dissolved in 1:1 mixture of concentrated nitric (65%, Avantor, Poland) and hydrofluoric (40%, VWR, UK) acid prior to analysis.

3. Leaching tests

3.1. Copper-cobalt cake

The first analyzed material was copper-cobalt cake. Two forms of the cake were investigated – wet and dried. The influence of process temperature and initial acid concentration on the leaching yield of selected elements, i.e., germanium, copper, zinc, and cadmium, was tested. Results of the investigation under non-oxidizing conditions for the dried copper-cobalt cake are presented in Figures 1-2, while for wet cake in Figures 3-4. Results of tests under oxidative conditions are shown in Figures 5-6.

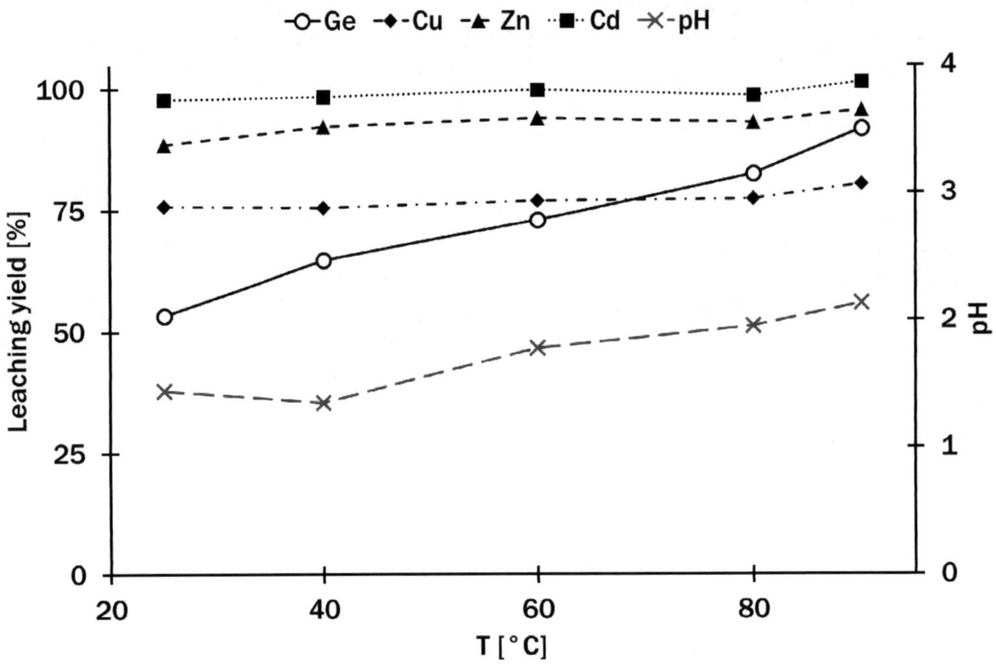

Figure 1. Leaching yields of germanium, copper, zinc, and cadmium from the dried copper-cobalt cake at different temperatures (tests for 2 hours, S:L 1:5, initial acid concentration 12.5 wt.%)

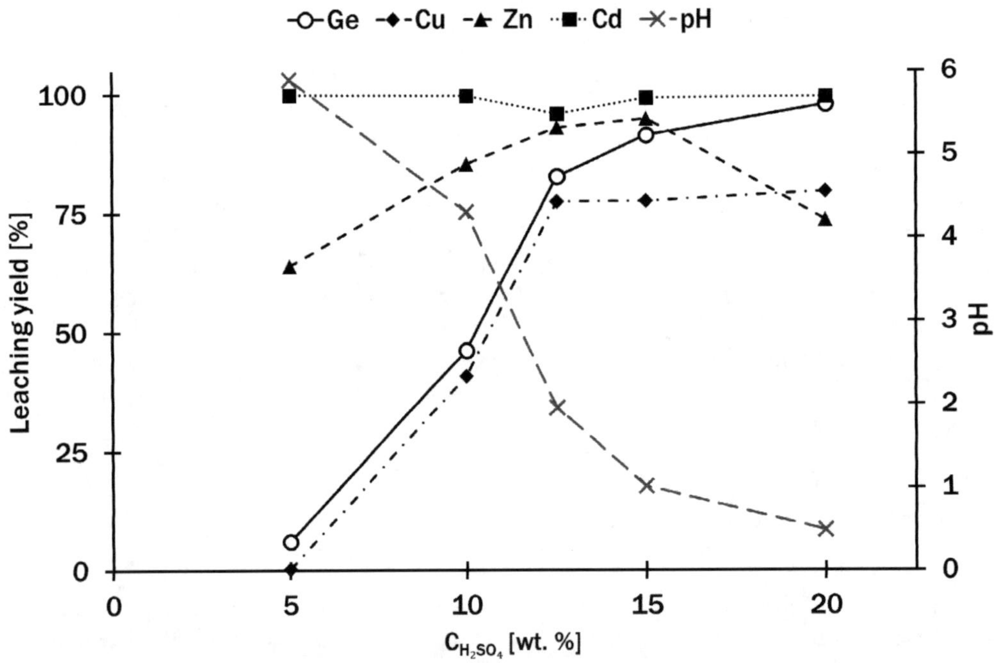

Figure 2. Leaching yields of germanium, copper, zinc, and cadmium from the dried copper-cobalt cake for different initial H_2SO_4 concentrations (tests at 80°C for 2 hours, S:L 1:5)

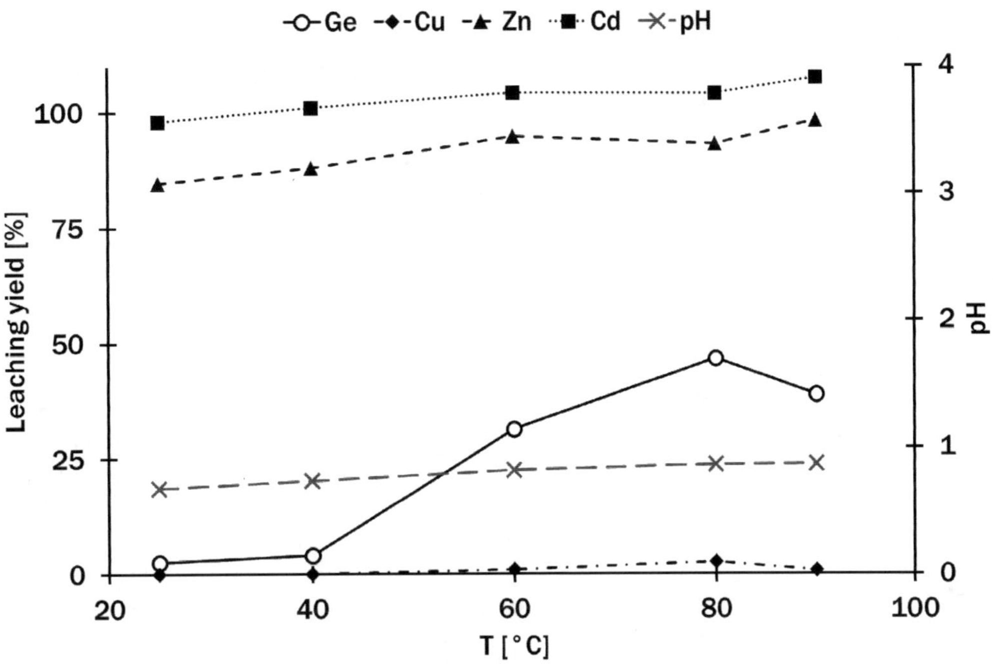

Figure 3. Leaching yields of germanium, copper, zinc, and cadmium from the wet copper-cobalt cake at different temperatures (tests for 2 hours, S:L 1:5, initial acid concentration 12.5 wt.%)

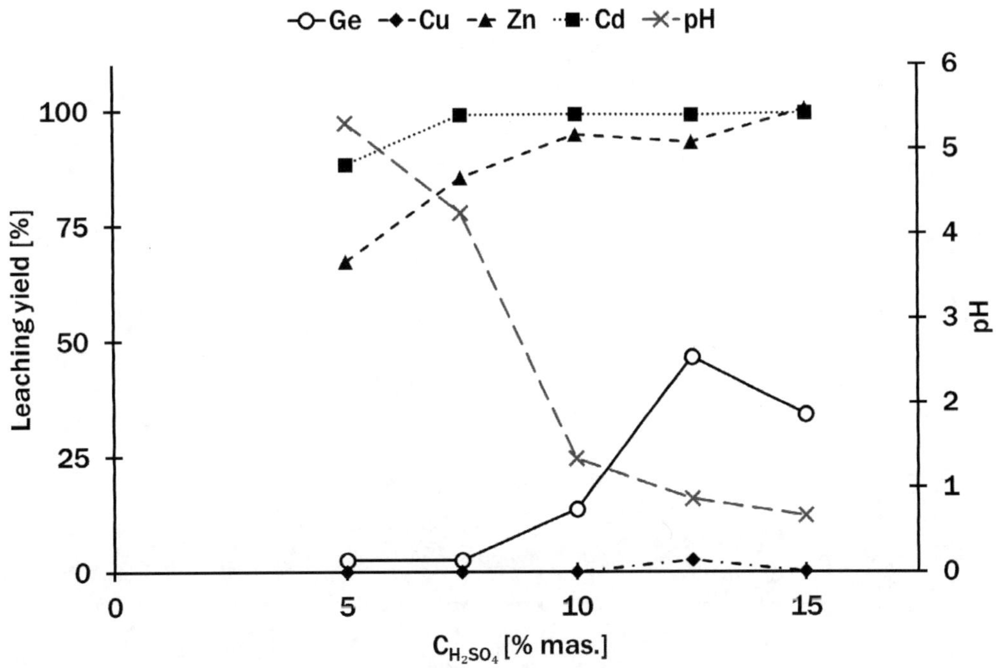

Figure 4. Leaching yields of germanium, copper, zinc, and cadmium from the wet copper-cobalt cake for different initial H_2SO_4 concentrations (tests at 80°C for 2 hours, S:L 1:5)

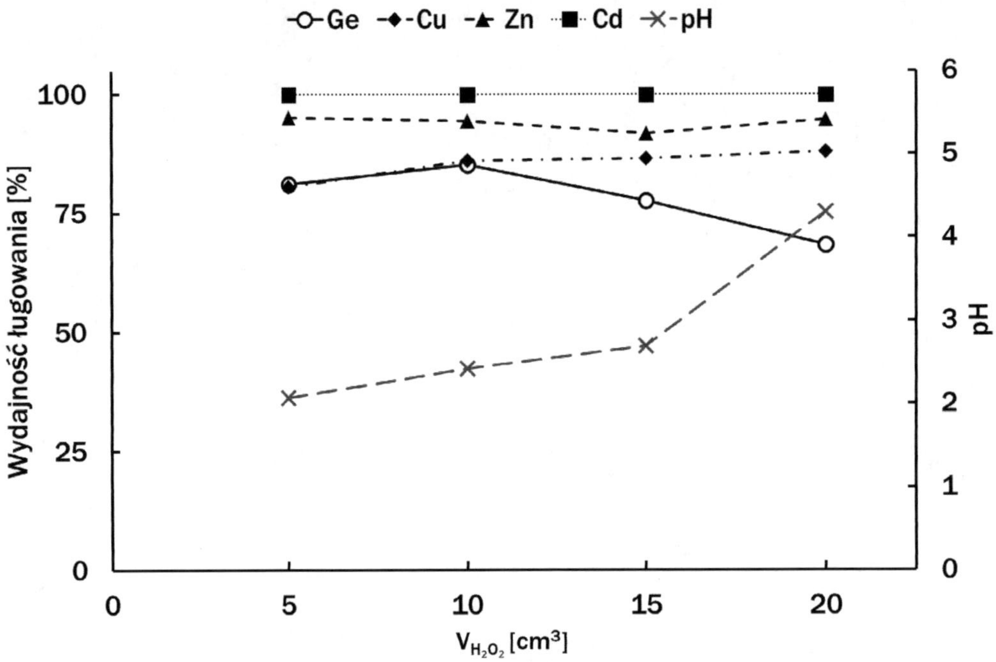

Figure 5. Leaching yields of germanium, copper, zinc, and cadmium from the dried copper-cobalt cake for different amounts of 30% H_2O_2 added (tests at 80°C for 2 hours, S:L 1:5, initial acid concentration 12.5 wt.%)

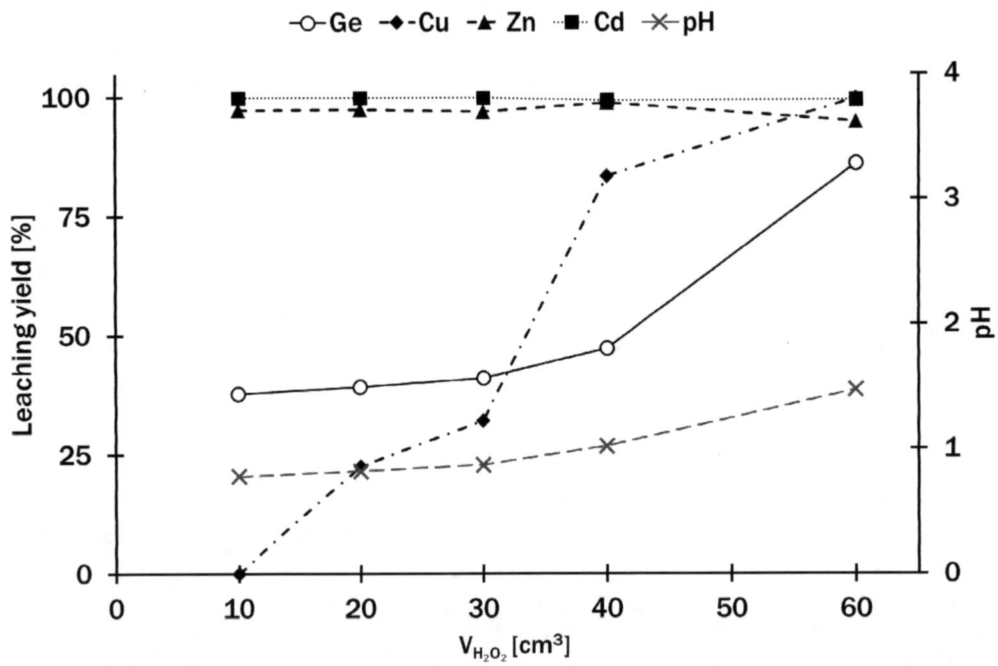

Figure 6. Leaching yields of germanium, copper, zinc, and cadmium from the wet copper-cobalt cake for different amounts of 30% H_2O_2 added (tests at 80°C for 2 hours, S:L 1:5 initial acid concentration 15 wt.%))

It is clear that germanium leaching yield depends on the initial material. Significantly higher leachability, even up to 99%, was achieved for dried cake (Figure 2, 20 wt.% initial H_2SO_4 concentration). It is associated with oxidation of the initial material during the drying process, even at a moderate temperature of 70°C). It is also supported by a significantly higher leaching yield of copper for dried material. In the case of the wet cake, almost no copper was leached. Leaching yields of zinc and cadmium are not significantly influenced by the kind of initial material. Germanium leaching yield is also significantly influenced by the final pH of suspension – it may be observed especially in tests with different initial acid concentrations (Figures 2 and 4). The general conclusion is that in the case of dried cake, it is possible to achieve almost complete leachability of germanium, while in the case of the wet cake, germanium is only partially leached.

Adding hydrogen peroxide as an oxidant increased the leaching yield of germanium from the wet cake. It may also be noticed that the addition of oxidant had, firstly, an influence on copper leachability, then, an increase in the amount of leached germanium. Oxidant reacts with copper at 0 and I oxidation state present in the cake and oxidize it to soluble Cu(II). Therefore, leachability of germanium, which might have been trapped in the undissolved solid is also increasing. On the other hand, germanium leaching yield was decreased when oxidant was used for leaching of dried cake. The leachabilities of cadmium and zinc were not significantly influenced by H_2O_2 addition. In this cake decrease of germanium leachability might be explained by formation of sparingly soluble oxides.

Summing up, the highest germanium leaching yield was achieved for the leaching of dried cake at 80°C, 2 hours, and 20 wt.% initial H_2SO_4 under non-oxidizing conditions. Under these conditions, the selectivity of leaching is moderate, as cadmium, zinc, and copper leaching yields are significant.

3.2. Liquation feeding furnace dross

The second investigated material was liquation feeding furnace dross. The influence of process temperature and initial acid concentration on the germanium and zinc leaching yield was examined. Results of the investigation under non-oxidizing conditions for the dried copper-cobalt cake are shown in Figures 7-8.

In the case of liquation feeding furnace dross, achieving a maximum of 55% germanium leaching yield was possible. On the other hand, almost complete zinc leaching was reached for most of the investigated parameters. The highest germanium leachability was obtained at 80°C, for 2 hours, with an S:L ratio of 1:10 and 30 wt.% H_2SO_4 initial concentration. It was also clear that temperature and initial acid concentration had some influence on germanium leaching yield. The yield increased as the leaching temperature increased – the highest (53%) for the test performed at 90°C. However, the difference between 80°C and 90°C was not significant. Also, the higher initial acid concentration results in higher germanium leachability. It was noticed that in the range of 5-20 wt.% germanium leaching yield increased significantly – from 1% to 54%. Further increase to 30 wt.% H_2SO_4 had a minor impact on germanium leachability – it increased to 55%. The highest germanium leachability was obtained at 80°C, for 2 hours, with an S:L ratio of 1:10 and 30 wt.% H_2SO_4 initial concentration.

The reason why it was not possible to achieve a higher germanium leaching yield may have been associated with the composition of the initial dross. Although the material was significantly oxidated, it might have also contained zinc and elements in metallic elementary form. Therefore, germanium might have been partially reduced and cementated during leaching. Moreover, germanium might have been partially enclosed in the material matrix; therefore, it was impossible to release Ge without complete dissolution of the dross fully.

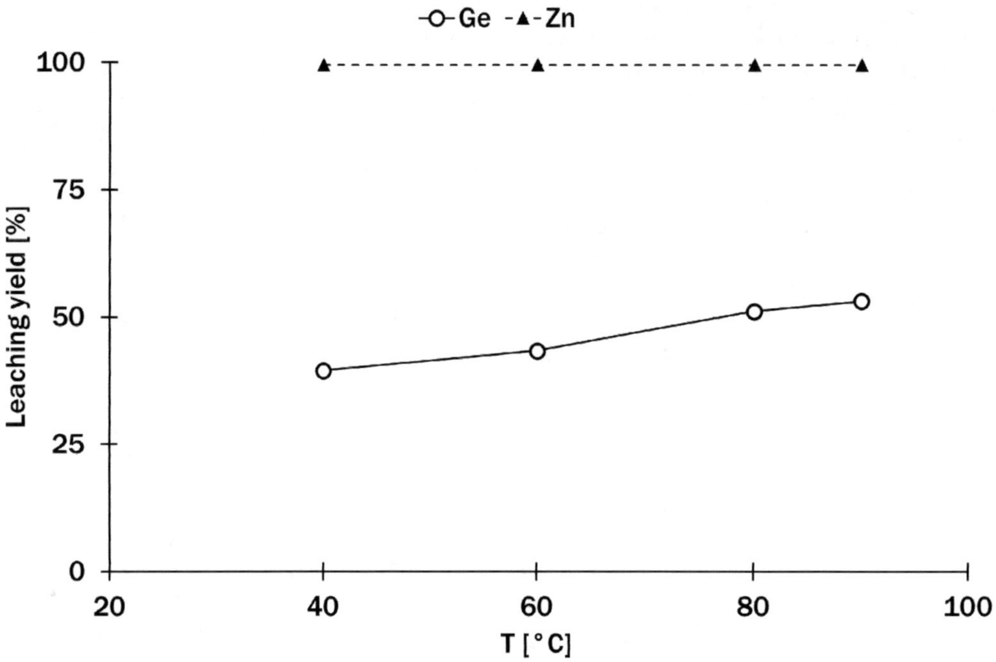

Figure 7. Leaching yields of germanium and zinc from liquation feeding furnace dross at different temperatures (tests for 2 hours, S:L 1:10, initial acid concentration 15 wt.%)

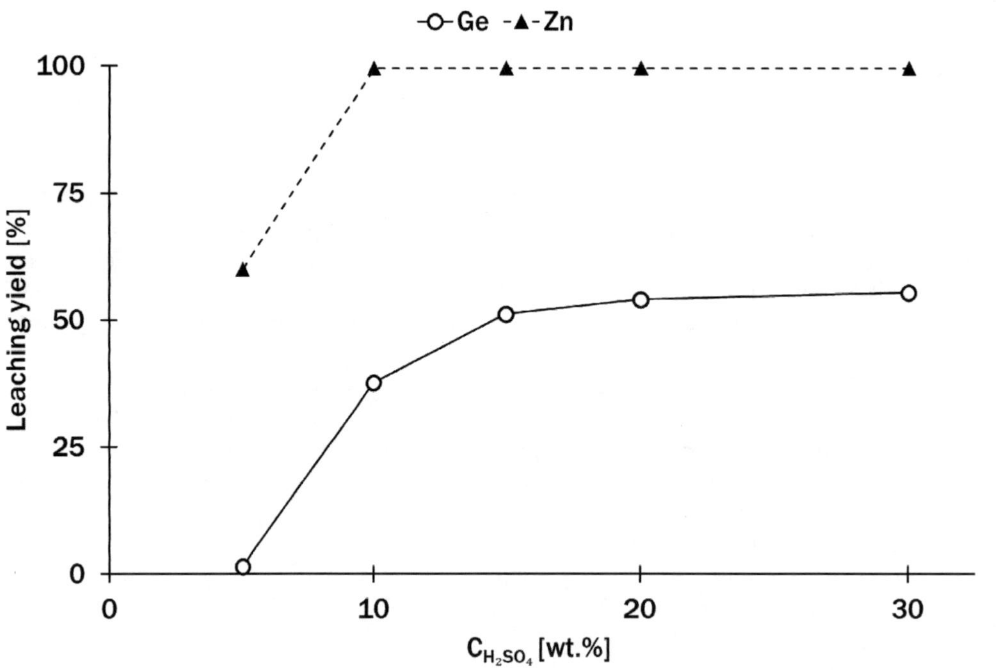

Figure 8. Leaching yields of germanium and zinc from liquation feeding furnace dross for different initial H_2SO_4 concentrations (tests at 80°C for 2 hours, S:L 1:10)

4. Further steps

Leaching of the cake resulted in a solution containing germanium along with other elements. The composition of the solution obtained after the leaching of copper-cobalt cake under oxidative conditions is presented in Table 5.

Table 5. Composition of the solution after oxidative leaching of copper-cobalt cake (80°C, 2h, 12.5 wt.% H_2SO_4, S:L 1:5, 40 cm3 30%H_2O_2/dm3)

Element	Cu	Cd	Zn	Ge
Content [g/dm^3]	34.8	33.4	17.2	36.4 mg/dm^3

The composition of the solution is quite complex, and the germanium concentration is low. Therefore, selective methods allowing the collection of germanium almost completely are highly desirable. It may be done using either precipitation with tannic acid or solvent extraction.

Tannic acid is a quite selective agent, forming an insoluble germanium complex used in some industrial applications. Using it is possible to achieve <0.5 mg/dm^3 Ge in the final solution [16-17]. Precipitate may be later burned to obtain germanium concentrate.

Solvent extraction was also investigated as a method for germanium separation. Several extractants were examined, including LIX 63, hydroxamic acid, D2EHPA, or Aliquat 336 [18]. Especially interesting is the way including complexation of germanium in sulfate solution with an organic acid, like tartaric acid, followed by extraction with an amine extractant, e.g., trioctylamine [19-20]. Germanium may be later stripped from organic acid using an aqueous sodium hydroxide solution and precipitated to obtain germanium concentrate. These operations will be further studied as the next steps of the developed germanium recovery process.

5. Conclusions

Leaching of copper-cobalt cake and liquation feeding furnace dross in aqueous sulfuric acid solutions was investigated. It was found that almost complete germanium leaching was achieved using dried cake at 80°C for 2 hours, S:L 1:5, and initial 20 wt.% H_2SO_4. Germanium leachability was influenced by temperature and initial H_2SO_4 concentration – more germanium was leached at 80-90°C and for higher acid concentrations. Also, the cake's form influenced leaching – the dried cake was more oxidized, and therefore more copper was leached. Similarly, adding an oxidant increased the wet cake's copper and germanium leaching yield. However, less germanium was leached under oxidative conditions than for tests with dried cake.

On the other hand, achieving complete germanium leaching from liquation feeding furnace dross was impossible. The highest germanium leaching yield was 55% for 80°C, 2 hours, S:L ratio 1:10 and 30 wt.% H_2SO_4 initial concentration.

Acknowledgements
This activity has received funding from the Polish National Centre for Research and Development within LIDER XI programme (grant no. 43/0159/L11/19/NCBR/2020) "Innovative technology for germanium recovery from wastes of Polish zinc metallurgy".

References

[1] Su Y, Hu D 2022 *E3S Web Conf.* **352** 03045
[2] U.S. Geological Survey 2022 Germanium *Mineral commodity summaries* (Reston, Virginia: U.S. Geological Survey) 72.
[3] Melcher F, Bucholz P 2014 Germanium *Critical Metals Handbook* ed G. Gunn (Chichester, UK: John Wiley and Sons), 177-203.
[4] Curtolo D C , Friedrich S, Friedrich B 2017 *J. Cryst. Process Technol.* **7(4)** 78624

[5] Dai S, Seredin V V , Ward C R, Jiang J, Hower J C, Song X, Jiang Y, Wang X, Gornostaeva T, Li X, Liu H, Zhao L, Zhao C 2014 *Int. J. Coal Geol.* **121** 79-97

[6] Font O, Querol X, López-Soler A, Chimenos J M, Fernández A I, Burgos S, García Peña F 2005 *Fuel* **84(11)** 1384-1392

[7] Arroyo Torralvo F, Fernández-Pereira C, García Villard E, Luna Y, Leiva C, Vilches L, Villegas R. 2018 *Miner. Eng.* **128** 106-114

[8] Liu F, Liu Z, Li Y, Wilson B P, Lundström M 2017 *Hydrometallurgy* **169** 564-570

[9] Zhang L, Guo W, Peng J, Li J, Lin G, Yu X 2016 *Ultrason. Sonochem.* **31** 143-149

[10] Kul M, Topkaya Y 2008 *Hydrometallurgy* **92(3-4)** 87-94

[11] Cengizler H, Eric R H 2011 Recovery of germanium from lead blast furnace slag, *Proceedings of the Second International Slag Valorisation Symposium* ed P T Jones, Y Pontikes, Y J Elsen, Ö Cizer, L Boehme, T Van Gerven, D Geysen, M Guo and B Blanpain (Leuven, Belgium: Katholieke Universiteit Leuven)

[12] González A, Font O, Moreno N, Querol X, Arancibia N, Navia R 2017 *Waste Biomass Valorization* **8** 2121-2129

[13] Chen W S, Chang B C , Chiu K L 2017 *J. Environ. Chem. Eng.* **5(5)** 5215-5221

[14] Kulawik S, Prajsnar R, Chmielarz A, Cybulski A, Michalski R, Klejnowska K, Drzazga M 2019 *Metals* **9(2)** 166

[15] Drzazga M, Prajsnar R, Chmielarz A, Benke G, Leszczyńska-Sejda K, Ciszewski M, Bilewska K, Krawiec G 2018 *Metals* **8(12)** 1041

[16] Liang D, Wang J, Wang Y, Wang F, Jiang J 2008 *Hydrometallurgy* **93(3–4)** 140-142

[17] Drzazga M, Chmielarz A, Benke G, Leszczyńska-Sejda K, Knapik M, Kowalik P, Ciszewski M, 2019 *Appl. Sci. (Basel)* **9** 966

[18] Tao J, Tao Z, Zhihong L 2021 *J. Clean. Prod.* **294** 126217

[19] Liang J, Fan L, Xu K, Huang,Y 2012 *Energy Procedia* **17** 1965-1973

[20] Drzazga M, Palmowski A, Benke G, Ciszewski M, Leszczyńska-Sejda K 2021 *Hydrometallurgy* **202** 105605

Recycling potential of secondary zinc materials

Martin van Leeuwen[1], Bruno Gay[2] and Sabina Grund[3]

[1,3] International Zinc Association, 1150 Brussels, Belgium

[2] B.CIRCLE, 1150 Brussels, Belgium

[1] mvanleeuwen@zinc.org

[2] bruno.gay1801@outlook.com

[3] sgrund@zinc.org

Abstract. Circularity is a decisive corner stone of a sustainable use of raw materials. This has been recognized at global level and is pushed forward specifically in Europe and e.g., Japan. Based on intensive desk top research and interviews with experts from the zinc industry, information on secondary raw materials for zinc recovery was compiled. This paper comprises of an overview of zinc bearing wastes, residues, and by-products occurring in all stages of zinc mining, production, as well as first-use and end-use production. These materials are quantified and characterized, and ways of recycling are described.

As a result, potentials for an increase of zinc recycling rates are identified. Mine tailings are a meaningful resource for zinc recovery. Industrial wastes occurring in zinc smelters, such as Jarosite, Goethite, and their likes are in the scope of research projects aiming at recycling zinc and other valuable metals. Residues, ashes, drosses, slags that are produced in first use operation of the zinc value chain, such as galvanizing, zinc die casting, or brass plants are recycled almost completely in Europe and in many other regions in the world. The biggest potential for increasing recycling rates for zinc still lies in steel mill dusts. Also, the majority of zinc die cast parts seem to end up in waste incineration and dump sites. This low end-of-life zinc recycling rate bears room for additional recycling opportunities via an improvement of the intelligent collection and sorting of waste and other innovative recycling processes utilizing residues from waste-to-energy plants.

1. Introduction

Over the past years, the importance of circularity has increasingly been acknowledged at European and at global level. In Europe, the EU New Green Deal [1] that was agreed on in December 2019, initiated and even more increased focus on the transition to a circular, European economy. Beyond Europe, China, Russia, India, Brazil, Indonesia, and other developing countries mark the regions in which the rapidly rising demand for metals is expected to occur. Metals are a major part of the materials that are needed to meet the demands caused by increasing world population and rising living standards as well as by changes in use patterns associated with greening the economy by e.g., renewable energy and e-mobility. The OECD projects the global metals use to more than double from 8 Gt in 2011 to 20 Gt in 2060 [2].

Increasing circularity of zinc in practical terms requires quantitative information on materials that are available for recycling, their chemical and physical characteristics, and technologies available for recycling. In an extensive desk top study combined with interviews with experts from the zinc industry, such information was compiled and may now serve as information hub for various questions and projects

Journal of Physics: Conference Series 2738 (2024) 012022 doi:10.1088/1742-6596/2738/1/012022

linked to secondary raw materials for zinc. This paper aims to provide a summary and high-level overview of the extensive desk top study done by the International Zinc Association in 2021.

2. Zinc losses in refined zinc production

Zinc production has long been studied throughout the world because this metal quickly found industrial uses around the beginning of the first industrial revolution. The volume treated increased with time thanks to its widespread use and new applications. Consequently, zinc production has increased gradually from the 19th century until the early 2000s.

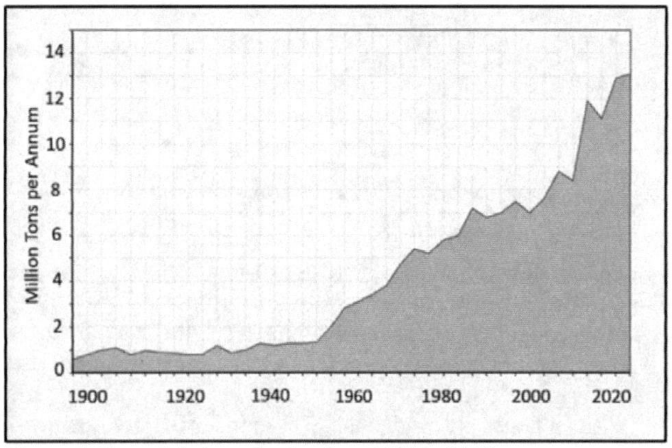

Figure 1: Zinc global production from 1900 to 2020

Since the use of zinc is linked to the industrialization of countries around the world, it is possible to draw a parallel between the global growth of zinc production and the economic growth of industrialized countries. Today, it seems that the refined zinc production level is in the range of 13 to 14 million tonnes (Mt) per year (Figure 1 [3]). That number includes the zinc production from minerals (primary or mined zinc) and the zinc production from steel dust reprocessing loops (refined secondary zinc) but doesn´t take into account direct zinc recycling loops from zinc industrial residues (directly re-used secondary zinc). Direct zinc remelting provided another 4 to 5 Mt of secondary zinc products per year to the market in between 2010 and 2019 [4].

2.1. Resources and Reserves

Mineral deposits can be classified as: 1) Mineral resources that are potentially valuable, and for which reasonable prospects exist for eventual economic extraction, or 2) Mineral reserves that are valuable and legally, economically, and technically feasible to extract. These two concepts can be further subdivided into different categories from proven ore reserve to probable inferred mineral resource.

Reserves and resources are not fixed once and for all. This notion of evolution is often misunderstood, and data published in the media is often misinterpreted as a fixed limit. This applies to many natural resources, including zinc. Of course, there is an ultimate limit, which would be related to the full exploitation of geologically feasible zinc ores at a very high theoretical cost.

In his report on the geological long-term availability of zinc, Pirard, University of Liege [5], concludes on zinc mining indicators:

- Accessible crustal content of zinc: 198,000,000 Mt
- Extractable global resources: 63,000 Mt
- Proven and probable reserves: 250 Mt

These tonnages are complemented by 247 Mt of zinc being in use today as anthropogenic stock with most of it becoming available for recycling one day (see Figure 2).

Journal of Physics: Conference Series **2738** (2024) 012022 doi:10.1088/1742-6596/2738/1/012022

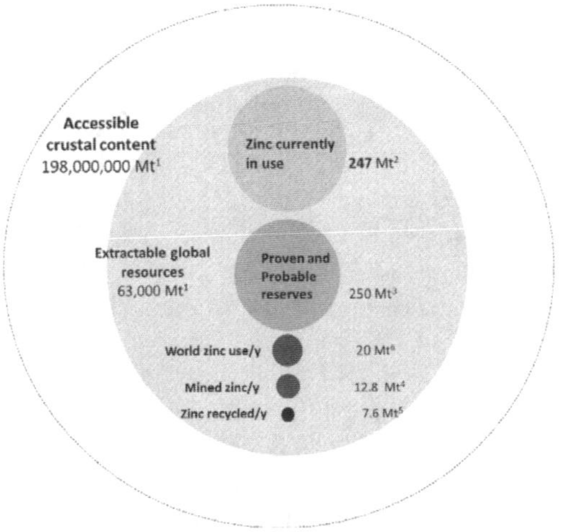

1. Pirard E, 2021 [5]
2. Rostek L et al., 2022 [4]
3. USGS Mineral Commodities Summary, 2021 [6]
4. International Lead Zinc Study Group, 2019 [7]
5. Rostek L et al., 2022 [4]
6. Rostek L et al., 2022 [4]

Figure 2: Evaluation of the Reserves and the resources of zinc

2.2. *Zinc losses in the Extraction and in the Minerals Processing*

Non-ferrous metal ores usually are polymetallic ores. Zinc in ores is typically associated with other metals of value, such as lead, copper, iron, silver, gold, indium, germanium, and sometimes gallium. Sometimes, zinc is mined as a by-product from ore bodies that are mined for their silver content in the first place. Other metals typically found in zinc ores are arsenic and cadmium.

A typical loss of 2.5% during ore mining is quoted, this is called the mine cut-off. This cut-off depends on many different factors such as the type of ore, the conformation of the deposit, the average zinc content and the technical means of processing the ore. Total zinc losses during mining for 2015 are estimated at about **350 kt Zn/year**. These losses stay behind in mines and are challenging to recover.

Losses during mineral processing (e.g., flotation) have been estimated at 15% on average, with a range between 10-30%. These losses depend on the type of ore, the type of ore processing facility and other technical and economic factors. The price of zinc compared to other non-ferrous metals is also an element to be considered, in particular in processing polymetallic minerals. 15% zinc losses during mineral processing for 2015 are estimated at an absolute volume of about **1,900 kt Zn/year**. These zinc losses report to the mine tailings after mineral processing. To recover part of these losses, tailings can be reprocessed in some cases. Reprocessing tailings is not always easy to implement. However, new paradigm shifts may influence tailings reclamation decisions. These include new environmental regulations, improved zinc recovery capacity, recovery of other metals that have recently increased in value and changes in mineral technology. New Century Resources' zinc tailings treatment in Australia is a good example. The tailings pond at the Century mine was converted into the 14th largest known zinc deposit in the world - 77 Mt of tailings, containing 3% recoverable zinc with a homogenous grade distribution, gave a deposit of 2.3 Mt of zinc. They have ramped up tailings throughput through the existing infrastructure. Another tailing reprocessing facility, in Poland, processes tailings from the Boleslaw mine (closed 1996), Olkusz mine (closed 2001) and Pomorzany mine (closed 2020) with an average Zn% of 1.0% and maximum Zn% of 1.5%. After reprocessing, the tailings have an average Zn% of 0.05% so the facility achieves a recovery of approx. 95%. Recovery can technically be as high as close to 100% in theory, but it is often not economically viable.

2.3. *Zinc losses in Refined Zinc Production*

The production of refined zinc is divided among various industrial groups operating in different countries around the world. Often, ore-producing countries are also producers of refined zinc so have integrated their mining and refining activities. Zinc losses in the main refining routes roast-leach-electrowinning (RLE), imperial smelting furnace (ISF) and solvent extraction-electrowinning (SX-EW) have been studied with a focus on residue production rates and related zinc losses.

RLE is a hydrometallurgical process and currently the most widely used process in the world for refining zinc ores and crude zinc oxides (CZO) to metallic refined zinc with high purity. It accounts for approximately over 90% of the volume of refined zinc produced. The RLE process is very efficient for treating zinc ores as it has a very high yield and produces high purity special high grade (SHG) zinc (min. 99.995%). However, the RLE process is limited to a feed with approx. 20% CZO originating from recycled electric arc furnace (EAF) dust, and even lower than 20% if no extra CZO washing and/or direct leaching capacity is available. In the RLE process, zinc losses to residues are estimated to be between 2% and 6%. These losses are mainly concentrated in iron residues of the type Jarosite, Goethite or Hematite, depending on the used residue precipitation process. An average of 3% for these zinc losses is assumed in this paper, which for 2015 is estimated at about **345 kt Zn/year**.

The precipitation of iron as Jarosite by hydrolysis generates sulphuric acid which must be neutralised in order to achieve near complete iron precipitation. Calcine from the roaster is added as neutralising agent, amongst others. Under the hydrolysis stage conditions, less than 5% of the zinc ferrite content in calcine is dissolved and 95% is retained in the Jarosite waste. This explains the presence of zinc in Jarosite and generates a limit in the amount zinc which is recovered in the leaching step, which can only be recovered by further treatment of Jarosite, for example in a Waelz Kiln. The more Fe is present in the concentrate feed to the roaster, the more Zn is lost to Jarosite. Jarosite contains 4.5% Zn on average, but can vary between 1.5 and 13% depending on the feed composition and leaching process conditions.

ISF is a pyrometallurgical process and is still used but its relative importance has been decreasing for the last 20 years. It accounts for approximately 7% of the volume of refined zinc produced. This process is generally considered to be less advantageous for treating zinc ores than the RLE process because the zinc product from this process has more and higher impurities. However, ISF is capable of processing zinc-lead ores and more secondary raw materials from the EAF dust reprocessing loop. It can produce zinc and lead at the same time. ISF is also able to process some secondary raw materials from the RLE process and it can treat more complex ores, polymetallic ores and other waste materials. In the ISF process, zinc losses occur to the slag which can have a zinc content of 6-10%. These zinc losses are estimated at about **42 kt Zn/year**.

The SX-EW process has been developed more recently and its importance has been increasing for the last 10 years. The main advantage of the SX-EW process is that it can treat secondary materials such as CZO from the EAF dust processing loop without the need of washing it first to remove the halides. The use of CZO as raw material can be up to 100% which is much higher than in the RLE process; it can be used to produce SHG zinc with a 100% recycled content, depending on CZO availability in the market. Total zinc losses to SX-EW residues in 2015 are estimated at **18 kt Zn/year**. At present, its importance is still low, but it will probably increase in the next few years depending on the need to treat secondary raw materials. China was planning to build three more SX-EW process plants between 2021 and 2023: Yongxin nonferrous phase II in Guangxi, Tenglong Recycling in Hebei and Yuguang Zinc in Henan.

Process	Mined Zinc (kt)	Relative importance (%)	Residues vs. Zinc ratio (%)	Zinc content in residues (%)	Zinc losses (kt)	WOX/CZO Zinc (kt)	Relative importance (%)	Residues vs. Zinc ratio (%)	Zinc content in residues (%)	Zinc losses (kt)
RLE	13,450	86%	65%	4.5%	338	890	50%	32.5%	4.5%	7
ISF		7%	50%	8.0%	38		25%	25%	8.0%	4
SX-EW		5%	50%	4.5%	15		25%	25%	4.5%	3
Other		2%	50%	8.0%	11		0%			
Total		100%					100%			
Grand Total	415									

Table 1: Estimated amount of zinc lost in residues from zinc smelting processes in 2015

Table 1 gives an overview of estimated total global zinc losses in refined zinc production in 2015. The lost zinc units are split by origin of the feed material, being either primary mined zinc or secondary crude zinc oxide (CZO). The total estimated losses for 2015 were **415 kt of zinc**. Some zinc form residues (including slags) from refined zinc production is recovered in recycling processes. Recycling of zinc residues is apparent today, in particular in China but also in other regions where these residues cannot be landfilled anymore. The following chapter will explain how zinc can be recovered from residues and shows that it is done on a smaller scale at some smelting operations.

3. Zinc recovery from secondary raw materials from refined zinc production

Reprocessing of residues from zinc refining (RLE and SX-EW) by hydrometallurgical processes was first used for the stabilization of Cd- and As-containing residues, to make them chemically inactive to allow for safe landfilling or use it road construction. The Jarofix process is operated by some smelting operations to stabilize residues (Table 2).

More recent processes are being developed to recover zinc, precious metals and other non-ferrous metal. Smelting residues such as Jarosite can still contain valuable amounts of silver, indium, lead and zinc, of which silver and indium together often represent more than 50% of the total value. Next to an economic incentive, new environmental regulations are a large driver for the development of new recovery and recycling technologies for smelter residues.

Existing process technologies to treat zinc residues from smelting and refining operations are the Ausmelt process which is a top submerged lancing (TSL) furnace (from Metso Outotec), ISASMELT which is a TSL furnace as well (Glencore technology), the Waelz process and the imperial smelting process (ISP). Table 2 gives an overview of zinc and lead producing companies operating such technologies for recycling of residues [8]. Other more experimental processes or technologies under development are the HIsarna process (from Tatasteel [9]) and the Jarogain process (from VTT Technical Research Center & Aalto University [10]).

	Technology	Company	Site	Country
	Stockpiled	Majority	Majority	Globally
Stabilisation	Jarofix	Xstrata	San Juan de Nieva Smelter	Spain
		Xstrata	CEZinc Operations	Canada
		Hindustan Zinc (HZL)	Chanderiya	India
	Jarosite sulphidation	Boliden	Kokkola	Finland
Pyrometallurgical	Ausmelt furnace	Korea Zinc	Onsan Operations	South Korea
		Young Poong Corp.	Young Poong Operations	South Korea
		Sun Metals Corp.	Sun Metals Zinc Refinery	Australia*/ South Korea
	Waelz kiln	Zhuzhou Smelter	Zhuzhou Smelter	China
		Votorantim	Juiz de Fora Smelter	Brazil
		Glencore	Portovesme	Italy
		KCM	Plovdiv	Bulgaria
	Pb smelter/ISF + fumer	Nyrstar	Hobart/Port Pirie Operations	Australia
		Mitsui Mining	Hachinoche Smelter	Japan
		TeckCominco	Trail Lead/Zinc Operations	Canada

* Zinc residues produced at Sun Metals Zinc Refinery in Australia are sent to South Korea Onsan refinery for treatment (Source: Based on data provided by Piret & Stolberg Partners [2])

Table 2: Examples of processes used for the reprocessing of RLE residues

In a TSL furnace, zinc is fumed out of the liquid phase in the furnace. This liquid phase is typically a lead or copper smelt. So, next to reprocessing secondary zinc residues, fumed zinc collected in zinc-rich dust can also be a by-product from lead or copper (alloy) production in a TSL furnace.

4. Zinc losses in zinc first use production

When zinc was first used, more than 150 years ago, the applications were initially related to objects made of solid zinc or zinc alloy such as household and industrial objects. Zinc sheets for roofing also appeared quite quickly. Then batch galvanizing was developed. The emergence of zinc alloys Zamak and brass allowed for a wider application range for zinc. During the last decades, continuous galvanizing has been an important growth engine for zinc. Other applications have also emerged such as zinc powders in oxide or metal form and zinc wires for thermal spraying. More recently, new applications are emerging such as rechargeable zinc batteries.

The distribution of zinc applications therefore takes into account its versatility of use. Typical uses of refined zinc, which account for more than 90% by volume of all metallic zinc uses, are galvanizing, die-casting, brass and production of zinc oxide powders. Other uses of zinc with minor volumes are production of zinc dust, zinc wire, fertilizer nutrients and zinc sheet. There are other applications of zinc that are produced in small volumes for niche markets.

4.1. Zinc losses in continuous steel sheet galvanizing

Continuous galvanizing is the largest first use of zinc and accounts for 35% of total zinc consumption (primary and secondary zinc). Continuous galvanizing allows steel sheets to be coated with zinc or zinc alloys in a continuous process. There are three technologies currently used: 1) Hot-dip galvanizing (Sendzimir process) by dipping the sheet in a molten zinc (alloy) bath; 2) Electro-galvanizing by depositing zinc to the steel surface by means of an electric current from a generally aqueous bath in which zinc ions and, possibly, other metals (Ni, Fe, Co) are present; 3) Vapour phase galvanizing in which the steel sheet surface is put in contact with a zinc vapor flux (Physical Vapour Deposition (PVD) and Jet Vapour Deposition (JVD) processes). With the evolution of technologies, it appears that, of these three technologies, hot-dip galvanizing has become the predominant technology for the automotive market, for the construction market and for other markets.

In continuous hot-dip galvanizing, different bath compositions exist and are detailed in Table 3. The zinc feed into these processes is often alloyed SHG zinc. Alloying elements are typically aluminium, magnesium, silicon, lanthanum and/or cerium in varying concentrations. During the process, the zinc bath is saturated with iron due to the reaction between the liquid zinc bath and the iron in the steel. Traces of other elements from the steel can be found in the coating, such as manganese or chromium.

	Composition				
Bath	**Al (Wt%)**	**La/Ce (Wt%)**	**Mg (Wt%)**	**Si (Wt%)**	**Fe**
Galvannealed (GA)	0,10 to 0,14	0	0	0	Over Satureted High
Galvanized (GI)	0,18 to 0,30	0	0	0	Over Satureted Medium
Galfan	4,7 to 6,00	0-0,05		Traces	Over Satureted Low
ZnAlMg	0,18-12,00	0	0,50-5,00	0-0,20	Over Satureted Low
Galvalume	50-60	0	0	1-2	Over Satureted Low

Table 3: Bath compositions used generally in the hot dip continuous galvanizing lines, the balance is zinc and trace elements

In continuous hot-dip galvanizing, top drosses and skimmings are formed in two ways: 1) Due to the reaction between the bath and the steel; and 2) Due to the oxidation of the bath alloy on the surface and during the wiping. When the aluminium content in the zinc bath is below 0.135%, bottom drosses are formed. The composition of the top skimmings and bottom drosses vary with the zinc bath composition.

The estimated global production of continuously galvanized steel in 2015 has been estimated at 160 Mt. It is estimated that in 2015 about 5.7 Mt of zinc was used for continuous galvanizing globally. About **635 kt of zinc** per year was lost to galvanizing drosses, galvanizing top skimmings, and drosses and ashes generated in zinc alloying & casting plants. Because of the very high zinc content, residues from continuous galvanizing are almost fully recycled into new secondary zinc products.

4.2. Zinc losses in batch galvanizing
Batch galvanizing is the second largest application for zinc. This represents about 25% of total zinc consumption in first use. Batch galvanizing is usually done by hot-dipping of steel parts in a bath of molten zinc. The batch galvanizing process typically exists out of the following process steps: cleaning, pickling, fluxing, galvanizing, and cooling. With some exceptions, the zinc bath composition is generally >99% zinc. The other elements are either small amounts of intentionally added elements (Bi, Ni, Sn, Al, Pb) or small amounts of associated elements (Fe, Pb, Cd, Cu).

Residues produced in batch galvanizing are of two types: bottom drosses and surface ashes or top ashes. Bottom drosses are formed as a result of the reaction between the liquid zinc and the iron in the workpieces, the steel in the tank and the steel in the working tools (suspension hooks). Bottom drosses contain >95% zinc, the remainder is mainly Fe and Pb. Surface ashes or top ashes are produced due to an exothermic reaction between the activating flux, the zinc, the iron and the oxygen from the air. The total composition of top ashes is a mix of solidified droplets of metallic zinc from the bath and the oxide-chloride powders (like ZnO, $Zn_xCl_y(OH)_z$, NaCl, KCl, …). Top ashes contain >80% zinc, of which about 50% is in the metallic form.

It is estimated that in 2015 about 4.4 Mt of zinc was used for batch galvanizing globally. About **1.15 Mt of zinc** per year was lost to galvanizing drosses, galvanizing ashes, and top skimmings generated in zinc alloying & casting plants. Because of the very high zinc content, residues from batch galvanizing are almost fully recycled into new secondary zinc products.

4.3. Zinc losses in zinc die casting
Zinc alloys for die casting (called Zamak) appeared between 1926 and 1929, when it was developed and patented by the New Jersey Zinc Company. Foundries needed a cheap, hard and resistant material easy to mould and mechanize. The aim was to obtain parts with lower price, higher precision, better finishing and higher quality than silver, which gets dirty and cannot be painted. Zamak alloys are cast in special hot-chamber casting machines. The hot-chamber casting machine allows for the manufacturing of pieces at very high precision and very high speed, so with a high productivity. Zamak alloys are very pure and have low levels of impurity elements Fe, Pb, Cd and Sn. Zamak alloys are generally produced from SHG zinc (min. 99.995% purity), pure aluminium, copper, and magnesium. Recycled "non-used cast Zamak" or new scrap coming directly from the casting process are also used. Those scrap pieces need to be very clean and without additional metallic coating or plating. In some cases, top skimmings from continuous galvanizing are used to produce Zamak. This secondary raw material requires reprocessing in order to lower the iron content, lower the iron-aluminium dross particles, eliminate the zinc and aluminium oxide particles and to precisely adjust the alloy composition according to international standards before it can be used as alloy for die casting. Aluminum is added to improve strength and fluidity, and Al% in Zamak alloys is typically 4.0% which is close to the Zn-Al eutectic composition. Copper makes the alloy stronger and harder; Cu% varies by alloy, and is 0.0% for Zamak alloy #3; 0.9% for #5 and 3.0% for #2. Magnesium is added as grain refiner and Mg% is typically 0.05% for high-pressure die cast alloys, but can be higher for zinc alloys for gravity casting. Zinc alloys for casting are specified in international standards such as EN 1774 (1997) and ASTM B86 (2022).

Global volumes of Zamak produced and used are difficult to estimate because the market fir zinc die castings is volatile and can change rapidly depending on the economic situation in a region or in the world. Zamak alloys are used in a wide variety of applications. In addition, the market represents an important part of total zinc first use (about 15%) and zinc die casting alloys are used in many countries by die cast companies of various sizes. The applications of Zamak are very versatile: buildings, appliances, clothing, electronics, electric, toys, automotive and various other applications. It is assumed that the global use of zinc for producing zinc die casting alloys is about 2 Mt of zinc per year.

For zinc die casting alloys, the big question revolves around end-of-life (EoL) recycling. Zinc die cast parts are generally small parts that are dispersed in many applications. Very often, they end up in municipal solid waste (MSW) going to landfills or incinerators. In addition, the equipment in which these parts are used is often imported or exported. Recycling at the EoL is therefore very difficult to implement and the estimation of the EoL recycling rate is complicated to evaluate. Europe is a region where the recovery of these metals may be well organized through effective waste collection and sorting schemes. In the rest of the world, it is likely to see much lower EoL recycling rates for post-consumer zinc die cast parts. It is estimated that about 475 kt of zinc is recycled per year from EoL zinc die cast scrap, the remaining **1.525 Mt of zinc** pet year is either abandoned in place or lost to landfilling sites and waste incinerators.

4.4. Zinc losses in brass production

The brass market is a large market in volume and represents about 8 Mt of brass per year. Typical brass products are tube, rolled brass sheet, wire, powders and castings. The zinc content in brass is between 32% and 45% (with an assumed average of 38%). It can also contain a few percent of lead, and the remainder is copper. Lead is considered necessary in brass because it gives better machinability to the alloys, so the brass industry is often considered as an attractive outlet for lead-containing secondary zinc from recycling operations. In recent years however, the use of lead in brass has also been questioned by brass users because of environmental concerns.

In 2016, the amount of primary brass produced was about 4.4 Mt. For producing this amount of primary brass, both primary and secondary zinc was required and estimated at 38% x 4.4 Mt = 1.7 Mt of zinc. The total zinc losses to residues generated during melting and alloying of primary brass, plus residues generated during manufacturing of primary brass products, are estimated at **100 kt of zinc** per year in total.

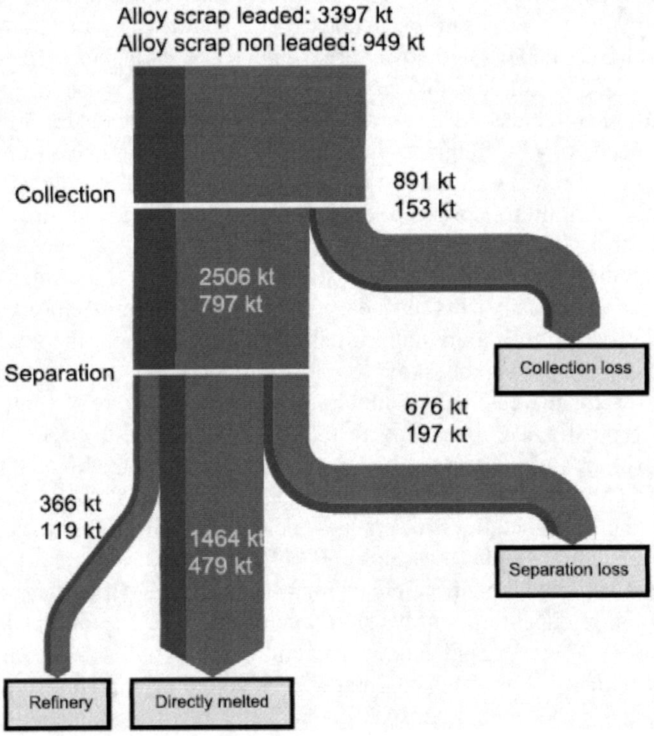

Figure 3: Copper alloy losses during the stages of collection and separation of post-consumer scraps from EoL phase

Similar to EoL zinc die castings, meaningful amounts of brass scrap are lost in the post-consumer phase. Figure 3 shows copper alloys (brass) losses during the stages of collection and separation stages of post-consumer scraps [11]. The figure shows that about 44% of post-consumer brass scrap is lost through dissipation or landfilling, caused by collection and separation inefficiencies. This 44%

represents about 1,917 kt x 38% = **730 kt of zinc** losses per year. The remaining 1,909 kt of copper alloys (brass) which is collected and sorted per year is recycled either through direct remelting into new brass, or treatment in a copper refinery to recover and refine the copper. Before the copper reaches the copper refinery for treatment, the zinc has been typically removed from the brass scrap, e.g., in a top submerged lancing (TSL) furnace in which it is fumed out. The so-called copper furnace dust from this TSL furnace, which is very high in zinc content, can be treated in a zinc refinery (e.g., in an ISF) to recover the zinc and produce refined zinc out of it.

4.5. Zinc losses in zinc sheet production

The manufacturing of rolled zinc is ancient and dates back to the beginning of the 19th century when the methods of extraction, purification and shaping of zinc began to be industrialized. Zinc sheet can be used to make roofs and facades of very different styles, often with very beautiful aesthetic views. The manufacturing method of rolled zinc consists of melting cathodes or ingots of SHG quality zinc, followed by adding and alloying with titanium, copper and aluminium in proportions generally lower than 1%. The melting and alloying yield is about 97%, so about 3% of residues are produced with a composition of pure SHG zinc or alloyed zinc. Generally, these residues are in the form of ashes which consists of oxides and metal. Once the zinc is alloyed, the liquid alloy is cast in a continuous or semi-continuous slab casting machine. The resulting slab is then rolled in several stages and coiled to obtain coils of zinc sheet. These coils are then cut to width. Coils can be sold as such, or they can be coated with different organic or inorganic coatings. Part of the produced material is for manufacturing of gutters, tubes, elements, and various downpipes for rain- and stormwater drainage.

Rolled zinc is mainly produced in Europe (France, Germany, Netherlands, Spain). There are also producers in Peru, the USA and China, and probably very small producers elsewhere in the world. Rolled zinc production is estimated between 250 kt and 290 kt per year. Rolled zinc is assumed to have the best end-of-life (EoL) recycling rates amongst other zinc applications. In Europe this rate is typically >95%. The recycling route exist of collection of scrap by roofing and building demolition companies, followed by sorting at metal recyclers, followed by remelting and casting at specialized recycling companies. The recycling market is estimated between 100 kt and 200 kt of zinc per year in Europe. Zinc losses to landfills are estimated to be **nil**.

Collected EoL rolled zinc ("roofing scrap") is typically recycled into alloys for the brass industry because it can contain small amounts of lead, tin and cadmium. Roofing zinc scrap can be very old, dating back more than 50 years when it was installed. At that time, rolled zinc was made with an alloy of zinc, lead and cadmium. Also, solders used in roofing installation are typically lead and/or tin alloys.

5. Recovery of zinc from secondary raw materials from first use production

It has been mentioned already that residues from galvanizing are almost fully recycled into new secondary zinc products, because of their very high zinc content. These secondary zinc products are typically zinc oxide powders, metallic zinc powders and metallic zinc dust. The manufacturing of zinc powders or zinc dust is carried out in different processes. Some processes manufacture only zinc powders and others only zinc dust, some only metallic zinc and some only zinc oxide. However, some processes can produce both metallic and oxidic type of products. Thanks to the processes used, often based on preferential evaporation of zinc at high temperature, the production of zinc powders or zinc dusts allows an efficient recycling of secondary material flows from galvanizing.

Journal of Physics: Conference Series **2738** (2024) 012022 doi:10.1088/1742-6596/2738/1/012022

5.1. *Zinc recovery from secondary raw materials with the French process*

The French process (FP) or indirect process is the simplest and most widely used process for making zinc oxide (ZnO) powders. Zinc is melted in a graphite crucible and vaporized at temperatures typically around or above 1,000°C. The high temperature drives the separation of zinc and iron in iron-zinc intermetallics, often without the help of reducing agents such as carbon. The zinc vapor reacts with the oxygen in the air to produce ZnO, accompanied by a drop in temperature and bright luminescence. The feed is generally bottom dross from batch galvanizing process but can also be a primary feed such as special high grade SHG or high grade (HG) zinc. This process is often used to treat this type of secondary raw materials because bottom drosses contain associated elements such as iron, lead, tin, and nickel. These elements are not very volatile around 1,000°C and therefore remain, for a large part, at the bottom of the crucible. This makes it possible to purify secondary materials in a very simple way. Depending on process parameters, a wide variety of zinc powder grades can be produced, see Table 4 [12].

Industrial grades of zinc oxide. Data are adapted from the product datasheets from industrial producers: PT. Indo Lysaght Indonesia, US Zinc in the USA, Umicore Zinc Chemicals and Silox in Belgium, IEQSA in Peru and Grillo Zinkoxid GmbH in Germany.

ZnO Grade	Nominal purity (%)	Specific surface area ($m^2 g^{-1}$)	Production process
Gold Seal	99.995	4–7	French Process
Pharma Grade	99.8–99.9	3–9	French Process
White Seal	99.8	3–5	French Process
Green Seal	99.6–99.7	4–10	French Process
Red Seal	99.5	3–5	French Process
American Grade	98.5–99.5	Max. 3	American Process
Active Grade	93–98	Min. 30	Wet process
Feed Grade	90–99	-	Various

Table 4: Zinc oxide powder commercialized grades

5.2. *Zinc recovery from secondary raw materials with the American process*

The American process (AP) or direct process is a more complicated process able to treat complex oxidized residues such as galvanizing top ashes, after mechanical separation of the metallic parts from the ashes. The zinc oxide in the residues is reduced via carbon sources in a rotary drum furnace at high temperature (about 1,200°C). When the zinc is reduced into metal, the zinc evaporates, and the fume is re-oxidized to make purer ZnO powders. This process is able to purify the zinc from different associated elements. Effectively, some associated elements are not reduced (Al, Mg) and stay in the oxide fraction. Other associated elements are reduced but the resulting metals are not volatile (Fe, Sn) so stay in the slag phase. It must be mentioned that the separation is not perfect because the oxides of some metals are also volatile at the process temperature and report to the ZnO fraction in the gas stream. Also, dust with other associated elements can be entrained in the ZnO fraction in the gas phase. Without addition of carbon, the American process can serve as a clinker process to remove halides (Cl, F) from material streams.

5.3. *Zinc recovery from secondary raw materials with other processes*

The Larvik process allows the treatment of different zinc materials to produce powders. The materials are melted in an electric resistance furnace equipped with various baffles. This furnace is topped by a distillation column which allows a certain purification to be achieved. The vapor is then condensed into zinc oxide powder or zinc metal dust depending on the atmosphere. The feed is usually SHG zinc ingots, HG zinc ingots and zinc drosses from the batch galvanizing process. Zinc dusts obtained by condensation are generally round in shape.

The Norzinco process (or New Jersey process) is also a process that uses an electric melting furnace topped by a distillation column. The electric furnace is heated by induction. Like the Larvik process, it is possible to produce oxidized zinc powders and zinc metal dusts. The feed for this type of furnace also consists of primary zinc (SHG and HG) and/or secondary zinc.

In 2015, the production of zinc oxide powders and zinc metal powders & dust was estimated at 1.28 Mt per year and 300 kt per year respectively, which represents a total of 1.58 Mt per year. Raw materials used for these products were 546 kt zinc in secondary raw materials (35%) and 1,034 kt zinc in primary raw materials (65%).

Journal of Physics: Conference Series **2738** (2024) 012022 doi:10.1088/1742-6596/2738/1/012022

6. Recovery of zinc from secondary raw materials from end-of-life recycling

6.1. Zinc recovery from galvanized steel scrap

Steel is primarily produced via: 1) The blast furnace (BF) / basic oxygen furnace (BOF) route, or 2) The electrical arc furnace (EAF) route. Other technologies exist such as the open hearth furnace (OHF), or are under development (e.g. gas direct reduced iron (GDRI) or coal direct reduced iron (CDRI)). The BF/BOF route represents about 70% and the EAF route represents about 25% of the total steel production volume globally. These two technology routes are the most used processes with about 95% of the total steel production volume. The ratio between EAF and BF/BOF varies as a function of time and region. In developing regions, BF/BOF production is predominant because the available of steel scrap is typically low and EAF technology is not well developed. Figure 4 shows a strong growth of EAF production technology in North America and, to a lesser extend, in Europe in the last 3 to 4 decades [13]. In the longer term, absolute steel volumes produced through the EAF route will increase in almost all regions, and probably, the relative volumes will also increase.

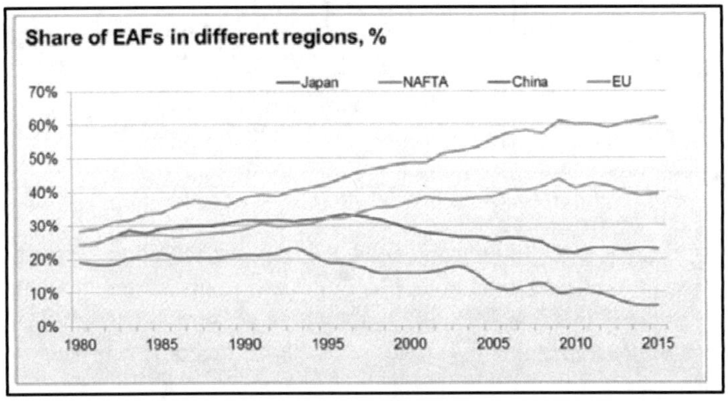

Figure 4: Variability of the EAF process share as a function of producing regions

New and old, non-galvanized and galvanized steel scraps are all recycled via the BF/BOF and EAF routes. The total scrap ratio in the BF/BOF process ranges between 5% to 35%, where scrap is used as coolant for the converter in the BOF process. The scrap ratio in the EAF process can be close to 100%, although direct reduced iron (DRI) can be used in EAF processes as well to substitute scrap, depending on availability of steel scrap in the marketplace. Zinc which is contained in the coating layer of galvanized scrap does not report to the new steel produced in an EAF. The operating temperatures of an EAF (generally >1,500°C) are much higher than the evaporation temperature of zinc (907°C). Zinc is therefore reporting to the steel dust residue, that is collected in filters.

Blast furnaces and basic oxygen furnaces also produce steel dust which is collected in filters. The typical zinc content in the 3 types of steel dust is very different:

- Blast Furnace (BF) flue dust consists of iron-oxide particles, oxides, carbon, and lime. It is generated by the blast furnace during smelting and collected in the gas exhaust and cleaning system (e.g., bag house filters). Those residues consist of mainly iron and 0-3% zinc on average.
- Basic Oxygen Furnace (BOF) flue dust consists of iron-oxide particles generated by the basic oxygen furnace during steel making and is collected in the gas exhaust and cleaning system (e.g., bag house filter). It typically consists of mainly iron oxide and 0.5-6% zinc on average.
- Electric Arc Furnace (EAF) flue dust consists of iron oxide particles generated in the electric arc furnace during steel making. It is collected in the gas exhaust and cleaning system (e.g., bag house filter). Most of the dust is iron oxide with 10-40% zinc and trace amounts of lead, cadmium, chromium and arsenic.

Figure 5 show the absolute weights and relative proportions of residues produced by BF/BOF and EAF processes [14]. Slags are the residues produced in larger quantities (150 kg to 300 kg per ton of steel produced). Dusts and sludges are generally produced in smaller quantities (20 kg to 30 kg per ton of steel produced). The latter quantities are nevertheless important in absolute terms if the very large quantity of steel produced in the world is taken into account.

Based on the estimates described above, a calculation of the quantities of zinc ending up in steel dusts per year can be made. The amount of crude steel produced in 2015 was 1,620 Mt per year. For each route, an estimate of the produced zinc volumes in the steel dusts can be assumed:

- BF (Blast Furnace): **340 kt zinc** per year in produced BF dust globally
- BOF (Basic Oxygen Furnace): **1,200 kt zinc** per year in produced BOF dust globally
- EAF (Electric Arc Furnace): **2,050 kt zinc** per year in produced EAF dust globally

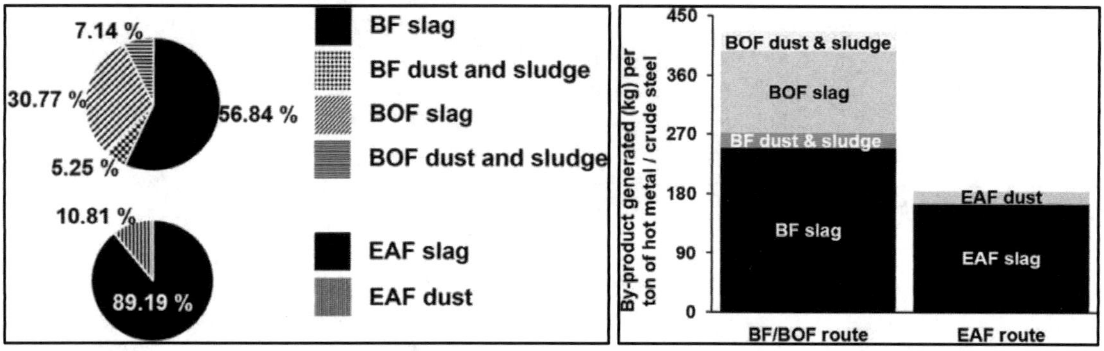

Figure 5: Absolute weights and relative proportions of residues produced by BF/BOF and EAF processes

The zinc concentration in BF and BOF dusts is too low to be economically processed in classical recycling processes for zinc. Consequently, at the time being, a very low amount of zinc is recycled and recovered from BF and BOF dusts. This quantity is below 10% of the total zinc present in BF and BOF dusts. An exemption is a BF process operated DK Recycling in Germany. This special BF process is able to process a variety of secondary raw materials, including a significant amount of BOF dusts. The unique BF process is able to accept up to 38% of zinc in the feed materials, which is extremely high compared to the feed limitations of classical BF processes which is typically up to 0.1% of zinc. Also, relatively small amounts of BOF dusts are used in the cement industry as BOF dust provides a source of Fe in cement. This is considered as a loss of zinc and not a use of zinc as the zinc has no meaningful function in cement apart from the fact that zinc in concrete is a corrosion inhibitor to reinforcing steel.

More constringent regulation for landfilling of BF and BOF residues can stimulate and increase the recycling of those materials. In some regions, like in some developed Asian countries where land area is scarce, innovative, or classical processes such as the rotary hearth furnace (RHF) and the electric smelting reduction furnace (ESRF) have been developed to recover both iron and zinc from these steel dusts. EAF dusts from these regions are recycled in these two processes.

The zinc concentration in EAF dusts is often high enough to be economically processed in classical recycling processes for zinc. There are many possible treatment technologies for EAF dust. Most of these treatments are pyro-metallurgical, but there are also hydro-metallurgical process options. However, the Waelz process is still by far the most used process and still representing, in volume, 85% of the market in 2015. Not all EAF dusts are recycled, a meaningful amount is still landfilled, even though EAF dust is considered a hazardous waste. It is mentioned above that 2,050 kt of zinc in EAF dusts was available for recycling in 2015. The actual recycled amount of zinc in 2015 from EAF dusts was a fraction of this, being 890 kt of zinc or 42% of all zinc in EAF being available for recycling. Consequently, **1,160 kt of zinc** was lost to landfill sites.

Pyrometallurgical processes are often used for recycling of steel dusts because they rely on the ability to reduce (at about 1,200°C) and vaporize the zinc metal at a relatively low temperature (the zinc boiling point is 907°C). Within this group of processes, some recover only zinc (such as the Waelz process) while other processes recover both zinc and iron. Within this last group, there are processes that recover iron in the form of DRI (e.g., RHF) and processes that recover iron in liquid form (e.g., ESRF). Table 5 presents the main commercial and experimental pyrometallurgical processes that are currently used in the world [15].

Pyrometallurgical processes for treatment of EAF dust.

Process	Temperature (°C)	Furnace type	Inputs	Products	Scale	Capacity (kt y^{-1})
Waelz kiln	1150–1200	Rotary kiln	Coal, lime/sand, natural gas, air	Crude zinc oxide and Fe-slag (waste)	Commercial	40–250
RHF	1250–1300	Rotary hearth furnace	Coke, binder, and natural gas	Crude zinc oxide and DRI/hot briquetted iron (HBI)	Commercial	100–300
PRIMUS	1000–1100	Multiple hearth furnace	Coal and air	Crude zinc oxide and DRI/iron concentrate	Commercial	100
OXYCUP	1500–1600	Cupola	Coke, scrap, and bricks (waste + cement + C)	Crude zinc oxide, pig iron, and slag	Commercial	200
Coke-packed bed	1500–1600	Shaft furnace	Coke, fluxes, and O_2-enriched air	Crude zinc oxide, molten metal, and slag	Commercial	10
Ausmelt	1250–1300	Top submerged lance furnace	Coal and O_2-enriched air	Crude zinc oxide, and Fe-slag (waste)	Commercial	100
ESRF	1300–1500	EAF	Binder, air, and electricity	Crude zinc oxide, pig iron, and slag	Commercial	36
Plasamadust	1400–1600	Shaft furnace equipped with plasma devices	Coke, coal, fluxes and superheated plasma gas	Crude zinc oxide, molten iron, and slag	Commercial	10–20
Plasma-arc	1400–1600	Plasma arc furnace	Coal, air, and electricity	Crude zinc oxide, molten iron, and slag	Commercial	5–40
EMPF	1400–1600	Slag-resistance furnace	Coal and air	Crude zinc oxide, molten iron, and slag	Commercial	40
Submerged plasma	1300–1400	Submerged plasma reactor	Coke, natural/petroleum gas, flux, air, and electricity	Crude zinc oxide and Fe-slag (waste)	Commercial	40–60
PIZO	1300–1500	Induction furnace	Coal, air, electricity	Crude zinc oxide, pig iron and slag	Commercial	50
Flame reactor	1600	Flame Reactor	Natural gas, O_2-enriched air	Crude zinc oxide, molten metal, and slag	Commercial	30
Thermal plasma reduction	1000–1400	Plasma arc furnace	Natural gas/CO, electricity	Crude zinc oxide and Fe-rich slag for steelmaking	Lab	–
Microwave processing	1000–1200	Microwave furnace	Coke/coal, air, electricity	Crude zinc oxide and DRI	Lab	–
Solar thermal reduction	800–1350	Solar furnace	Graphite, silica, and sunlight	Zinc, crude zinc oxide, and slag	Lab	–
Iron bath smelting	1400–1500	–	Coal/silicon, fluxes, and heat	Crude zinc oxide, molten metal, and slag	Lab	–
Calcification	900–1100	–	$CaCO_3$/lime, O_2-enriched air, and heat	Crude zinc oxide and $Ca_2Fe_2O_5$ for use in blast furnace	Lab	–
Halogenation	875–1050	–	PVC and heat	$ZnCl_2$ and Fe + C pellets	Lab	–

Table 5: Pyrometallurgical processes for treatment of EAF dust

Hydrometallurgical processes for recycling of steel dusts have disadvantages linked to the intensive use of water, and advantages such as the possibility of recovering valuable metals present in the steel dusts (copper, nickel, etc.).

Once zinc is separated from the other elements, the resulting material, which is called crude zinc oxide (CZO), contains typically 50-70% zinc. If CZO is produced with the Waelz process, it is typically called Waelz oxide (WOX). CZO or WOX is generally further processed by existing hydro-metallurgical processes to produce special high grade (SHG) zinc (min. 99.995%). Other process for treating CZO or WOX can be used as well, like the pyro-metallurgical imperial smelting process (ISP) or the solvent extraction & electrowinning (SX-EW) process. Other processes like the ZincOx process purify CZO or WOX to manufacture marketable zinc oxide powders directly, thus avoiding the energy spent (3,100-3,500 kWh/t of electricity) to produce metallic refined zinc metal.

6.2. Zinc recovery from municipal solid waste

According to the International Solid Waste Association (ISWA), the total volume of municipal solid waste (MSW) generated worldwide is about 2,000 Mt per year. It is estimated that about 260 Mt of this waste is incinerated. Many countries in Europe, North America and part of Asia incinerate a higher proportion of their waste. Incineration, composting and recycling have become increasingly important over time. At the time being, the MSW incinerated in Europe is estimated at 70 Mt per year.

The incineration of municipal wastes is often conducted at around 850°C for normal waste (1,100°C for hazardous waste). Figure 6 shows the incineration process with the formation of incinerator fly ashes (IFA) and incinerator bottom ashes (IBA) [16]. In this figure, the (potential) recovery of metals and energy is also mentioned. In some incinerators, IFAs are treated in order to sort out certain elements like metals. At 850°C, zinc is melted and partially oxidized. Part of it can be vaporized and/or oxide particles can be carried to the filter. Figure 7 shows the repartition of the metals between the IFA and the IBA [17].

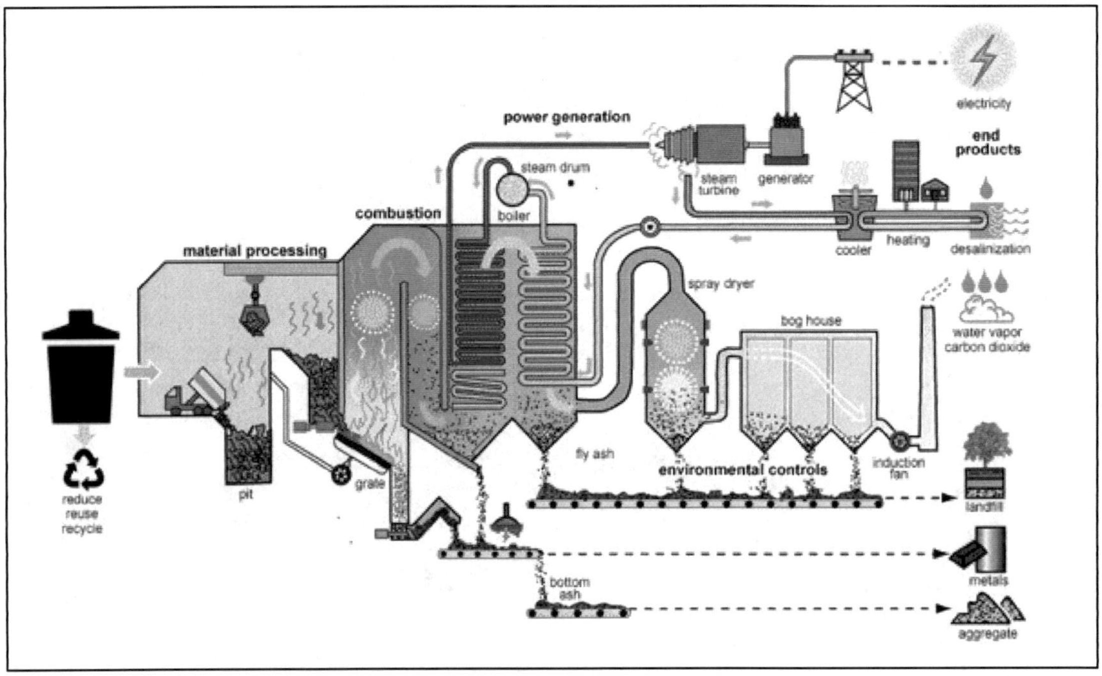

Figure 6: Incinerator with energy and metals recovery [16]

It is estimated that an average of **200 kt of zinc** in IBA and **400 kt of zinc** in IFA is available in Europe for recycling from waste incinerator residues, however most of these zinc volumes are currently not recovered in zinc recovery processes. If all MSW produced in the world would be incinerated, there is a potential source of **1.8 Mt of zinc** to be recovered per year.

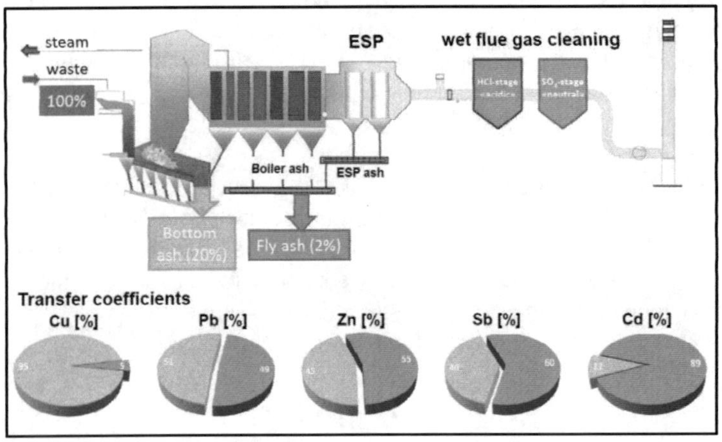

Figure 7: IBA and IFA volume fractions. Repartition of the metals between IBA and IFA

7. Conclusions

Zinc losses occur in the refined zinc production chain, in zinc first use production processes and at the end-of-life phase of products that contain zinc. Some of these zinc losses are recycled, in particular in case when the collection and sorting of zinc scrap or zinc waste is possible, when zinc scraps or wastes contain a high amount of zinc and low concentrations of other impurities, and when the recycling of zinc is economical.

The largest recycling loops for zinc are the recycling and recovery of zinc from batch galvanizing residues (1,150 kt Zn/year); pre- and post-consumer galvanized steel scrap (890 kt Zn/year); brass scrap (874 kt Zn/year); continuous galvanizing residues (635 kt Zn/year) and end-of-life zinc die casting scrap (475 kt Zn/year), all with reference to the year 2015. Recycling of rolled zinc (for roofing and building) has the highest end-of-life recycling rate of >95%. Recycling of manufacturing or prompt scrap from galvanizing (galvanizing residues) has a very high recycling rate also due to the very high zinc content in these residues and relatively low concentration of impurities in these pre-consumer scrap materials, which makes it economically viable to recycle them in commercial grade secondary zinc products.

At the other hand, many opportunities still exist to increase zinc recycling and reduce losses of zinc in abandoned products, to landfilling sites and to waste incinerators. Various zinc containing waste materials are not yet recycled and zinc is not recovered because the collection and sorting is a challenge, or recycling is not economical. Examples of opportunities to increase recycled zinc volumes are losses of zinc in the form of tailing losses from mineral processing (up to 1,900 kt Zn/year); not collected end-of-life zinc die cast parts (up to 1,525 kt Zn/year); landfilled BF and BOF steel dusts (up to 1,540 kt Zn/year) including BOF dust now used in cement production; landfilled EAF steel dusts (up to 1,160 kt Zn/year); not collected end-of-life brass parts (up to 730 kt Zn/year); not recycled bottom and fly ashes from municipal solid waste incinerators in Europe (up to 600 kt Zn/year) and residue losses from zinc smelting and refining (up to 415 kt Zn/year), all with reference to the year 2015.

8. Acknowledgements

Mr. Bruno Gay from B.CIRCLE, who co-authored this paper, is acknowledged for his expertise, knowledge and experience, and network in the zinc industry. He has done most of the desk top research and writing of the full report for the International Zinc Association.

References

[1] European_Commision. (11 December 2019). *The European Green Deal*. Retrieved from European Commission: https://ec.europa.eu/info/sites/info/files/european-green-deal-communication_en.pdf

[2] OECD_Publishing. (12 February 2019). *OECD iLibrary*. Retrieved from Global Material Resources Outlook to 2060: Economic Drivers and Environmental Consequences: https://doi.org/10.1787/9789264307452-en

[3] Metalprodex. (2021). *The global metal product exchange*. Retrieved from Zinc: https://www.metalprodex.com/metals/zinc.html

[4] Rostek, L., Tercero Espinoza, L., Goldmann, D., & Loibl, A. (2022). A dynamic material flow analysis of the global anthropogenic zinc cycle: Providing a quantitative basis for circularity discussions. *Resources, Conservation & Recycling 180*, 1-10.

[5] Pirard, E. (2021). *Zinc Resources – A State of Knowledge* (IZA project report). Brussels: International Zinc Association.

[6] USGS. (2021). *Mineral Statistics and Commodity Information*. Retrieved from USGS: https://www.usgs.gov/programs/mineral-resources-program

[7] ILZSG. (2020). *Press Release April 2020*. Lisbon: ILZSG

[8] Creedy, S., Glinin, A., Matusewicz, R., Hughes, S., & Reuter, M. (2013). Outotec Ausmelt Technology for Treating Zinc residues. *World of Metallurgy – ERZMETALL 66 No. 4*, 230-235.

[9] Vanderheyden, B. (2019). RECLAiming valuable METals from process residues through HIsarna ironmaking (RECLAMET) - focus on pre-processing. *Sixth PROMETIA Annual Seminar*. Kristiansand: CRM Group.

[10] Kangas, P., Koukkari, P., Wilson, B., Lundström, M., Rastas, J., Saikkonen, P., Hintikka, V. (n.d.). *Hydrometallurgical processing of jarosite to value-added products*. VTT – Technical Research Centre of Finland, Aalto University, School of Chemical Technology, Finland & Promeca Consulting Oy.

[11] Tercero Espinoza, L. (2018), *A global, dynamic stock-and-flow model for leaded brass*, Fraunhofer ISI.

[12] Moezzi, A., McDonagh, A. M., & Cortie, M. B. (2012). Zinc oxide particles: Synthesis, properties and applications. *Chemical Engineering Journal*, 1– 22.

[13] Ciftci, B. B. (2017). Global steel industry: outlook, challenges and opportunities. *5th International Steel Industry & Sector Relations Conference*. Istanbul: World Steel Association.

[14] Rieger, J., & Schenk, J. (2020). State-of-the-art Processing Solutions of Steelmaking Residuals , February 2020. *BHM Berg- und Hüttenmännische Monatshefte, Issue 5.*

[15] Lin, X., Peng, Z., Yan, J., Li, Z., Hwang, J.-Y., Zhang, Y., Jiang, T. (2017). Pyrometallurgical recycling of electric arc furnace dust. *Journal of Cleaner Production 149*, 1079-1100

[16] Cindrigo Ltd. (2021). Cindrigo Ltd. Retrieved from Technical Flow Chart: https://www.cindrigo.com/industry/wte/technical-flow-chart/

[17] Schlumberger, S. (2020). Urban mining - Zinc recovery from municipal solid waste incineration fly ashes in Switzerland. *GDMB Zinc Expert Meeting 2020*. Virtual: KEBAG / ZAR.

Application of Oxygen-enriched Side-blowing Technology in Treating Zinc Leaching Residue in China

Xu Liang [1,*], Zhang Ge[2] and Wang Bo[3]

1. China ENFI, Beijing, China, 100038. Email: xuliang@enfi.com.cn

2. China ENFI, Beijing, China, 100038. Email: zhangg@enfi.com.cn

3. China ENFI, Beijing, China, 100038. Email: wbo@enfi.com.cn

Keywords: Zinc residue, Side-submerged combustion smelting, Vanukov furnace

Abstract: Zinc hydrometallurgy is the main process of zinc smelting in the world. The leaching residue produced by different hydrometallurgical processes have different components, but they contain valuable metals such as zinc, silver, gold, indium and germanium, and heavy metals such as lead, cadmium and arsenic. In China, leaching residue is defined as hazardous waste, which requires harmless treatment and recovery of valuable metals. At present, China's main zinc leaching residue treatment technologies include Welz kiln, fuming furnace and Top-submerged lance Smelting furnace. These technologies have been applied for many years and are mature and reliable, but they need further improvement in energy conservation and environmental protection. In recent years, with the continuous progress of China's smelting technology, the oxygen-enriched side-blowing technology has been successfully applied to the treatment of zinc leaching residue, mainly including the Side-submerged combustion smelting technology of China ENFI and the Vanukov side-blowing technology. At present, the above two side blowing technologies have been put into industrial practice in many zinc smelters in China, which are characterized by low energy consumption, high recovery rate of valuable metals and high sulfur recovery rate. This paper introduces the application of the above two side blowing technologies in China's zinc smelters, and compares them from furnace structure, production operation, and technical indicators. Oxygen-enriched side-blowing technology will become the mainstream treatment process of zinc residue in China and even in the world in the future.

1. Current status of zinc leaching residue treatment technology

The zinc smelting process is mainly divided into two types: pyrometallurgy and hydrometallurgy, with hydrometallurgy accounting for over 80% of the total zinc smelting capacity worldwide. The leaching residues produced by the zinc hydrometallurgy process contains heavy metals such as lead, arsenic, and cadmium. Zinc leaching residue has been classified as hazardous waste in China (waste codes 321-004, 005, 006, 007-48). Before 2020, some zinc smelting factories constructed hazardous waste disposal sites to reduce production costs. In 2020, the Ministry of Industry and Information Technology of China issued the "Standard Conditions for the Lead and Zinc Industry", which stipulated that the zinc hydrometallurgy process must be equipped with a harmless treatment system for leaching residue and sulfur residue treatment facilities[1]. Therefore, the new zinc factories need to be equipped with harmless disposal facilities for zinc leaching residue. For existing zinc factories that use hazardous waste disposal sites to store leaching residues, limited by the strict environmental policies, they will also build harmless treatment facilities for zinc leaching residues. The selection of harmless disposal technology for zinc leaching residue has become a top priority for the sustainable development of zinc factories.

The treatment of zinc leaching residue usually adopts the high-temperature volatilization process by pyrometallurgy, which not only achieves harmless treatment of the residue but also achieves the recovery of valuable metals. At present, the main technology of zinc residues treatment in China is Waelz kiln, box fuming furnace,top-submerged lance smelting furnace (TSL) and side-blowing furnace & box fuming furnace.

1.1 Waelz kiln

Waelz kiln is the most widely used in the world. Its advantages are mature and reliable technology, and simple operation. However, the coke rate of this technology is as high as 40%~50%, and it needs to use coke or high-quality anthracite, which has high processing costs. At the same time, there are also problems such as low silver recovery rate, large amount of flue gas, low concentration of sulfur dioxide in the flue gas, and easy formation of kiln deposits in the kiln[2].

1.2 Box fuming furnace

Box fuming furnace was first applied in China at the beginning of this century and is currently being used by two factories. Its advantage is that the fuel can be bituminous coal, with a low price. But this technology can only use air or low oxygen enriched air for smelting, with low energy efficiency and coal consumption of up to 50%~60%. Due to the intermittent feeding process in its smelting process, the fluctuation of sulfur dioxide in the flue gas is significant, resulting in high flue gas treatment costs[3]. This technology is suitable for lead-zinc joint smelting factory, which can use lead smelting hot slag to treat zinc leaching residue.

1.3 TSL

TSL is an imported technology that was first applied in China in 2015.At present, only one factory in China is using it. Its advantage is the use of oxygen rich smelting, which has a high smelting intensity. However, this technology has drawbacks such as high coal consumption, frequent replacement of lance, high construction investment, low sulfur dioxide concentration in flue gas[4]. Due to its drawbacks of this technology, this technology has not been widely applied in China.

1.4 Side-blowing furnace & box fuming furnace

In view of the problems existing in the above technologies, ENFI developed the technology of treating zinc leaching residue with side-submerged combustion furnace & box fuming furnace, and it was successfully put into production in Chihong Zn & Ge Huize Smelter in July 2019. This device is the world's first production line to use side blowing furnace to treat zinc leaching residue. With the successful application of this technology, ENFI has promoted the process to Gansu Baiyin Nonferrous and Sichuan Shengtun Mining, and has made continuous optimization and improvement in technology. The above two factories were successfully put into operation at the end of 2021 and achieved production and standard in 2022, with all indicators exceeding the design values.

At the same time, a design institute in Northwest China designed a 200000 ton Jarosite residue treatment system in Hanzhong Zinc Industry in September 2020, using a Vanukov type side-blowing furnace & box fuming furnace.

At present, this technology has been applied in multiple zinc smelting factories in China, with the advantages of using high oxygen rich smelting, high smelting intensity, low coal consumption, high recovery rate of precious metals such as gold and silver, high sulfur dioxide concentration in side-blowing furnace flue gas, which can directly produce acid and achieve sulfur resource utilization at low cost.

1.5 Summary

Practice has proven that side-blowing furnace & box fuming furnace technology has significant advantages, such as simple operation, low energy consumption, high recovery rate of valuable metals, and good environmental protection. The following text introduces the application of this technology and its application in zinc smelting factories in China, and analyzes and compares the two side blowing technologies from aspects such as furnace structure, production operation, and technical indicators.

2. Introduction of side-blowing furnace & box fuming furnace technology

2.1 Reaction principle of zinc residues smelting

Zinc in zinc residues mainly exists in the form of $ZnO{\bullet}Fe_2O_3$, ZnO, ZnS, $ZnSO_4$, and $ZnO{\bullet}SiO_2$, with $ZnO{\bullet}Fe_2O_3$ and ZnO accounting for 90% of the total zinc in the residue. Pb is mainly $PbSO_4$. The compounds of lead and zinc are mainly oxides and sulfates, both of which are non heating materials. The boiling point of zinc is 907℃, and lead and its compounds are easily volatile at high temperatures. This technology utilizes the volatile characteristics of lead and zinc metals at high temperatures to achieve the reduction and volatilization. The lead and zinc vapor is re oxidized in the gas phase zone in contact with oxygen to generate small particles of zinc oxide and lead oxide, which enter the dust collection facility with the flue gas and are collected. The main chemical reactions are shown in equations (1) to (7) [5].

$$ZnO{\bullet}Fe_2O_3+SiO_2+CO_{(g)}=ZnO+2FeO{\bullet}SiO_2+CO_{2(g)} \quad \triangle H=-23\ kJ/mol \quad (1)$$

$$ZnS+3/2\ O_{2(g)}=ZnO+SO_{2(g)} \quad \triangle H=-444\ kJ/mol \quad (2)$$

$$ZnSO_4=ZnO+SO_{2(g)}+1/2O_{2(g)} \quad \triangle H=+333\ kJ/mol \quad (3)$$

$$ZnO{\bullet}SiO_2+CaO=ZnO+CaO{\bullet}SiO_2 \quad \triangle H=-88\ kJ/mol \quad (4)$$

$ZnO + CO_{(g)} = Zn_{(g)} + CO_{2(g)}$ △H= +198 kJ/mol （5）

$PbSO_4 = PbO + SO_{2(g)} + 1/2O_{2(g)}$ △H= +405 kJ/mol （6）

$PbO + CO_{(g)} = Pb_{(g)} + CO_{2(g)}$ △H= +131 kJ/mol （7）

The smelting of zinc leaching residue is mainly divided into two stages: residue melting and reduction volatilization. The melting stage deals with the cold material of zinc leaching residue, which absorbs a large amount of heat during the melting and decomposition of various substances and consumes a large amount of heat. The reduction and volatilization stage deals with hot slag, and the reduction and volatilization of lead and zinc require some heat, which accounts for a relatively low proportion of the total energy consumption. Therefore, during the residue melting stage, the side-blowing furnace adopts high oxygen rich combustion to achieve the goal of saving energy consumption. The reduction and volatilization stage adopts a box fuming furnace, which provides strong stirring and controls a strong reducing atmosphere to achieve the reduction and volatilization of lead and zinc.

Zinc leaching residue is melted in a side-blowing furnace and reduced and volatilized in a box fuming furnace. Elements such as zinc, lead, silver, indium, germanium, and gold in the slag evaporate into the flue gas at high temperatures and are collected through boilers and dust collectors. The purified flue gas is sent for acid production and desulfurization. The tailings are general solid waste, stored or sold out.

2.2 Zinc residue melting in side-blowing furnace

As mentioned earlier, the main energy consumption and the removal of sulfur elements are also in the melting stage. The side-blowing furnace is a fixed type furnace with straight elliptical or rectangular edges, which is heated by a spray gun or crushed coal is added as fuel from the top of the furnace. Oxygen rich air is sprayed into the side spray gun or tuyere of the furnace to support combustion, and the concentration of oxygen rich air injected into the furnace is 60%~90%. The high concentration of oxygen rich air reduces the amount of gas generated during combustion, thereby reducing the heat carried away by the flue gas and effectively reducing fuel efficiency. The smelting temperature of the side-blowing furnace is controlled between 1200℃ and 1500℃. The sulfate in the zinc leaching residue decomposes to produce sulfur dioxide, and some zinc and lead volatilize into the flue gas. The slag is discharged into the box fuming furnace through a chute. The flue gas is cooled by a waste heat boiler and purified by a dust collector to obtain zinc oxide dust. The purified flue gas is sent to the acid production system to produce sulfuric acid.

2.3 Reduction and volatilization in box fuming furnace

The box fuming furnace is a rectangular fixed furnace type, where air and pulverized coal are injected through the side tuyere. The coal is burned to provide heat to the molten pool, while the excess coal plays a reducing role. The excess oxygen coefficient in the molten pool area is~0.6 [6]. The reduction and volatilization of zinc in the molten pool must ensure dynamic conditions. Therefore, the box fuming furnace uses air to support combustion, with a large gas volume and sufficient stirring strength. The temperature of the melting pool is controlled to 1300℃. After most of the lead and zinc are reduced and volatilized into the flue gas, they are oxidized again in the gas and ultimately collected in the form of dust. The tailings contains<2.0% zinc and<0.2% lead. The flue gas is cooled by a waste heat boiler and purified by a dust collector to obtain zinc oxide dust. The purified flue gas is sent for desulfurization.

The process flow chart[7] is as follows:

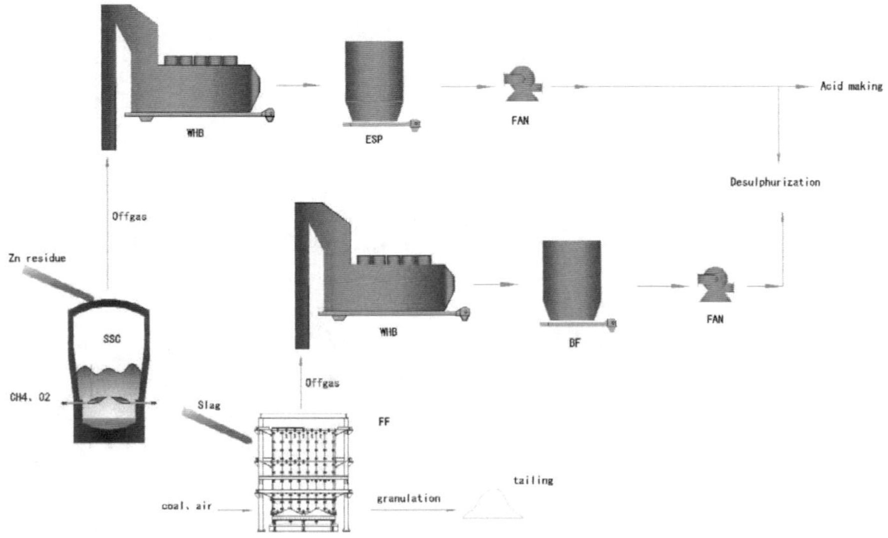

Fig. 1 Process Flow

3. Engineering cases

3.1 Case A

The factory has built one side-submerged combustion furnace to treat zinc leaching residue and oxidized ore. The slag produced by the side-submerged combustion furnace enters the existing box fuming furnace through a chute. The project was completed and put into operation at the end of July 2019, and the side-submerged combustion furnace processes approximately 500 tons of zinc containing materials on a dry basis per day.

(1) Raw materials

The amount of leaching residue and oxidized ore processed by side-submerged combustion furnace is about 20t/h dry basis. The components of leaching residue and oxidized ore are shown in Table 1.

Table 1 Leaching residue and oxidized ore

Item	Element	Zn	Pb	Ag	Ge
leaching residues	wt%	13～16	5～7	0.014～0.018	0.016～0.022
oxidized ore	wt%	12～18	4～6	～0.01	～0.01

(2) Output

The composition of tailing is shown in table 2.

The volatilization rates of Zn,Pb, Ag and Ge are shown in the table 3.

Table 2 Tailing

Element	Zn	Pb	Ag	Ge
wt%	1 ~ 2	0.02 ~ 0.2	0.001 ~ 0.002	0.001 ~ 0.002

Table 3 Volatilization rates

Element	Zn	Pb	Ag	Ge
%	90 ~ 95	95 ~ 98	92 ~ 98	90 ~ 95

(3) Consumption

The fuel consumption for processing each ton of residue is equivalent to 310kg~330kg of standard coal.

The equivalent oxygen consumption per ton of residue is about 300Nm^3/t to 330Nm^3/t.

3.2 Case B

The factory has built one Vanukov type side-blowing furnace, one insulation furnace, and two box fuming furnaces. The function of the insulation furnace is to distribute slag to two box fuming furnaces. The project was completed and put into industrial production at the end of September 2020, with a side-blowing furnace processing 600 tons of residue on a dry basis per day.

(1) Raw materials

The amount of leaching residue processed by side-blowing furnace is about 24t/h dry basis. The components of leaching residue is shown in Table 4.

Table 4 Leaching residue

Element	Zn	Pb	Ag
wt%	4～6	2～5	0.01～0.03

(2) Output

The composition of tailing is shown in table 5.

The volatilization rates of Zn,Pb, Ag are shown in the table 6.

Table 5 Tailing

Element	Zn	Pb	Ag
wt%	1～2	0.05～0.2	0.001～0.003

Table 6 Volatilization rates

Element	Zn	Pb	Ag
%	85～90	92～95	95～98

(3) Consumption

The fuel consumption for processing each ton of residue is equivalent to 320kg~340kg of standard coal.

The equivalent oxygen consumption per ton of residue is about 350Nm^3/t to 380Nm^3/t.

3.3 Case C

The factory has built one side-submerged combustion furnace, one box fuming furnaces. The project was completed and put into industrial production at the end of December 2021, with a side-blowing furnace processing 500 tons of residue on a dry basis per day.

(1) Raw materials

The amount of leaching residue processed by side-submerged combustion furnace is about 20t/h dry basis. The components of leaching residue is shown in Table 7.

Table 7 Leaching residue

Element	Zn	Pb	Ag
wt%	5~8	3~5	0.01~0.03

(2) Output

The composition of tailing is shown in table 8.

The volatilization rates of Zn,Pb, Ag are shown in the table 9.

Table 8 Tailing

Element	Zn	Pb	Ag
wt%	1~2	0.05~0.2	0.001~0.003

Table 9 Volatilization rates

Element	Zn	Pb	Ag
%	85~90	93~96	90~95

(3) Consumption

The fuel consumption for processing each ton of residue is equivalent to 270kg~290kg of standard coal.

The equivalent oxygen consumption per ton of residue is about 270Nm^3/t to 300Nm^3/t.

3.4 Case D

The factory has built one Vanukov type side-blowing furnace, one box fuming furnaces. The project was completed and put into industrial production at the end of December 2021. After nearly half a year of production in the factory, there are problems with temperature control difficulty, high energy consumption, large fluctuations in flue gas sulfur dioxide, unstable acid production, and high start-up costs in the side-blowing furnace. In May 2022, ENFI transformed it into side-submerged combustion. At present, the side-blowing furnace processes approximately 600 tons of residue on a dry basis every day.

(1) Raw materials

Before the renovation, the treatment capacity of residue was 20t/h dry basis, and after the renovation, the treatment capacity was increased to 25t/h dry basis. The components of leaching residue is shown in table 10.

Table 10 Leaching residue

Element	Zn	Pb	Ag	Ge
wt%	11 ~ 14	5 ~ 7	0.02 ~ 0.03	0.005 ~ 0.01

(2) Output

The composition of tailing is shown in table 11.

The volatilization rates of Zn,Pb, Ag and Ge are shown in the table 12.

Table 11 Tailing

Element	Zn	Pb	Ag	Ge
wt%	0.5～1.5	0.02～0.2	0.002～0.004	0.0005～0.0015

Table 12 Volatilization rates

Element	Zn	Pb	Ag	Ge
%	90～95	95～98	94～98	85～90

(3) Consumption

Before the renovation, the fuel consumption for processing each ton of residue is equivalent to 300kg~320kg of standard coal; after the renovation, the fuel consumption for processing each ton of residue is equivalent to 270kg~290kg of standard coal.

Before the renovation, the equivalent oxygen consumption per ton of residue is about 320Nm^3/h to 350Nm^3/h; after the renovation, the equivalent oxygen consumption per ton of residue is about 270Nm^3/t to 300Nm^3/t.

3.5 Summary

Based on the data from the above cases, make a summary as shown in table 13.

Table 13 Comparison of main indicators

Item	CASE A	CASE B	CASE C	CASE D
Treatment capacity(t/h)	~20	~24	~20	~25
Standard coal consumption in Tons of residue（kgcet/t）	310~330	320~340	270~290	270~290
Oxygen consumption(Nm^3/t)	300~330	350~380	270~300	270~300

4. Comparison of side-blowing technology for treating zinc leaching residue

At present, there are two types of side-blowing furnaces for treating zinc leaching residue in China, with significant differences in furnace structure, production operation, and technical indicators. The following text compares the two types of furnaces.

4.1 Furnace structure

(1)Furnace type

The distance between the tuyeres of the Vanukov type side blowing furnace is 1.5－2 meters; The distance between the tuyeres of the side-submerged combustion furnace is about 2－2.5 meters. The lower part of the two types of furnaces is equipped with a shallow hearth; Copper water jackets are used in the upper melting pool area of the furnace hearth, which are protected by hanging slag.

(2)Blowing method

The Vanukov type side blowing furnace is a single channel tuyere that only passes through oxygen rich air, and the valve group is simple. The side-submerged combustion furnace is a multi-channel spray gun that injects coal powder, oxygen rich air. The valve group is relatively complex and requires a matching coal powder injection device.

The distance between the tuyeres of the Vanukov type side blowing furnace is small and there are many tuyeres. It is necessary to change the tuyeres according to the furnace condition, and 50% of the tuyeres should be used normally. The distance between the spray guns of the side-submerged combustion furnace is large and the quantity is small. Generally, more than 70% of the spray guns are used.

(3) Heating method

Vanukov type side blowing furnace uses high calorific value granular coal (washed coal or Anthracite) as fuel and reducing agent. After being metered together with zinc leaching residue in the bin, it is fed into the furnace from the furnace top through the belt, and then the oxygen enriched air is blown through the side tuyere to support combustion and supplement heat. By adding crushed coal from the furnace top for combustion and heating, the temperature of the slag is uneven, and the temperature of the upper part of the molten pool reaches 1400℃ or even higher. Meanwhile, due to the high smelting temperature, a large amount of nitrogen oxide is produced.

The side-submerged combustion furnace uses a small amount of granular coal (with no special requirements for coal quality) as a reducing agent, measured together with zinc leaching residue in the bin, and then added to the furnace through a belt from the furnace top. The side spray gun injects powder coal and oxygen rich air to support combustion, and powder coal is used as the main fuel for heating. Due to the fact that only powder coal is injected from the middle of the molten pool for combustion and heating, the temperature of the molten pool is uniform and the slag temperature is low, with a temperature of less than 1300℃.

4.2 Production operation

(1) Oven

The Vanukov type side blowing furnace is operated by inserting multiple oxygen+natural gas burners from the tuyere for heating, and requires a separate natural gas drying system.

The side-submerged combustion furnace is mainly dried by inserting natural gas burner into the slag discharge port and the upper burner hole of the furnace.

(2) Furnace startup

The Vanukov type side blowing furnace does not have a burner and adopts the traditional blowing furnace opening method. It requires workers to add a large amount of wood, block coke, block slag, lead plates to create a molten pool. Taking a 13m^2 side blowing furnace as an example, 60 to 70 tons of materials are manually added to each melting pool.

The method of creating a molten pool in a side-submerged combustion furnace is relatively simple.

In the initial stage of the molten pool, a small amount of wood and crushed coke needs to be added. After the wood and coke are fully burned, automatic heating operation is carried out through powder coal spray gun.

(3) Tuyere operation

During normal production, the operation of the side blowing furnace mainly includes ingredient control and blast control. Both types of furnaces are automatically adjusted through DCS. However, due to the differences in tuyere and heating methods, the main difference in operation between the two types of furnaces lies in the tuyere operation.

The Vanukov type side blowing furnace uses tuyere to directly blow oxygen-rich air into the molten pool. However, due to the form of the tuyere, manual tuyere blowing operations need to be carried out on a regular basis for each shift to prevent tuyere blockage and observe the slag situation, which is labor-intensive and risky.

The side-submerged combustion furnace adopts a multi-channel spray gun to directly inject powder coal and oxygen-enriched air into the molten pool, without the need for normal operation. The gas flow and pressure of all spray guns, as well as the injection amount of powder coal, are monitored in real-time online, achieving fully automated operation of the spray gun.

4.3 Technical indicators

Based on the production practice of each factory, the following analysis and comparison were conducted on the indicators of using different side blowing furnaces to treat zinc leaching residue. The results are shown in table 14.

Table 14 Technical indicators

Item	Side-submerged combustion furnace+box fuming furnace	Vanukov type side-blowing furnace+box fuming furnace
Zn recovery rate	>90%	>90%
Pb recovery rate	>90%	>90%
Ag recovery rate	>90%	>90%
Fuel type	soft coal	anthracite+soft coal
Standard coal consumption in Tons of residue（kgcet/t）	270～290	300～340
O_2 (v%)	70%～85%(side blowing furnace)，21%(fuming furnace)	75～85%(side blowing furnace)，21%(fuming furnace)
SO_2 in flue gas(v%)	3%~4%	2%~4%
Zn in dust products(w%)	>50%	>50%
Smelting temperature（℃）	1150~1250	1350~1450
Smelting capacity of bed ($t/m^2.d$)	40~50	40~50
Nitrogen oxide content in side blowing furnace flue gas (mg/Nm^3)	500~1500	2000~8000

Based on the data in table 14, the following conclusions can be drawn:

(1) The volatilization rates of lead, zinc, and precious metals in the two side blowing technology are basically the same.

(2) Side-submerged combustion furnace, injecting powder coal effectively improves slag properties, lowers melting temperature, and reduces the generation of nitrogen oxides.

(3) Side-submerged combustion furnace with low coal consumption.

(4) Vanukov side blowing furnace needs to use high-quality anthracite, and the production cost is high.

5. Conclusion

In recent years, with the continuous progress of China's smelting technology, the treatment of zinc leaching residue using oxygen rich side blowing technology has been put into industrial practice in multiple zinc smelting factories in China, gradually replacing other zinc leaching residue treatment technologies. Practice has proven that the oxygen rich side blowing technology has the following advantages in treating zinc leaching residue.

(1) The melting of zinc leaching residue adopts high concentration oxygen rich side blowing smelting, and the actual standard coal consumption per ton of residue is 270~340kg, which reduces the standard coal consumption by more than 30% compared to other existing pyrometallurgical residue treatment processes. The energy-saving and carbon reduction effect is significant.

(2) The recovery rate of zinc, lead, silver and germanium is high, up to 90~98%, especially the recovery rate of silver is far higher than that of Welz kiln.

(3) The flue gas from the side blowing furnace has a high SO_2 content, which can directly produce acid. The flue gas treatment cost is low, achieving the resource utilization of sulfur.

(4) The side blowing furnace+fuming furnace process, using high-temperature melting, truly realizes the harmless treatment of zinc leaching residue, and has obvious environmental advantages.

At the same time, this article introduces the application of Vanukov type side blowing technology and side-submerged combustion technology in zinc smelting factories in China, and analyzes and compares them from the aspects of furnace structure, production operation, and technical indicators. Based on the above comparison, it can be seen that the side-submerged combustion technology developed by ENFI has advantages such as low energy consumption, controllable melting temperature, simple operation, high automation, low start-up cost, stable flue gas, and low nitrogen oxide content.

ENFI's side-submerged combustion technology has been successfully applied to many factories in the treatment of zinc leaching residue, such as Chihong Zn & Ge Huize Smelter, Gansu Baiyin Nonferrous and Sichuan Shengtun Mining, and will be widely used in global zinc smelting factories in the future. Through continuous engineering practice and continuous optimization and improvement of this technology, the zinc leaching residue treatment technology will be elevated to a higher level.

References

1. Standard Conditions for the Lead and Zinc Industry(2020)
2. FEN Ji-ping, Wang min,WEI Shan-xin (2014). Optimization of zinc volatilization kiln operation and production practice. MINING & METALLURGY, 23(2), 62-64.
3. JIANG Rong-sheng，CHAI Li-yuan,JIA Zhu-hong,WANG Qing-wei(2014). Production Practice and Discussion on Treating Lead and Zinc Smelting Slag by Fuming Process. YUNNAN METALUURGY ,43(1), 58-61.
4. WU Wan-gang,LI Wen-long,WANG Peng-fei (2018). Commercial production practice of zinc leaching residue treatment process by top-blowing furnace. CHINA NONFERROUS METALLURGY, 47(5), 6-9.
5. Lead Zinc Metallurgy[M]. CHINA SCIENCE PUBLISHING & MEDIA LTD, 2003.
6. Peng Rongqiu. Zinc metallurgy [M]. CENTRAL SOUTH UNIBERSITY PRESS, 2005.
7. XU Liang,Ma Shao-bin (2021). Zinc residue treatment technology by side-submerged combustion furnace + fuming Furnace. CHINA NONFERROUS METALLURGY, 50(2), 40-44.

Journal of Physics: Conference Series **2738** (2024) 012024 doi:10.1088/1742-6596/2738/1/012024

Low Carbon and High Efficient Utilization of Zinc-bearing Dust in Iron & Steel Industry by Rotary Kiln Process

Hengdi Ye[1], Fuqiang Zheng[1, *], Qian Li[1], Weimin Shen[2], Xingfu Lu[1], Bing Hu[1], Chen Liu[1]

1 *Zhongye Changtian* International *Engineering Co., Ltd., Changsha, Hunan, China. 410205.*

2 *Hunan Zhongye Changtian Energy Conservation and Environmental Protection Technology Co., Ltd. Changsha, Hunan, China. 410205.*

Corresponding author's e-mail: fq_zheng@126.com.

Abstract: More than 50 million tons per year of zinc-bearing dust with large amounts of zinc and iron resources is produced by iron and steel smelting in China, which is classified as hazardous solid waste. The zinc-bearing dust is an important secondary resource, in which the zinc and iron can be recycled as raw materials for the iron and zinc industries after separating them. Rotary kiln reduction is one of the main methods for separating zinc and iron from zinc-bearing dust. In this study, the 5 mm composite pellets containing zinc-bearing dust is prepared by forced disturbed pelleting as reduction charge. The rotary kiln system with controllable temperature field and atmosphere field by multi-point quantitative air suppling is used as the main equipment of reduction. This technology can realize the treatment of raw materials with high iron and zinc content in rotary kiln, but also can effectively improve the production efficiency and the quality of zinc oxide powder, reduce the kiln-ringing. Compared with the conventional rotary kiln method, the processing capacity of zinc-bearing dust increase by 30%. The zinc removal ratio reaches 95% and the ZnO content of zinc oxide powder is 53.93 wt.%. The TFe content and ZnO content of reduction product are 60.56 wt.% and 0.25%, respectively. The annual operation ratio of zinc-iron separation by rotary kiln with zinc-bearing dust is 92%. The metallization rate of reduction product is controllable. The reduction product with metallization rate of 65.97% can be used as a sintering raw material, and the comprehensive energy consumption is 187kgce/t. The reduction product with metallization rate of 88% can be used as a converter steelmaking raw material, and the comprehensive energy consumption is 253.1kgce/t.

1. Introduction

The iron and steel production process produces a large number of solid wastes, most of which can be used as building materials for overall utilization[1-3]. The other solid wastes cannot be directly utilized due to the high content of heavy metals and alkali metals[4, 5]. The requirements on transportation, storage and disposal of solid wastes with heavy metals and alkali metals are much higher than those of ordinary solid wastes[6, 7]. The difficulty and high cost of complex solid waste disposal are the key factors that restrict whether iron and steel enterprises can realize "solid waste not leaving the factory"[8]. Zinc-bearing dust is the main solid waste containing heavy metals produced in iron and steel process[9]. The output of zinc-bearing dust is about 3~5% of the output of crude steel. The output of zinc-bearing dust in iron and steel enterprises in China is about 30 ~50 million t/a. Zinc-bearing dust rich in iron, zinc and other valuable resources, has high recovery value[10, 11]. The reserves of zinc and iron are insufficient in China[12]. Therefore, the recovery of valuable metal from the zinc-bearing dust is an important way to solve the shortage of zinc and iron resources. However, zinc-bearing dust disposal process has great environmental risk because of its high pollution, which is also the key and difficult point of solid waste disposal in iron and steel plant.

At present, zinc-bearing dust are generally disposed by rotary hearth furnace or rotary kiln process[13, 14]. Rotary hearth furnace process has high primary investment, high energy consumption and poor product uniformity[15, 16]. In comparison, the rotary kiln method has wide adaptability of raw materials[17]. The material rolls forward in the kiln and is evenly heated. The uniformity of physical mixing and chemical reaction is better. So, the stability and technical economy of rotary kiln products are better than those of rotary hearth furnace[18, 19]. However, the problems of easy ring forming and low metallization ratio in traditional rotary kiln process need to be solved urgently[19, 20]. It is of great significance to develop a low-carbon and efficient recycling technology of zinc-bearing dust in green iron and steel process.

In this paper, the new technology and equipment for efficient cleaning and separation of valuable components in zinc-bearing dust are investigated to reduce energy consumption, extend the service life of equipment and improve product value. Finally, the full quantitative recovery of zinc-bearing dust can be achieved by sintering and other metallurgical processes.

2. Experimental

2.1. Raw Materials

Three kinds of zinc-bearing dust were used as raw materials in this study. The chemical compositions of zinc-bearing dust are listed in Table 1. The ZnO content of zinc-bearing dust A is the highest of 8.51 wt.%, and the Fe_2O_3 content is 16.11 wt.%. The ZnO content of zinc-bearing dust B is 4.73 wt.%, and the Fe_2O_3 content reaches 51.50 wt.%. The ZnO content of zinc-bearing dust C is 0.99.

Table 1 Chemical composition of zinc-bearing dust (wt.%).

Zinc-bearing dust	ZnO	Fe_2O_3	CaO	MgO	Al_2O_3	SiO_2	FC
A	8.51	16.11	14.01	0.01	0.90	1.48	0
B	4.73	51.50	11.78	3.44	0.33	2.34	1.98
C	0.99	22.03	7.92	1.24	3.17	7.25	42.83

The XRD pattern of zinc-bearing dust are shown in Figure 1. The major phases in the zinc-bearing dust A are hemihydrate gypsum and hematite. The major phases in the zinc-bearing dust B are magnetite, zinc ferrite, metallic iron, and a small amount of zinc oxide and calcium carbonate. The major phases in the zinc-bearing dust C are complex, mainly containing hematite, magnetite and a small amount of manganese-zinc ferrite with crystalline water, whose crystal structure is similar to magnetite. The other phases also include carbon, a small amount of quartz, calcium carbonate and calcium ferrite.

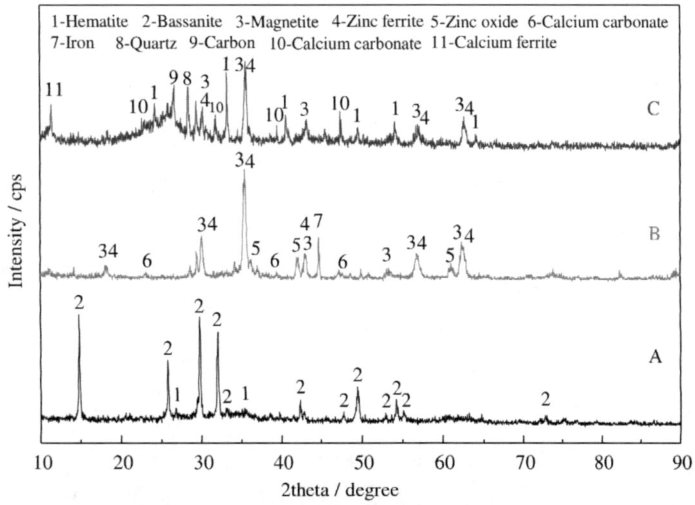

Fig. 1 XRD pattern of the zinc-bearing dust

2.2. Experimental Method

The composite pellets containing zinc-bearing dust were prepared by cylindrical pelletizing machine (Φ1000 × H200 mm) and forced disturbed pelletizing machine (Φ1000 × H500 mm). The reduction experiments were carried out in a muffle furnace. The high-temperature reduction products were cooled to room temperature in a nitrogen atmosphere. The reduction production were carried out in rotary kilns with a diameter of 2.5m and a length of 38m. The reduction products were prepared for analyzing the elemental composition, mineral phases and microstructure.

2.3. Analysis Methods

The chemical compositions of the samples were determined by Inductively coupled plasma spectrometer（ICP, Spectro blue sop).

The reduction samples were mounted into epoxy resin, mechanically ground and polished. Then the polished surface was characterized by an electron microscope for microstructure.

The mineral phases of the samples were characterized by X-ray diffraction (Cu K_a radiation, λ=0.154056 nm, 40 kV, 250 mA, SCAN: 10.0/70.0/0.02/.15 sec, D/max2550PC, Rigaku Co., Ltd., Japan).

3. Results and Discussion

3.1. Preparation of composite pellets containing zinc-bearing dust

3.1.1. Effects of pelletizing methods on pellet formation behavior

The effect of cylindrical pelletizing and forced disturbed pelletizing on pellet formation was simulated by EDEM, and the results were shown in Fig. 2 and Fig. 3. Under the same conditions, the agglomeration obtained by forced disturbed pelletizing is larger and the adhesion of fine particles is more. The strength of aggregate force chain increases rapidly with time and then decreases gradually in the process of forced disturbed pelletizing. At the early stage of the stirring paddle movement, the contact force of mixture increases rapidly, so that the strength of aggregate force chain increase. And then, the mixture accumulated under the feed port gradually tends to be distributed smoothly in the action of stirring force. The mixture acted by the stirring paddle decreases and tends to be stable, making the contact force gradually decrease and tends to be stable, and the aggregate further increases. The small contact force between the fine mineral powder attached to the surface will gradually reduce the strength of the right-hand chain.

The growth of particle aggregates can be divided into three stages: rapid growth stage, slow growth stage and steady change stage. When the agglomerates no longer grow, the strength of the particle agglomerates obtained from the forced disturbed pelletizing is greater than that of cylindrical pelletizing. The fluctuation of fine powder amount adhering to the surface of particle aggregates by forced disturbed pelletizing is less than that of cylindrical pelletizing. The particle size distribution of particle aggregates is more uniform.

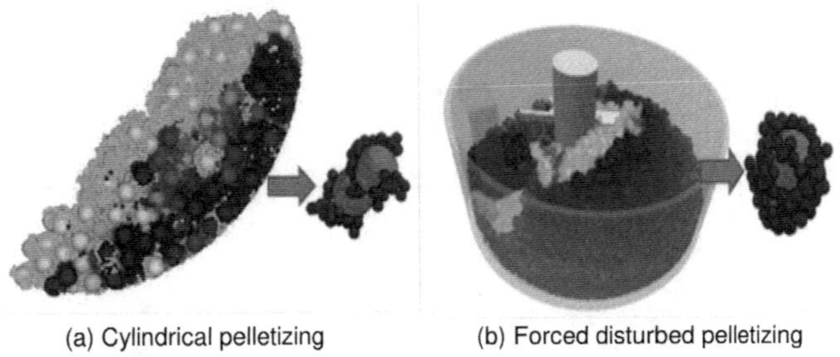

Fig. 2 Simulation of cylindrical pelletizing and forced disturbed pelletizing

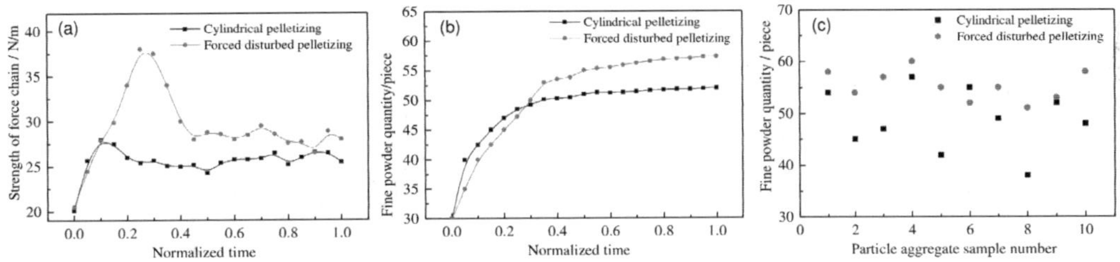

Fig. 3 Effect of cylindrical pelletizing and forced disturbed pelletizing on pellet formation

3.1.2. Effects of pelletizing methods on pellet size

The zinc-bearing dust A, B, C were mixed at 30%, 45%, 25% to pellet. The effect of forced disturbed pelletizing and cylindrical pelletizing of the pellet size under different moisture is shown in Table 2. When the moisture is 8.25 wt.%, the content of +3mm pellets by forced disturbed pelletizing is 10.4% more than that by cylindrical pelletizing. The average pellet size of pellets by forced disturbed pelletizing is 5.08 mm. When the proportion of +3mm pellets is at the same level, the moisture of forced disturbed pellets is 0.5% lower than that of cylinder pellets, and the proportion of 3-5mm pellets is 59.28%, which is 12.71% higher than that of cylinder pellets. The particle size of forced disturbed pellets are more uniform than that of cylinder pellet. The effect of different pelletizing methods on the volatilization of zinc and the reduction of iron is not significant. However, the moisture of the forced disturbed pellets is lower and the strength is higher, which is beneficial to improve the strength of the green pellet drying process and reduce the drying energy consumption. The pellet produced by forced disturbed pelletizing has high strength and produces less powder during the reduction process, which can alleviate the agglomeration in the kiln and prevent the reoxidation of reduction products.

Table 2 Effects of pelletizing methods on pellet size

Pelletizing methods	Moisture /wt.%	+3mm wt./%	3-5mm wt./%	average pellet size /mm
Cylindrical pelletizing	8.25	74.13	46.57	4.41
Forced disturbed pelletizing	7.75	73.46	59.28	4.54
	8.0	77.62	70.09	4.97
	8.25	84.53	79.65	5.08

3.2. Reduction of composite pellets containing zinc-bearing dust

3.2.1. Effect of temperature and time on reduction

The effect of temperature and time on zinc removal and iron reduction is shown in Fig. 4 and Fig. 5. The zinc removal ratio and metallization ratio increased gradually with the increase of reduction temperature. When the reduction temperature is 1150℃, the zinc removal ratio and metallization ratio reach the highest value. After the reduction time exceeds 35min, the zinc removal ratio and metallization ratio do not increase. The experimental results show that the metallization ratio of pellets can reach more than 96% and the zinc removal ratio can reach more than 99% as the reduction temperature is 1150°C and the reduction time is 35min.

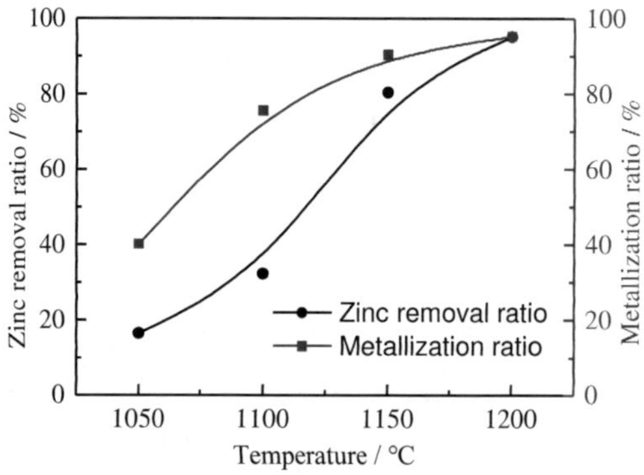

Fig. 4 Effect of temperature on zinc removal and iron reduction (reduction time of 30min)

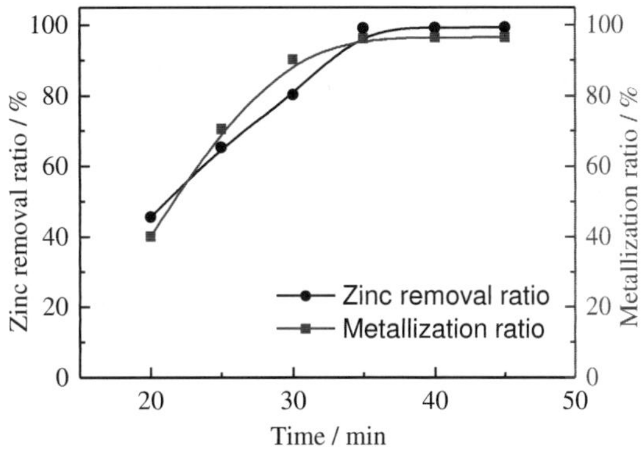

Fig. 5 Effect of time on zinc removal and iron reduction (reduction temperature of 1150°C)

The microstructure of composite pellets and reduced pellets reduced at 1150℃ for 35min is shown in Fig.6. The mineral particles in the composite pellets are dispersed and almost have no pore structure. The iron element in reduction pellets mainly exists in the form of metallic iron. The number of metallic iron particles in reduced pellets were well developed. The metal iron particles are connected with each other, and some iron particles enriches and grows to about 100 μm. At the same time, the carbon in the composite pellets is consumed by the reduction reaction, resulting in the increase of pore structure and pore size. The pellet internal structure is relatively loose.

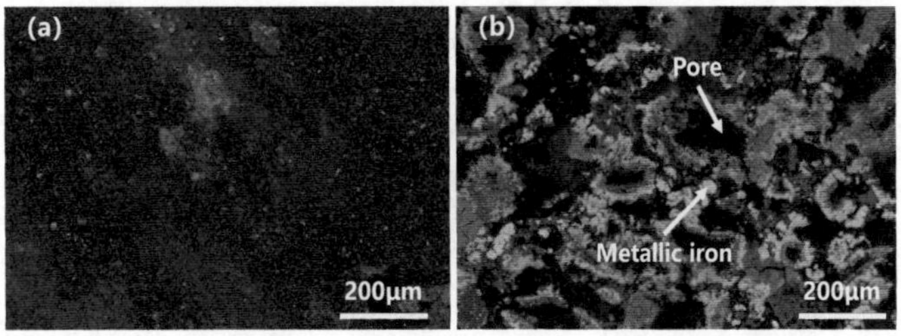

Fig.6 Microstructure of composite pellets (a) and reduced pellets reduced at 1150 ℃ for 35 min (b)

3.2.2. Kinetics of zinc removal by compound pellet reduction

In the reduction process, the carbon contained in the composite pellets will gasification and generate CO reducing gas, which will enhance the reducing atmosphere inside the pellets and contribute to the rapid progress of the reduction reaction. The reduction kinetics of composite pellets was studied by isothermal kinetics method. Isothermal kinetic models including Avrami-Erofeev equation, chemical reaction model, shrinkage nucleus model and diffusion model were used to analyze the reduction process of composite pellets. The linear relationship between different reaction models and reduction time was calculated. $G(\alpha)=[-\ln(1-\alpha)]^{1/2}$ has the highest regression coefficient r, which is greater than 0.9400. The linear relationship of $G(\alpha)$ with reduction time is shown in Fig.

7(a). The results show that the zinc removal efficiency of composite pellets reduction conforms to the model functions of nucleation and growth. The removal of zinc from complex pellets was controlled by random nucleation and the mechanism function was the Avrami-Erofeev equation (m=2). Kinetic parameters such as apparent activation energy and preexponential factor were calculated using the Arrhenius equation. The relationship between -lnk and $10^4/T$ in the kinetic model of reduction pellets is shown in Fig. 7(b). The kinetic parameters of zinc reduction and removal from compound pellet are shown in Table 3.

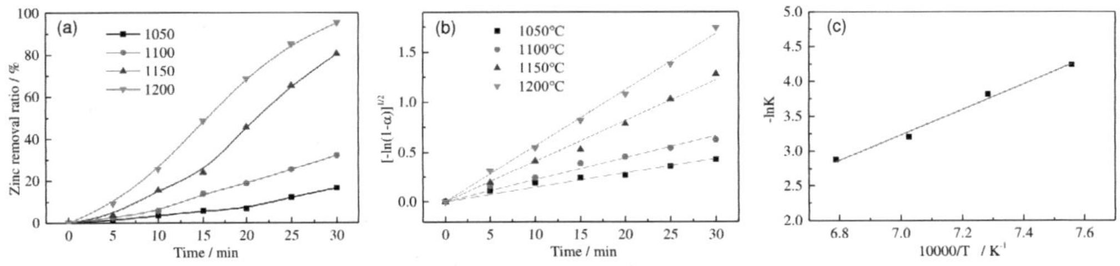

Fig.7 Reduction kinetics analysis of 5 mm reduction pellets

Table 3 Kinetic parameters of zinc reduction and removal from compound pellet

Kinetic model	-lnK-10^4/T equation	R^2	E/(kJ·mol^{-1})
A_2: $[-\ln(1-\alpha)]^{1/2}$	-lnk=-9.5133+1.8209(10^4/T)	0.9841	151.39

The reduction kinetics model of compound pellet conforms to the nucleation and growth models. After the carbon inside the pellets undergoes gasification reaction, a high reduction atmosphere is formed inside the pellets. The reduction reaction of iron and zinc components occurs synchronously and uniformly inside the pellets. This avoids the low reduction rate caused by the reduction gas passing through the product layer from outside during the conventional pellet reduction process, improves the reduction efficiency.

3.3. Mechanism of ring formation in rotary kiln during reduction process

3.3.1. Thermodynamic analysis of ring formation in rotary kiln

The isothermal phase diagram of CaO-FeO-SiO_2-0.2% ZnO slag system is shown in Fig. 8. The liquid phase region gradually expands as the temperature increases. The liquid phase region is mainly distributed in areas with high silicon content and high ferrous content. When the temperature increases from 1200 ℃ to 1250 ℃, the liquid phase region in the system significantly expands. The liquid phase region expands as CaO/SiO_2 decreases. Therefore, in order to achieve the reduction and separation of zinc while avoiding the melting and kilning of reaction materials, the reduction temperature should be controlled below 1200 ℃, and CaO/SiO_2 should be increased as much as possible.

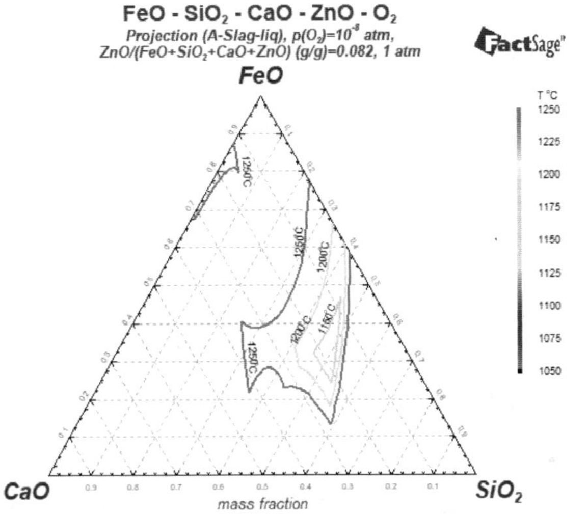

Fig.8 FeO-SiO2-CaO-0.2%ZnO isothermal phase diagram

3.3.2. Bonding between zinc-bearing dust powder and refractory materials

The zinc-bearing dust powder will react with the refractory material of rotary kiln during the reduction process. The effect of reduction temperature on the bonding between zinc-bearing dust powder and refractory materials is shown in Fig. 9. When the reduction temperature increases from 1050 ℃ to 1100 ℃, there is less reduced zinc-bearing dust powder adhered to the refractory material. During the process of increasing the reduction temperature from 1100 ℃ to 1200 ℃, the reduced zinc-bearing dust powder adhered to the refractory material increases sharply. The adhesion ratio increased to 45.9% when the reduction temperature was 1150 ℃. When the reduction temperature reaches 1200 ° C and 1250 ° C, the bonding phenomenon between the reduced zinc-bearing dust powder and refractory material is severe, and the adhesion ratio is close to 100%. Strictly controlling the high-temperature range of the reduction process will be beneficial for reducing the erosion of refractory materials by reduced zinc-bearing dust. Extending the high-temperature reduction time at 1100 ℃ can achieve efficient reduction and dezincification of zinc-bearing dust composite pellets, while avoiding ring formation in the rotary kiln.

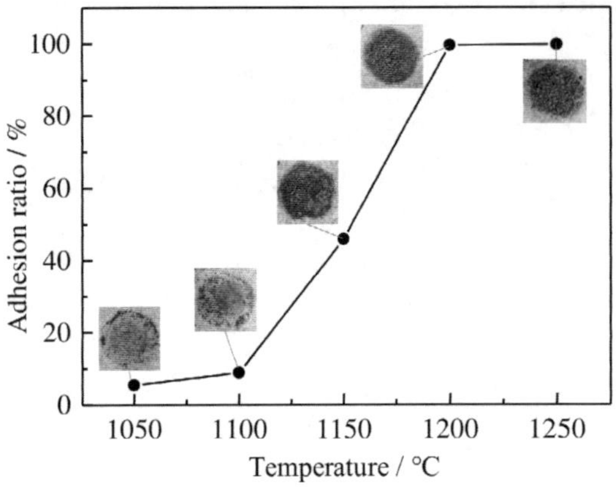

Fig. 9 Effect of temperature on the bonding between zinc-bearing dust powder and refractory materials

3.4. Industrial application of zinc-bearing dust reduction by rotary kiln method

The low-temperature direct reduction pilot production line that includes forced disturbed pelleting, rotary kiln system with controllable temperature field and atmosphere field has been built by Zhongye Changtian International Engineering Co., Ltd., which is shown in Fig. 10. The traditional rotary kiln method for treating zinc-bearing dust involves blowing all the air from the kiln head, which will form a high-temperature and oxygen rich zone inside the kiln. High temperature and oxygen rich areas will generate local high temperatures, exacerbating the tendency of rotary kilns to form rings and the re-oxidation of reduced products. The rotary kiln of this technology adopts multi-point air intake on the kiln body, reducing the oxygen intake at the kiln head, alleviating the re-oxidation of metallic iron. At the same time, controlling the multi-point air intake of the kiln body achieves precise temperature control inside the kiln, eliminates local high temperature points, and greatly alleviates the ring formation in rotary kiln.

Fig. 10 Low-temperature direct reduction pilot production line

The pilot study on the reduction and dezincification of zinc-bearing dust was carried on the low-temperature direct reduction pilot production line of Zhongye Changtian International Engineering Co., Ltd.. The 5 mm composite pellets containing zinc-bearing dust is prepared by forced disturbed pelleting as reduction charge. The ϕ 1m×10m rotary kiln system with controllable temperature field and atmosphere field by multi-point quantitative air suppling is used as the main equipment of reduction. When the reduction temperature is 1100 ℃ and the material stays at high temperature for 30 min, the dezincification rate reaches 99.09%. The ZnO content of the reduction product is 0.14%, which is only half of that when the multi-point air supply is not turned on. The metallization rate of the reduction product is 79.15%. The ZnO content in the zinc oxide powder reached 77.19%. There is no obvious ring formation phenomenon during the stable operation of the rotary kiln. The reduction product and collected dust is shown in Fig. 11.

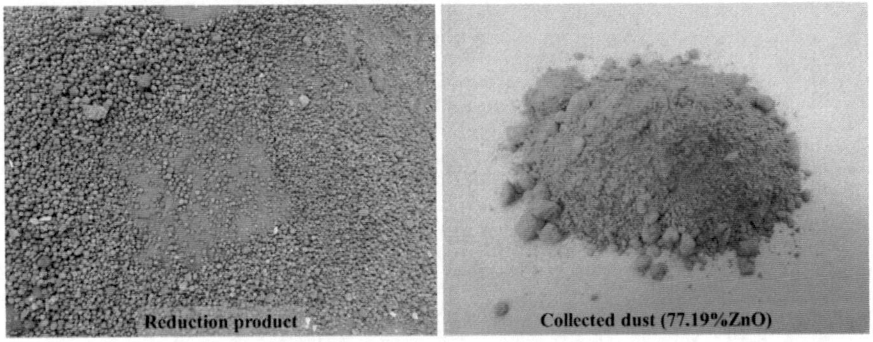

Fig. 11 The reduction product and collected dust

The key technologies of this study have been applied industrially in the production lines of 150000 t/a and 60000 t/a zinc-bearing dust reduction and dezincification. The 150000 t/a production line adopts a kiln head burner heating combined with carbon reduction inside the pellet. The 60000 t/a production line adopts the method of carbon distribution inside the pellet for heating and reduction. Compared with the conventional rotary kiln method, the processing capacity of zinc-bearing dust increase by 30%. The zinc removal ratio reaches 95% and the ZnO content of zinc oxide powder is 53.93 wt.%. The TFe content and ZnO content of reduction product are 60.56 wt.% and 0.25%, respectively. The annual operation ratio of zinc-iron separation by rotary kiln with zinc-bearing dust is 92%. The metallization rate of reduction product is controllable. The reduction product with metallization rate of 65.97% can be used as a sintering raw material, and the comprehensive energy consumption is 187kgce/t. The reduction product with metallization rate of 88% can be used as a converter steelmaking raw material, and the comprehensive energy consumption is 253.1kgce/t.

4. Conclusion

(1) The forced disturbed pelletizing can achieve efficient preparation of zinc-bearing dust composite pellets. The particle size of forced disturbed pellets are more uniform than that of cylinder pellet. The moisture of the forced disturbed pellets is lower and the strength is higher, which is beneficial to improve the strength of green pellet drying process and reduce the drying energy consumption.

(2) The composite pellets with carbon inside are conducive to the rapid reduction of iron and zinc. Lowering the reduction temperature and increasing the CaO/SiO_2 ratio of composite pellets can effectively inhibit the adhesion of reduction products to refractory materials. Extending the high-temperature reduction time at 1100 ℃ can achieve efficient reduction and dezincification of zinc-bearing dust composite pellets, while avoiding ring formation in the rotary kiln.

(3) The industrial production of zinc-bearing dust disposal shows that efficient reduction and dezincification of zinc-bearing dust can be achieved by using forced disturbed pelletizing, multi field collaborative controllable rotary kiln reduction. Compared with the conventional rotary kiln method, the processing capacity of zinc-bearing dust increase by 30%. The zinc removal ratio reaches 95%. The TFe content and ZnO content of reduction product are 60.56 wt.% and 0.25%, respectively. The annual operation ratio of zinc-iron separation by rotary kiln with zinc-bearing dust is 92%. The metallization rate of reduction product is controllable.

References

[1] Lin WL, Wang H, Liu DH, Xing HW, Liu C, Zhang YZ, Li S: Research progress on granulation process of iron and steel slag. *Metallurgical Research & Technology* 2022, 119(1):120.

[2] Chung SH, Kim KH, Sohn I: DRI from recycled Iron bearing wastes for lower carbon in the blast furnace. *ISIJ International* 2015, 55(6):1157-1164.

[3] Abdelwaheb, Ben, Slama N, Feki D, Levacher M, Zairi: Valorization of harbor dredged sediment activated with blast furnace slag in road layers. *International Journal of Sediment Research* 2021, v.36(01):132-140.

[4] Wang X, Xiuying LI, Yan X, Cheng TU, Zhaoguo YU: Environmental risks for application of iron and steel slags in soils in China: A review. *Pedosphere* 2021, 31(1):28-42.

[5] Wang D, Zhu D, Pan J, Guo Z, Yang C, Tian H, Xue Y: Integration of preparation of K,Na-embedded activated carbon and reduction of Zn-bearing dusts. *International Journal of Mining Science and Technology* 2022, 32(3):10.

[6] Ollonqvist P, Riihimaki T, Koukkari P: New hydrometallurgical treatment to recover zinc and iron from EAF dust. *Erzmetall* 2022(1):75.

[7] Bouckley E: Russia's USM Urban Mining to recycle EAF dust to reclaim zinc for steel industry. *SBB Steel Markets Daily* 2022(200):16.

[8] Amin AMM, Soliman YMM, El-Dek SI, Ahmed YMZ, Zaki AH: Valorization of industrial iron and zinc sludges for the synthesis of $ZnFe_2O_4$ ceramics. *Journal of Magnetism Magnetic Materials* 2022, 544:168681-.

[9] Mamyachenkov SV, Anisimova OS, Toporkova YI: Studies on the regularities of the steelmaking zinc-bearing dusts leaching in ammonium chloride solutions. *Non-Ferrous Metals* 2019, 47(2):1-6.

[10] Bayat O, Sever E, Bayat B, Arslan V, Poole C: Bioleaching of zinc and iron from steel plant waste using Acidithiobacillus ferrooxidans. *Applied Biochemistry Biotechnology* 2009, 152(1):117-126.

[11] Kazinczy B, Kótai L, Gács I, Sajó IE, Sreedhar B, Lázár K: Study of the preparation of zinc(II) ferrite and ZnO from zinc- and iron-containing industrial wastes. *Industrial & Engineering Chemistry Research* 2003, 42(2):318-322.

[12] ZHANG Z, JIANG G, WANG X, ZHANG J: Development and utilization of the world's and China's bulk mineral resources and their supply and demand situation in the next twenty years *Acta Geologica Sinica* 2016, 90(4):1370-1417.

[13] Wu YL, Jiang ZY, Zhang XX, Wang P, She XF: Numerical simulation of the direct reduction of pellets in a rotary hearth furnace for zinc-containing metallurgical dust treatment. *International Journal of Minerals, Metallurgy, Materials* 2013, 20:636-644.

[14] She XF, Xue QG, Yang HX, Ding YG: Formation mechanism of soot in zinc-bearing dusts direct reduction process. *Advanced Materials Research* 2011, 156-157:817-823.

[15] Yao H, Ma H, Mao R, Qiu J, Chen C, Ju D: Preparation of ZnO nanoparticles from Zn-

containing rotary hearth furnace dust. *Journal of Wuhan University of Technology-Mater Sci Ed* 2022, 37(1):32-37.

[16] Mao R, Wang F, Xu Y, Ren K, Wang GJMR, amp, Technology: Preparation process of cold bonded pellets with iron-bearing dust and sludge from steel production process. *Metall Res Technol* 2021, 118(6):602.

[17] Matthes J, Waibel P, Keller HB: A new infrared camera-based technology for the optimization of the Waelz process for zinc recycling - ScienceDirect. *Minerals Engineering* 2011, 24(8):944-949.

[18] Suetens T, Klaasen B, Van Acker K, Blanpain B: Exergy based comparison of the Waelz Kiln process and RHF for the treatment of EAF dust. *Chinese Science Bulletin Kexue Tongbao* 2013, 58(18):2262-2268.

[19] Vogelbacher M, Keller S, Zehm W, Matthes J: Advanced methods for kiln-shell monitoring to optimize the waelz process for zinc recycling. *Processes* 2021, 9(6):1062.

[20] Tan P, Vix P: Modeling and optimization of rotary kiln treating EAF dust. *International Journal of Minerals, Metallurgy Materials*, 12(4):308-312.

Journal of Physics: Conference Series **2738** (2024) 012025 doi:10.1088/1742-6596/2738/1/012025

Application of side-submerged combustion smelting(SSC) process in recovery of lead smelting by-products

Siyao Peng

China ENFI Engineering Corporation, Beijing, People's Republic of China, 100038.
Email: pengsy@enfi.com.cn

Abstract. The treatment of lead smelting by-products , such as anode slime and antimony-containing dust and slag is an important process, which is not only related to improvement the rate of comprehensive recovery of resources, but also plays a very important role in improving the economic benefits of plant. The existing by-product treatment technology and current situation were introduced in first. Combined with the current treatment process and characteristics of by-products, this paper focuses on the characteristics of the oxygen-enriched submerged combustion side-blown smelting technology and the application of by-products treatment. Based on the engineering practice and relevant technical and economic indicators of treating 5812t anode slime annually, compared with the conventional technology, this technology has obvious advantages in operation cost, treatment efficiency and environmental protection.

1. Introduction

Lead resources are mainly distributed in Australia, China, Russia, Peru, etc. China's lead reserves account for nearly 20% of the total, and like other important metals, there is a problem of excessive exploitation and utilization of lead. China is the largest producer of refined lead, with a production of approximately 60% of the world refined lead[1,2]. Under the tight supply of lead concentrate in China, smelter is seeking complex resources, imported lead concentrate, silver concentrate, etc. to expand the source of raw materials. In recent years, lead smelters often get low TC/RC (Treatment and refining charges for Processing concentrates), thus the smelters are on the edge of loss. Facing this situation, smelters need to compensate for their benefits by increasing the income of by-products. The revenue from rare and precious metals has always been the main source of profit for lead smelter, especially the recovery of precious metals such as gold, silver, antimony, and bismuth.

In order to improve the resource recycling efficiency and benefits, lead smelters are maximizing the amount of valuable metals associated with raw materials, resulting in a sharp increase in the output of anode slime, copper dross, antimony dust, etc. during the lead refining process. Although there has been a significant improvement in economic benefits due to the increase in by-products, the impact on existing comprehensive recovery systems is gradually becoming prominent. Mainly reflected in the outdated equipment, insufficient processing capacity, deteriorating indicators, and backlog of precious metals in the anode slime treatment system.

The yield of lead anode slime is usually 1~3%. For a lead smelters with an annual output of 100,000 tons, the annual output of anode slime is 3000 tons at a yield of 3 %. At present, the yield of anode slime in lead smelter enterprises is generally 3 ~ 6 %. For smelter that deal with low lead-containing materials in large quantities, the yield of anode slime is greater. The conventional pyrometallurgical-hydrometallurgical combined process is widely used in most lead smelter, includes two steps: pyrometallurgical reduction, oxidation refining and electrolytic refining[3,4]. There are some main advantages of the process: wide adaptability of raw materials, simple and reliable equipment, moderate processing capacity, and low technical requirements.

However, the processes still have shortcomings, such as long working period, producing excessive slag, more precious metal backlog, large dust collection systems, dust pollution and poor working conditions. The main purpose of the pyrometallurgical process is to produce qualified silver alloys for further refining in the electrolysis system. After the successful application of bottom blown smelting and side blown smelting in lead smelting, Chinese metallurgical researchers have developed new processes for treating lead anode slime using oxygen bottom blown smelting and oxygen-enrich side blown smelting. In this paper, the treatment process widely used in China is briefly analyzed, the characteristics of the new treatment process are introduced, and the application prospect of the new process is prospected.

2. Current status of lead anode slime treatment technology

2.1. Conventional smelting process

The conventional pyrometallurgical treatment process is the wet anode slime is piled up for oxidation for 7~10 days, then add reducing coal and flux, and add to the reduction smelting furnace in batches. Heat is provided in the natural gas or diesel burner to maintain the furnace temperature above 1100 °C. PbO, CuO, Bi_2O_3, part of Sb_2O_3 and As_2O_3 in anode slime were reduced by reducing coal to obtain molten metal lead and enrich gold and silver. Flux is usually soda, which can form stable slag with oxides such as arsenic and antimony in anode slime, and reduce the melting point of slag, improve the fluidity of slag, and make it easy to separate from silver-rich lead. During the reduction, some Sb and As volatilize into the dust, which can be used as raw materials for recovering Sb. The silver-rich lead produced by the smelting furnace is transferred to a converter where it is blown with air to further oxidized and refined to produce qualified gold and silver alloys.

The smelting process is carried out in two special converters. There are many procedures and long operation period in converter operation, which are divided into feeding, melting, precipitation, slag discharge, antimony blowing and lead discharge. Usually, the operation period of the smelting furnace is generally 15~24h, and the operation period of the refining furnace is generally 60~78h. During operation, the feeding and discharge needs to be completed manually.

The natural gas or oil burner is used to heat the smelting furnace. The heat transfer mode is mainly radiation, which is low efficiency, and most of heat loss taken away by the off-gas. Usually gas consumption of 1t anode slime smelting section is about 100~200 m^3/t, and gas consumption of 1t silver-rich lead refining section is about 1500~2500 m^3/t, which equivalent to gas consumption of 1t anode slime smelting section is about 1000~2000 m^3/t.

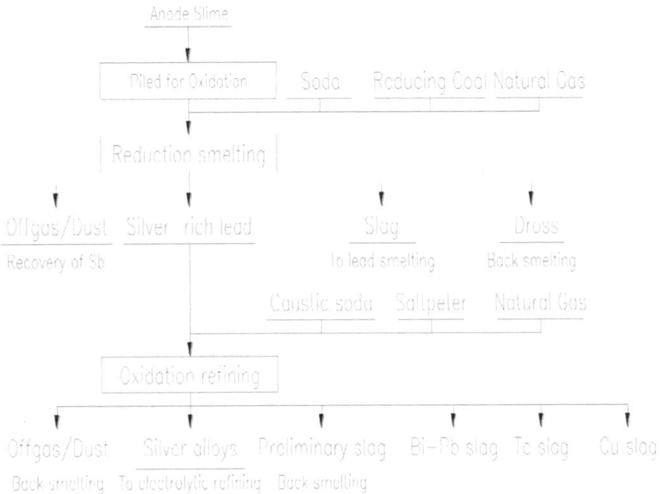

Figure 1. Process flow diagram of conventional anode slime smelting

2.2. Bottom blowing smelting process

A lead smelter in Henan Province built a new anode slime process system with a capacity of about 6000 tons in 2006, which is the first oxygen bottom blowing smelting system. The new process system consists of 5 smelting furnaces, which including an oxygen bottom blowing smelting furnace, an oxygen bottom blowing refining furnace, two refining furnaces, and a slag smelting furnace. After a period of operation and modification, the indexes of the new smelting system have reached or exceeded the design indexes. Practice has proved that the process of oxygen bottom blowing smelting anode slime is feasible, and the energy consumption and cost of anode slime smelting are significantly reduced[5,6].

The bottom blowing smelting process is mainly divided into three stages, and the process flow is shown in Figure 2:

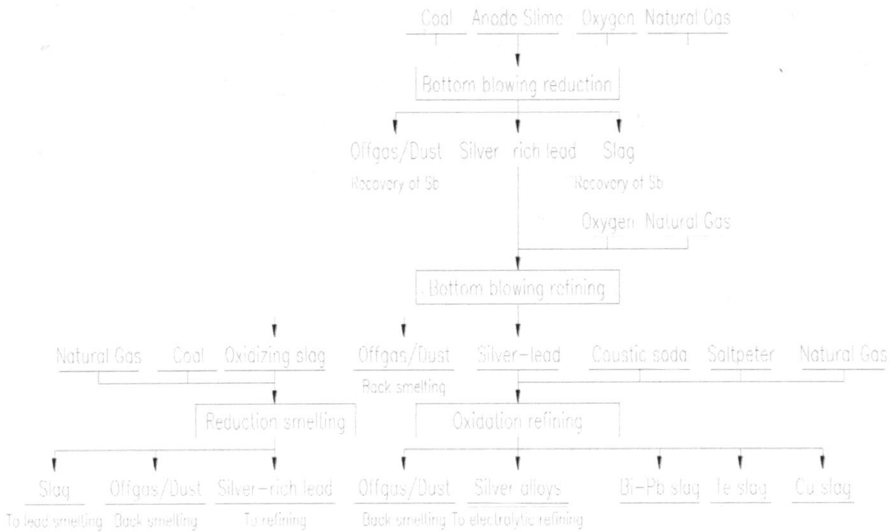

Figure 2. Process flow diagram of bottom blowing smelting process

First stage: The anode slime is continuously fed into the oxygen bottom blown smelting furnace with coal and other return. Under the heating of the natural gas burner, and the oxidation and stirring of the bottom oxygen lance, most of the arsenic, antimony and some lead volatilized into the off gas, and a small amount of antimony, lead and quartz react to form slag, and bismuth, copper and some lead were reduced to form molten lead, which enriches a large amount of rare and precious metals such as gold, silver and tellurium. The smelting slag is sent to the recovery system to further recover antimony.

Second stage: the preliminary oxidation refining of silver-rich lead in the bottom blowing refining furnace to improve the grade of gold and silver in lead, and prepare for the final refining. In the refining furnace with bottom oxygen supply, arsenic, antimony, lead, bismuth and other impurities in silver-rich lead are oxidized. The refining furnace with oxygen supply from bottom has strong oxidation ability and has a good removal effect on impurities such as arsenic, antimony and lead in precious lead.

Third paragraph: It is difficult to oxidize impurities such as bismuth, tellurium, copper in the alloy, and the final impurity control is relatively low. So Intermittent refining is adopted to remove all impurities in gold-silver alloy, meeting the requirements of subsequent electrolytic refining.

Process characteristics:

- Chutes are used to connect different melting furnaces in different heights, reducing the workload of melt transfer.
- The processing capacity is greatly improved, more than doubled.
- The bottom blowing has strong impurity removal ability and high precious metal grade in lead, which is beneficial to improve the efficiency of refining process. Arsenic and antimony are enriched in dust, which is beneficial for the recovery of arsenic and antimony.
- Fresh and wet anode slime can treated directly, which can shorten the operation period and reduce the intermediate occupation.

- The energy utilization efficiency is improved, the heat transfer efficiency is improved by bottom blowing stirring, and the metallurgical process reaction is strengthened. The natural gas consumption of the bottom blowing smelting process is only 60~70 % of that of the conventional converter smelting process.

2.3. Side-submerged combustion bath smelting process

2.3.1. Technical description

Side-submerged combustion bath smelting technology(SSC) is a kind of enhanced bath smelting technology for treating non-exothermic materials, which inject oxygen-enriched air and fuel (natural gas, producer gas, pulverized coal) into the molten bath with the multi-channel side-submerged lance at high speed and pressure. The materials on the molten bath are rapidly mixed into the bath due to the strong agitation of the injected gas, and the physical and chemical reactions are completed in bath.

This SSC technology is particularly suitable for the treatment of non-exothermic materials, and also suitable for the treatment of small-scale materials such as anode slime. The blowing injected gas in the molten bath and the rolling molten slag create a huge reaction interface, where the material is heated, melted and reduced, and the reduced metal settle and separate from slag. Under the action of gravity, a relatively static metal layer below the lance is formed. Thus, a layer of slag is formed on the upper part of the metal layer, which collects some impurity elements and gangue.

2.3.2. Technical feature

1) Simple preparation process and environmentally friendly.

The basic characteristic of the molten bath smelting is that the preparation is relatively simple, and the raw material does not need drying, and batching processes, which reduces the dust pollution.

The conventional converter smelting process has strict requirements on preparation. The wet anode slime needs to be stored and naturally oxidized for 7-10 days, which is convenient for removing impurities during the smelting process. At the same time, the converter smelting needs to limit the moisture of the material to avoid causing the furnace accident.

2) High temperature smelting, high heat utilization rate, and stable heat balance in the furnace.

The oxygen-enriched air concentration reaches 60~80%. The amount of smelting off gas is small, which take away the less heat loss. Through the side blowing lance directly inserted into the molten bath, the fuel is injected into the bath to replenish heat. The fuel burns in the bath directly, and the heat is released fully and absorbed by the bath. The heating transfer is fast and the heat utilization rate is high. The temperature of the molten bath can be adjusted quickly by change fuel, and the heat balance can be controlled.

3) Control oxygen potential flexible in furnace

By controlling the amount of oxygen and fuel, the redox atmosphere of the molten bath in the furnace can be flexibly adjusted, and the on-line control of the furnace can be realized.

4) High recovery rate and less backlog of precious metals is

The submerged bath smelting technology injects and blows the slag layer of the molten bath, and the position is higher than the bottom blowing smelting process. So less base metal in the silver-rich lead can be oxidized and the base metal content in the silver-rich lead is higher, then the ability to enrich precious metals is stronger, and the loss of precious metals in dust and slag is less, which is beneficial to improve the yield of gold and silver in the silver-rich lead. The silver content of side-submerged smelting dust is less than 1000g/t, and less than 700g/t in the slag, while the silver content of bottom-blown smelting dust is less than 1600g/t, and 1300g/t in the slag.

The discharge frequency of SSC technical for silver-rich lead is higher, once every two hours; While the bottom blowing smelting technology has a long period of discharge, which usually takes 12~24 hours to discharge once, and there are many precious metals in the furnace.

2.3.3. Process flow

Submerged combustion smelting anode slime system mainly includes side blowing reduction smelting and slag settlement, preliminary refining and final refining. And the process flow is shown in Figure 3.

Anode slime, returned high-silver dust and a small amount of reduced coal were blended and mixed, then added to the reduction furnace continuously. Under the action of oxygen-rich air and natural gas supplied by the side-blown lance, violent combustion and agitation, part of the arsenic and antimony in the anode slime volatilized into the dust as low-valence oxides, and some of them reacted with lead oxide and gangue components to form slag. Most of bismuth, copper and some lead are reduced and form the alloy in the molten state, and the rare and precious metals such as gold, silver in the anode slime are enriched in the alloy.

The slag produced by the reduction furnace enters the slag converter for insulation through the chutes, further settling and separating the silver-rich lead mixed in the slag. The discharged slag is sent to the antimony system for further recovery of valuable metals.

The refining process is the same as the traditional smelting process, that is, the oxidation is used to remove the impurities except gold and silver in the alloy, and the gold and silver alloy with a gold and silver content of more than 98 % is obtained for further electrolytic refining. The oxidation order of metals in alloy is: Sb, As, Pb, Bi, Cu, Te, Se, Ag.

The silver-rich lead produced by the reduction furnace is initially removed in the preliminary refining furnace, mainly to remove arsenic, antimony, and some lead, which is preparing for further refining and reduce the period of final refining. By blowing air to the surface of silver-rich lead, the easily oxidized arsenic, antimony, lead are oxidized and volatilized into dust, and the grade of gold and silver in the alloy is improved further. The high-grade silver-lead alloy, which removes arsenic, antimony and some lead, complete the final refining in the refining furnace and produces qualified anode alloy. From the perspective of comprehensive recovery, the refining process is divided into two sections, which can not only improve the efficiency of the refining furnace, but also preliminarily separated arsenic and antimony dust as raw material for comprehensive recovery of antimony.

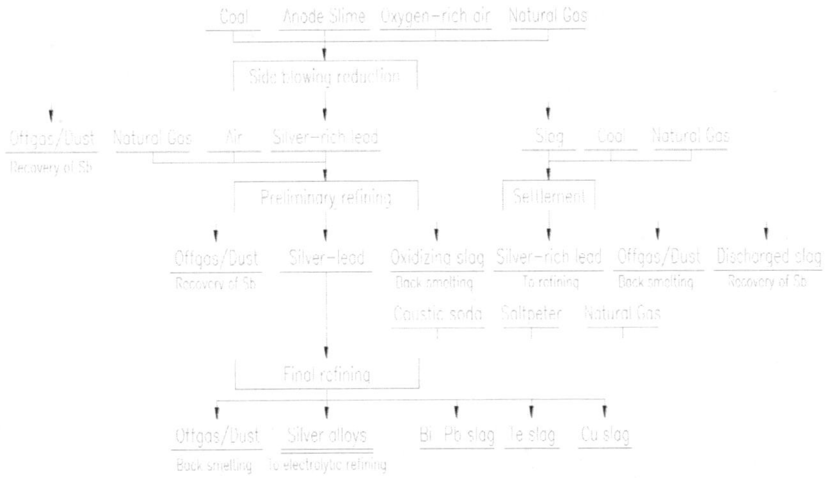

Figure 3. Process flow diagram of side-submerged combustion bath smelting process

3. Industrial application

3.1. Application of SSC in anode slime smelting process

Taking the project with an annual processing capacity of 5812 t anode slime as an example, the operation of the project is briefly introduced[7]. The lead anode slime produced by lead smelter, and copper smelter anode slime (copper removal) are raw materials. The main components are shown in Table 1. The moisture of anode slime is about 30 %.

Table 1 Chemical composition of anode slime (Dry basis, Wt %)

Material	t/a	Pb	Cu	As	Sb	Bi	Te	Ag	Au
Self-produced anode slime	3843	15.00	1.64	18.07	30.88	14.53	0.64	9.29	0.03
Copper smelter anode slime	600	21.85	1.35	1.69	3.09	1.01	1.04	13.45	0.56
Purchased lead anode slime	1369	10.00	2.22	7.86	33.02	22.40	0.53	3.17	0.02

The main smelting equipment of the anode slime smelting project is shown in Table 2. The main technical indexes of treating anode slime by side-submerged combustion bath smelting process are shown in Table 3.

Table 2 Equipment table of anode slime smelting project

No.	Equipment	Types and Specifications	QTY
1	Side- submerged reduction furnace	$2m^2$	1
2	Slag settling furnace	Φ 2000mm×2600mm	1
3	Pre-refining furnace	Φ 2000mm×2600mm	2
4	Final refining furnace	Φ 2000mm×2600mm	2

Journal of Physics: Conference Series **2738** (2024) 012025 doi:10.1088/1742-6596/2738/1/012025

Table 3 Main technical and economic indicators of anode slime smelting

No.	Indicators	Unit	Design value	Actual value	Remark
1	Annual Processing capacity	t	5812	5113	Lack of anode slime
2	Daily Processing capacity	t	25~30	50~60	
3	Coal consumption	t/a	397	130	
4	Natural gas consumption	$10^3 \cdot Nm^3/a$	3220	2791	
5	Natural gas consumption (reduction smelting)	$10^3 \cdot Nm^3/a$	581	460	
6	Oxygen consumption	$10^3 \cdot Nm^3/a$	1295	1075	
7	Direct recovery rate of silver	%	>97	>97	

3.2. Application of SSC in the smelting of antimony-containing materials

Antimony-containing dust and slag produced in the process of anode slime smelting are used as raw materials for comprehensive recovery of antimony. Taking the project with an annual processing capacity of 4685 t antimony-containing materials as an example, the operation of the project is briefly introduced. The main components antimony-containing dust and slag were Sb 35.5 % and Pb 7.9 %.

The antimony comprehensive recovery process is shown in figure 4.

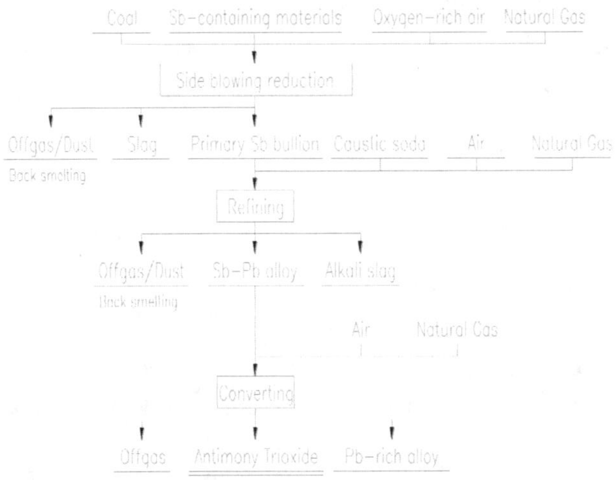

Figure 4. Process flow diagram of antimony comprehensive recovery process

The main smelting equipment of antimony comprehensive recovery project is shown in Table 4.

Table 4 Equipment table of antimony comprehensive recovery project

No.	Equipment	Types and Specifications	QTY
1	Side- submerged reduction furnace	$2m^2$	1
2	Refining kettle	Φ1500mm, Q=10t	3
3	Converting kettle	Φ1500mm, Q=10t	3

The main technical indexes of antimony comprehensive recovery by side-submerged combustion bath smelting process are shown in Table 5.

Table 5 Main technical and economic indicators of antimony comprehensive recovery

No.	Indicators	Unit	Design value	Actual value
1	Annual Processing capacity	t	4634	4851
2	Sb_2O_3 capacity	t	1439	1450
3	Daily Processing capacity	t	20	30~40
4	Coal consumption	t/a	611	388
5	Natural gas consumption	10^3·Nm^3/a	1841	1741
6	Natural gas consumption (reduction smelting)	10^3·Nm^3/a	371	292
7	Oxygen consumption	10^3·Nm^3/a	1056	628
8	Direct recovery rate of Sb	%	88	90

4. Conclusion

Side-submerged combustion bath smelting process has many successful practices in the non-ferrous metal smelting, and also has been applied in the field of anode slime and antimony-containing material smelting, which has showed the special advantages and characteristics.

At present, after years of technical improvement and upgrading, the overall technical and equipment of main non-ferrous metals smelting in China have been significantly improved. However, the technical and equipment of rare and precious metal comprehensive recovery plants are uneven. The new technology is suitable for lead and copper anode slime smelting projects, other rare and precious metal smelting and comprehensive recovery, has broad market and application prospects.

References

[1]GU Y, WANG JP. Situation and sustainable development proposals for lead resource development in china. Resources & Industries, 20.1(2018):39-46.

[2]YANG RL. Analysis of currant situation of lead and zinc mine resources development and suggestions for sustainable development in China. World Nonferrous Metals, 1(2018):148,150.

[3]ZHOU JQ,ZHANG SH. The processing Technology and application practice on the mixture of copper and lead anode slime. World Nonferrous Metals, 2(2018):20-21.

[4]DENG LJ, ZENG ZY. Improvement of lead anode slime production process. China Nonferrous Metallurgy, 6(2016):47-48.

[5] Yuan YF, Liu SH. Process design and production practice of continuously treating lead anode slime by bottom-blowing bath smelting. China Nonferrous Metallurgy, 41.4(2012):3.

[6] Wang GZ, Chen HJ. Practice on smelting of lead anodic slime with oxygen bottom blowing technique. Hunan nonferrous metals, 28.1(2012):3.

[7] Li D, Cao KF. Design and application of anode slime side-blowing reduction furnace. Nonferrous Metallurgical Equipment, 36.6(2022):4.

Journal of Physics: Conference Series **2738** (2024) 012026 doi:10.1088/1742-6596/2738/1/012026

The transformation behaviour of $ZnFe_2O_4$ spinel particles in the $CaO-Fe_xO-SiO_2-ZnO$ slag system

Lihua Wang[a], Kehua Xu[b], Huanwen Wang[a], Shaogang Chen[b], Zhiqian Yu[a], Kai Ma[a], Tao Wu[b], Longgong Xia[a],*

a: Refractory Material Metallurgy and Slag Chemistry Research Group, School of Metallurgy and Environment, Central South University, South Lushan Road 932, Changsha, P.R.

b: Danxia smelter, Shenzhen Zhongjin Lingnan Non-ferrous metal Company Limited, Renhua, P.R. China

* Contacting person and Presenter: longgong.xia@csu.edu.cn

Abstract. Treating zinc hydrometallurgical residues and other Zn-bearing solid waste in direct lead smelting processes has been recognized as the preferred solution for the sustainable development of the lead and zinc industry. However, new technical problems occurred when changing the starting material of lead-making processes, the pyrometallurgical process for treating zinc hydrometallurgical residues and zinc bearing solid waste usually had the $CaO-Fe_XO-SiO_2-ZnO$ slag. Along with the increasing of Zn content in the raw material, high ZnO slag formed, and the $ZnFe_2O_4$ spinel particles intended to precipitate from the molten slag. The understanding of the particle behaviour of $ZnFe_2O_4$ spinel was the core part of monitoring the slag properties. This study investigated the transformation behaviour of zinc ferrite spinel particles and the changes in glassy slag phase through the high-temperature phase equilibrium experimental technique, and the temperature was 1100 °C-1300 °C and the oxygen potential was 10^{-8} atm – 10^{-4} atm. It is found that the particle size of $ZnFe_2O_4$ spinel in the molten slag increased along with the increasing of the oxygen potential. The Zn/Fe mass ratio in spinel particles decreased when increasing of temperature and it increased when increasing of oxygen potential. The Fe/SiO_2 mass ratio and the Zn content in glassy slag phase gradually increased along with increasing temperature and they decreased along with increasing oxygen potential. The growing and decomposition of $ZnFe_2O_4$ spinel was essentially a process of material exchange with the glassy slag phase.

1. Introduction

Global lead and zinc production in 2022 reached 12.30 million tons and 13.286 million tons, respectively. Lead and zinc consumption was about 12.40 million tons and 13.59 million tons in recent years (sources from International Lead and Zinc Study Group). In the past three decades, significant progresses had been made in the lead and zinc metallurgy. [1-3] Advanced technologies such as direct smelting of lead sulfides, [4, 5] direct reduction of high lead containing slag and direct leaching of complex zinc sulfide concentrates have been widely used all over the world. In addition, the Pb & Zn integrated co-smelting technology has been proved as a high resource utilization and excellent environmental performances model, which is a consensus in the field of Pb-Zn smelting industry and scientific research.

Along with the depletion of the high-quality Pb-Zn mineral resources, there was more pressure in raw material supplying, and new technical and environmental problems occurred. Sustainable development has turned out to be the key idea in this field. Complex resources and secondary materials were having larger proportion in starting materials. Secondary resources, such as lead-silver residues, iron precipitations and lead past, have been co-treated with lead concentrates. The co-smelting technology can effectively utilize the heat releasing from the oxidation of lead sulfide concentrate, and it can also achieve the harmless treatment and resource utilization of solid wastes, especially the hazardous. [6] The development of modern lead smelting technologies such as Kivcet, oxygen-enriched side-blowing, oxygen-enriched bottom-blowing and QSL, have already started to co-smelt secondary materials with lead concentrates. All of them had industrial practices, for example, the Canadian Trail Smelter, [7] Jiangtong Lead and Zinc Company, [8] Zhongjin Lingnan Danxia Smelter.[9], etc. The pyrometallurgical processes owned the advantages of fast reaction rate, large capacity, low pollution emission and stable final products, which showed a good potential in co-treating complex Pb concentrate and Zn hydrometallurgy residues [10]. Researches and practices have shown that the co-processing of hydrometallurgical zinc smelting residues in the pyrometallurgical lead smelting process was the preferred method in line with the production reality and industrial development direction, there were significant environmental benefits. With its Kivcet furnace, Jiangxi Copper Lead and Zinc Company dealt with lead-containing materials such as zinc leaching residue, lead sulfate sludge, and neutralization residue. Moreover, it had realized good performance for more than six years of practice. [8] The lead-silver residue from zinc hydrometallurgy has also been successfully treated in the furnace, which made full use of resources and reduces the environmental risks. However, the current co-processing of zinc hydrometallurgical residues in direct lead smelting process caused prominent problems, such as the high zinc content in slag, deterioration of slag properties, difficult separation of slag and metal phase, etc. At the early stage of Trail plant operation, it produced many furnace knots in the period of treating high Zn charge with the Kivcet furnace, resulting in the reduced reduction capacity of the coke filter layer, which made the lead content of the slag increase and the capacity of the kiln to process zinc smelting slag decrease. [7] Slag-metal separation was difficult in the Kivcet furnace of the Danxia smelter when processing high proportion of goethite residues, and one sticky slag layer has been found in the molten bath. [9] The increase of zinc content in the raw material was the major reason for the above industrial difficulties, and zinc mainly entered the slag phase in the oxidation smelting stage. While under the currently direct lead smelting process conditions, high zinc slag was prone to have saturated zinc ferrite spinel, leading to the increase of slag viscosity, poor slag-metal separation and so on, which seriously affected the industrial practice. The analysis about the behavior of spinel in high Zn smelting slag and the formation of slag performance control mechanism was the way to solve above problems.

The behavior of $ZnFe_2O_4$ spinel in CaO-Fe_XO-SiO_2-ZnO slag system is the core knowledge of slag, which is also the bottleneck of technology development. Literature review showed that phase diagram studies of pyrometallurgical slag systems were predominant. [10-21] But there was a lack of systematic understanding about the transformation behavior of $ZnFe_2O_4$ spinel. In this paper, the transformation behavior of $ZnFe_2O_4$ spinel particles in CaO-Fe_XO-SiO_2-ZnO slag system has been systematically studied with high-temperature phase equilibrium experimental technique. Obtained results can help in optimizing these processing parameters, and it was also useful for the treatment of zinc bearing solid waste and the synthesis of related spinel materials.

2. Experimental section

2.1 Experimental apparatus

The schematic picture of the high temperature phase equilibrium setup has been depicted in Figure 1.

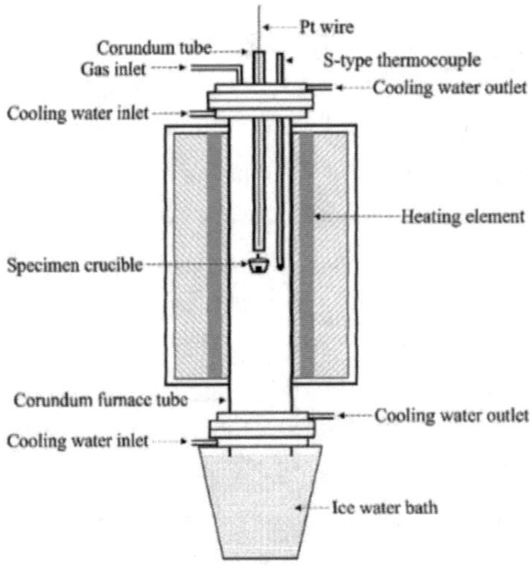

Figure 1. Schematic of the experimental equipment.

It is consisted of one vertical tube furnace (Hefei Kejing KSL-1700) with Si-Mo heating element, water cooling system, gas supplying system and a connected quenching device. Zirconia crucible (purity 99.9 wt %, inner ⌀, outer ⌀ and hight = 8, 10 and 20 mm, respectively), Cornell wire/platinum wire have been used as sample reaction vessel and suspension wire, respectively. The dissolution of zirconia or platinum into the slag can be neglected in these experiments. Samples were kept at the hot zone of the vertical furnace for a certain period. The temperature profile of the working tube has been pre-checked with a S-type thermocouple, and the temperature accuracy was estimated to be ± 3 °C. When the inquired equilibrium time reached, the suspension line was cut, and the sample dropped and quenched into an ice-water bath immediately.

2.2 Preparation of initial Material

Based on the industrial slag composition [7-10], master slag was pre-synthesized from chemical pure oxides. Keeping the mass ratio of Fe/SiO_2 and CaO/SiO_2 as constants, and the value was 0.85 and 0.33, respectively. Zn, Fe, Si and Ca contents in the starting slag were 20.0 wt %, 23.91 wt %, 11.89 wt % and 5.83 wt %, respectively.

Accurately balanced oxides were well ground and mixed in an agate mortar. After that, the mixture was pressed into flakes in a pressing tool with a pressure of 1.5 MPa/cm^2 and followed by loading them into the zirconia crucible. The crucibles containing the samples were suspended in the high temperature zone of the furnace through platinum wires. Pre-mixed N_2-O_2 gas or CO-CO_2-Ar (with the oxygen partial pressure of 10^{-6} atm) has been used as carrying gas in master slag preparation process. The raw material sintering conditions were determined from the sample characterization, melting performance, and chemical reactions, and processing parameters were 1200 °C, 8h and 10^{-6} atm. When the reaction time reached, the crucible was quenched and cooled in the low-temperature zone by adjusting the length of the platinum wire, and the experimental raw material was obtained by physically breaking it into several pieces, then it was fully ground in a mortar and pestle to obtain a powder material. Ultimately it was dried and put into a storing sample bag.

2.3 Sample characterization

The synthesized slag was examined by electron microscopy scanning and X-ray diffraction, the results were shown in Figure 2. The calculation of the CaO-Fe_xO-SiO_2-ZnO slag system phase diagram was carried out through FactSage 7.1 and the results were obtained and shown in Figure 3.

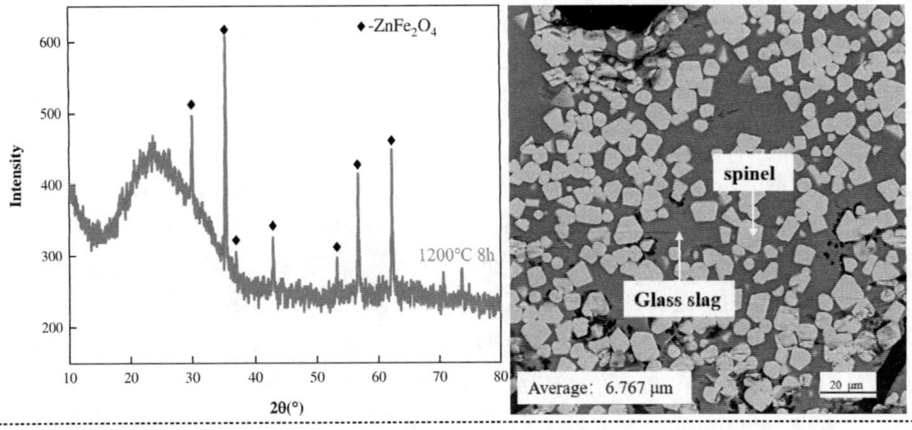

Figure 2. XRD result and microscopic morphology of synthesized slag.

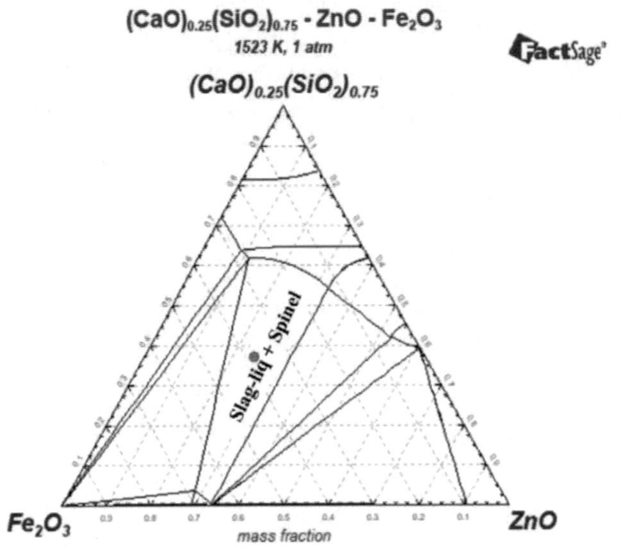

Figure 3. Phase diagram of CaO-Fe_XO-SiO_2-ZnO slag system at 1200 °C, CaO/SiO_2 = 1/3 (Red point represents the starting slag).

From the microscopic morphology, the synthesized slag had different contrasts, which referred to two phases. The bright and regular shaped area was the spinel phase, the dark gray area was the glass slag phase, and the black area was the SiO_2 precipitates, which has also been confirmed by the energy spectrum scan. The size of the spinel particles was statistically analyzed from obtained electron microscope images, and the result was 6.77 μm. The mean statistical analysis of the composition of glass slag and spinel in the synthesis slag was carried out from EDS results. The Zn/Fe mass ratio in spinel was 0.486, the Zn content in glass slag was 20.338 wt %, and the Fe/SiO_2 mass ratio in glass slag was 0.135. The XRD results showed that the precipitated phase was $ZnFe_2O_4$.

Combining XRD detection, SEM results and the calculated phase diagram of CaO-Fe_XO-SiO_2-ZnO slag system, it can be determined that the slag is consisted of precipitated solid phase and glass slag liquid phase, and the precipitated solid phase was $ZnFe_2O_4$.

2.4 Experimental program

The successfully synthesized raw material, zirconia crucible, and 1700 °C vertical tube furnace were selected to experimentally determined the spinel conversion behavior by high-temperature phase equilibrium experiments. High ZnO fayalite slag was used to adjust two influencing factors: temperature and oxygen partial pressure. The temperature interval was 1100 °C - 1300 °C, with five temperature points of 1100 °C, 1150 °C, 1200 °C, 1250 °C, 1300 °C, and oxygen partial pressure of 10^{-8} atm, 10^{-7} atm, 10^{-6} atm, 10^{-5} atm, 10^{-4} atm standard gas (O_2+N_2), a total of 5*5 = 25 sets of experiments need to be completed, and the samples were quenched and cooled, dried, and stored in samples. The above 25 sets of experimental samples were characterized by using XRD detection method and SEM-EDS.

3. Result and discussion

3.1 Effect of temperature

The experimental results were obtained by controlling a certain oxygen potential condition and investigating the phase transformation behavior of spinel particles in high ZnO fayalite slag with the change of temperature. The crystallinity of spinel was evaluated by X-ray diffraction, and the effect of temperature on crystallinity can be reflected in the variation of X-ray diffraction peak intensity under different oxygen potential conditions. The result was shown in Figure 4-6.

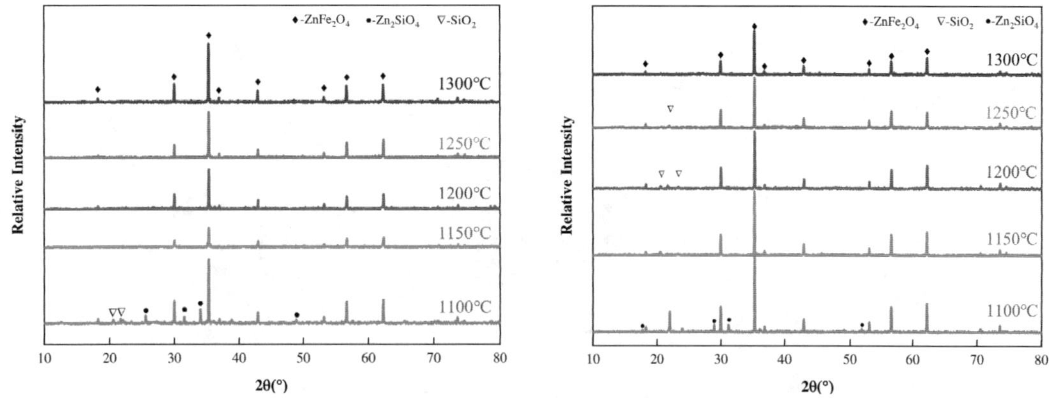

Figure 4. Recorded XRD patterns of slag samples at different conditions.

Left: at an oxygen potential of 10^{-8} atm; Right: at an oxygen potential of 10^{-4} atm.

As shown in Figure 4, it suggested that the crystallinity of $ZnFe_2O_4$ spinel increased with temperature increasing at an oxygen potential of 10^{-8} atm. However, when slag samples were treated at 10^{-4} atm, the conclusions were different, it decreased with temperature increasing. In terms of microscopic morphology, the average particle size of $ZnFe_2O_4$ spinel particles in equilibrium slag varied with temperature at different oxygen potentials shown in Figure 5.

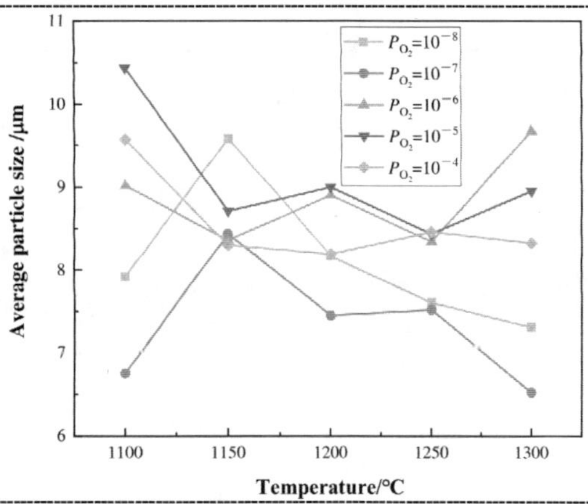

Figure 5. Variation of average size of $ZnFe_2O_4$ spinel particles in equilibrium slag with temperature, at different oxygen potentials.

According to the data shown in Figure 5, the average size of $ZnFe_2O_4$ spinel particles tended to decrease with the increasing of temperature at different oxygen potentials. Solid spinel particles nucleated or precipitated in the molten slag phase, and the particle size had the correlation with temperature, hold time and oxygen partial pressure. Small particles might grow big with the mechanism of Ostwald Ripening or Nucleation-Growing. Samples prepared at low temperatures, such as 1100 °C, more Fe（III） in the system at high p_{O2}, and that was a privilege for big particles formation. Along with increasing the temperature or decreasing the p_{O2}, the Fe (III) / Fe $_{total}$ ratio in the system decreased, which was a favor for spinel decomposition. Taking the oxygen potential of 10^{-8} atm as an example, the backscattered electron microscopy results of the samples obtained at different temperature conditions were shown in Figure 6.

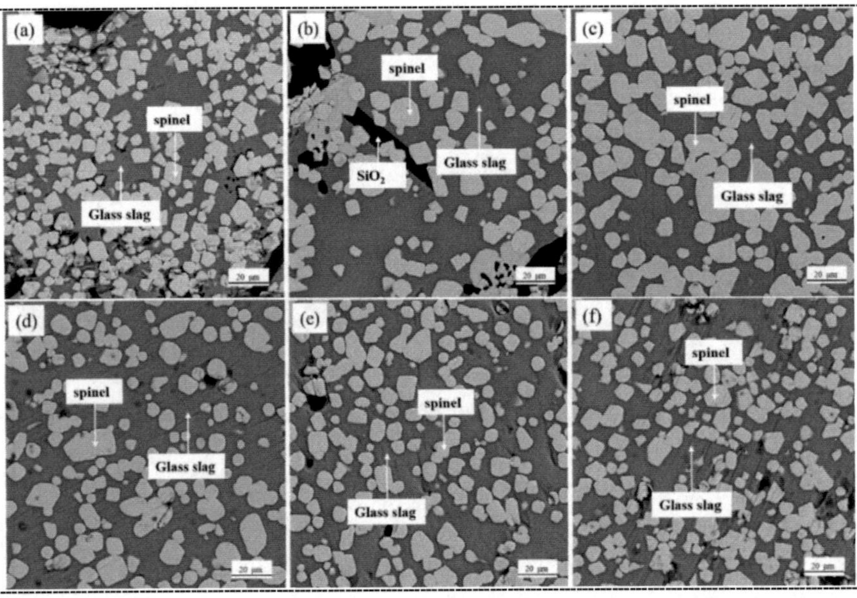

Figure 6. The BSE pictures and EDS results of the samples at different temperature conditions, at 10^{-8} atm (a) row material (b) 1100 °C (c) 1150 °C (d) 1200 °C (e) 1250 °C (f) 1300 °C

The information of the picture in Figure 6 showed a clear support for the variation of the size of $ZnFe_2O_4$ spinel particles tends. The effect of temperature also reflected in the chemical composition. The Zn/Fe mass ratio of $ZnFe_2O_4$ spinel in slag samples with different oxygen potential was shown in Figure 7, and the variation of Fe/SiO_2 mass ratio and Zn content in glass slag of samples with temperature was shown in Figure 8.

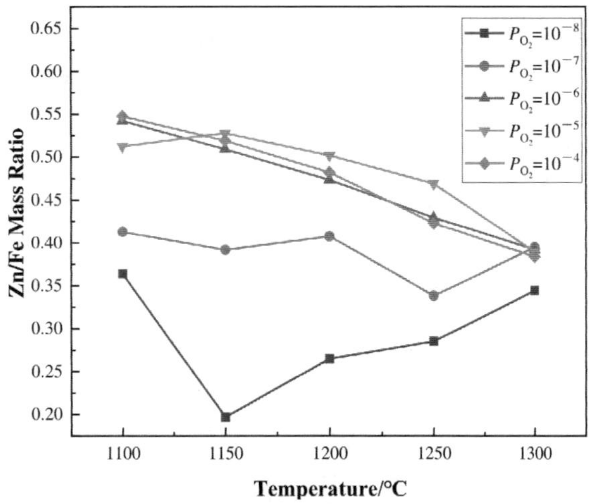

Figure 7. The Zn/Fe mass ratio variation of spinel with temperature in samples, at different oxygen potentials.

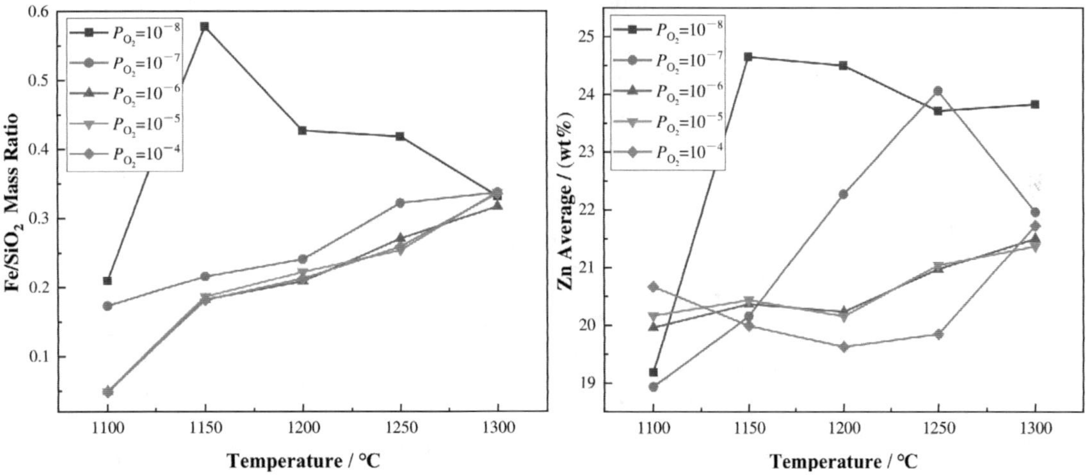

Figure 8. The Fe/SiO_2 mass ratio and Zn content of glass slag with the variation of oxygen potential and temperature.

From Figure 7 and Figure 8, it has been demonstrated that the Zn/Fe mass ratio in spinel gradually decreased and the Fe/SiO_2 mass ratio in glass slag phase gradually increased with temperature increasing under different oxygen potential conditions; the Zn content in glass slag phase showed an increasing trend with temperature increasing.

3.2 Effect of oxygen potential

The experimental results were obtained by controlling a certain temperature condition and investigating the phase transformation behavior of spinel particles in high ZnO fayalite slag with the change of oxygen potential. The effect of oxygen potential on crystallinity can be reflected in the variation of X-ray diffraction intensity under different oxygen potential conditions. The results were shown in Figure 9.

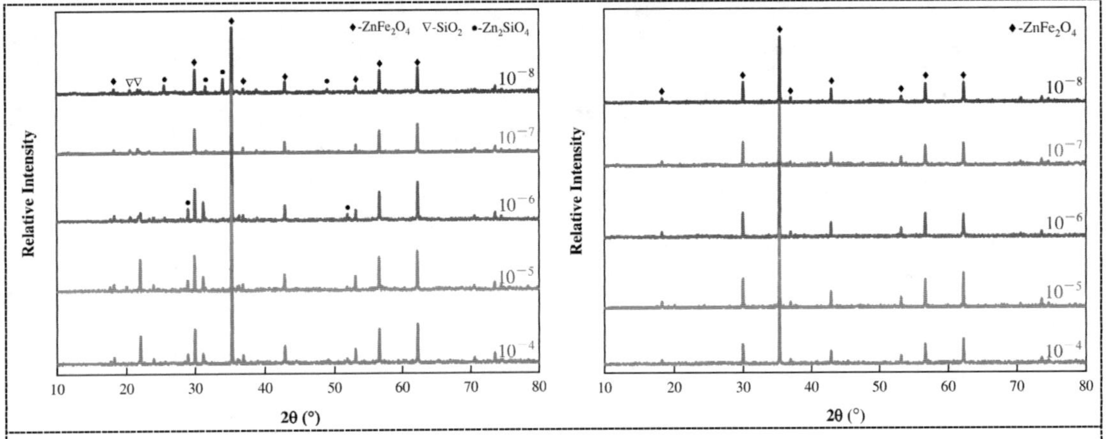

Figure 9. XRD results of slag samples at different oxygen potential conditions.

Left: at 1100 °C; Right: at 1300 °C.

From Figure 9, It has been suggested that the crystallinity of $ZnFe_2O_4$ spinel increased with increasing oxygen potential at all temperature conditions. Combined with Figure 4, we found that the fraction of high ZnO fayalite slag does not change too much in the range of all conditions of this study. In terms of microscopic morphology, the average particle size of $ZnFe_2O_4$ spinel particles in equilibrium slag varied with oxygen potentials at different temperature shown in Figure 10.

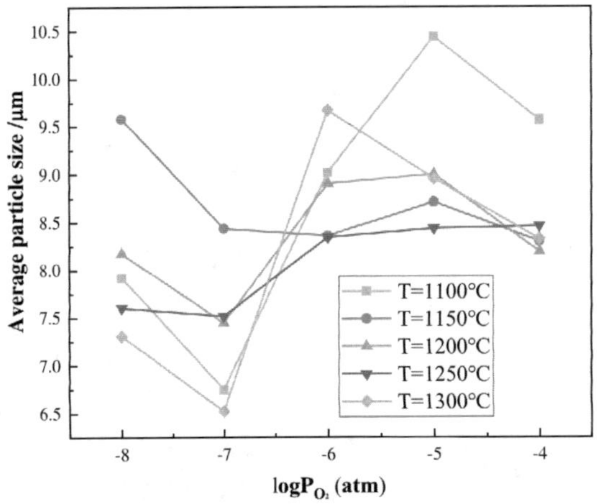

Figure 10. Variation of the average size of $ZnFe_2O_4$ spinel particles in equilibrium slag with the variation of oxygen potential and temperature.

According to the information shown in Figure 10, the average size of $ZnFe_2O_4$ spinel particles tended to increase with the increase of oxygen potential at different temperatures. Taking an example at 1300 ℃, the backscattered electron microscopy results of the samples obtained at different temperature conditions were shown in Figure 11.

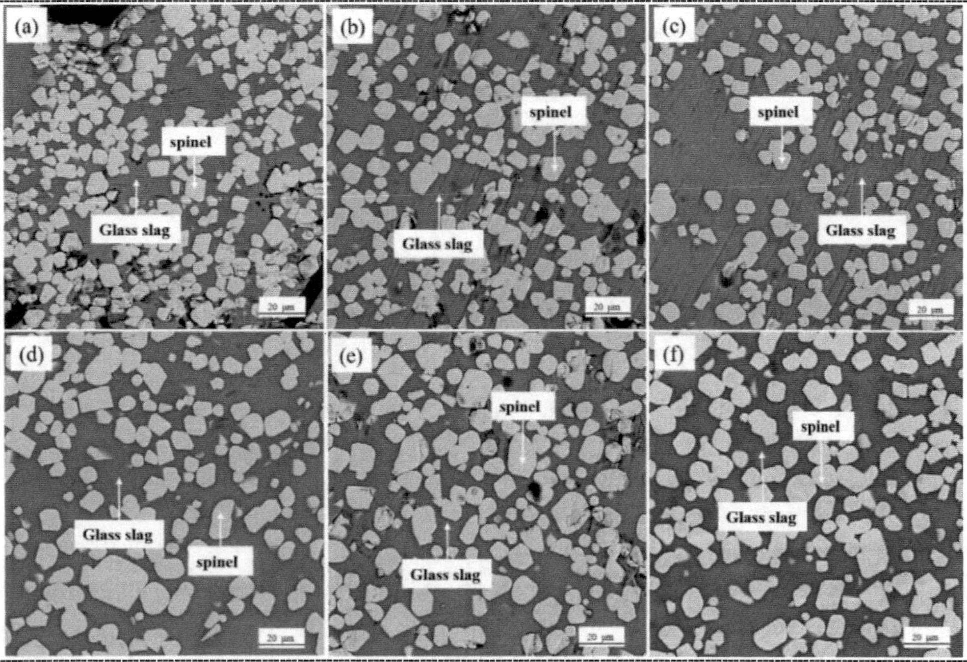

Figure 11. The BSE pictures and EDS results of the samples at 1300 °C.
(a) row material (b) 10^{-8} atm (c) 10^{-7} atm (d) 10^{-6} atm (e) 10^{-5} atm (f) 10^{-4} atm.

As shown in Figure 11, $ZnFe_2O_4$ spinel particles grown larger along with increasing the oxygen partial pressure. The Zn/Fe mass ratio of $ZnFe_2O_4$ spinel in these samples was shown in Figure 12, and the Fe/SiO_2 mass ratio and Zn content in glassy slag have also been shown in Figure 13.

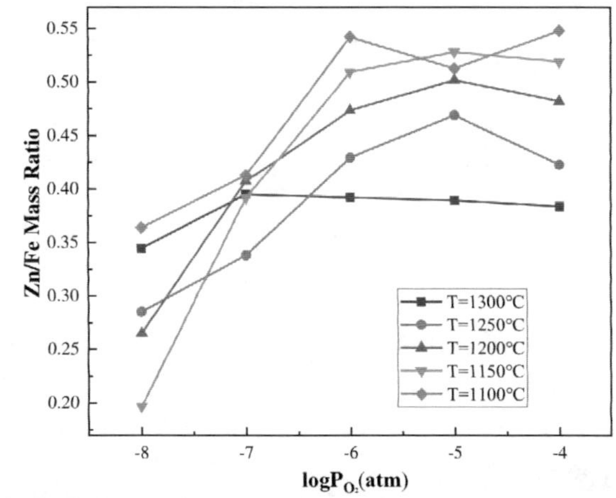

Figure 12. The Zn/Fe mass ratio variation of spinel with the variation of oxygen potential and temperature.

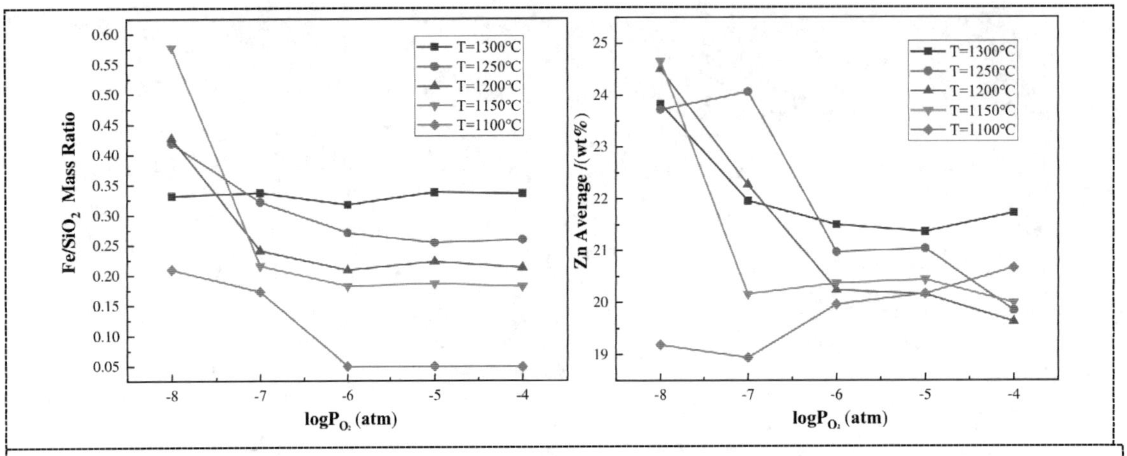

Figure 13. The Fe/SiO_2 mass ratio and Zn content of glass slag with the variation of oxygen potential and temperature

From Figure 12 and Figure 13, it has been demonstrated that the Zn/Fe mass ratio in spinel gradually increased with increasing oxygen potential under different temperature conditions; the Fe/SiO_2 mass ratio in glass slag phase gradually decreased with increasing oxygen potential; the Zn content in glass slag phase had a decreasing trend with increasing oxygen potential.

4. Summary and conclusions

The phase equilibrium and the spinel particle transformation have been experimentally studied in the high ZnO contain fayalite slag by controlling different oxygen potential conditions and temperatures. The individual experimental data has been summarized and the variation pattern has been organized, and the following conclusions has been obtained:

(1) the average size of spinel particles Variation pattern. The average size of $ZnFe_2O_4$ spinel particles tended to decrease with the increase of temperature, and gradually increased with the increase of oxygen potential.

(2) Spinel phase components variation pattern. The Zn/Fe mass ratio of $ZnFe_2O_4$ spinel phase gradually decreased with temperature increasing and gradually increased with increasing oxygen potential.

(3) The change of components in the glass slag phase. In $ZnFe_2O_4$ fayalite slag, the variation pattern of Fe/SiO_2 mass ratio in the glass slag phase remained consistent, gradually increased with temperature and gradually decreased with oxygen potential; the Zn content increased with temperature and decreased with oxygen potential.

(4) In the high ZnO fayalite slag, the variation of Zn/Fe mass ratio in $ZnFe_2O_4$ spinel particles was exactly opposite to the variation of Fe/SiO_2 mass ratio and Zn content in the glass slag. Therefore, the growth and decomposition process of $ZnFe_2O_4$ spinel is essentially a process of material exchange with the glass slag phase.

Acknowledgments

The research was funded by the National Solid Waste Project (No. 2022YFC3901604). The financial support from "The science and technology innovation Program of Hunan Province" (No. 2022RC1084) and the "111" project (Green and Value-Added Metallurgy of Non-ferrous Resources) is also acknowledged.

5. References

[1] Kania H, Saternus M. Evaluation and Current State of Primary and Secondary Zinc Production-A Review[J]. Applied sciences-basel, 2023,13(3).

[2] Hyk W, Kitka K, Rudnicki D. Selective Recovery of Zinc from Metallurgical Waste Materials from Processing Zinc and Lead Ores[J]. Molecules, 2019,24(12).

[3] Agrawal A, Sahu K K, Pandey B D. Recent trends and current practices for secondary processing of zinc and lead. Part I: lead recovery from secondary sources[J]. Waste management & Research, 2004,22(4):240-247.

[4] Liu J, Liu Y T. The research and application of oxygen-enriched side blowing smelting technology[J]. Chinese Non-ferrous Metallurgy, 2013,42(01):34-36. (In Chinese)

[5] Song X C, Gu H L. The industrial practice of top-blowing smelting technology in direct lead smelting [J]. The Design and Research on Non-ferrous Metals, 2013,34(05):18-21. (In Chinese)

[6] Xu C Q, Xin P F, Xu L. The co-smelting of lead-silver residue from zinc hydrometallurgy process and lead concentrate [J]. Chinese Non-ferrous Metallurgy,, 2022,51(04):37-42. (In Chinese)

[7] You L H. The new lead smelting process in Trail smelter of Cominco company [J]. Chinese Non-ferrous Metallurgy,2004(03):40-46. (In Chinese)

[8] Jiang J X, Li X R, Guo H J. The practice of Kivcet smelting technologu [J]. Global Non-ferrous Metals, 2018(18):7-9. (In Chinese)

[9] Zhang J L. The treatment of zinc hydrometallurgy residues for safe disposal and value recovery in Danxia smelter [J]. Metallurgy Management, 2023(02):100-103. (In Chinese)

[10] Wen P. The study of the sub-merged blowing technology in treating lead-zinc solid wastes [D]. Hunan University of Industry, 2016. (In Chinese)

[11] Hidayat T, Shishin D, Decterov S A, et al. Critical thermodynamic re-evaluation and re-optimization of the CaO-FeO-Fe_2O_3-SiO_2 system[J]. Calphad-computer coupling of phase diagrams and thermochemistry, 2017,56:58-71.

[12] Hidayat T, Fallah-Mehrjardi A, Hayes P C, et al. Experimental Investigation of Gas/Slag/Matte/Spinel Equilibria in the Cu-Fe-O-S-Si System at 1473K (1200°C) and P_{SO_2} =0.25atm[J]. Metallurgical and Materials Transactions B, 2018,49(4):1750-1765.

[13] Sineva S, Hidayat T, Fallah-Mehrjardi A, et al. Experimental investigation of gas-matte-spinel and gas-slag-matte-spinel equilibria in the Cu-Fe-O-S-Si system at 1200°C: effect of SO_2 partial pressure[J]. Mineral Processing and Extractive Metallurgy, 2022,131(3):195-202.

[14] Shevchenko M, Jak E. Experimental liquidus studies of the Zn–Fe–Si–O system in air[J]. 2019,110(7):600-607.

[15] Shevchenko M, Jak E. Experimental study and thermodynamic optimization of the ZnO–FeO–Fe_2O_3–CaO–SiO_2 system[J]. Calphad, 2020,71:102011.

[16] Hidayat T, Shishin D, Decterov S A, et al. Experimental Study and Thermodynamic Re-optimization of the FeO-Fe_2O_3-SiO_2 System[J]. Journal of phase equilibria and diffusion, 2017,38(4):477-492.

[17] Henao H M, Kongoli F, Itagaki K. High temperature phase relations in FeO_X (X=1 and 1.33)-CaO-SiO_2 systems under various oxygen partial pressure[J]. Materials transactions, 2005,46(4):812-819.

[18] Chen M, Avarmaa K, Taskinen P, et al. Investigation on the Matte/Slag/Spinel/Gas Equilibria in the Cu-Fe-O-S-SiO_2-(CaO, Al_2O_3) system at 1250 °C and P_{SO_2} of 0.25 atm[J]. Mineral Processing and Extractive Metallurgy Review, 2023,44(4):313-324.

[19] Nicol S, Jak E, Hayes P. Effect of Cooling Rate on the Controlled Solidification of "Fe_2O_3"-CaO-SiO_2 Liquids in Air in Synthetic Iron Ore Sinter[J]. Isij international, 2020,60(5):865-875.

[20] Mou Q, Li J, Zeng Q, et al. Effect of Fe_2O_3 on the size and components of spinel crystals in the CaO-SiO_2-MgO-Al_2O_3-Cr_2O_3 system[J]. International journal of minerals metallurgy and materials, 2019,26(9):1113-1119.

[21] Zhang L, Gao F, Deng T, et al. Phase equilibria in the FeO–Fe_2O_3–SiO_2 system: Experimental measurement and thermodynamic modeling[J]. Calphad, 2022,79:102459.

A novel leaching process of zinc ferrite and its application in the treatment of zinc leaching residue

Limin Zhang[1], Yunyan Wang[1,2,3], Yong Ke[1,2,3*], Yongjian Luo[1], Xiaobo Min[1,2,3], Cong Peng[1,2] and Yun Li[1,2]

[1] School of Metallurgy and Environment, Central South University, Changsha 410083, China
[2] State Key Laboratory of Advanced Metallurgy for Non-ferrous Metals, Changsha 410083, China
[3] Chinese National Engineering Research Center for Control & Treatment of Heavy Metal Pollution, Changsha 410083, China

Email: keyong000ke@csu.edu.cn

Abstract. Zinc ferrite is a refractory phase generated in the pyrometallurgical process of zinc and steel production. Much energy is invested in the decomposition of zinc ferrite to recycle zinc since zinc ferrite is difficult to leach. In this work, a novel leaching process targeted at decomposition of zinc ferrite was proposed to save energy and improve metal recovery efficiency. The key of this novel leaching process was the use of copper powder as the reductant. Leaching of zinc ferrite in the presence of copper powder was investigated. The extraction of zinc was 100% when molar ratio of copper to zinc ferrite was 1.5 while the extraction of zinc was only 19.3% without copper. Effects of leaching temperature, acid concentration, the ratio of liquid to solid and reaction time were studied. Under the conditions: 60 °C, ≥ 70 g·L^{-1} H_2SO_4, the ratio of liquid to solid ≥ 25 mL/g and the molar ratio of copper to zinc ferrite ≥ 1.5 at the open system, the synthetic zinc ferrite was dissolved completely within 60 min. Besides, the zinc ferrite-bearing zinc leaching residue from a roast-leach-electrowin plant was analyzed and leached under the optimal conditions. It was found that most of zinc ferrite in the zinc leaching residue was removed, remaining unreacted lead sulphate in the leached residue. The leaching efficiency of zinc in zinc leaching residue reached 94.2%. The results suggest that copper powder enhanced facile and efficient zinc extraction from zinc ferrite without concentrated acid, high temperature, long reaction time and specific investment. The recyclability of copper and its intrinsic commodity value showed the potential application in industrial zinc and iron containing wastes.

1. Introduction

Zinc and iron are important metals with various applications in vehicles, buildings, machinery and infrastructure [1]. In the processes of production and regeneration of zinc or iron, pyrometallurgical methods are necessary. Since zinc and iron tend to be companions in common use and zinc-bearing minerals, co-treatment of massive zinc and iron containing materials at the furnace brings about zinc ferrite, which is a stable phase causing trouble in recycling zinc [2, 3]. Many solid wastes such as zinc-bearing flue dust and zinc leaching residue are reported to contain considerable amounts of zinc ferrite [4]. To realize resource recycle, carbothermal reduction method has been long used, followed by obvious drawbacks including coal consumption, slag stockpile and loss of non-volatile metals [5]. Hot-acid leaching is another industrial practice targeted at zinc ferrite destruction while this intensive investment plays a limited role because of kinetics retard [6].Thus, developing a new method to decompose zinc ferrite is key to sustainable and effective treatment of zinc and iron containing wastes.

Many methods are proposed to extract zinc from zinc ferrite. Among them, the pyrometallurgical processes combined with leaching are effective on the phase transformation of zinc ferrite. For instance, reducing gases including CO and H_2 change zinc ferrite into weak acid soluble zinc oxide and magnetically separated iron oxides [7, 8]. Roasting by sulphate makes zinc in zinc ferrite be water soluble zinc sulphate [9,10]. However, these ways have no apparent advantages over carbothermal reduction method because the resulting residue or slag demands further safe and effective utilization. Besides, they are energy-consuming. Hydrometallurgical processes have the potential to save energy and reach complete extraction. In particular, reductive leaching has attracted wide interest in enhancing leaching efficiency of zinc ferrite with shorter time. As shown in Table 1, in the presence of the reductant, the leaching efficiency of metals in the zinc ferrite could reach more than 85% within 120 min. Without the reductant, the leaching efficiency reduces by more than 20% under the same leaching conditions. However, the mentioned reductants rarely succeed in practical application due to material cost, impurity input, new residue generation and other reasons. Besides, hot acid seems to be a must though the reductant exists, which improves the leaching conditions slightly.

Table 1. Previously reported reductive leaching conditions and results of zinc ferrite.

Material	Lixiviant	Condition	Leaching efficiency	Reference
Zinc neutral leaching residue	33.3 g/L $N_2H_4 \cdot H_2SO_4$, 80 g/L H_2SO_4	95 °C, 120 min, 10 mL/g, 400 rpm	Fe: 94.19%, Zn: 95.83%, Cd: 90.81%	[11]
Zinc neutral leaching residue (NLR)	ZnS in zinc concentrate: Fe in NLR=0.95, 160 g/L H_2SO_4	90 °C, 4 h, 11.4 mL/g, 400 rpm	Fe: 92.7%, Zn: 94.7%, In: 93.7%	[5]
Zinc residue	SO_2 200 kPa, 56 g/L H_2SO_4	100 °C, 120 min, 7 mL/g, autoclave	Fe: 90%, Zn: 96%, Ga: 91%	[12]
Zinc leaching residue (ZLR)	PbS in galena: zinc ferrite in ZLR=1, 140 g/L H_2SO_4	85 °C, 90 min, 40 mL/g, 500 rpm	Fe: 91%, Zn: 87%, In: 88%	[13]

To further improve the economics of the reductive leaching process, copper which much decreases the reaction temperature as well recyclable and reusable was proposed as the reductant in the sulfuric acid leaching system. In this study, the high and fast destruction efficiency of the synthetic zinc ferrite under mild conditions was achieved. Effect of parameters including temperature, acid concentration, dosage of the reductant, ratio of liquid-solid and leaching time was investigated in detail. Moreover, zinc leaching residue containing zinc ferrite was leached coupled with copper to verify the performance.

2. Materials and methods

2.1. Materials

The zinc ferrite was synthesized via roasting the mixture of iron oxide (AR, purity >99%, Sinopharm Chemical Reagent Co., Ltd.) and zinc oxide (99%, MACKLIN reagent. co., Ltd.) [14]. Then, the produced zinc ferrite was crushed and ground to fractions, which were weighed and stored for further studies. Leaching reagent, sulfuric acid (GR, 95%~98%, Sinopharm Chemical Reagent Co., Ltd.) was diluted before use. Reductant, copper powder (electrolytic, purity>99.7%, 200 mesh, Sinopharm Chemical Reagent Co., Ltd.). All solutions were prepared using pure water.

Zinc leaching residue was collected from a zinc smelter using the traditional roast-leach-electrowin process. The sample was dried at 105 °C for 12 hours in a draught drying cabinet then ground. The zinc ferrite and zinc leaching residue were screened using a 200-mesh sieve to obtain fractions below 74 μm size as the experiment materials.

2.2. Leaching experiments

Batch tests were conducted in 50 ml centrifuge tubes with perforated membranes in a thermostatic water-bath oscillator with a speed of 260 r/min for 60 min. 6 ml of the leaching solution was put in a centrifuge tube and heated to the set temperature followed by adding the mixture of zinc ferrite and copper powder. The leached residue at 60 min was collected via a centrifuge and the cake was washed and dried in a vacuum freeze dryer for 12 h. The addition of copper powder (k) was evaluated by the formulas:

$$ZnFe_2O_4 + Cu + 8H^+ \rightarrow Zn^{2+} + 2Fe^{2+} + Cu^{2+} + 4H_2O \quad (1)$$

$$m_k = \frac{M_{Cu}}{M_{ZnFe_2O_4}} \times m_{ZnFe_2O_4} \quad (2)$$

where M is the molecular weight and m is the mass.

For the experiment of time, 0.5 ml sample of the solution was extracted using the pinhole filter membrane of 0.22 μm at 10, 20, 30, 40, 50, 60 min from 12 ml of the leaching solution. The solution was chemically analyzed to calculate the leaching rate (w). The computational formula was as follows:

$$w = \frac{V_L \times C_L}{G_S \times W_S} \quad (3)$$

where V_L is the calibrated volume of the leaching solution; C_L is the concentrations of metals in the leaching solution; G_S is the mass of the solid used in the experiment; W_S is the content of metals in the solid.

All experiments were conducted at the open system.

2.3. Analytical methods

X-ray diffraction (XRD, D8 Advance, Bruker) and scanning electron microscopy (SEM-EDX, JSM-IT300, JEOL) were utilized to study the characteristics of the solids. Inductively Coupled Plasma Optical Emission Spectroscopy (ICP-OES, Agilent 5100) was used to determine the total Zn, Fe and Cu in the solution.

3. Results and discussions

3.1. Characterization of zinc ferrite, zinc leaching residue and copper powder

As indicated in XRD pattern of Figure 1-a, the synthetic solid was totally zinc ferrite. The zinc ferrite was constituted by piled particles in both nano size (~500 nm diameter) and micro size (~1.5 μm diameter) as illustrated in Fig. 1d, which agreed with the published study [15]. The main phases in zinc leaching residue are zinc ferrite and lead sulphate as shown in Figure 1-b. The particle size of zinc ferrite in zinc leaching residue was similar to that of the synthetic one including nano size and micro size particles in Figure 1-e. X-ray diffraction pattern and SEM image of copper powder were presented as Figure 1-c and Figure 1-f, respectively. It could be seen that the copper powder was composed with independent coral-like particles, which were stacked by smaller granules (~5 μm diameter). The micro particle size of the copper powder was over 20 μm.

The zinc leaching residue sample contains 16.9% Zn and 24.8% Fe analyzed by ICP-OES. The mass of zinc ferrite in zinc leaching residue was calculated as 53.4% according to the ratio of Fe in zinc ferrite and in zinc leaching residue.

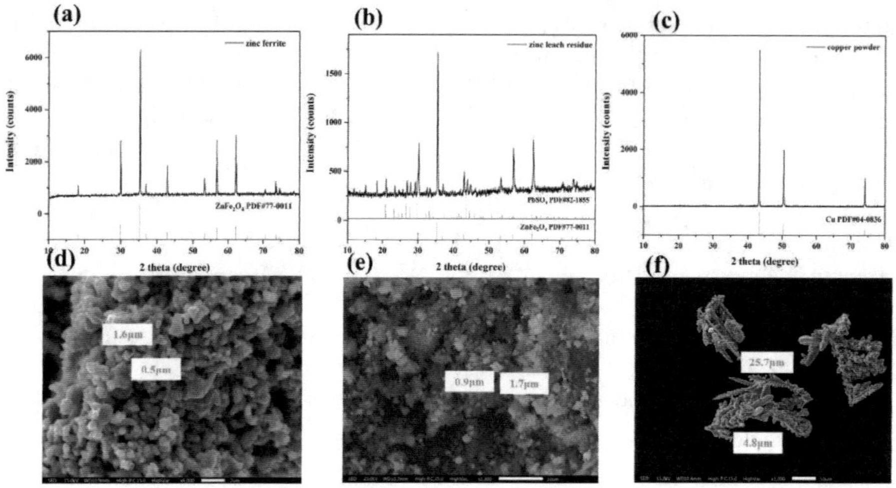

Figure 1. X-ray diffraction patterns and SEM images of the synthetic zinc ferrite (a) (d), zinc leaching residue (b) (e) and copper powder (c) (f).

3.2. Leaching of zinc ferrite in the presence of copper

The effect of copper powder dosage on the leaching of zinc ferrite was studied under the conditions: liquid-solid ratio of 30, sulfuric acid concentration of 100 g/L and 60 °C. As shown from Figure 2-a, the leaching efficiency of zinc were strongly dependent on the dosage of copper. When zinc ferrite was leached without copper, zinc was extracted only 19.3%. The leaching efficiencies of zinc increased to 43.2% and 76.4% as molar ratio of copper to zinc ferrite was 0.5 and 1.0, respectively. Zinc ferrite was completely dissolved when molar ratio of copper to zinc ferrite was 1.5 and 2.0. Figure 2-b illustrated the phase transformation of the residues. Without copper, the peak intensity of the residual zinc ferrite was stronger, indicating that pure acid leaching made more crystalline zinc ferrite stay in the leached residue. With copper, the peak intensities of the residual zinc ferrite were weaker as the of copper was more, demonstrating that copper helped destruction of zinc ferrite. When molar ratio of copper to zinc ferrite was 1.5, the residue was only copper. Notably, a little copper was found in the residue along with the main phase zinc ferrite when molar ratio of copper to zinc ferrite was 1.0. It could be speculated that excess copper was a must to make zinc ferrite disappear.

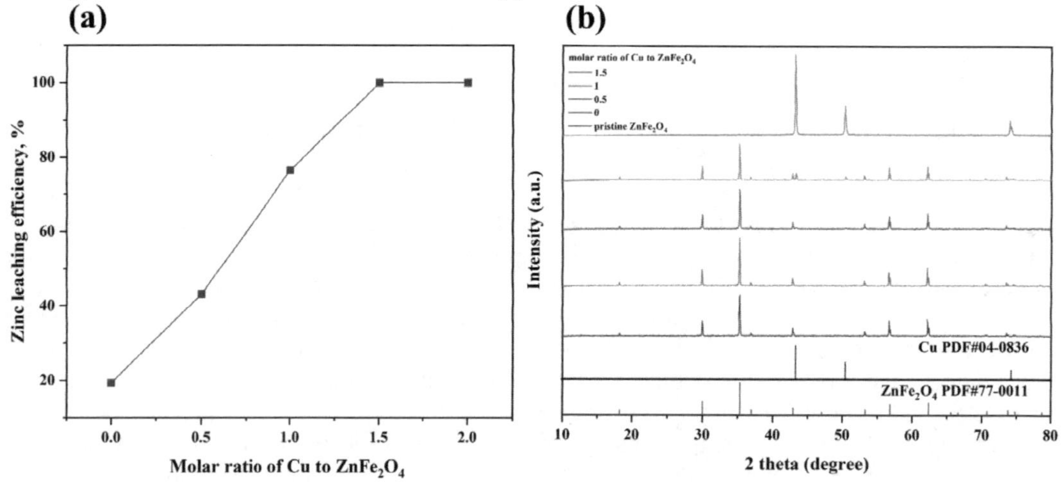

Figure 2. Leaching of zinc ferrite in the presence of copper (a) zinc leaching efficiencies; (b) X-ray diffraction patterns of the leached residues

3.3. Effects of factors

3.3.1 Temperature. The effect of temperature on the leaching of zinc ferrite was studied under the conditions: liquid solid ratio of 30 mL/g, sulfuric acid concentration of 100 g/L and molar ratio of copper to zinc ferrite = 1.5. As shown in Figure 3-a, the temperature put a positive effect on zinc leaching. The leaching efficiencies of zinc increased as the temperature increased from 28.8% at 30 °C to 78.6% at 50 °C. When the temperature was over 60 °C, the zinc extraction reached 100%.

3.3.2 Acid concentration. The effect of acid concentration on the leaching of zinc ferrite was studied under the conditions: liquid-solid ratio of 30 mL/g, molar ratio of copper to zinc ferrite = 1.5 and temperature of 60 °C. It could be seen that in the weak acid solution like 30 g/L sulfuric acid, facile leaching of zinc ferrite was realized. The leaching efficiency of zinc was 51.4% and 85.2% as the sulfuric acid concentration was 30 g/L and 50 g/L, much higher than that without copper when the sulfuric acid concentration was 100 g/L in Figure 3-b. Zinc ferrite was totally dissolved as the sulfuric acid concentration was over 70 g/L.

3.3.3 Ratio of liquid to solid. The effect of liquid to solid ratio on the leaching of zinc ferrite was studied under the conditions: molar ratio of copper to zinc ferrite = 1.5, sulfuric acid concentration of 100 g/L and temperature of 60 °C. As shown in Figure 3-c, zinc leaching efficiency increased proportionally to the ratio of liquid to solid when the ratio of liquid to solid was below 20 mL/g. The leaching efficiencies of zinc were 25.6%, 47.7% and 91.3% as ratio of liquid to solid was 5 mL/g, 10 mL/g and 20 mL/g. When ratio of liquid to solid was more than 25 mL/g, zinc ferrite was completely dissolved. It is indicated that sufficient acid is a must for the dissolution of zinc ferrite, which is an acid-consuming process. However, excessive acid is unnecessary because zinc ferrite has been completely dissolved, from results of acid concentration (Figure 3-b) and liquid-solid ratio (Figure 3-c).

3.3.4 Time. The experiment of time was conducted under the conditions: molar ratio of copper to zinc ferrite = 1.5, sulfuric acid concentration of 100 g/L, ratio of liquid to solid=30 mL/g and 60 °C. As shown in Figure 3-d, the rate of zinc extraction maintained at a high level within 30 min as the reaction proceeded. The leaching efficiencies of zinc was 13.6%, 43.3% and 79.5% at 10 min, 20 min and 30 min, respectively. When the reaction time was 40 min, the extraction of zinc was 97.1%. Zinc extraction reached 100% after 50 min, indicating that zinc ferrite could be completely dissolved within 60 min.

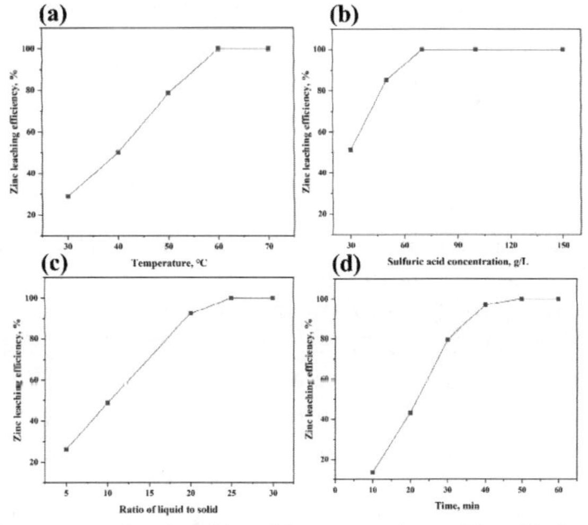

Figure 3. Effects of factors on zinc leaching (a) temperature; (b) sulfuric acid concentration; (c) ratio of liquid to solid; (d) time

3.4. Leaching of zinc leaching residue

The experiment of zinc leaching residue was studied under the conditions: molar ratio of copper to zinc ferrite = 1.5, ratio of liquid to solid (zinc ferrite in zinc leaching residue) = 30 mL/g, sulfuric acid concentration of 100 g/L and 60 °C. As shown in Figure 4-a, the leaching behavior of zinc in zinc leaching residue was similar to that in the synthetic zinc ferrite. The leaching efficiency of zinc climbed up within 30 min from 40.7% at 10 min to 83% at 30 min and reached a plateau after 40 min at 94.2%. Figure 4-b compared the constituents of the leached residue with those of the pristine zinc leaching residue. It could be seen that zinc ferrite almost disappeared after leaching and lead sulphate was the main phase in the leach residue, demonstrating that copper worked for the industrial zinc ferrite and very efficient for zinc extraction from zinc ferrite-bearing zinc leaching residue.

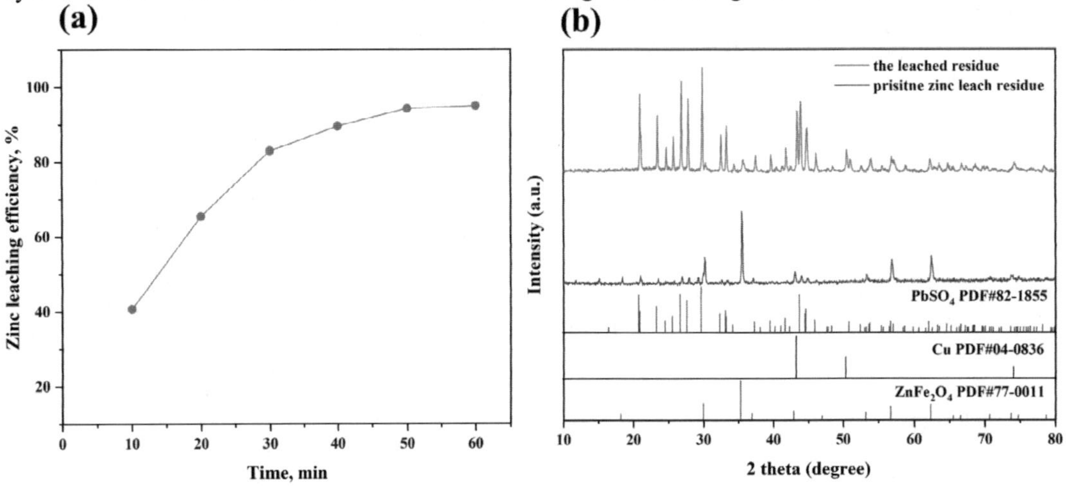

Figure 4. Leaching of zinc leaching residue (a) zinc leaching efficiencies; (b) X-ray diffraction patterns of the leached residue

4. Conclusions

In this study, copper powder was used as a reductant in the sulfuric acid system to realize the facile and efficient dissolution of the man-made zinc ferrite as well as zinc ferrite-bearing zinc leaching residue. Zinc ferrite could be completely dissolved within 60 min at 60 °C without concentrated acid. Under the conditions: molar ratio of copper to zinc ferrite = 1.5, ratio of liquid to solid (zinc ferrite in zinc leaching residue) = 30 mL/g, sulfuric acid concentration of 100 g/L and temperature of 60 °C, the extraction of zinc from zinc leaching residue reached 94.2%. Lead sulphate was the main phase in the leached residue.

Acknowledgments

We thank for constructive comments on this manuscript from the reviewers. This work was financially supported by Major program Natural Science Foundation of Hunan Province of China (2021JC0001), Foundation for Innovative Research Groups of the National Natural Science Foundation of China (52121004), National Science Fund for Distinguished Young Scholars (51825403), Science and Technology Innovation Program of Hunan Province, China (2021RC3013), Science and Technology Project of Gansu Province, China (21ZD4GD033) and National Natural Science Foundation of China (22276219), Hunan Provincial Innovation Foundation for Postgraduate (CX20230175), and the Fundamental Research Funds for the Central Universities of Central South University.

References

[1] T. Watari, K. Nansai and K. Nakajima 2021 *Resources, Conservation and Recycling* 164 105107.
[2] D. K. Xia and C. A. Pickles 1999 *Minerals Engineering* 12 693-700.
[3] N. Leclerc, E. Meux and J.-M. Lecuire 2003 *Hydrometallurgy* 70 175-183.

[4] V. Kashyap and P. Taylor 2021 *Mining, Metallurgy & Exploration* 38 27-36.
[5] F. Zhang, C. Wei, Z.-g. Deng, C.-x. Li, X.-b. Li and M.-t. Li 2016 *Transactions of Nonferrous Metals Society of China* 26 2495-2501.
[6] D. Filippou and G. P. Demopoulos 1992 *Canadian Metallurgical Quarterly* 31 41-54.
[7] H. Yan, L.-y. Chai, B. Peng, M. Li, N. Peng and D.-k. Hou 2014 *Minerals Engineering* 55 103-110.
[8] V. Kashyap, P. Taylor, E. Tshijik Karumb and M. Cheshire 2021 *Minerals Engineering* 171 107078.
[9] Y. Li, H. Liu, B. Peng, X. Min, M. Hu, N. Peng, Y. Yuang and J. Lei 2015 *Hydrometallurgy* 158 42-48.
[10] P. Grudinsky, D. Pankratov, D. Kovalev, D. Grigoreva and V. Dyubanov 2021 *Materials* 14.
[11] C. Zhang, X.-b. Min, J.-q. Zhang, M. Wang, B.-s. Zhou and C. Shen 2015 *Transactions of Nonferrous Metals Society of China* 25 4175-4182.
[12] X. Wu, S. Wu, W. Qin, X. Ma, Y. Niu, S. Lai, C. Yang, F. Jiao and L. Ren 2012 *Hydrometallurgy* 113-114 195-199.
[13] Y. Fan, Y. Liu, L. Niu, W. Zhang and Z. Zhang 2021 *Minerals Engineering* 163 106777.
[14] X. Min, B. Zhou, Y. Ke, L. Chai, K. Xue, C. Zhang, Z. Zhao and C. Shen 2016 *Applied Surface Science* 371 67-73.
[15] Y. Zhang, X. Li, L. Pan, Y. Wei and X. Liang 2010 *Hydrometallurgy* 102 95-100.

Journal of Physics: Conference Series **2738** (2024) 012028 doi:10.1088/1742-6596/2738/1/012028

Increasing of the unit capacity of MF process at Miike Smelter

H Kubo[1], F Nishiyama[2] and S Torikai[3]

Miike Smelting Company Limited
2-1 Shinkai-cho, Omuta-city, Fukuoka, Japan, 836-0017

1. Email: hiroshi_kubo@mitsui-kinzoku.com
2. Email: f_nishiyama@mitsui-kinzoku.com
3. Email: storikai@mitsui-kinzoku.com

Abstract. MF (Mitsui Furnace) is a half shaft blast furnace developed by Mitsui Miike Smelter in the 1960's to treat vertical retort residue. Currently MF mainly treats various secondary wastes (EAF dust, Secondary Fly Ash, etc.) and recovers zinc and lead as crude zinc oxide, copper and silver as matte, meanwhile other elements such as FeO, CaO, SiO_2, Al_2O_3 are concentrated in slag.

There are two steps in the MF process. The first step is pretreatment. In this process, various waste materials are blended in a certain proportion with reductant and dried adding binder in a rotary dryer, then grinded in a rod mill. The grinded materials are made into an oval briquette by briquetting machine. The second step is treatment process. The briquettes fed into the MF are smelted and reduced by blowing air through the tuyeres. By enriching the tuyere air with oxygen generated by the VSA (Vacuum Swing Adsorption) oxygen generator in 2005 and 2012, we have improved the throughput of wastes.

By the improvement of oxygen enrichment in the MF, the unit capacity increased significantly. However shortage of the drying capacity in the pretreatment process became a problem. To ensure sufficient drying capacity, we installed one more rotary drier in series in 2016. Additionally in 2018, we installed a granulator after the rod mill to improve the bulk density in briquette. As a result of these efforts, the treatment amount in the whole MF process has increased.

The MF can treat various kinds of materials containing various non-ferrous metal-valuables. Through these process improvements, we promote the treatment of wastes containing heavy metals and contributes significantly to the recycling of metal resources.

1. Introduction

Miike Smelting Co., Ltd. is recycling waste materials by its own technology using the half shaft blast furnace called MF Process, and produces the crude zinc oxide. EAF dust was the main composition of conventional raw materials, and Miike Smelting Co, Ltd. started treating secondary fly ash which contain zinc and lead in 2002 newly. Miike Smelting Co., Ltd. produces the crude zinc oxide mainly by treating these two kinds of raw materials. And raw materials treating capacity was improved 30%, by enrichment of air with oxygen from 2005 to 2012. Currently, Miike Smelting Co., Ltd. is recycling about 140,000 tons of waste materials per year and production capacity of crude zinc oxide is 28,000 t/y.

2. Outline of MF Process

2.1. History of Residue Treatment Process at Miike Smelter

Since it started zinc smelting using horizontal retort furnaces in 1913, the Miike Smelter of Mitsui Mining and Smelting Company has equipped electrolysis plants and vertical retort furnaces in turn and produced 108,000 tons of zinc per year at its peak. The residue produced by these processes was partly treated in fuming furnaces.

However, for the purpose of treating increasing residue and former piled-up residue, MF Process was developed, and the first MF furnace started operations for practical use in 1965. During the 20 years from 1965 to 1985, the residue treated by MF Process reached 2,100,000 tons, and the treatment of almost all piled-up residue was completed. After the production of zinc metal was stopped at Miike by the down-sizing of its zinc smelting department in 1985, it was decided to treat the recycle material by which it's for EAF dust from outside our company, only one MF furnace is treating waste material. And then Miike Smelting Co., Ltd. started treating secondary fly ash in 2002, enrichment oxygen operation started in 2005.

2.2. Features of MF Process

2.2.1. First step is Pretreatment process. Today the two main raw materials which MF is treating are EAF dust and secondary fly ash. Table 1 shows the typical construction of MF materials. Figure 1 shows the flow sheet of MF Process. In MF process, lump materials that are difficult to pulverize in a rod mill are fed straight to the MF furnace, and powder materials after being briquette.

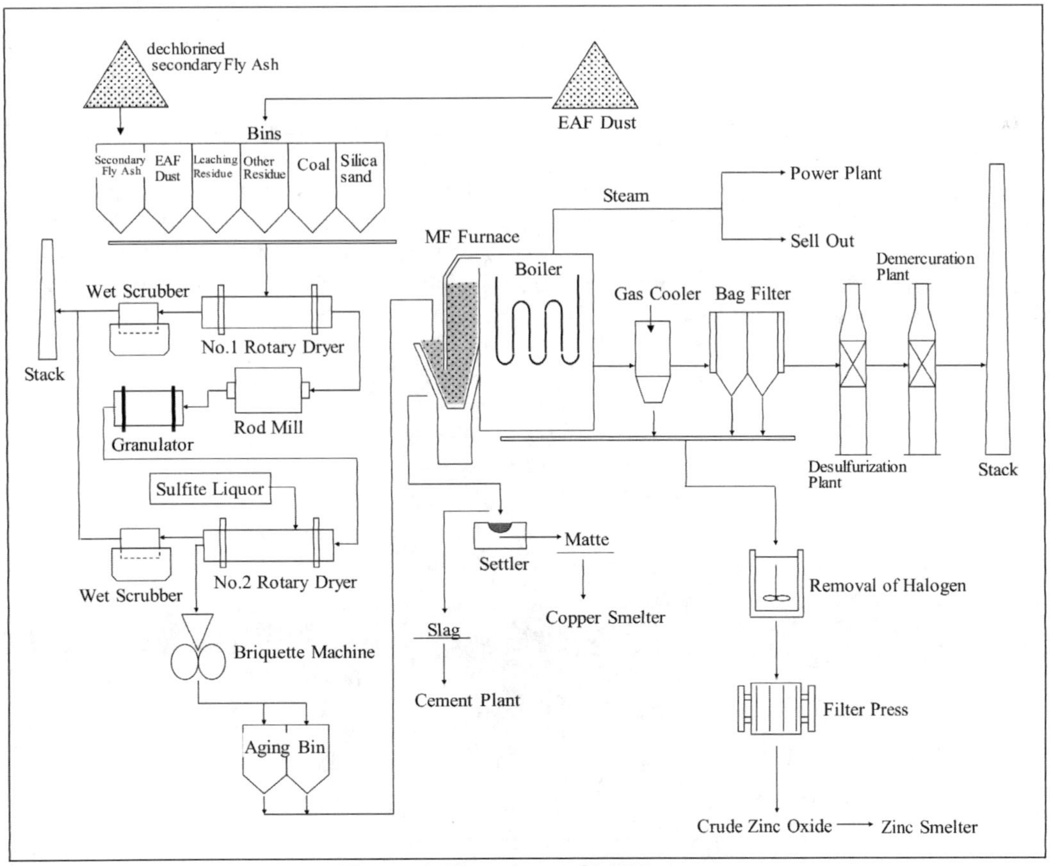

Figure 1. Flowsheet of MF Process.

Table 1. MF Feed Materials (t/yeat).

EAF Dust	Secondary Fly Ash	Zn Leaching Residue etc.	PCB waste	Total
40,000 ~ 60,000	30,000 ~ 50,000	40,000	400 Cap:10,000	140,000

In the briquetting process, such materials as EAF dust and secondary fly ash, silica as flux and coal are constantly fed and transmitted to the rotary dryer, where blended with the materials, and the moisture is regulated. These are milled and mixed in the rod mill, granulated in the granulator, and then sulfite liquor as binder is added in another rotary dryer. This mixture is briquetted and constantly fed to the furnace.

2.2.2. Second step is MF process. In the MF furnace, zinc and lead oxide dust, matte and slag are produced. The oxide dust, carried out from the furnace with the exhaust gas, is collected in the boiler, cooler and the bag filter. The exhaust gas, after the dust has been collected, contains Hg and SOx, which is decreased by the exhaust gas washing facility and emitted into the air. The oxide dust, after being washed, filtered in order to remove chlorine and fluorine, is transmitted to Zinc Smelter, as crude zinc oxide. From the MF furnace slag and matte are taken out constantly; the matte, separated and recovered in the settler by its specific gravity, is sold as the material for copper. This slag, stable and non-toxic, is sold as the material for cement. Table 2 shows the product amount of MF process.

Table 2. The Product Amount of MF Process (t/yeat).

Crude Zinc Oxide	Slag	Matte
28,000	70,000	1,000

Figure 2 shows the conceptual figure of MF furnace. The briquettes fed from the feeding shoot, after being dried and preheated in the coking zone, enter the smelting zone. Zinc and lead are deoxidized there, and after deoxidization turned into oxide dust; copper and precious metals are concentrated in the matte. It is an important feature of MF furnace that these metals can be recovered in it at the same time. Fed into the furnace, 85-90% of the zinc, 95-97% of the lead are recovered as oxide dust, and the copper and precious metals as matte. In addition, the furnace and the waste heat boiler are constructed in one unit so that efficient recovery of heat is possible.

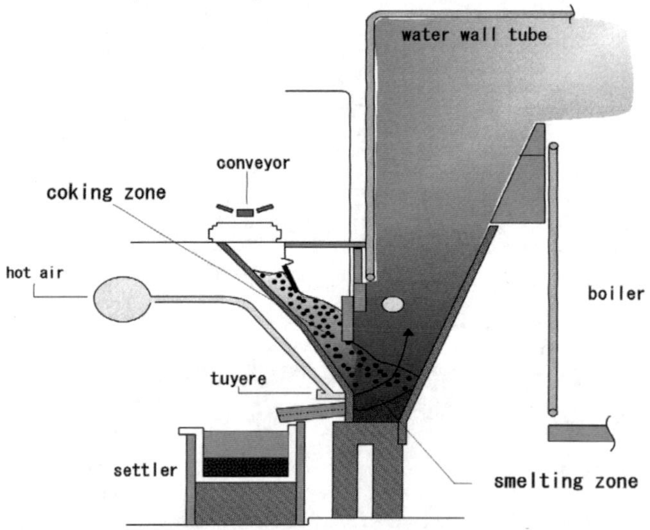

Figure 2. Conceptual Figure of MF Furnace.

2.3. Comparison of MF process with other process

Currently the main method used for EAF dust treatment is the Waelz process. The Waelz process utilizes a rotary kiln, where carbon material is added to EAF dust. The mixture is heated to around 1300°C using heavy oil burners etc., zinc is vapored after zinc oxide is reducted. As the generated zinc vapor is re-oxidized by atmospheric O_2, zinc is finally recovered in crude zinc oxide. On the other hand, the residue left after zinc extraction, called as clinker, is discharged from the kiln. While a portion of it is recycled as electric furnace feedstock, the majority is processed for use as roadbed material or cement raw material [1]. The clinker is not fully melted in the kiln, it exists in semi-melted state. Therefore, it cannot be separated into slag and matte. Consequently, if precious metals are present in the waste, they would be lost in the clinker. Whereas, in the MF process, the raw materials are completely melted in the furnace, so it can be separated into slag and matte. Therefore, valuable metals such as gold and silver contained in materials can be recovered as matte, which is highly economical.

Furthermore, the MF furnace has characteristics such as the furnace temperature exceeds 1300°C, the ability to recover heavy metals. Taking advantage of them characteristics, Miike Smelting Co, Ltd. has started the detoxification treatment for low-concentration PCBs (polychlorinated biphenyls) waste in 2015. Because of the characteristics of PCBs, such as low solubility, high boiling point, resistance to thermal decomposition, non-flammability, and high electrical insulation properties, it was extensively produced in Japan during the 1960s to 1970s [2]. PCBs were primarily used as electrical insulating fluid in capacitors and transformers, as well as in hydraulic systems, heat transfer systems, and lubricants. PCBs were also blended with other chemicals, finding applications in various products like sealants, adhesives, and carbonless copy paper [3]. However, due to the recognized environmental and human health hazards posed by PCBs, their production and importation have been prohibited. Incinerating products containing PCBs or PCBs waste at inadequate temperatures and residence times can lead to the generation of dioxins and the dispersion of PCBs into the environment.

At Miike Smelting Co, Ltd., certification for the harmless disposal facility for low-concentration PCBs waste was obtained in 2015, enabling the treatment of waste with PCBs concentration below 0.5%. For advantage in heavy metal recovery techniques in MF furnace, the company has become capable of addressing the challenging needs of processing PCBs waste, containing lead. Additionally, in 2021, certification was acquired for waste treatment with PCBs concentration below 10% or less, handling a wider range of PCBs waste. Figure 3 shows the flow sheet of MF process at the point of initiating treatment of PCBs waste. PCBs waste that transferred to a small-sealed plastic container feed from the side of MF furnace.

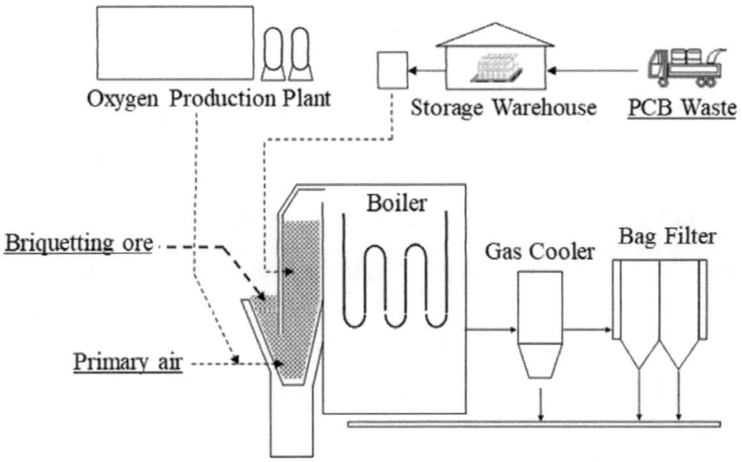

Figure 3. Flowsheet of MF Process at the point of Initiating Treatment of PCB waste.

3. Technological Progress in MF Process

3.1. Enhancement in pretreatment process

For increase of the treatment capacity at the MF furnace, the oxygen production plant (VSA: Vacuum Swing Adsorption) with the capacity of 1,700 Nm^3/h and 90% O_2 was introduced in 2005, which improved the melting speed inside the MF furnace. Furthermore, in 2012 (1,700 Nm^3/h, 90% O_2), additional oxygen enrichment was installed, and two VSA units were operated [4]. Currently, high oxygen concentration air with the preheating temperature of 200°C and O_2 concentration: 27% is blown into the furnace at the rate of 30,000 Nm^3/h. As a result, the reaction speed inside the MF furnace improved, leading to the 30% increase in the treatment capacity.

With the aforementioned improvements, the MF treatment capacity, which was around 100,000 tons in the early 2000s, gradually increased, and reached 140,000 tons after the introduction of VSA in 2012. Figure 4 shows the trend in treatment amount of waste material from 2002 to 2022. However, increasing the treatment capacity, the drying capacity in the pretreatment process became insufficient apparently. Therefore, in 2016, the additional rotary dryer was installed, and the existing line, which was previously operated with one unit, was upgraded to operate with two units in series, thus increasing the drying capacity. Table 3 shows the specifications of the drying equipment.

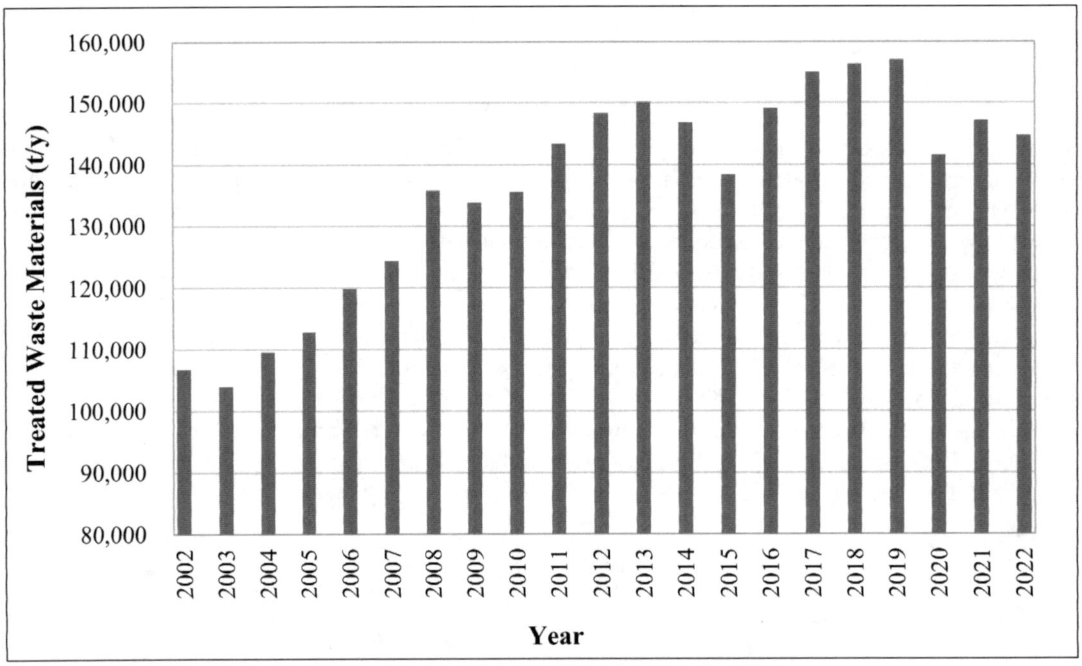

Figure 4. Transition of the Treating Amount in MF process.

Table 3. Changes in Rotary Dryer Equipment specifications.

Period	Units	Number of burners	Flow rate of heavy oil
Before April 2016	1	2 x 1 unit	700 L/h x 1 unit
After April 2016	2 (series)	2 x 2 units	700 L/h x 2 units

Previously, granulated EAF dust accounted for a significant portion of the treatment materials in Miike Smelting Co,. Ltd.. However, since 2002, the proportion of powdered secondary fly ash has gradually increased. When the larger proportion of powdered waste materials is fed into the briquetting machine, the bulk density decreases, resulting in the decrease for briquette strength. To solve this problem, the granulator with water injection was installed before the second rotary dryer to increase the proportion of granulated waste materials fed into the briquetting machine and increase the bulk density of the briquette ore. Figure 5 shows the change in waste material density before and after granulation. The increase of the bulk density of the raw material, making it possible to suppress volume reduction during briquetting compared to only powdered material. Therefore, it is thought that harder briquetting ore can be produced.

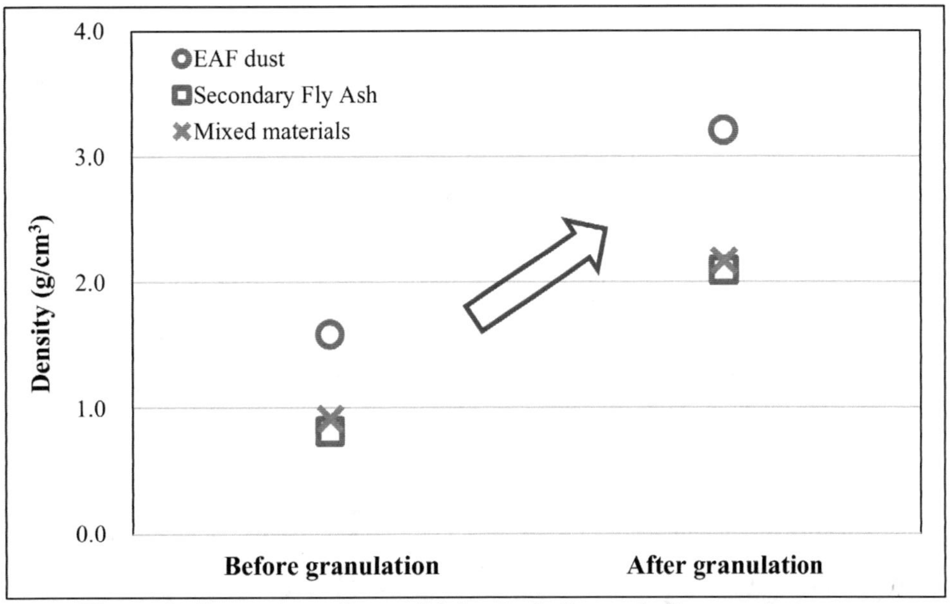

Figure 5. Comparison of material density before and after granulation.

3.2. Flexibilty of MF Treatment

The treatment of secondary fly ash has continued since 2002 at Miike Smelter, as a result, the composition of slag has gradually changed, with increases in calcium content. Table 4 shows the transition of slag composition up to the present. Since SiO_2 content in the slag is relatively decreased and the viscosity of the slag is lowered, it is thought that the fluidity of the slag can be improved. Figure 6 shows the transition of slag composition in the FeO-SiO_2-CaO phase diagram. In this figure, it is confirmed that the melting point of the slag gradually increased, the melting point of the slag exceeded 1100°C in 2018. If the melting point of slag rises, the risk of clogging the slag port increases when cooling around the slag port or insufficient heat in the furnace occurs. However, it can be operated without problems since the temperature inside MF furnace is over 1300°C.

Table 4. Transition of slag composition.

Year	CaO (%)	SiO_2 (%)	Al_2O_3 (%)	Fe (%)
2000	6.7	20.3	5.2	36.2
2012	16.1	17.6	5.8	35.7
2018	18.9	16.1	5.8	32.5

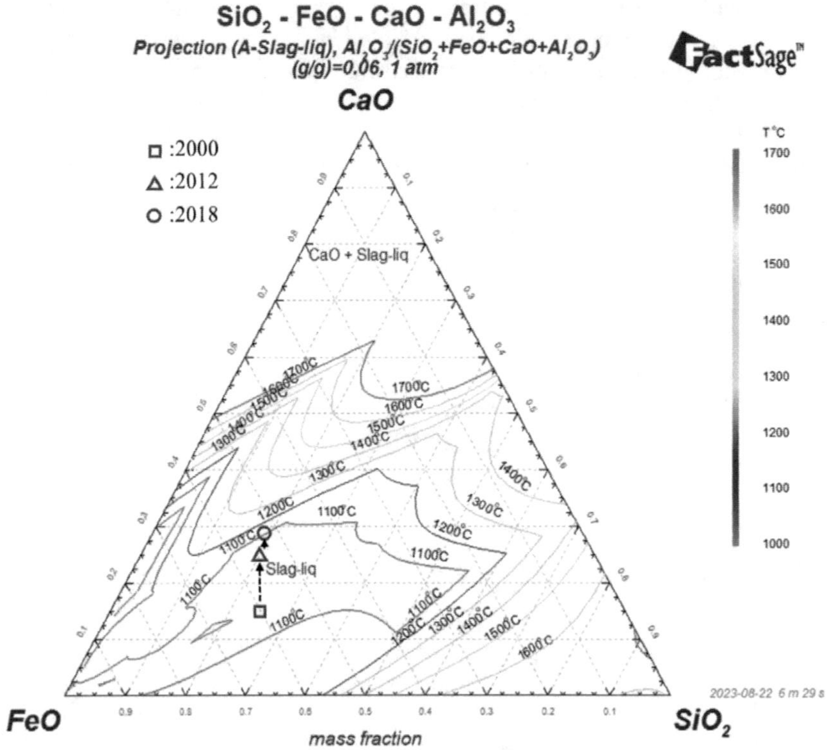

Figure 6. Transition of Slag Composition in the FeO-SiO_2-CaO phase diagram.

In this way, MF furnace is able to perform stable raw material processing even with changes in the slag composition, allowing for flexible adaptation to a wide range of materials. Since 2016, approximately 150,000 tons of waste materials have consistently treated.

3.3. Enhancement in exhaust gas washing facility

With the increase in the treatment amount of waste material in the MF furnace, the absolute amount of sulfur content also increased, prompting the enhancement of the desulfurization plant. Sodium hydroxide is used as the alkaline agent in the desulfurization plant, and the exhaust gas is subjected to countercurrent washing using spray. To reduce the exhaust gas velocity during washing, the tower diameter was expanded from 5.1 m to 10 m, allowing for increased exhaust gas volume. The bottom liquid of the desulfurization plant is mostly discharged, but the portion is circulated within the tower to effectively utilize the chemical liquid.

Furthermore, to comply with the mercury emission regulations implemented in 2018, the demercuration plant was installed after the desulfurization plant. Table 5 shows the details of the washing plant. In the demercuration plant, an oxidizing agent is used to collect mercury from the gas into the liquid, and countercurrent washing has performed in the similar method.

Table 5. Specifications of desulfurization and demercuration plant.

Form	Filling	Diameter (m)	Height (m)	Flow rate (Nm^3/h)
Packed tower	Tellerette	10	17	140,000

These equipment enhancements have ensured stable operation of mercury and SOx concentrations below the regulatory limits, enabling the treatment of raw materials containing mercury even after the implementation of mercury regulations.

4. CONCLUSIONS

MF Process, by its original technology, will contribute greatly to the recycling treatment of waste materials, especially those containing nonferrous metals. Our company earnestly hopes to develop MF Process into a more efficient and more economical process. Also, Mitsui Mining and Smelting Group including Miike Smelting Co., Ltd. is actively promoting initiatives aimed at achieving carbon neutrality by 2050. By 2030, the group aims to reduce CO_2 emission by 38% compared to 2013 fiscal year. As environmental concerns intensify year by year, the group must address to minimize the environmental impact of our operations.

References

[1] P I Grudinskya, D V Zinoveeva, V G Dyubanova and P A Kozlovb 2019 *Inorganic Materials: Applied Research* **10** pp 1220–1226
[2] P D Voogt and U A Th Brinkman 1989 *Halogenated biphenyls, terphenyls, naphthalenes, dibenzodioxins and related products* (Topics in environmental health vol 4) eds R D Kimbrough and A A Jensen (Amsterdam, New York, Oxford: ELSEVIER) pp 3–46
[3] M D Erickson and R G Kaley II 2011 *Environmental Science and Pollution Research* **18** pp 135–151
[4] K Fujioka and S Hatae 2015 *Proc. of Lead-Zinc Conference Pb-Zn 2015 (Pb-Zn 2015* vol 2*)* ed U Waschki (Clausthal-Zellerfeld: GDMB) pp 791–797

Journal of Physics: Conference Series **2738** (2024) 012029 doi:10.1088/1742-6596/2738/1/012029

Evaluation of recycled galvanised steel as a resource for Zn production

N Uys[1], M Becker[1,2] and J Petersen[1,3]

1 Mineral to Metals Initiative, Department of Chemical Engineering, University of Cape Town, Cape Town, South Africa.
2 Centre for Minerals Research (CMR), Department of Chemical Engineering, University of Cape Town, Cape Town, South Africa.
3 HydroMet Research Group, Initiative, Department of Chemical Engineering, University of Cape Town, Cape Town, South Africa.

uysnic002@myuct.ac.za , megan.becker@uct.ac.za , jochen.petersen@uct.ac.za

Abstract. The high reduction potential of zinc has enabled its extensive use in the steel industry as a protective galvanising coating. Galvanising, which consumes approximately 50-60% of the global production of zinc, significantly increases the life span of steel products and contributes towards addressing the costly global problem of corrosion. While there does not appear to be a readily available and agreed-upon annual quantity of galvanised steel produced, it has been estimated by the authors to be 320-390 Mt/a (equating to 15-20% of the annual steel production). Due to the various methods employed to galvanise steel and different steel shapes which are coated, there is no clarity on the average percentage of zinc which galvanised steel is comprised of. However, a zinc content of between 1.5-2.5% appears to correlate with the estimated annual galvanised steel production.

The reprocessing of secondary zinc streams is becoming increasingly important due to legislation changes and increased consumer demand for recycled and low carbon metals. Within the steel industry this will include the recycling of galvanised scrap steel, using primary process routes, as well as the re-processing of waste streams, such as steel dusts, using additional technologies such as the Waelz kiln. Zinc associated with galvanised scrap steel would exclusively report to these dusts, and thus represents a key resource for recycled zinc.

Although there are two main steel process routes preceding the galvanising process, the focus here is on the Electric Arc Furnace (EAF) process route due to its ability to incorporate high levels of secondary (zinc containing) steel feeds, and the dust produced typically contains economically recoverable levels of zinc. EAF dust production is estimated between 5-10 Mt/a at a zinc content of 10-36%. Failure to recover this zinc would represent a loss of 0.5-3.6 Mt Zn/a to waste. The Waelz kiln is the most prominent technology utilised to recover Zn from EAF dust. However, this process requires additional energy, and thus the associated carbon footprint needs to be included in the evaluation of the potential of Waelz oxide as a secondary Zn resource. To this end, model analysis of a simplified Waelz kiln using the HSC simulation software shows that the carbon footprint associated with the processing of EAF dust ranges from 4-12 kg CO_2/kg Zn dust at 10 wt.% Zn in the dust to 1-4 kg CO_2/kg Zn at 35 wt.% Zn in the dust.

Keywords: Secondary Zinc, Galvanised Steel, Electric Arc Furnace Dust, Waelz Kiln

1. Introduction

Zinc's most significant property is its high reduction potential. This property has enabled its extensive use in the steel industry as a protective galvanised coating. While the growth in demand for steel as a ubiquitous engineering and construction material is due to increased urbanisation and population growth, an additional demand relates to the costly global problem of material loss due to corrosion. The addition of zinc in the form of galvanising contributes towards addressing corrosion and thus significantly increases the life span of steel products. As a result, approximately 50-60% of the global production of zinc (13 Mt/a) is used to galvanise steel [1] which equates to 6.5-7.8 Mt/a.

The reprocessing of secondary zinc streams is becoming increasingly important, due to legislation changes and increased consumer demand for recycled and low carbon metals [2–5]. As zinc's primary usage is in galvanised steel, the recycling of this product and the associated by-product and waste streams represents the biggest opportunity for zinc recovery. The recycling of galvanised scrap steel would occur using primary production process routes while the re-processing of waste streams, such as steel dusts, would occur using additional technologies (e.g. Waelz kiln). During reprocessing, zinc associated with galvanised scrap steel almost exclusively reports to these dusts, and thus they represent a key resource for recycled zinc. Although there are other secondary zinc streams, such as the neutral leach residue from the dominant Roast-Leach-Electrowinning (RLE) primary process route, these are not considered in the present study which is focussed on recycling of Zn used in galvanised steel.

The current global recycled content of zinc is approximately 25% [6]. While this may appear low, it is due to zinc's primary function in extending the lifespan of products. Therefore, a large portion of Zn historically used in galvanising steel remains in service and is thus not yet available for reprocessing and recycling.

Recovering the zinc from secondary resources requires additional energy, typically supplied to a pyrometallurgical process through the addition of coke (carbon) as a reductant or through the addition of electrical energy. Given the relatively large recycling rate of steel (± 85%) [7], the carbon footprint of recycling technologies warrants investigation. This additional carbon footprint needs to be determined to more comprehensively assess the metrics with which the "greenness", "sustainability" and "circularity" of metals are evaluated.

The objective of this desktop study was therefore to address gaps in available literature with respect to quantifying potential zinc secondary resources and the additional carbon (CO_2) footprint associated with returning this secondary zinc back into primary production. To this end, a preliminary model analysis of a simplified Waelz kiln using simulation software is presented to provide this data.

2. Methodology

The methodology for this desktop study involved a literature review of the technologies associated with galvanised steel production using a combination of journal papers and open access reports from the steel industry. As galvanised steel is seen as the major source for recycled Zn, it will be the steel type primarily considered in terms of demand, supply, and associated technologies and processing routes (EAF and Waelz kiln). Based on the collected information, an estimate of the quantity (Mt/a) of galvanised steel produced was made. Similarly, the quantity (Mt/a) and zinc content (%) associated with the production of galvanised steel was also estimated using extrapolated data from reports. It should be noted that many of the values presented have a wide range due to conflicting pieces of readily available information, resulting in uncertainty. Finally, the Waelz kiln was reviewed as a technology for the recycling of electric arc furnace (EAF) dust, and the CO_2 footprint (kg CO_2/kg Zn) was estimated using a high-level Waelz kiln model with Metso's HSC simulation software [8] in combination with carbon footprint associated with different types of energy sources available from literature.

3. Results

3.1. Global Steel and Galvanised Steel Production and Key Technologies

3.1.1. Overview: Steel and galvanised steel. Although there are over 3 500 different grades of steel, each with their own physical and chemical properties, steel is principally an iron alloy which is commonly combined with other elements to improve the properties of steel, such as carbon (<2%), manganese (<1%), silicon, phosphorus sulphur and oxygen [9]. There are many different standards which are used to classify the grades and types of steel. Common classifications include carbon steel, alloy steel (low and high alloy steel), stainless steel (>10% chromium) and tool steel, each of which can be further divided into grades.

The embodied energy and the carbon footprint per tonne of steel produced is reported to be low compared to other metals [10]; however, the sheer quantity of crude steel produced globally (> 1950 Mt in 2021) results in high annual energy consumption and carbon emissions [11]. Although, the steel industry has reduced the energy intensity of its processes, with the average energy needed to produce one tonne of steel decreasing by 60% since the 1960s and the carbon footprint (per tonne of steel) decreasing by 16% in the last 10 years, the industry still accounts for approximately 7-8% of total global carbon emissions [11,12].

Corrosion is an underestimated and costly global problem. It is estimated that it costs the global economy US $2.5-3 trillion annually (in the U.S.A. it accounts for ± 3-3.5% of GDP) as a result of corroded steel infrastructure and buildings [13,14]. This is not only an economic cost, but also adds to the demand for primary resources and therefore has environmental implications [15,16]. Corrosion prevention and management strategies are therefore of vital importance to maximise resource efficiency.

One method to protect steel from corrosion is the addition of a thin zinc coating on top of the steel, through galvanisation. The most common method is hot-dip galvanising (HDG), although other processes such as electroplating can also be applied [17,18]. The versatility of galvanised steel has allowed its application in construction, telecommunications, automotive, agriculture and energy (solar and wind) projects [18].

Stainless steel is also often used due to its corrosion resistance properties. However, stainless steel usually costs 4-5 times more than galvanised steel. Consequently, the demand for galvanised steel is highest in certain industries such as construction, automotive, and home-appliances [19].

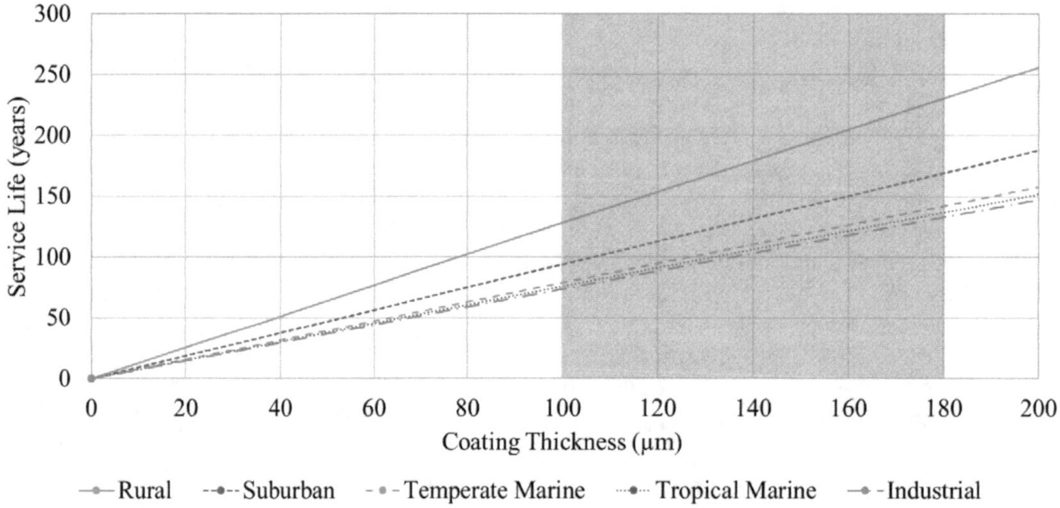

Figure 1. Life expectancy of galvanised steel based on thickness of zinc coating (µm) [20]. Service life = time to 5% rusting of steel surface. Grey area = typical range of coating thickness.

As the lifespan of steel products is measured in decades, longer if galvanising finishing treatments are applied (Figure 1), there is a significant time delay between production and end of service life at which recycling would occur. Although this optimises the in-service value of the material, it does result in the sustained primary production of steel. Steel currently has an average of 37% recycled content [11]. Therefore, both primary and secondary production methods of steel production will continue to be prevalent in the industry.

3.1.2. Steel production processing routes and technologies. There are two main process routes for steel, namely the blast furnace-basic oxygen furnace (BF-BOF) and the electric arc furnace (EAF), which account for 70-72% and 27-29% of global steel production, respectively (Figure 2), with variations and combinations of technologies in existence [11]. The key difference between the two routes is the incorporation of coke as a raw material within the BF-BOF process (Figure 3). While BF-BOF primarily uses iron ore and coke as a feed and has more limited option of incorporating recycled steel, the EAF process typically uses recycled steel as its primary feed and electricity as its energy source [11]. The large carbon footprint associated with steel production therefore primarily originates from the BF-BOF's use of coke as a raw material, while for EAF the carbon footprint would be dependent on the local source of electricity.

In the BF-BOF process, the iron ores need to be first reduced to iron (pig iron / hot metal) in the BF, and this is followed by its conversion to steel within the BOF. The steel is then cast and rolled into various shapes (tubes, sheets, rods etc.) [11]. Within the EAF process, electricity is used to melt recycled steel, while additives are used to adjust the steel to the required chemical composition. Oxygen injection can also be used to supplement the electrical energy supply to the EAF. The downstream process steps are similar to those of the BF-BOF process [11].

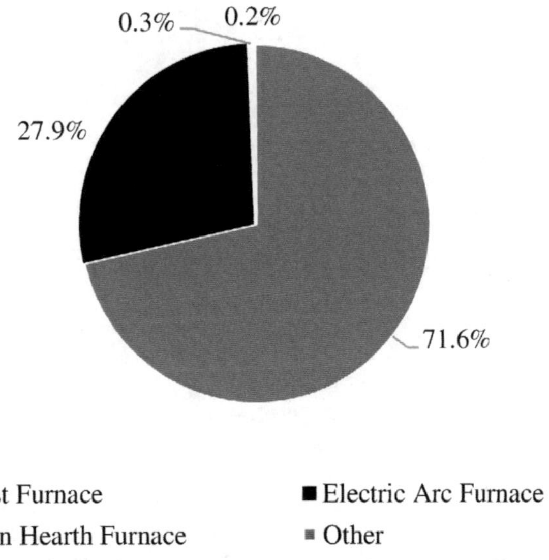

Figure 2. Technology options associated with global steel production in 2019 [21].

While the BF-BOF process can use up to 30% recycled steel as a feed, 100% of the EAF processes feed can use recycled steel. Although steel can be infinitely recycled as its properties are preserved during reprocessing, the increasing demand will require the continued usage of both BF-BOF and EAF routes due to need to process fresh iron ore for the foreseeable future [11].

With reference to Zn recycling, the EAF process route can incorporate high levels of secondary, zinc containing steel feeds, and the dust produced typically contains economically recoverable levels of zinc.

The Waelz kiln is the most prominent technology utilised to subsequently treat EAF dust. EAF dust production is estimated between 5-10 Mt/a, at a zinc content of 10-36%. The recycling of this dust is and should continue to be a priority as failure to recover this zinc would represent a loss of 0.5-3.6 Mt Zn/a, or up to 60% of the Zn consumed in galvanising.

Other technologies include the open-hearth furnace (OHF), otherwise known as reverberatory furnace, which accounts for approximately 0.4% of crude steel production, and the usage of which is in decline. As of 2012, there were only 7 plants that use reverberatory furnaces (of which OHF is the main type) left globally [22]. Since then, 6 OHFs in Mariupol, Ukraine, closed in 2015 [23] and 2 further OHFs in Vyksa, Russia, were closed in 2018 [24], with any remaining furnaces likely to be to be located in Ukraine, India, or Uzbekistan [24–26]. The reason for this decline is that the OHF is uneconomical due to its excessively energy-intensive and long process times, in addition to other environmental concerns [11]. Another known technology is the induction furnace (IF); however, this technology was banned in China in 2017 for producing environmentally unfriendly and sub-standard quality steel products [27,28], and it is unclear whether any IF facility remains in operation.

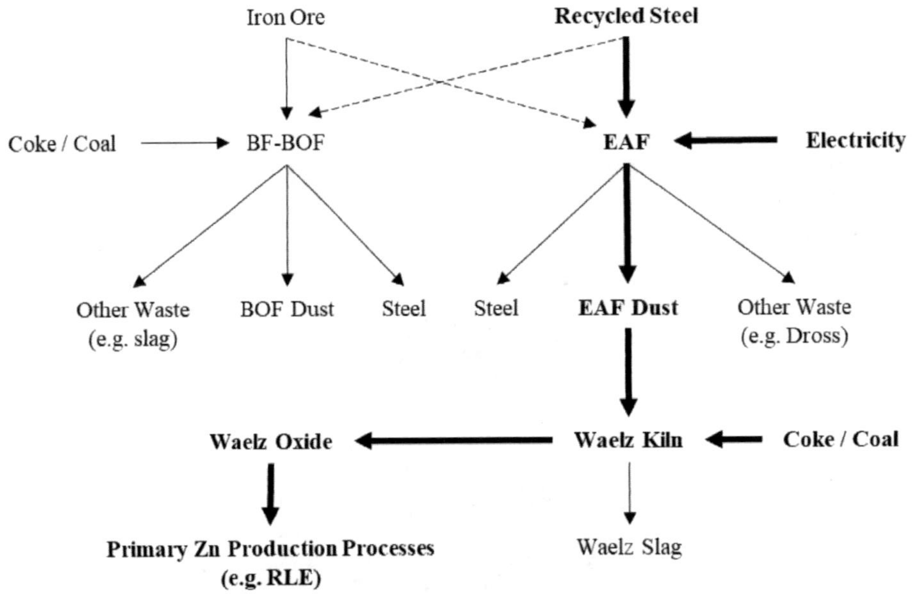

Figure 3. Process flow diagram for the incorporation of secondary zinc streams (EAF dust) back into primary zinc production. Dotted lines represent feasible, but less commonly incorporated feeds.

3.1.3. Regional distribution of steel technologies. With regards to the regional contribution that each of these technologies make to global crude steel production, Asia accounts for the combined majority (72%), and China represents 74% of Asia's production (i.e. China represents 53% of global production of crude steel). With respect to global BF-BOF and EAF steel production, Asia represents 81% and 48% respectively, with China representing 82% (China produces 67% of world BF-BOF steel) and 41% (China produces 20% of world EAF steel) of Asia's production [21].

Figure 4 summarises the global contribution that the BF-BOF and EAF technologies make to global crude steel production and is divided by region. Asia represents 48% of global EAF steel production with China representing 41% (i.e. China produces 20% of world EAF steel) of Asia's production. North America and Europe produce a combined total of 28% of world EAF production (compared with 12% of BF-BOF production), which highlights the focus of the Western world on secondary (recycling) rather than primary production.

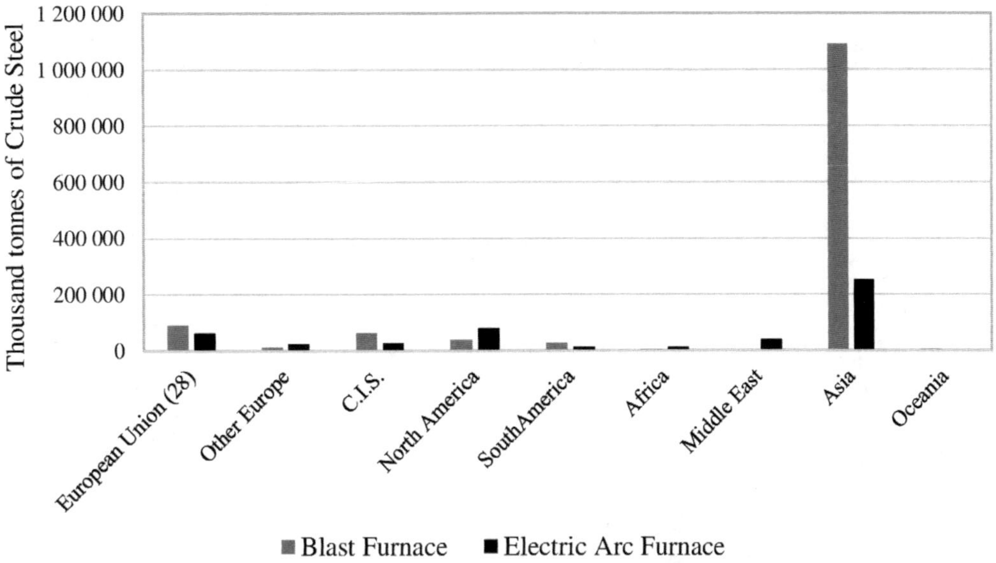

Figure 4. Global distribution of crude steel using the BF-BOF, EAF, OHF and other process routes in 2019 [21]. C.I.S = Commonwealth of Independent States.

3.2. Quantifying annual galvanised steel production

3.2.1. Galvanised Steel. There does not appear to be a readily available and agreed-upon annual production quantity of galvanised steel. Multiple methods were therefore used to arrive at a consistent value. Firstly, a recent market forecast report placed the global production as 337.3 Mt/a [29] in 2017. Secondly, another source reported that China consistently produced ± 65 Mt/a of HDG sheets during the 2017-2020 period [30]. Assuming that sheeting represents ± 40-41% [31] of the global galvanised steel finished product market, this places China's total HDG steel production at ± 160 Mt/a. It should be noted that this would be an under-estimate as other galvanising techniques are not included in this value. Given China's dominance within the zinc industry (± 50% market share) as well as within the steel industry (53%), a similar dominance (± 50%) is expected within the galvanising steel market, thereby conservatively placing the global annual production close to 320 Mt/a. A third attempt to corroborate these values and to establish a more recent (2022) value was made using values found within one of the market forecast reports [29]. The Compound Annual Growth Rate (CAGR) (Formula 1 in Table 1) was restructured (Formula 2 in Table 1) using $V_i=V_f - D$, and applied. The annual production in 2022 was then back-estimated as 388.2 Mt/a respectively. The values used within the calculations can be found in the Table 2.

Table 1. Compound Annual Growth Rate (CAGR) formula.

$CARG = (\frac{V_f}{V_i})^{1/t} - 1$	(1)
$V_f = \frac{D\ (CARG+1)^t}{((CARG+1)^t-1)}$	(2)

where CAGR=compound annual growth rate (%); t = time (yrs); V_f = final value; V_i = initial value; D= difference ($V_f - V_i$).

Thus, the three methods appear to produce relatively consistent values ranging from 320-390 Mt/a between 2017 and 2022 (equating to 15-20% of the annual steel production). As the zinc used in galvanisation only represents approximately 6.5-7.8 Mt/a, the values remain the same irrespective of whether it is assumed that the above values are inclusive or exclusive of the zinc coating.

These same reports, forecasting the global galvanised steel market, project the galvanised steel market to grow by over 80 Mt between 2022 and 2027, primarily due to demand from China and India in the pre-engineered buildings sectors [29,31,32]. The United States of America and Japan also represent large consumers of galvanised steel [19].

Table 2. Galvanised steel market forecast [29].

		2022 Report	Units
Reported	**CAGR**	4.11	%
	Time (t)	5 [1]	yrs
	Difference (D)	86.6	Mt
Calculated	**$(CAGR + 1)^t$**	1.2	
	Final Value (V_f)	474.8	Mt/a in 2027
	Initial Value (V_i)	388.2	Mt/a in 2022
	% of galvanised steel of total crude steel	19.9	%

Underlined values provided in reports; CAGR = compound annual growth rate; t = time in years; D = Difference ($V_f - V_i$).

In addition to the uncertainty surrounding the size of the galvanised steel market, there is no average percentage of zinc which galvanised steel is comprised of. This is due to the various methods used to coat steel with zinc, the different shapes of steel which are coated (sheets, tubes etc.) as well as the numerous standard categories which each specify minimum thicknesses of zinc coating. For instance HDG sheets produced from cold rolled steel have an average thickness of 0.25-0.2 mm which sheets made form hot rolled steel are typically in the 2-3 mm thickness range [33]. It has been claimed that zinc usage is up to 5-7% [2] of the mass of coated steel [34] (assuming no zinc losses nor wastage this equates to only 80-130 Mt/a galvanised steel production), however in contrast, zinc content of between 1.5-2.5% [3] (realistically averaging between 1.6-2% if taking into account up to 10% Zn loss) would correlate with the reported galvanised steel production quantities of 320-390 Mt /a.

Table 3. Estimated Zinc content of Galvanised Steel at different coating thicknesses (μm) and different steel thicknesses (assuming Zn coating included in steel thickness and steel density of 7.75 g/cm^3) [33].

	Galvanised steel thickness (mm)				
Zn coating/side (μm)	**0.25**	**2**	**3**	**6**	**20**
7	5.2%	0.6%	0.4%	0.2%	0.1%
10	7.4%	0.9%	0.6%	0.3%	0.1%
14	10.4%	1.3%	0.9%	0.4%	0.1%
19	14.2%	1.8%	1.2%	0.6%	0.2%
32	24.0%	3.0%	2.0%	1.0%	0.3%
43	32.5%	4.0%	2.6%	1.3%	0.4%

[1] 2027-2022 = 5 yrs

[2] This is the total zinc usage meaning zinc coating and zinc wastage (dross etc). However, the quantity / percentage of zinc wastage during galvanising is unknown.

[3] This is assumed to represent average zinc content of galvanised steel.

The exact density of the ungalvanised steel (7.75-8.05 g/cm^3) and whether the zinc coating is included or excluded, appears to make very little difference to the overall Zn content (%). The most significant factor is the thickness of the steel to be galvanised and secondly is the thickness of the Zn coating (Table 3). Based on these numbers, assuming a zinc content of between 1.5-2.5% in galvanised steel, for the modelling of zinc content within recycled steel processed within EAF furnaces, does not seem unreasonable.

3.3. Quantifying production of waste from galvanised steel production

Both the BF-BOF and EAF processes produce wastes (Figure 3). These wastes are primarily in the form of slag (90% w/w), dust and sludge and cumulatively represent 400 kg (BF-BOF) and 200 kg (EAF) worth of co-products per tonne of steel produced [4] [35].

With over 400 Mt/a of iron (BF) and steel (BOF) slag (calculated as >500 Mt in 2019), it is reported that the slag recovery rate / usage is >80% for steel and ± 100% for iron which would leave 25-34 Mt of BOF slag remaining [35]. The BF-BOF dust generated appears to vary from 7-30 kg/t steel produced (1-3%) which translates into between 9-40 Mt of BF-BOF dust generated annually worldwide [36–38]. EAF dust production is between 1-2% of EAF steel production (5-10 Mt) as summarised by region in Figure 5 [21,39,40].

As can be seen from the above numbers, BF-BOF dust represents a significantly higher contribution (2-4 times) to the quantity of steel dust compared to EAF dust. Due to the chemical nature of these dusts, they are regarded as hazardous and must be chemically / thermally treated prior to disposing in landfills, which is expensive [41]. Alternative methods for (economically) processing these BOF dusts are therefore important areas of research. In addition, there needs to be increased focus on producing "quality" waste which would be better suited for further processing [42].

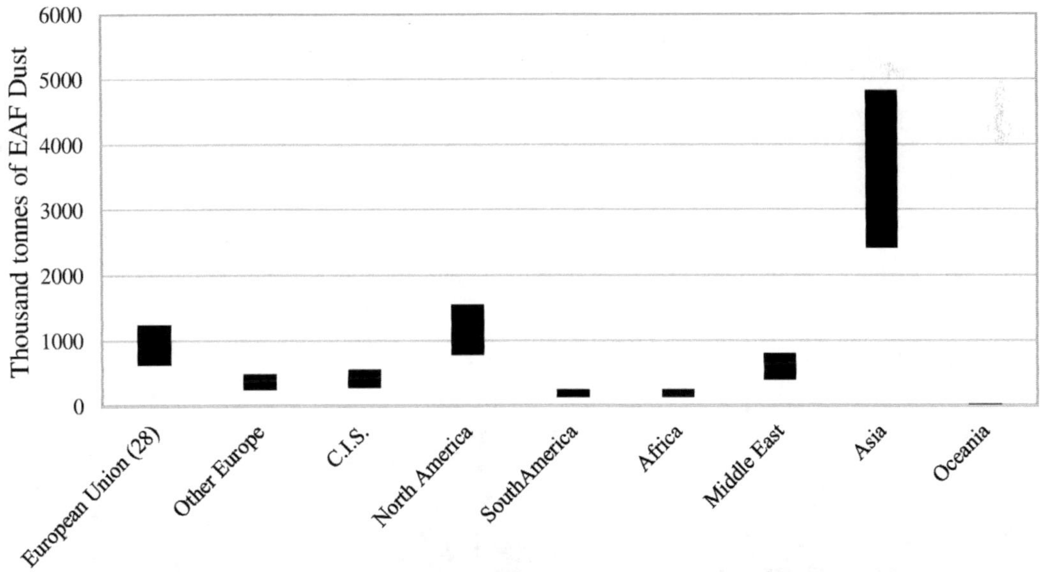

Figure 5. Global estimated production range of EAF dust (EAFD) [21].

Currently the economic threshold for the recycling of EAF dust is determined by the zinc content being higher than 15% [41,42]. EAF dusts containing less than this threshold of zinc are currently sent to landfill, which in the EU reportedly makes up two thirds of their EAF dust production [41]. The reason for the low Zn content in the EU in unknown and may simply be as a result of the types of scrap

[4] Per tonne of crude steel

steel available for recycling (i.e., non-galvanised steel). However, the costs of disposal are no longer negligible and continue to rise with increasing legal responsibilities being placed on producers of hazardous material [42,43].

Given that 30-40% of the steel produced by the steel industry is from recycled scrap steel, which may not necessarily be galvanised steel scrap, the composition of EAF dust depends strongly on which types of steel scraps which are processed in the EAF [44]. Only when galvanised steel scraps are included in the feed material to EAF will it result in the presence of zinc in the EAF dust. For this study only EAF dust which results from the processing of galvanised steel is considered.

Iron and zinc oxides represent the major constituents in EAF dust. Depending on the source of galvanised scrap, which is increasingly introduced into EAFs, the content of zinc in EAF dust can vary from 10-36% (Table 4) [42,44,45]. The most common range, however, is between 20-30% [5], while the Fe content can range from 20-30% [40]. In comparison, the zinc content of BOF dust can vary between 0.1-3.2% while Fe ranges from 30-85% [36]. The low Zn content in BOF dusts have made it difficult to economically recover the zinc, however, the presence of Zn and other deleterious metals still presents a hazard which needs to be treated prior to the disposal of the BOF dust.

Table 4. Summary of Chemical compositions of EAF dust [40,42]

Element Wt.%	Fe	Zn	Pb	Mn	Si	Cu	Cr	Ca	Cl	Al
Min	20	10	1.24	2.74	1.50	0.18	0.10	2.2	0.87	0.00
Max	30	36	2.2	4.44	1.74	0.4	0.47	4.35	3	0.30

The exact percentages of EAF dust and BOF dust which are recycled is unknown. While some authors have reported that 75% of the EAF dust produced is treated using Waelz kilns [40], others claim that over 50% of EAF dust is landfilled [46]. In the absence of other information this range (50-75% recycling of EAF dust) will therefore be assumed (see Figure 6).

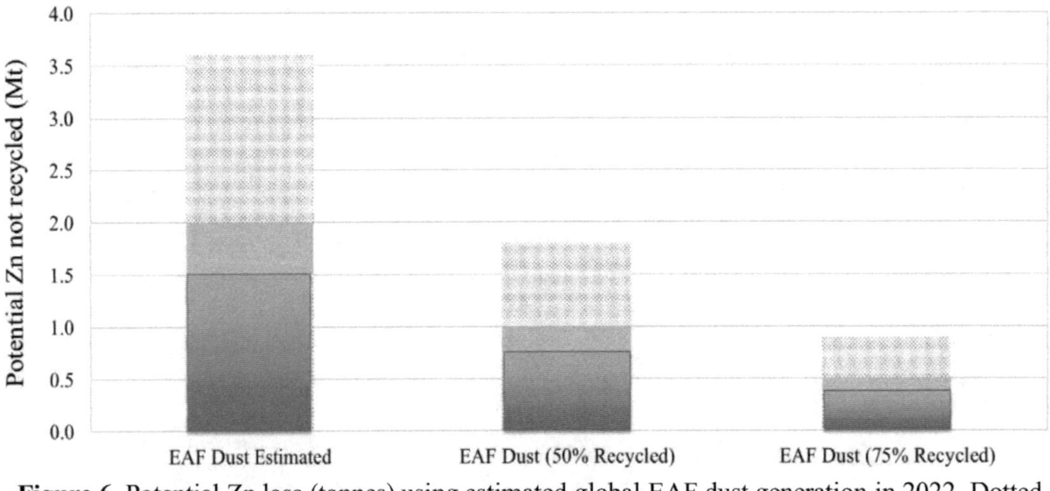

Figure 6. Potential Zn loss (tonnes) using estimated global EAF dust generation in 2022. Dotted area represents the maximum range of Zn loss, yellow colour represents the realistic range with blue representing the minimum expected loss. Left bar assumes no recycling of EAF dust; middle bar assumes 50% recycling rate; right bar assumes 75% recycling rate [40,46].

[5] It has been found to be economical to recover zinc from EAF dust where zinc content is >15% [42,67].

If there is no recycling of EAF dust this equates to a maximum loss of Zn of 3.5 Mt/a, which is the maximum height of the dotted area on the far-left bar of Figure 6. However, realistically the maximum zinc loss would be closer to 1.5-2 Mt/a (11-15% of annual Zn production), which is represented in yellow in Figure 6. If 50% was recycled it would represent a maximum loss of about 1.75 Mt/a, while realistically being closer to 0.75-1 Mt/a (5.5-7.5% of annual Zn production). This corresponds to another author's estimation of annual Zn loss to landfilled EAF dust [46]. Lastly at 75% recycling there would be a maximum loss of ±1 Mt/a, and realistically 0.4-0.5 Mt/a (3-4% of annual Zn production). In the absence of additional information, BOF dust is assumed to have limited or no recycling. Therefore, it is assumed that the zinc in BOF dust is lost to landfill (±0.5 Mt Zn/a lost via BOFD).

3.4. Review of Waelz kiln for EAF dust recycling and Carbon Footprint

Although some hydrometallurgical processing routes have been proposed and demonstrated, only High Temperature Metals Recovery (HTMR) processes, which are pyrometallurgical, have been commercially successful [42]. The following pyrometallurgical processes are or have been in operation to treat EAF dust: Rotary kiln (Waelz kiln), Rotary Hearth Furnace (RHF), Multi-stage furnace (PRIMUS), and the Electric Smelting Reduction Furnace (ESRF) [47]. 80-90% of the EAF dust which is recycled (approximately 2-2.1 Mt) is treated using the Waelz kiln, or a derivate of this technology, as it is currently deemed the most efficient processing route [42,44,48].

3.4.1. Waelz kiln. In principle, the Waelz kiln process involves the carbothermic reduction of zinc compounds (typically ZnO), followed by the volatilisation of metallic zinc and the subsequent re-oxidation of metallic zinc to ZnO. The selective volatilization of certain metals in the zinc waste (EAF dust or leaching residue) results in a more concentrated final zinc oxide product which can then be returned to traditional zinc processing routes such as the RLE for further zinc recovery.

The following reactions (3(8) represent the key reactions occurring in the Waelz kiln, assuming that the Zn in the feed (EAF dust) is primarily present in the form of ZnO. Further details regarding the processing of sulphide feeds and additional operations of the Waelz kiln can be found through the following references: [44,49–52]. The first stage is the evaporation of moisture at temperatures of approximately 600ºC.

$$H_2O_{(l)} \rightarrow H_2O_{(g)} \tag{3}$$

This is followed by the reduction and immediate volatilization of Zn after which the zinc vapour is re-oxidised and removed as fume.

$$ZnO + C = CO_{(g)} + Zn_{(g)} \tag{4}$$

$$Zn_{(g)} + CO_{(g)} + O_{2(g)} = ZnO + CO_{2(g)} \tag{5}$$

$$C + O_{2(g)} = CO_{2(g)} \qquad (4) + (5) = (6)$$

Zn is also often present in ferrite ($ZnFe_2O_4$) form and is reduced via the following two routes:

$$3ZnFe_2O_4 + CO_{(g)} = 3ZnO + 2Fe_3O_4 + CO_{2(g)} \tag{7}$$

$$ZnFe_2O_4 + CO_{(g)} = ZnO + 2FeO + CO_{2(g)} \tag{8}$$

Most metals and compounds do not achieve a perfect split between the Waelz oxide fume or to the slag. The volatility, amongst other aspects, of each metal and/ compound at the operating temperature, plays an important role in where they report. For instance, the majority (>90%) of the Zn, Pb, Cd and Cl report to the fume, while the majority (>90%) of Fe, SiO_2, Cu, Al, Ag, and Au report to the slag.

The Waelz kiln is a type of rotary kiln / furnace, that typically consists of the kiln and auxiliary equipment such as a mixer, a fan, and a fume-collecting system (settlers). The Waelz kilns, for zinc operations, can reach a 1250ºC, however the temperature is difficult to measure and varies over the length of the kiln. The largest kilns require 18-22 kW of power (motors etc.). The fume-collecting equipment consists of different types of settlers, namely cyclones, baghouses, and Electrostatic Precipitators (ESP). As the dust particles are heavier than the zinc fume, a simple settler is often sufficient [50]. The zinc enriched product is referred to as Waelz oxide with the by-product being the

Waelz slag. Waste gas, dust and Waelz oxide (fume) leave at the upper (feed) end of the kiln, while the slag leaves from the lower end of the kiln.

Like many pyrometallurgical processes the Waelz kiln allows for greater operational flexibility in terms of the feed streams. The Waelz kiln's ability to treat ZnO (zinc oxide) and $ZnFe_2O_4$ (ferrites) has added to its greater success over hydrometallurgical processes [42]. It is also unnecessary to control the roasting conditions as at the operating conditions of the Waelz kiln ferrites, ferrates, sweet roast and sulphate roast are all easily reduced [50].

The Waelz process can process material with low zinc concentrations, however, the higher amounts of inert material result in significant increase in energy consumption. As a result, recycling companies will charge producers a higher price for the processing of low zinc dusts. A minimum threshold of 15% Zn is typically required to achieve financial feasibility [42].

Unfortunately, like many pyrometallurgical processes, the Waelz kiln requires a high Capital (CAPEX) investment and operating (OPEX) expenditure. It benefits from scaling, which means that larger operations are more economically viable than small scale. This poses a challenge for processing operations which would normally incur high transportation costs due to their remote locations [42].

Another disadvantage of the Waelz process is that it leaves Waelz slag as a residue which contains oxides of iron and calcium [42]. This Waelz slag is not typically of a quality that makes it saleable or useful [47] and is often landfilled. To address this shortcoming, significant research has been put into other technologies such as MINTEK's Enviroplas process [53].

The most commonly used options for reduction of the Waelz kiln feed streams are metallurgical coke and coke breeze [50]. Coke usage can range from as low as 10-12% [50] to treat calamine waste, to 25-40% [44] of the feed (assumed on a dry basis), with the average being 30% [50]. The moisture content of the coke can range from 15-20% [50]. The incoming air is assumed to be at STP and consist of 78-79% nitrogen and 20-21% oxygen. The flowrate of air into the Waelz kiln is reported to be ±3000-3500 m^3/h (35-50 cu. ft/s) [50]. The main heat losses are in the waste gas stream. The exit gas compositions have been reported as 18-28% CO_2, 0-3% O_2 and 0-2% CO [50].

3.4.2. Carbon Footprint and reduction options. The zinc industry's average carbon footprint, across all processing routes, appears to range from 2.5-4 kg CO_2 eq/ kg Zn, however, this can vary widely reaching as much as 6 kg CO_2 eq/ kg Zn in extreme cases [54–57]. Although the embodied energy and the carbon footprint per tonne of steel produced is considered low compared to other metals, the enormous quantity of steel produced results in significant energy usage and CO_2 footprint considerations.

While the average energy consumption of steel is 21.31 GJ/t crude steel, the technology choice plays a significant factor. While the BF-BOF route requires 24.43 GJ/t, the EAF only requires 10.04 GJ/t, which is less than half of that associated with BF-BOF [58–60]. As coke is a significant feed source for BF-BOF route, the associated CO_2 footprint is also significantly higher than that of EAF. EAF has an average global carbon footprint of 0.67 kg CO_2/kg steel, as opposed to 2.32 kg CO_2/kg BF-BOF steel. It is unclear, which exact energy mix was used for this calculation, however; 10.04 GJ/t equates to 2789 kWh/t which in turn results in a carbon footprint of ±0.24 kg CO_2/kWh, which is significantly lower than coal while being higher than any of the usual low carbon green sources of energy (wind, solar, hydro). It shows, however, that the CO_2 footprint for EAF steel can be further reduced if other low carbon green sources of energy such as those listed are used for the electricity generation [59,61,62].

With respect to the EAF, the reported energy (electricity) requirement is only 440 kWh/t EAF steel produced (1.61 GJ/t EAF steel) [66], although this directly equates to approximately 0.36 kg CO_2/kg EAF steel (assuming 820 g CO_2/kWh using coal as an electricity source). This is significantly less than the above reported values and it is assumed that there is additional auxiliary equipment which has been taken into account to form the above significantly larger energy requirement and CO_2 footprint. As EAF processes are reported to have a total CO_2 footprint of 0.67 kg CO_2/kg EAF steel, the values in Table 5 were adjusted to account for this discrepancy by normalising them with reference to coal.

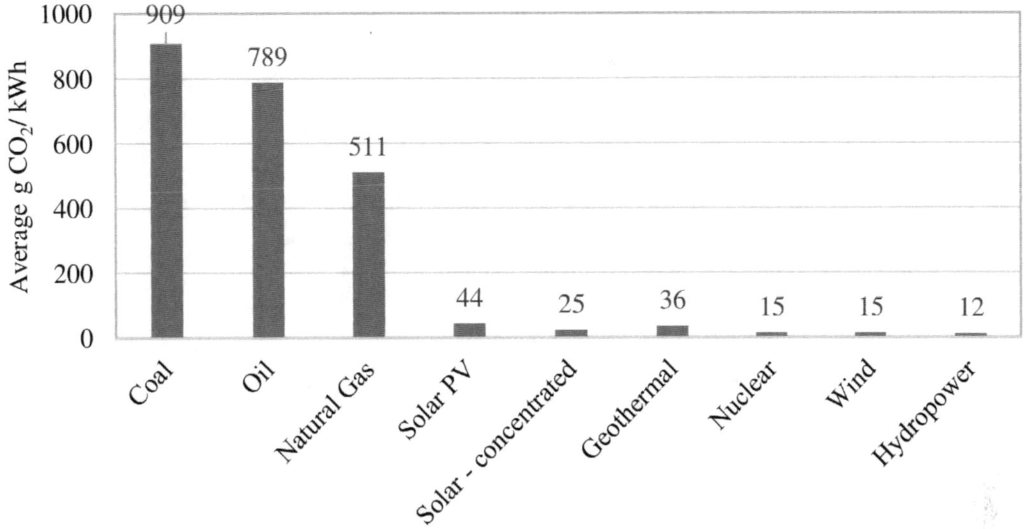

Figure 7. Carbon footprint (g CO_2/kWh) associated with different energy sources for electricity generation [64,65].

As can be seen in Table 5, green energy options (solar, wind, hydropower) may result in the footprint decreasing by up to 95%. However, it should be noted that these values do not take into account the footprint of any auxiliary equipment (energy storage) which may also need to be installed when changing energy sources.

Table 5. Estimated CO_2 footprint of EAF steel production based on different energy sources.

	CO_2 footprint as % of coal	**CO_2 Footprint (kg CO_2/kg EAF steel - adjusted)** [a]
Coal	100%	0.67
Oil	81%	0.54
Natural Gas	55%	0.37
Solar PV	5.0%	0.03
Solar - concentrated	2.7%	0.02
Geothermal	4.0%	0.03
Nuclear	1.6%	0.01
Wind	1.7%	0.01
Hydropower	1.3%	0.01

[a] The CO_2 footprint associated with other energy sources for EAF steel have been calculated by ratios using the reported CO_2 footprint of EAF steel (using coal-based electricity) as a basis.

3.1.3. HSC Modelling. A simplified Waelz kiln model was created using Metso's HSC simulation software package. Three different Zn wt.% (10%, 26%, 35%) were used to create simplified EAF dust compositions. Each of these Zn wt.% were used to investigate the carbon footprint associated with the Waelz kiln (see Table 6). For each of the 3 chosen wt.% which span the expected Zn content within EAF dusts (Table 4), the mineralogy was varied to determine whether the presence of ferrites significantly impacted the carbon footprint.

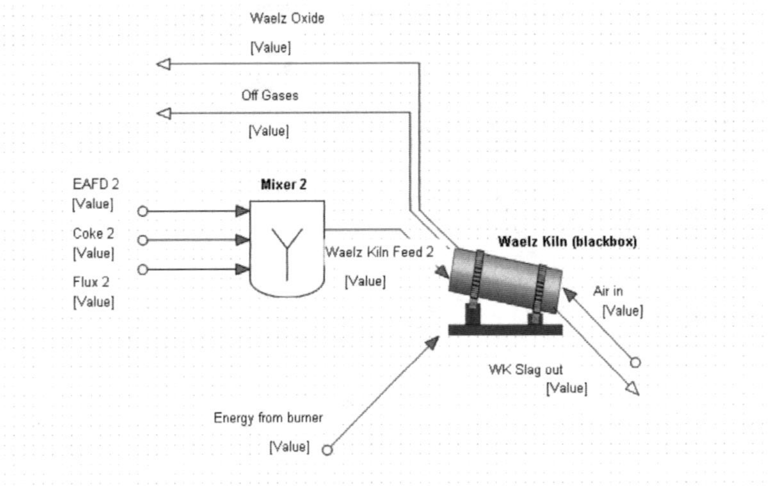

Figure 8. Waelz kiln model using HSC.

The following assumptions and ranges were used. Inlet temperatures of EAF dust, coke, flux and air were at 25ºC. Exit temperatures of the off gas and Waelz oxide was 600ºC while the Waelz slag was 1300ºC. Flux (SiO_2) was added as 20% of the EAF dust feed (wt.%) and the coke (assumed pure carbon) was varied between 10 and 40% of the EAF dust feed (wt.%). The moisture content of the EAF dust was 15%. The exit gas composition was maintained at approximately 15-25% CO_2, 0.5-1% O_2 and <2% CO which is in line with reported values [50]. 100% of the water was assumed to have evaporated and reported to the off-gas stream, 99% of Zn was assumed to have reacted and volatilised, while 100% of Fe was assumed to have reacted to form FeO. An unknown factor is the extent to which the carbon reacts and therefore the extent to which it reports to the Waelz Slag. From literature carbon content in the slag typically ranges from 10-20% and this range was thus used [49].

Table 6. EAF dust compositions used as feed to the Waelz kiln.

Composition (wt.%)	1	2	3	4	5	6
Zn	10		26		35	
ZnO	12.90	1.76	31.90	6.60	43.55	23.58
Fe_3O_4	71.60	30.06	53.10	5.20	41.45	4.09
$ZnFe_2O_4$	0.00	46.95	0.00 [a]	73.40 [b]	0.00	57.53
FeO	0.00	0.00	0.00	0.00	0.00	0.00
$H_2O_{(l)}$	15.50	21.23	15.00	14.80	15.00	14.81

[a] Zn was assumed to be 100% present as ZnO and Fe as Fe_3O_4 i.e. no ferrites ($ZnFe_2O_3$) were assumed.
[b] The same Zn and Fe wt.% were maintained while it was assumed that 90% of Fe was present as ferrites ($ZnFe_2O_3$), with the balance as ZnO and Fe_3O_4.

As can be seen in Figure 9, the carbon footprint associated with processing of EAF dust in a Waelz kiln appears to range from 4-12 kg CO_2/kg Zn dust (10 wt.% Zn in EAF dust) to 1-4 kg CO_2/kg Zn (35 wt.% Zn in EAF dust). In this simple model, the presence of ferrites appeared to have no significant impact on the carbon demand and thus carbon footprint.

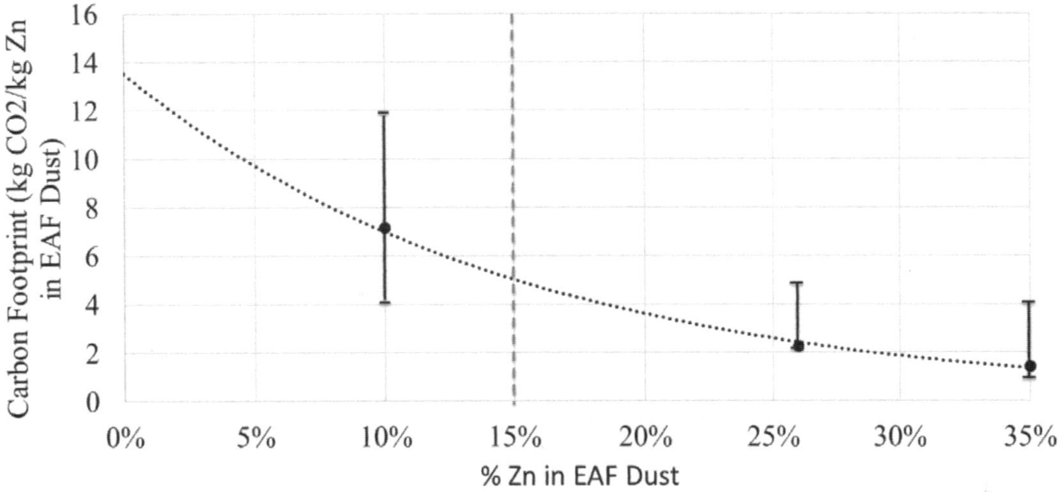

Figure 9. Carbon footprint (kg CO_2/kg Zn in EAF dust) ranges associated with different concentrations of Zn in EAF dust. Dotted line represents minimum viable zinc content.

4. Conclusions

The addition of zinc in the form of galvanising contributes towards addressing corrosion and thus significantly increases the life span of steel products, with approximately 50-60% of the global production of zinc being used to galvanise steel. Almost all processes produce waste, and the reprocessing of secondary zinc streams is becoming increasingly important, due to legislation changes and increased consumer demand for recycled and low carbon metals. However, the recovering of zinc from these secondary resources requires additional energy which needs to be quantified.

The objective of this desktop study was to address gaps in available literature with respect to quantifying potential zinc secondary resources and the additional carbon (CO_2) footprint associated with returning this secondary zinc back into primary production. To this end the authors have estimated the annual galvanised steel production between 2017 and 2022 ranges from 320-390 Mt/a (equating to 15-20% of the annual steel production).

The application of zinc for galvanising can be considered dispersive in nature due to the low quantities of zinc coating required per ton of steel. This makes economically recycling zinc from these sources challenging. For modelling purposes, it has been determined that a zinc content of 1.5-2.5 % for galvanised steel is reasonable. However, it is acknowledged that these values will vary widely depending on the type of steel being recycled.

Although there are two main process routes for steel making, the EAF process route is predominantly discussed with reference to zinc due to its ability to incorporate high levels of secondary (zinc containing) steel feeds and the dust produced typically contains economically recoverable levels of zinc (15 wt.%). EAF dust production is estimated between 5-10 Mt/annum and at a zinc content of 10-36%. If this zinc is not recovered it represents a loss of 0.5-3.6 Mt/annum or up to 60% of primary consumption for galvanising. Due to conflicting information the recycling rate of EAF dust appears to range from 50-75%. While the CO_2 footprint for EAF steel is reported as 0.67 kg CO_2/kg EAF steel and can therefore be significantly reduced (up to 95%) if low carbon green sources of energy (wind, solar, hydro) are used for the electricity generation which is required for the EAF operation.

The Waelz kiln is the most prominent technology utilised to treat EAF dust and further extract the zinc, in the form of Waelz oxide, which can subsequently be incorporated into primary zinc processing routes such as roast-leach-electrowinning (RLE). Unlike the EAF, the Waelz kiln is reliant on coke as a reducing agent. This represents a significant contributing factor to the carbon footprint associated with the recycling of secondary zinc streams.

Model analysis of a simplified Waelz kiln using the HSC simulation software was presented with the carbon footprint primarily dependent on the wt.% of Zn in the EAF dust, the ratio of coke to EAF dust, as well as the extent of reaction of the coke within the kiln. The carbon footprint associated with processing of EAF dust in a Waelz kiln appears to range from 4-12 kg CO_2/kg Zn dust (10 wt.% Zn in EAF dust) to 1-4 kg CO_2/kg Zn (35 wt.% Zn in EAF dust). The carbon footprint associated with the Waelz kiln could be lowered if the zinc content is kept at the highest possible level through selective processing of high zinc secondary waste streams in EAF and through an increased efficient utilisation of coal which would reduce the feed rate of carbon per tonne EAF dust.

A future study to investigate the carbon footprint associated with the primary zinc processing routes will allow for a comparison between the primary and secondary routes. However, it is anticipated that the carbon footprint associated with primary Zn processing routes such as RLE will be significantly lower than the of recovery of Zn through the Waelz kiln. It is therefore anticipated that while the reprocessing of zinc dust is important from an environmental and material efficiency perspective, the industry may be unable to achieve the consumers' demand for recycled "green" zinc at a lower carbon footprint than primary zinc.

5. Acknowledgements

The authors would like to acknowledge the following funders for supporting this research: the South African Minerals to Metals Research Institute (SAMMRI), the South African National Research Foundation (NRF) through its SARChI Chair in Minerals Beneficiation (Grant UID 64829) and the South African Department of Science and Innovation (DSI) as well as input through discussions with the International Zinc Association (IZA).

References

[1] Statista. Global end use of zinc as of 2017, by category [Internet]. 2018 [cited 2018 Jun 22]. Available from: https://www.statista.com/statistics/240626/share-of-zinc-consumption-by-category/
[2] Fortune Business Insights. Recycled Metal Market Report [Internet]. Fortune Business Insights. 2022 [cited 2023 Feb 2]. Available from: https://www.fortunebusinessinsights.com/recycled-metal-market-105328
[3] Leotaud VR. Demand for green metals from recycling expected to grow [Internet]. Mining.com. 2022 [cited 2023 Feb 10]. Available from: https://www.mining.com/demand-for-green-metals-from-recycling-expected-to-grow-report/
[4] Davis C. Low carbon still a high priority for metals markets [Internet]. S&P Global Commodity Insights. 2022 [cited 2023 Feb 4]. Available from: https://www.spglobal.com/commodityinsights/en/market-insights/blogs/metals/092322-low-carbon-aluminum-billet-decarbonization#
[5] Valckx N, Stuermer M, Seneviratne D, Ananthakrishnan P. Metals Demand From Energy Transition May Top Current Global Supply [Internet]. International Monetary Fund. 2021 [cited 2023 Feb 10]. Available from: https://www.imf.org/en/Blogs/Articles/2021/12/08/metals-demand-from-energy-transition-may-top-current-global-supply
[6] IZA. Zinc Recycling- Stocks and Flows. International Zinc Association. 2015.
[7] IEA. Iron and Steel [Internet]. International Environmental Agency. 2021 [cited 2022 Apr 26]. Available from: https://www.iea.org/reports/iron-and-steel
[8] Metso. HSC Chemistry [Internet]. Metso. 2023 [cited 2023 Jun 1]. Available from: https://www.hsc-chemistry.com/hscchemistry
[9] World Steel Association. About steel [Internet]. 2023 [cited 2023 Apr 10]. Available from: https://worldsteel.org/about-steel/about-steel/
[10] Rankin J. Energy Use in Metal Production. High Temp Process Symp [Internet]. 2012 [cited 2023 Apr 10]; Available from: http://www.iea.
[11] World Steel Association. About steel [Internet]. 2022 [cited 2022 Sep 22]. Available from: https://worldsteel.org/about-steel/about-steel/
[12] Hoffmann C, Van Hoey M, Zeumer B. Decarbonization in steel [Internet]. McKinsey. 2020 [cited 2023 Apr 10]. Available from: https://www.mckinsey.com/industries/metals-and-mining/our-insights/decarbonization-challenge-for-steel#/
[13] NACE Inernational. Corrosion Costs. Corrosion Costs and Preventive Strategies in the United States. 2015.
[14] NACE International. Economic Impact - Assessment of the Global Cost of Corrosion [Internet]. IMPACT. 2016 [cited 2023 May 16]. Available from: http://impact.nace.org/economic-impact.aspx
[15] Alltimes Coatings. What Is the True Cost of Corrosion? [Internet]. Alltimes Coatings. 2021 [cited 2023 May 16]. Available from: https://www.alltimescoatings.co.uk/news/what-is-the-true-cost-of-corrosion
[16] MINTEK. The high cost of corrosion [Internet]. MINTEK. 2011 [cited 2023 May 18]. Available from: https://www.mintek.co.za/2011/11/15/the-high-cost-of-corrosion/
[17] American Galvanizers Association. Zinc Coatings: A comparative analaysis of process and performance characteristic. Am Galvaniz Assoc. 2009;14(2):184–270.
[18] National Material Company. Galvanized Steel: Types, Uses, Benefits [Internet]. National Material Company. 2019 [cited 2023 May 18]. Available from: https://www.nationalmaterial.com/galvanized-steel-types-uses-benefits/
[19] Precision Reports. Global Galvanized Steel Market Growth 2023-2029 [Internet]. 2023 [cited 2023 Apr 11]. Available from: https://www.precisionreports.co/global-galvanized-steel-market-22516721

[20] Silver City Galvanizing. Steel Composition [Internet]. Silver City Galvanizing Inc. 2013 [cited 2023 May 22]. Available from: https://silvercitygalv.com/technical-support/steel-composition/#prettyPhoto
[21] World Steel Association. Steel statistical yearbook [Internet]. worldsteel. 2020 [cited 2022 Sep 22]. Available from: https://worldsteel.org/steel-topics/statistics/steel-statistical-yearbook/
[22] World Steel Association. The White Book of Steel [Internet]. 2012. 53 p. Available from: https://www.worldsteel.org/en/dam/jcr:7b406f65-3d94-4e8a-819f-c0b6e0c1624e/The%2520White%2520Book%2520of%2520Steel_web.pdf
[23] MetInvest. Ilyich Iron and Steel Works, Mariupol [Internet]. Metinvest Holding. 2015. Available from: https://www.viktormacha.com/galerie/ilyich-iron-and-steel-works-mariupol-308/
[24] Organisation for Economic Co-operation and Development. Latest developments in steelmaking capacity [Internet]. 2019. Available from: https://one.oecd.org/document/DSTI/SC(2018)10/en/pdf
[25] knoema. India - Open Hearth Furnace - Steel. knoema. 2019.
[26] Industrial Metal Castings. Open Hearth Furnace,Open Hearth Furnace Application,Open Hearth Furnace Manufacturers,Open Hearth Furnace Process. Industrial Metal Castings. 2009.
[27] Serapio M. China's outcast steel machines find unwelcome home in Southeast Asia [Internet]. Reuters. 2018 [cited 2023 Apr 12]. Available from: https://www.reuters.com/article/us-steel-asia-china-analysis-idUSKBN1OG06E
[28] Cordero T. Iron and steel group urges ASEAN to ban 'obsolete' furnaces from China [Internet]. GMA News Online. 2018 [cited 2023 Apr 12]. Available from: https://www.gmanetwork.com/news/money/companies/641855/iron-and-steel-group-urges-asean-to-ban-obsolete-furnaces-from-china/story/
[29] Technavio. Galvanized Steel Market Size, Share & Trends to 2027 [Internet]. 2022 [cited 2023 May 9]. Available from: https://www.technavio.com/report/galvanized-steel-market-industry-analysis
[30] van Leeuwen M. Current Status and Future Expectations for the Zinc Market. In: InterZAC [Internet]. 2022. Available from: https://www.researchgate.net/publication/343350539_Current_Status_and_Future_Expectations_for_the_Zinc_Market
[31] Research and Markets. Global Galvanized Steel Market 2023-2027 [Internet]. 2022 Dec [cited 2023 Apr 10]. Available from: https://www.researchandmarkets.com/reports/5504426/global-galvanized-steel-market-2023-2027?utm_source=GNOM&utm_medium=PressRelease&utm_code=m95l6x&utm_campaign=1639320+-+Global+Galvanized+Steel+Industry%3A+Trends+and+Opportunities+2021-2025&utm_exec=joca220prd
[32] Technavio. Overview Galvanized Steel Production - Statistics and Analysis 2021-2025 [Internet]. 2021 [cited 2023 May 9]. Available from: https://statistics.technavio.org/statistics/galvanized-steel-production
[33] SuperTec. GALVANIZED STEEL SPECIFICATIONS. SuperTec Ceilings & Board (Pty) Ltd.
[34] AZoM. Galvanized Coatings - Surface Area Calculation Tables [Internet]. AZO Materials. 2003 [cited 2023 May 9]. Available from: https://www.azom.com/article.aspx?ArticleID=1852
[35] World Steel Association. Steel industry coproducts [Internet]. World Steel Association. 2021 [cited 2023 Apr 12]. Available from: https://worldsteel.org/wp-content/uploads/Fact-sheet-Steel-industry-co-products.pdf
[36] Hoang-Trung Z, Kukurugya F, Takacova Z, Havlik T. Acid leaching of BOF (basic oxygen furnace) sludge. XVIII Int Student's Day Metall. 2011;241–7.
[37] da Rocha EP, de Castro JA, Araujo GM, Martins L. Characterization of BOF dust for pellets production used in blast furnace. Mater Sci Forum. 2014;798–799:611–6.
[38] Ray SK, Chattopadhyay G, Ray AK. Evaluation of Dust Generated from Basic Oxygen

Furnace Steel Making. J Air Waste Manag Assoc. 1997;47(6):716–21.
[39] Kozicki C, Carlson C. A Look at the Process of EAF Dust Recycling [Internet]. FEECO International. 2022 [cited 2022 Sep 22]. Available from: https://feeco.com/a-look-at-the-process-of-eaf-dust-recycling/
[40] Małecki S, Gargul K, Warzecha M, Stradomski G, Hutny A, Madej M, et al. High-performance method of recovery of metals from eaf dust—processing without solid waste. Materials (Basel). 2021;14(20).
[41] Cifrian E, Dacuba J, Llano T, Díaz-Fernández MDC, Andrés A. Coal fly ash-clay based geopolymer-incorporating electric arc furnace dust (EAFD): Leaching behavior and geochemical modeling. Appl Sci. 2021;11(2):1–17.
[42] de Buzin PJWK, Heck NC, Vilela ACF. EAFD – a Thermodynamic Analysis and Classification of Dust Types. In: 45th Steelmaking Seminar [Internet]. Porto Alegre, Brazil; 2014. p. 689–98. Available from: issn: 1982-9345
[43] de Buzin PJWK, Heck NC, Vilela ACF. EAF dust: An overview on the influences of physical, chemical and mineral features in its recycling and waste incorporation routes. J Mater Res Technol [Internet]. 2017;6(2):194–202. Available from: http://dx.doi.org/10.1016/j.jmrt.2016.10.002
[44] Antuñano N, Cambra JF, Arias PL. Hydrometallurgical processes for Waelz oxide valorisation – An overview. Process Saf Environ Prot [Internet]. 2019;129(July):308–20. Available from: https://doi.org/10.1016/j.psep.2019.06.028
[45] Omran M, Fabritius T, Heikkinen EP. Selective Zinc Removal from Electric Arc Furnace (EAF) Dust by Using Microwave Heating. J Sustain Metall [Internet]. 2019;5(3):331–40. Available from: https://doi.org/10.1007/s40831-019-00222-0
[46] Antrekowitsch J, Rösler G, Steinacker S. State of the Art in Steel Mill Dust Recycling. Chemie-Ingenieur-Technik. 2015;87(11):1498–503.
[47] Nakayama M. New Eaf Dust Treatment Process: ESRF [Internet]. 2009. Available from: http://www.steelplantech.co.jp/wp-content/uploads/2012/08/201108_EAF_Dust Treatment_fortlightMetalRecovery.pdf
[48] Yakornov SA, Pan'shin AM, Grudinsky PI, Dyubanov VG, Leont'ev LI, Kozlov PA, et al. Thermodynamic Analysis of Zinc Ferrite Decomposition in Electric Arc Furnace Dust by Lime. Russ J Non-Ferrous Met. 2017;58(6):586–90.
[49] Kashiwada M, Kumagai T. Waelz Process for Leach Residues at Nisso Smelting Company Ltd. AIME World Symp Min Metall Lead Zinc. 1970;2:409–22.
[50] Harris WE. The Waelz Process. Metall Lead Zinc. 1936;121:702–20.
[51] Chen T., Poirier G. Mineralogical characterisation of Waelz kiln products. In: Rao SR, Harrison JA, Kozinski JA, Amarantunga LM, Cheng TC, Richards GG, editors. Waste Processing and Recyling in Mineral and Metallurgical Industries - Fith International Symposium. Hamilton, Ontario, Canada; 2004. p. 681–96.
[52] Peng N, Peng B, Chai L, Liu W, Li M, Yuan Y, et al. Decomposition of Zinc Ferrite in Zinc Leaching Residue by Reduction Roasting. Procedia Environ Sci. 2012;16:705–14.
[53] Stewart DJC, Barron AR. Pyrometallurgical removal of zinc from basic oxygen steelmaking dust – A review of best available technology. Resour Conserv Recycl. 2020;157:1–51.
[54] Ellis D. Boliden starts sales of low carbon zinc [Internet]. Mining Digital. 2022 [cited 2022 Oct 4]. Available from: https://miningdigital.com/sustainability/boliden-produces-low-carbon-zinc
[55] GreenSpec. Zinc production & environmental impact [Internet]. GreenSpec. 2022 [cited 2022 Oct 4]. Available from: https://www.greenspec.co.uk/building-design/zinc-production-environmental-impact/
[56] Nilsson AE, Aragonés MM, Torralvo FA, Dunon V, Angel H, Komnitsas K, et al. A review of the carbon footprint of Cu and Zn production from primary and secondary sources. Minerals. 2017;7(9):1–12.

[57] Rezinal. Rezinal on its way to produce carbon-neutral zinc! [Internet]. Rezinal. 2016 [cited 2022 Oct 4]. Available from: https://www.rezinal.be/en/node/77
[58] Steel Dynamics. EAF vs BlastFurnace [Internet]. 2022 [cited 2022 Sep 22]. Available from: https://stld.steeldynamics.com/eafvsblastfurnace/
[59] World Steel Association. Sustainability indicators [Internet]. World Steel Association. 2023 [cited 2023 Apr 12]. Available from: https://worldsteel.org/steel-topics/sustainability/sustainability-indicators/
[60] Dan Carbon. Difference between electric arc furnace vs blast furnace [Internet]. 2020 [cited 2022 Sep 22]. Available from: https://www.dancarbon.com/q/eaf/105.html
[61] Steel Supply. Electric Arc Furnace vs. Blast Furnace [Internet]. Steel Supply L.P. 2022 [cited 2022 Sep 22]. Available from: https://www.steelsupplylp.com/blog/electric-arc-furnace-vs-blast-furnace
[62] World Steel Association. Sustainability Indicators 2022 report. 2022.
[63] Bellona Europa. Hydrogen in steel production: what is happening in Europe – part one [Internet]. 2021 [cited 2023 May 8]. Available from: https://bellona.org/news/eu/2021-03-hydrogen-in-steel-production-what-is-happening-in-europe-part-one
[64] World Nuclear Association. Carbon Dioxide Emissions From Electricity [Internet]. World Nuclear Association. 2022 [cited 2023 Jun 2]. Available from: https://www.world-nuclear.org/information-library/energy-and-the-environment/carbon-dioxide-emissions-from-electricity.aspx
[65] Abdallah L, El-Shennawy T. Reducing Carbon Dioxide Emissions from Electricity Sector Using Smart Electric Grid Applications. J Eng [Internet]. 2013;8. Available from: https://www.hindawi.com/journals/je/2013/845051/
[66] Islam M. Assessment of torsional resonance in turbine-generator shaft due to multiple EAF operation. 2013.
[67] Schoukens AFS, Shawt F, Chemaly EC. The Envroplas process for the treatment of steel-plant dusts. J South African Inst Min Metall. 1993;93(1):1–7.

Removal of Lead and Zinc from Blast Furnace Sludge Through High Temperature Carbochlorination

Y. Xiao[1] *, P. Beentjes[1], M. Soefihara[2] and Y. Yang[2]

[1] Tata Steel, IJmuiden, 1970 CA, The Netherlands. Email: yanping.xiao@tatasteeleurope.com
[2] Delft University of Technology, 2628 CD Delft, The Netherlands

Abstract. A new dual kiln processing route has been proposed and investigated for removal of heavy metals from the zinc-rich fraction of blast furnace (BF) sludge and recovery of ferrous materials. The zinc in BF sludge exists both as ZnO and ZnS. The thermodynamic analysis indicates that the carbochlorination of the sludge shows a selective heavy metal removal. Lead can be removed at a temperature below 900°C followed by the zinc removal at higher temperatures, producing an iron-rich clean residue free of heavy metals in the end feasible for recycling back to ironmaking. High temperature carbochlorination experiments have been conducted with designed mini-pellets. The experimental results prove that via the proposed dual kiln route, a valuable zinc-rich fraction and a clean iron-bearing residue can be achieved with addition of iron chloride, leaving a small stream of lead-rich residue for further processing. This newly developed recycling flowsheet provides a sequential carbochlorination in two inter-connected rotary kilns, incorporated with efficient dust-capturing systems. In total, over 98% heavy metal removal has been achieved. The Zn-rich dust contains more than 60% of zinc, and the quality of the iron-rich residue is sufficient for re-use in the integrated steel works.

1. Introduction

Steelmaking industry contributes greatly to the development of global economy, but at the same time generates significant amount of emissions to the environment. Today CO_2 reduction and materials circularity become even more important than the economic growth. The development and application for sustainable and more efficient process technology will be crucial for the further improvement of the steelmaking industry. According to World Steel Association [1], global crude steel production has reached 1,878 million tons in 2022 with 4.2% decrease from 2021. The European Union produces 136.7 million tons of steel in 2022 with 10.5% decrease from 2021, with approximately 60% BF-BOF route and 40% EAF route at 500 steel production sites across 27 EU member states. Approximately 56% of EU steel is made from scrap, with around 100 million tonnes of scrap steel recycled every year [2]. Increasing efforts have been made in recent years inside the European Union for developing CO_2-mitigating or even carbon-free H_2-based technologies, improving resource efficiency and fostering sustainable and green transition. In 2020, the EU published a Circular Economy Action Plan, an important step in developing a truly circular economy in Europe, which helps to clarify the policies regarding recycling and the absolute minimum of waste generation from the steel production cycle.

Blast Furnace (BF) operation generates on average about 10 - 15 kg/thm (hot metal) dry flue dust worldwide. It contains enriched levels of heavy metals, which can contaminate soil, water and air if not properly processed, leading to adverse environmental and ecological impacts. It is therefore essential to develop efficient methods with advanced clean-production techniques for heavy metal removal from blast furnace dusts. Tata Steel in IJmuiden of the Netherlands is one of the Europe's leading integrated steel production plants, with around 7 million tons annual rolled steel production. Two blast furnaces are in operation, with self-produced sinter, pellets, and coke. Steel scrap is recycled via steel converter with a controlled zinc input, and the steelmaking dust is recycled to BFs via sinter plant. There is a site limit for recirculating zinc in the integrated steel works. It is known that the heavy metals have a detrimental impact on the BF process stability. For example, zinc can circulate and deposit inside the furnace, and will cause scaffold in the upper shaft of blast furnace, decreasing the lining lifetime via wearing, and even creating issues during tapping. The heavy metals (Zn, Pb, Cd etc.) leave the blast furnace together with flue gas, and are enriched through a dry classifier followed by wet scrubber. The dry coarse fraction containing lower amount of zinc and lead is recycled internally in the sinter plant. The finer fraction is further processed via the hydro-cyclones into three fractions with low, middle and high zinc contents. Depending on the input limitation of zinc in the blast furnace, the low and partially the medium zinc containing fractions are recycled internally, and the high zinc fractions are currently landfilled as an external zinc exit stream. The landfill of high-zinc fractions has a higher environmental impact, and constant effort has been given in finding a more sustainable solution since 1980's at IJmuiden site, with various processing routes tested and evaluated, including both hydro- and pyrometallurgical processes [3 - 6].

Chlorination is one of the effective techniques for the heavy metal extractions. It is known that zinc is present in the oxide form (ZnO) and sulfide form (ZnS), both of which can be converted into metal chloride. Metal chlorides are generally very volatile at relatively low temperatures, and often have different vapour pressure from one another [7]. These characteristics lead to good selectivity in the metal separation via volatilization at a specific temperature utilizing the difference in the vapour pressure. A recent research was conducted by Hamann et al. in 2021 [8] in treating the BF sludge through direct reduction with the addition of spent pickling acid ($FeCl_2$) to remove the heavy metals in a mixture of chlorides. A recycling solution to selectively remove the heavy metals is required, and preferably Pb could be removed from Zn at lower temperatures, leaving a zinc-rich fraction for further extraction and utilization. The small amount of Pb-rich fraction will be subjected to further processing, or landfilled after proper immobilization. In the present study, a new carbochlorination processing route was explored and tested, and the lab scale experimental results will be presented.

2. Raw materials and characterisations

The zinc rich BF dust is used in the present study, which is generated as filter cake after heavy metal enrichment via the hydro cyclones. Four mini-pellet samples, with and without $FeCl_2$ addition, were prepared for the carbon-chlorination experiments. Table 1 lists the pellet size and its chemical compositions determined with Inductively Coupled Plasma Optical Emission Spectroscopy (ICP-OES) and ELTRA for carbon and sulphur analyses, and the mini-pellets are shown in Figure 1.

Table 1. Chemical compositions of the prepared mini-pellets

Sample	Size, mm	Cl, wt%	C, wt%	Ca, wt%	Cd, wt%	Fe, wt%	Mg, wt%	Mn, wt%	Pb, wt%	S, wt%	Ti, wt%	V, wt%	Zn, wt%	Si, wt%
1	<3.15	0.14	40.80	1.63	0.01	24.45	0.62	0.10	0.51	1.10	0.09	0.02	2.77	2.70
2	<3.15 (with $FeCl_2$)	2.22	38.00	1.63	0.01	25.15	0.58	0.09	0.49	1.04	0.08	0.02	2.48	2.60
3	3.15 - 5	0.13	40.65	1.65	0.01	24.65	0.63	0.10	0.56	1.11	0.09	0.02	2.83	2.70
4	3.15 - 5 (with $FeCl_2$)	2.10	38.95	1.59	0.01	24.75	0.57	0.10	0.51	1.06	0.08	0.02	2.60	2.50

Figure 1: Mini-pellets of BFD (Blast Furnace Dust) with or without $FeCl_2$ addition

The composition of each sample was the average of a duplicate analysis. The samples 1 and 3 have almost the same chemical compositions, and the only difference is the size. The samples 2 and 4 are with the same addition of $FeCl_2.4H_2O$ (in two size ranges), calculated based on the stoichiometry required to remove lead and zinc present in the dust samples. The highest element present in the pellets is carbon (~38 - 40 %), followed by iron (~25%). Zinc, lead, and cadmium content are around 2.6%, 0.5%, and 0.01%, respectively. The initial Cl level for the samples 1 & 3 is at the level of 0.13 - 0.14% and it increases to the level of 2.1 - 2.2% with $FeCl_2$ addition to facilitate the heavy metal chlorination reactions. The zinc in the BF dust is mainly present as ZnO and ZnS identified with XRD, and the other possible zinc phase as zinc ferrite ($ZnFe_2O_4$) can hardly be quantified. Based on the XRD analyses, ZnO was found as zincite and ZnS was found as wurtzite and sphalerite in both samples. It is found that the zinc phases consist of approximately 1/3 of ZnO and 2/3 of ZnS. The mineralogical form of lead is not confirmed by XRD due to its very low content, but PbS was identified via Raman spectrometer previously.

3. Experimental

The experiments were carried out using two different furnaces. The first was the TGA furnace for the small-scale experiment focusing on mass changing during the heating and removal rate of zinc and other heavy metals. The second furnace was a horizontal tube furnace for a larger scale experiment focusing on the heavy metal removal and separation, and on generating secondary dust rich in heavy metals. The initial and the treated samples were analysed by using various analytical techniques.

3.1. Small scale TGA tests - thermal behaviour of BF dust

A Netzsch STA 409 TGA/DSC furnace was used for understanding the thermal behaviour of the mini pellets prepared from BF dust. The experiments in the TGA furnace were conducted to a maximum temperature of 800, 1000, 1200, and 1400°C for each of the four samples, respectively. The mini-pellets samples were first dried overnight at 105 °C. Each sample weighing between 100-200 mg was put in a crucible, placed next to a reference crucible inside the heating chamber and heated at 40°C/min starting from room temperature. After reaching the designated maximum temperature, the isothermal condition was maintained at the temperature level for 60 minutes. The nitrogen gas was introduced at 20 ml/min during the experiment for keeping an inert atmosphere. After the holding period of 60 minutes, the sample was then cooled through the water cooling system attached internally. Each test was repeated to ensure the reproducibility.

3.2. Selective heavy metal removal from BF dust in a horizontal tube furnace

The heavy metal removal experiments were carried out in a horizontal tube furnace, with an installed glass filter to capture the flue dust from the off gas, as illustrated in Figure 2. The stage-wise tests were designed to investigate the feasibility for selective removing and separating the heavy metals (in particular zinc and lead) under carbochlorination conditions. Samples 1 and 2 of smaller sized mini-pellets without and with $FeCl_2$ addition were selected for the experiments, and about 40 grams of each sample was put in a stainless steel boat, and the boat was placed into the constant temperature zone of the tube furnace. The furnace was closed gas tight and flushed with dry nitrogen.

Two sets of experiments were designed. Table 2 and Table 3 give an overview of the stage-wise experimental design with the samples and the achieved conditions for both sample 1 (series A, without $FeCl_2$ addition) and sample 2 (B series, with $FeCl_2$ addition). For tests No. 1-8 (A1-A8, B1-B8), the same mini-pellets sample (A or B) went through 8 stage-wise heating, with a local holding of 30 minutes, at the temperature intervals specified in Table 2. Only filter dust samples were collected stage-wise following thermal treatment profile, with just one last residue pellet sample in the boat. In each stage a new glass filter was used. For the test No. 9-16 in Table 3, 8 individual experiments were conducted for

Journal of Physics: Conference Series **2738** (2024) 012030 doi:10.1088/1742-6596/2738/1/012030

both sample 1 and sample 2, each started at 150°C and ended at the pre-set temperatures for 30 minutes. Each experiment will generate an accumulative dust and the residue (after reaction) samples.

Prior to the experiments the pellet samples were dried at 150°C for one hour in the same furnace. The glass filter was applied at the exit of the off-gas pipe and held at the position by load. At a given temperature, the sample was reacted for 30 minutes, then cooled quickly at the low temperature zone of the furnace by pulling out of the hot zone The residues after reactions and the captured secondary flue dust on the filters were subjected for further analyses.

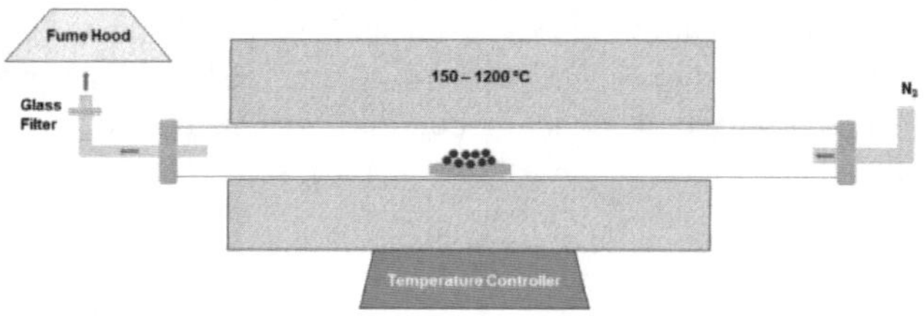

Figure 2: Schematic diagram of horizontal tube furnace for carbochlorination tests

The residue samples and the captured secondary flue dusts were analysed with various analytical methods including ICP-OES, X-ray Fluorescence (XRF), X-ray Diffraction (XRD), Scanning Electron Microscopy and Energy Dispersive X-ray Spectroscopy (SEM/EDS), ELTRA for carbon, as well as wet chemical analysis.

Table 2: An overview of experiment design and the samples collected for analyses (A series for sample 2, and B-series for sample 1): stage-wise experiments

Test no.	Initial T, °C	Final T, °C	Sample type	Heating profile
A0/B0	20	150	Residue	Initial composition dried
A1/B1	150	350	Filter	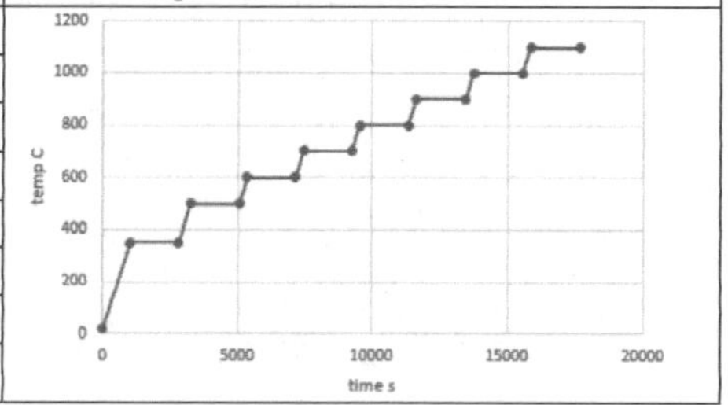
A2/B2	350	500	Filter	
A3/B3	500	600	Filter	
A4/B4	600	700	Filter	
A5/B5	700	800	Filter	
A6/B6	800	900	Filter	
A7/B7	900	1000	Filter	
A8/B8	1000	1100	Filter	

Table 3: An overview of experiment design and the samples collected for analyses (A series for sample 2, and B-series for sample 1): continuous experiments

Test no.	Initial T, °C	Final T, °C	Sample type	Heating profile
A9/B9	150	350	Filter + residue	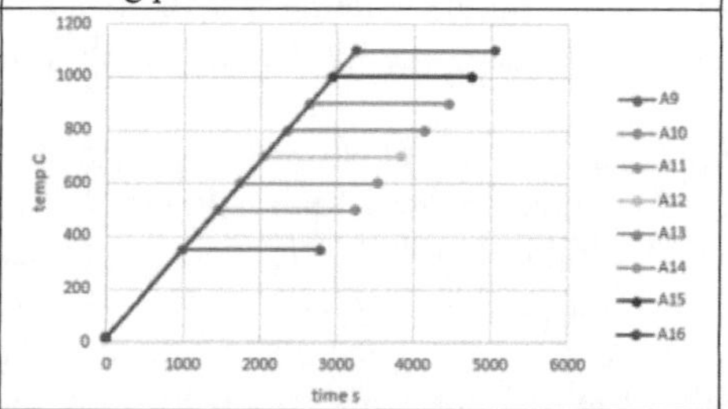
A10/B10	150	500	Filter + residue	
A11/B11	150	600	Filter + residue	
A12/B12	150	700	Filter + residue	
A13/B13	150	800	Filter + residue	
A14/B14	150	900	Filter + residue	
A15/B15	150	1000	Filter + residue	
A16/B16	150	1100	Filter + residue	

4. Results and discussion

4.1. Impact of chloride addition on the thermal behavior of BF dust - TGA experiments

TGA experiments were conducted at four different temperatures of 800°C, 1000°C, 1200°C and 1400°C with a constant heating rate of 40 °C/min in an inert atmosphere.. When the temperature was reached, the isothermal heating was held for another 60 mins. The results of the mass change during the non-isothermal and isothermal heating periods for samples of 1-4 are compared and shown in Figure 3.

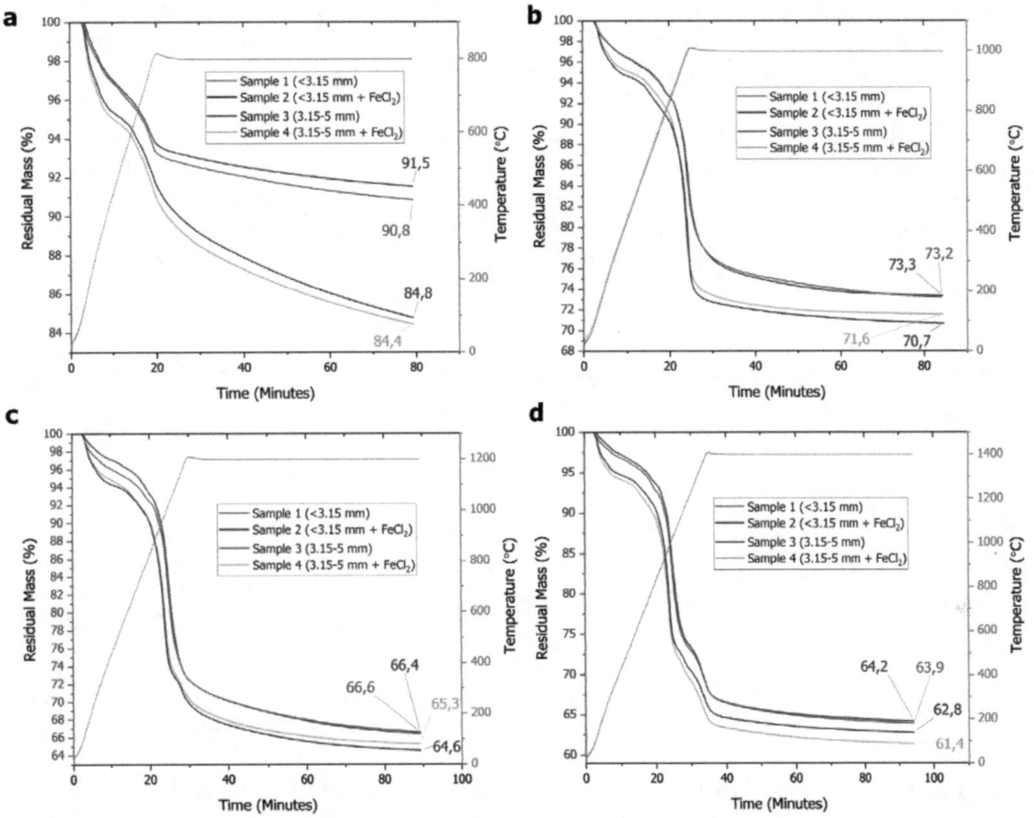

Figure 3: Evolution of mass loss of BF dust samples under different thermal conditions of maximum temperatures: (a) 800°C, (b) 1000°C, (c) 1200°C (d) 1400°C.

It can be seen that the significant mass change occurred during the non-isothermal heating period. In addition, for all the samples heated up to 800°C (Figure 3(a)), there were continued considerable mass loss within the isothermal heating period. On the other hand, for all the samples heated up to 1000, 1200, and 1400°C, the kinetics of the carbochlorination reactions in the blast furnace sludge were sufficiently fast to be almost completed within the non-isothermal heating period. As the maximum temperature increases, the sample's residual mass becomes smaller, which means that more elements were converted into the gas phase. It is expected that the most generated gases were the mixtures of carbon monoxide and carbon dioxide during processing at temperature range from 800°C to 1400°C, due to the high presence of carbon and iron oxide in the dust samples. Another contribution would arise from the volatilization of zinc, lead, and cadmium compounds which could account for up to 5% when heated up to 1400°C, based on FactSage simulation.

The largest influence of chloride addition is shown by the samples heated to the maximum temperature of 800°C and 1000°C, with further increase in the maximum temperature, the influence of chloride on the mass loss is less significant. The samples with chloride addition (samples 2 and 4) have higher mass loss than those without chloride addition (samples 1 and 3). This difference is caused by the vaporization of crystal water in $FeCl_2$ and by the chlorination reactions which promote the removal of heavy metals from the blast furnace dust with faster reaction kinetics. In this study, the difference of mass loss between sample 1 and 3, and between sample 2 and 4 are insignificant which reflects the insignificant effect of the pellets size.

Figure 4 illustrates the zinc and lead removal efficiency from the TGA experiments. The results show that all the sample reacted at temperatures higher than 1000°C led to a high removal rate of Zn and Pb, up to 98 and 99%, respectively. Furthermore, it is interesting to note that up to 98% of Pb has already been removed at 800°C from the samples with iron chloride addition (samples 2 and 4), however, only less than 40% Pb was removed from the samples without chloride addition (samples 1 and 3). This indicates that the carbochlorination reactions could lower the temperature required for the complete removal of Pb. Thermodynamic analyses of this reaction system were conducted with the FactSage simulations to support the experimental results, which is shown in Figure 5. The thermodynamic calculation results show a consistent agreement with the TGA results, and both show the chloride addition leads to a relatively lower temperature of lead removal, subsequently followed by zinc removal at higher temperatures.

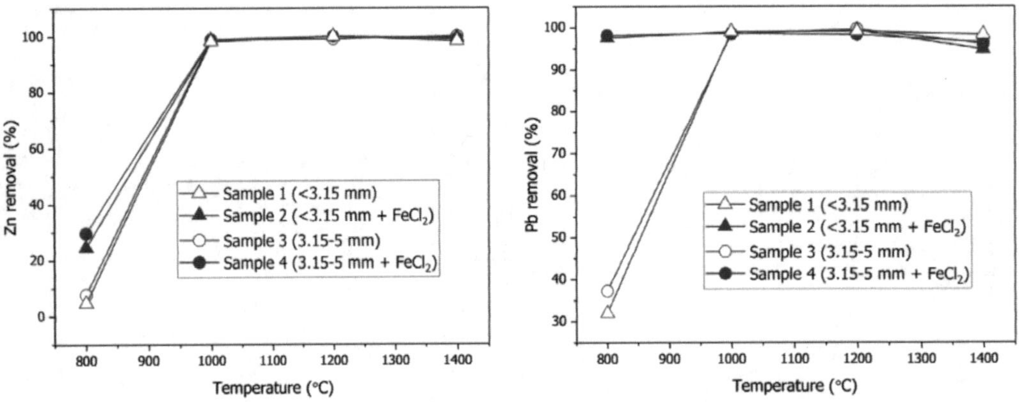

Figure 4: Removal of zinc and lead from the BF dust during the TGA experiments: effect of maximum temperature.

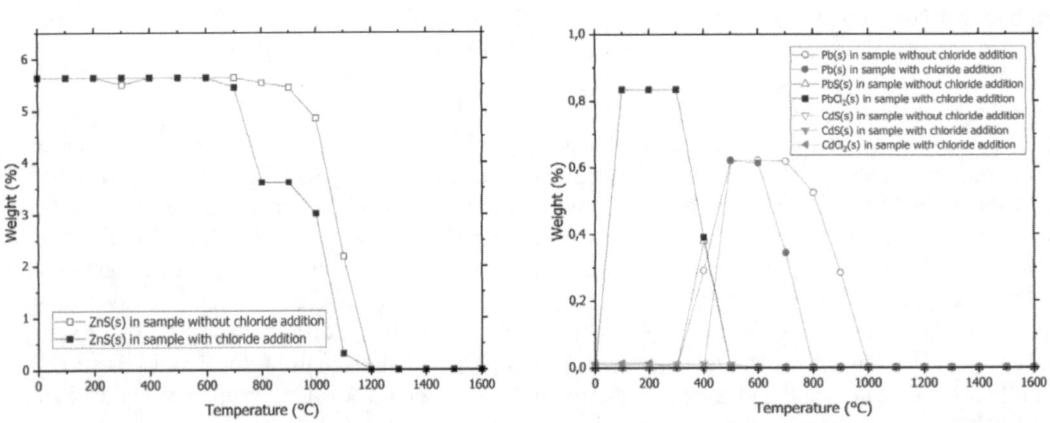

Figure 5: Equilibrium compositions of zinc and lead in the solid residues based on the thermodynamic analyses with FactSage's equilibrium module.

No significant impact from the pellet size could be observed by comparing the differences from samples 1 with 3 and 2 with 4, regardless of the effect of chloride addition. Usually the sample size would influence the removal rate, especially when the sample is heated at a high temperature. In this case, the heating temperature is sufficient to trigger the removal, and the sample of less than 5 mm is sufficiently small not to hinder the diffusion of the zinc and heavy metal volatiles. It demonstrates that zinc oxide, zinc sulfide, lead oxide/sulfide are thermodynamically favourable to be converted into chloride gas at a temperature higher than 800°C. A small sample (diameter <10 mm) could increase the diffusion rate from the inner core to the bulk, increasing the removal rate of heavy metals. Moreover, Zhu (2021) [12] reported a reduced removal rate of zinc when the sample size was increased from 12 mm to 18 mm, which is far greater than the diameter of the pellets used in this study.

4.2. Selective heavy metal removal via carbochlorination in the horizontal furnace

The pre-mixed mini-pellets were first dried in the same horizontal furnace and then reacted at given temperature for 30 minutes in N_2 atmosphere, following the schemes described in Table 2. Due to the limited amount of secondary flue dust captured on the glass filters from the stage-wise experiments (Table 2), it is difficult to accurately quantify the total chemical compositions. Instead the flue dust deposited on the filters were dissolved following the certified procedures and analyzed for the relative ratio of Pb to Zn in mass with ICP-OES, and the results are given in Figure 6. Figure 7 shows the impact of chloride additions on the Pb to Zn ratio in the captured flue dust and solid residues following the scheme in Table 3, which illustrates the potential of an earlier removal of Pb starting from the temperature of 600 °C. At 900 °C, over 97% of Pb has been removed from the solid residue with $FeCl_2$ addition, where the majority of zinc still remains in the solid residues, as seen in Figure 8. Figure 9 (a-f) shows the composition changes of various elements of Zn, Pb, Fe, V, Cd and Sn in the remaining residue as a function of temperature, respectively.

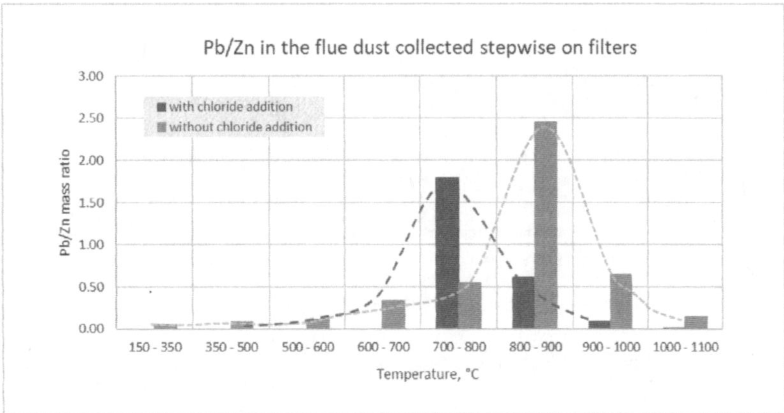

Figure 6: Impact of $FeCl_2$ addition on the removal of Zn and Pb from the BF dust at different temperature range *(following experimental scheme Table 2-A)*

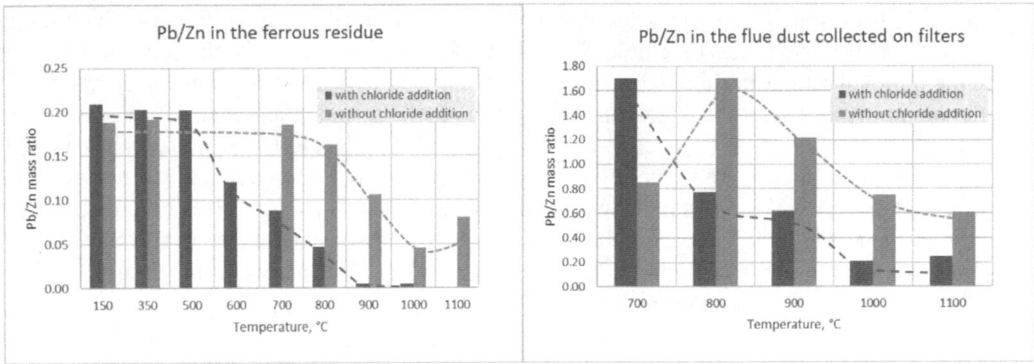

Figure 7: The mass ratio of Pb to Zn in the residues and accumulated filter dust
(following experimental scheme Table 2-B).

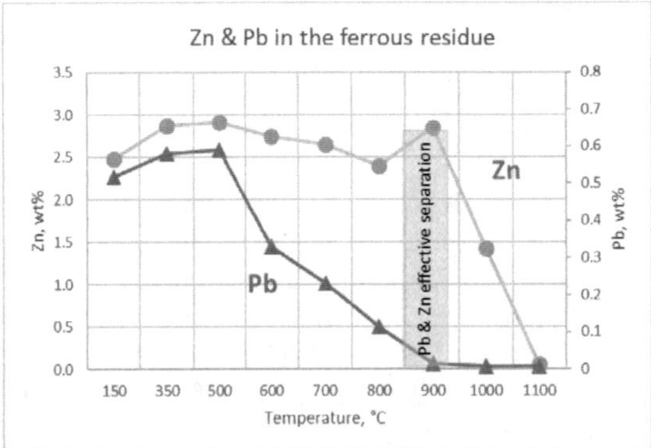

Figure 8: Zinc and lead in the remaining solid residue in the presence of $FeCl_2$ addition

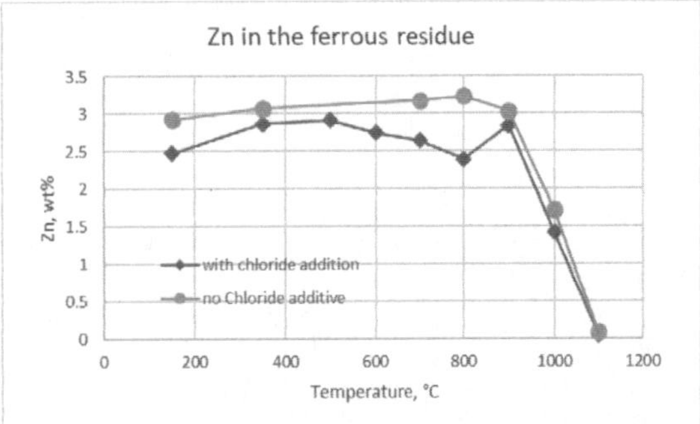

(9a): Impact of chloride addition on Zn in the residue

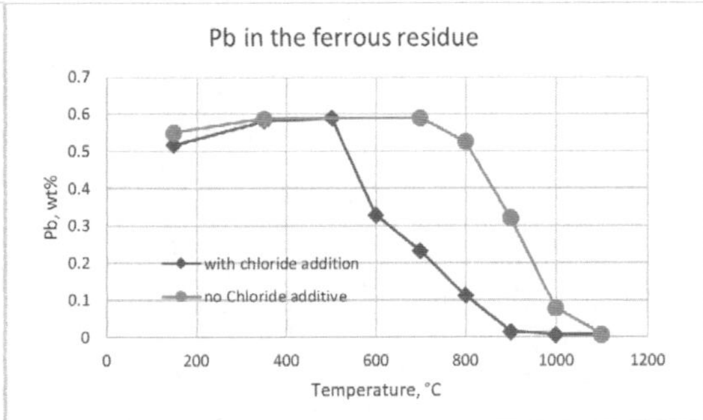

(9b): Impact of chloride addition on Pb in the residue

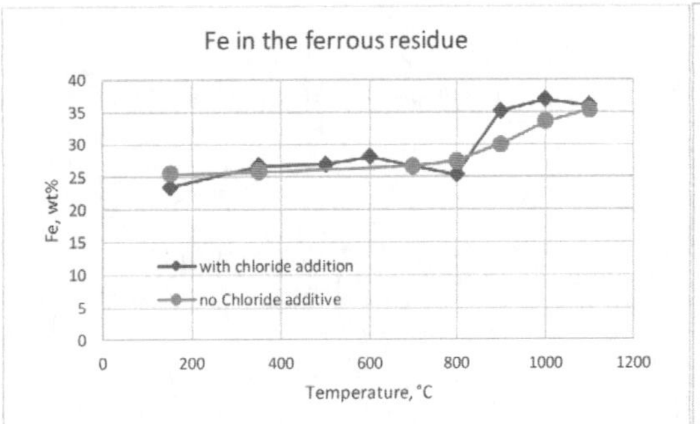

(9c): Impact of chloride addition on Fe in the residue

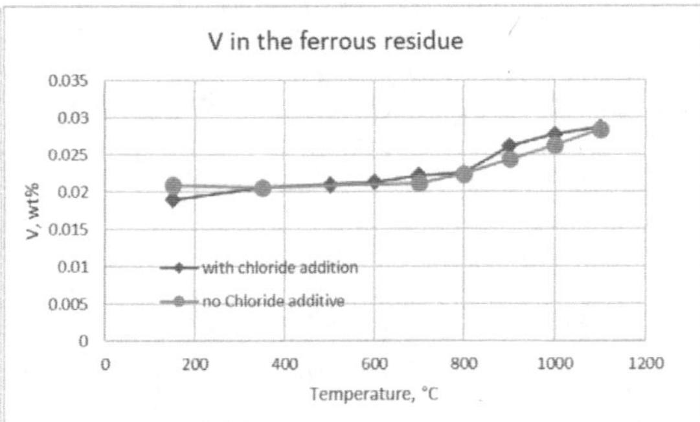

(9d): Impact of chloride addition on V in the residue

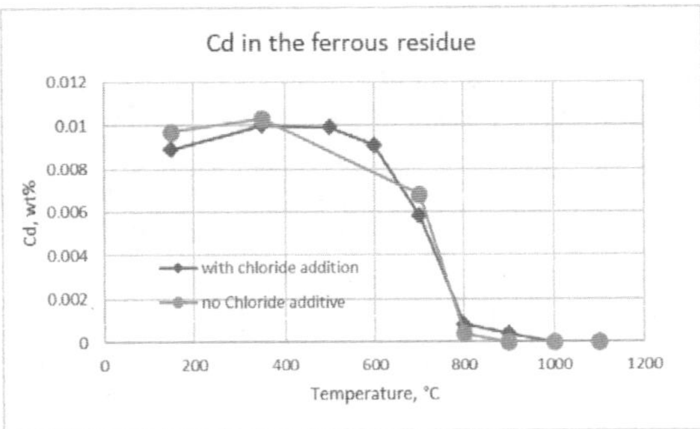

(9e): Impact of chloride addition on Cd in the residue

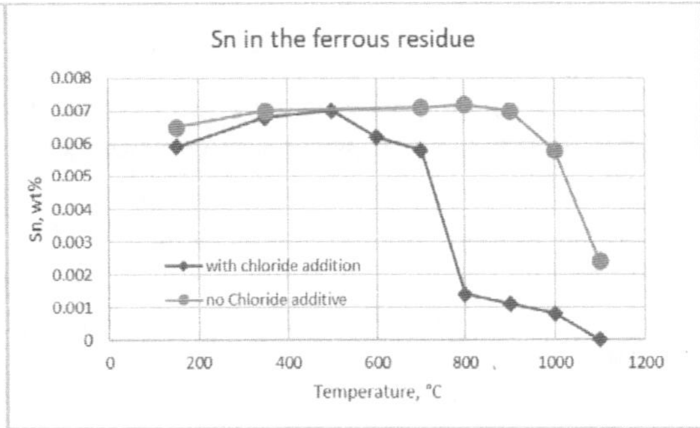

(9f): Impact of chloride addition on Sn in the residue

Figure 9: Impact of chloride addition on the composition changes of various elements in the residue

It can be seen that the samples with chloride addition have a slight higher zinc removal rate than the samples without chloride addition (Figure 9a), probably due to the presence of ZnS as thermodynamically predicted in Figure 5. For lead removal, more considerable difference can be seen as the samples with chloride addition lead to significant lead decrease in the residue starting at 500°C (Figure 9b), and have a removal efficiency of higher than 97% already at 900°C while zinc loss from the residues is insignificant (Figure 8). In contrast, the samples without chloride addition have only a lead removal efficiency of 42%, which also agrees with the results from the TGA measurements. At higher temperatures at 1100°C, the addition of chloride is no longer of influence on the removal efficiency of lead and zinc, and nearly all lead and most of the zinc have been removed from the residue.

Figures 9(c) and 9(d) show the variations of iron and vanadium in the residue with slightly increased concentrations at higher temperatures, and this is mainly due to the consumptions of carbon and the loss of heavy metals. However, some amount of iron chloride might be lost to the gas during chlorination, based on the thermodynamic assessment, but this requires further quantifications. Adding chlorides shows no impact on cadmium removal since its removal rate has already reached nearly 100% at 800°C for all samples, as shown in Figure 9(e). Figure 9(f) shows the similar behaviour of tin to that of lead, and this may create an option for Sn coated scrap recycling in the integrated steel works.

Upscaling the experiments with larger samples is essential to mimic the industrial case, especially to have a more representative composition of the generated secondary dust. In this study, the highest content of Pb in the lead-rich dust is about 30%, and Zn in zinc-rich dust is 68%. This quality is a baseline for the further utilization of the by-products from the recycling process. The ferrous residue after the thermal treatment contains very low heavy metals and can re-enter into the ironmaking route.

The metallization of iron was also confirmed by the wet-chemical analyses of the reacted samples (the residue), and the highest metallization degree of 44% was achieved for the samples without chloride addition. With increasing temperature, the iron metallization increases and stabilized after 1100 °C. The samples with chloride addition has a relative lower metallization degree. It is clear that addition of ferrous chloride from pickling liquor as source of chloride can lower the onset temperature and increase the lead removal through chlorination reactions, which is consistent with the thermodynamic simulations.

5. A flexible dual kiln route and perspectives

Compared to the hydrometallurgical routes, the pyrometallurgical processes have been used in the industry to treat zinc-bearing wastes, including blast furnace sludge. It offers a high levels of zinc and heavy metals removal through roasting, direct reduction, fuming or smelting. The pyrometallurgical route can remove zinc compounds, including the phases resistant to leaching (e.g., ZnS, $ZnFe_2O_4$), usually being a hindrance to the hydrometallurgical route [6, 9]. The direct reduction of the sludge/dust results often in good quality products for direct use in the relevant industry, consisting of an iron-rich fraction as the residue and a zinc-rich fraction as flue dust (containing heavy metals) from off-gas filtering. The high carbon content of ironmaking flue dust, especially from the blast furnace could promote a self-reducing behaviour for direct reduction without the need for reducing agent for both iron and heavy metal constituents. The limitation lies within the minimum zinc content based on the economic perspective. Some pyrometallurgical techniques require a high content of zinc input with specific input tonnage. For example, Rotary/Waelz Kiln is mainly used to recycle the dust with more than 15% of zinc, commonly contained in the EAF dust. Moreover, pyrometallurgy involving a high temperature furnace requires a substantial initial capital investment, commonly used to extract high-grade ores. Blast furnace sludge with low zinc content (1.5 - 8%) is limited to particular pyrometallurgical techniques to achieve a positive economic benefit. The established smelting-based processes consume significant amount of energy to melt the input materials entirely. Slag is then generated as a by-product that needs additional processing for utilization.

Most pyrometallurgical techniques can remove over 90% of zinc. The direct reduced iron or the iron-rich residue contains a high amount of metallic iron which can be directly sent to the sinter plant or even blast furnace after briquetting. As discussed, BF sludge contains a large amount of carbon compared to the dust from EAF and BOF steelmaking. The carbon within the dust or sludge can lead to self-reduction of the materials without the need for coke or reducing gas (CO) which has been shown effective in producing high-grade (partly) metallic iron and zinc-rich secondary dust.

Direct reduction in the rotary kiln has been well established in the industry, with a high removal rate of zinc up to 94% [10 - 13]. Around 83% of EAF dust is recycled through this process, with the capacity in the range of 100 - 350 kt/year. [10]. The typical input for Rotary Kiln or Waelz Kiln is pelletized mixture of dust, binder, pulverized coal, and water. At a temperature of 1100 to 1200°C, the reduction of metal oxides starts to occur while zinc and alkali metals will be selectively volatilized. The reactions in the Waelz kiln occur in the bottom area for reduction and upper area for oxidation, with a counter-current movement of the solid feed and injected hot air in a rotating chamber. Rotary kiln has the advantages of the relatively straightforward process of low environmental pollution and has been widely used in treating EAF dust with high zinc content (>16%) [10]. The problem of the rotary or Waelz kiln is the solid ring formation on the furnace lining due to the adhesion of elements with a low melting point as the kiln rotates. Compared to the rotary kiln mainly used for EAF dust (high zinc) processing, RHF technology has been widely used to treat BF and BOF dust with the final product of DRI containing high Fe, below 0.004% of Zn with almost no lead [10]. However it requires higher capex and has no selectivity for heavy metals.

Recycling of blast furnace sludge has been theoretically feasible and proved by the high removal efficiency of the zinc and other heavy metals *selectively* through the present experimental works. The proposed flowsheet as shown in Figure 10 is based on a dual-kiln *selective* heavy metal removal process, which launches a new recycling route with higher flexibility in waste processing. The recycling process starts with raw materials preparation using a designed recipe for pelletizing. The addition of chlorination agent together with bentonite as binder promotes heavy metal removal at lower temperatures, in particular for lead compounds. The carbochlorination reaction creates the selective removal of Pb at 900°C or below (together with Cd), before zinc is substantially removed. Thus the preliminary direct reduction takes place at 800 – 900°C in the first reactor for removing the environmental harmful elements like Pb and Cd, the dust from which requires further investigations for safe disposal. The solid residue will be transported hot to the second reactor for further processing at 1100°C to capture zinc as a valuable secondary raw material for the zinc smelter. The maximum operating temperature will be preferably lower than the softening temperature of the residue material to reduce the liquid formation. Finally, nearly heavy metals-free residue can be recycled back to the ironmaking process internally.

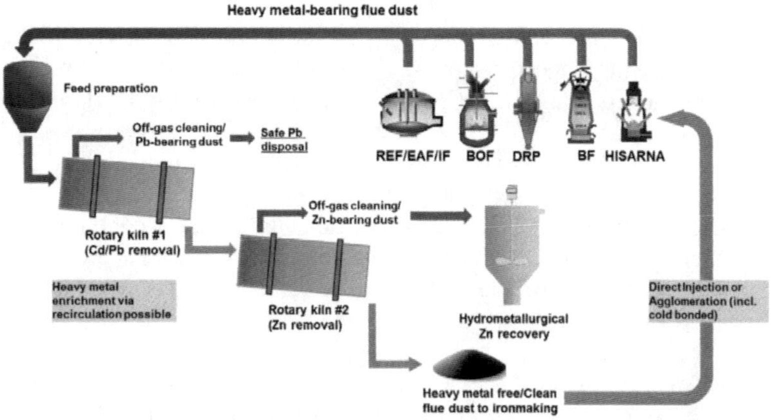

Figure 10: Proposed flowsheet for BF dust processing and heavy metal separation

The selection and design of a new process, such as the proposed dual kiln process, for waste minimization in an integrated steel works offers several benefits. One of the key advantages is the potential for a significant reduction in waste generation for external treatment or landfill. The rotary kiln process is known for its ability to efficiently process various types of waste materials, by converting the waste streams into useful resources. The dual kilns increase the flexibility and selectivity in the process operation and targets control. This minimizes the amount of waste sent to landfills, therefore reducing environmental pollution and the associated costs of waste management and disposal. Additionally, the use of this new process can enhance the overall waste management and reduce the risk of spread of environmental harmful elements. On the other hand, implementing a new technology like the dual kiln process involves substantial upfront investment and operational costs. The installation and maintenance of the required equipment along with the operational investment can be expensive and time consuming. Moreover, transition to a new green production route requires modifications to the existing infrastructure and production system, may cause potential disruptions to the existing operations and create risks to the new market situation.

It is therefore important to consider the potential risks associated with the new process, such as emission control related to the carbochlorination, the influence of the added chloride. The final processing of the heavy metal chlorides requires further investigations in the follow-up research. Careful evaluation and planning are necessary to ensure that the benefits outweigh the drawbacks and that the chosen process aligns with the long-term goals of the business and the local regulatory requirement.

6. Conclusions

The present study aims to investigate the efficiency and viability of selective removal of the heavy metals in the BF flue dust or sludge via carbochlorination reactions. Through this research, the following key findings have been obtained.

(1) Carbochlorination offers a new route to selectively remove the heavy metals from BF dust at relative lower temperatures. The experimental results reveal a high removal rate of heavy metals (Zn, Pb, Cd) at a relatively lower temperature and could produce nearly a heavy metal-free ferrous residue along with the partial self-reduction of iron oxides.

(2) Adding chloride reduces the onset reaction temperature and thus lowers the temperature of the carbochlorination reactions. High removal efficiency of lead is achieved at the temperatures of below 1000°C. The addition of chlorides improves the removal kinetics of the heavy metals, completed at 800°C (mainly for Cd and Pb), leaving a cleaner Zn-rich secondary dust and a ferrous residue nearly free of heavy metals.

(3) A high Pb/Zn ratio in the secondary flue dust during early removal stage at 800 – 900 °C has been achieved, which requires further investigation for proper processing and disposal. In

contrast, zinc is concentrated in the dust generated at a distinctively higher temperature of around 1000°C. Zinc concentration of up to 60% with very low Pb was captured in the secondary dust generated at 1100°C, which can be extracted more efficiently to close the zinc cycle.

(4) The proposed flowsheet provides a sequential double-step reduction in two inter-connected kilns or other reactors with separately attached dust capturing system, resulting in the final products of iron-rich ferrous residue and zinc-rich dust as new secondary resources. With using a chlorination agent such as $FeCl_2$, the selective removal of lead and cadmium (even tin) over zinc at lower temperatures of 800 - 900°C create a strong business case for easier zinc recovery as secondary resource.

In conclusion, the recycling of blast furnace sludge with a novel double-step direct reduction and carbochlorination is a favourable and attractive route. The developed process could be implemented with low risks, subject to a further economic feasibility study.

References

[1] World Steel Association, retrieved June 16, 2023 from www.worldsteel.org.

[2] The European Steel Association, retrieved March 18, 2019 from www.eurofer.eu.

[3] Honingh S, van Weert G. and Reuter M A 2000, *Fourth Intl. Symposium on Recycling of Metals and Engineered Materials*, edited by D.L. Steward et al., TMS, 313-327.

[4] Meijer K, Borlee J, Skorianz M, Feilmayr C, Goedert P and Dry R 2015 *METEC-ESTAD*. Duesseldorf.

[5] ReclaMet, retrieved July 22, 2022 from https://eitrawmaterials.eu/project/reclamet.

[6] Peeters T, Xiao Y, Yang Y. 2018 Process for selective removal of zinc from metallurgical plant waste. European Patent EP 3333272 A1.

[7] Peek E 2011 *Chloride Metallurgy: Process Technology Development*, EPD Congress John Wiley and Sons, Ltd, 467–501.

[8] Hamann C, Spanka M, Stolle D, Auer G, Weingart E, Al-Sabbagh D, Ostermann M and Adam C 2021 *Journal of Hazardous Materials* **402** 123511.

[9] Binnemans K, Jones P T, Manjon Fernández A and Masaguer Torres V 2020 *Journal of Sustainable Metallurgy* **6** 505– 540.

[10] Wang J, Zhang Y, Cui K, Fu T, Hussain S and AlGarni T S 2021 *Journal of Cleaner Production* **298** 126788.

[11] Ye L et al. 2021 *Waste Management* **135** 389–396.

[12] Zhu D, Wang D, Pan J, Tian H and Xue Y 2021 *Powder Technology* **380** 273–281.

[13] Jiang L, Li P, Du C and Yu J 2022 *China Nonferrous Metallurgy* **51** 81 – 90.

Production increase of high-grade crude zinc oxide pellets at Shisaka Smelting Co., Ltd.

S. Egawa, S. Takaya, K. Yamamoto and H. Fujiyama

SHISAKA SMELTING CO., LTD. 3-5-3 Nishibara-cho, Niihama, Ehime 792-0011 Japan

satoshi.egawa.y5@smm-g.com

Abstract. Shisaka Smelting Co., Ltd. has recovered crude zinc oxide pellets from an electric arc furnace dust (EAF dust) by the Waelz kiln process since 1977. The EAF dust processing line consists of a reduction roasting process, a Waelz oxide washing process, and a drying/calcination process. In late years, about 90,000 tons of EAF dust is processed annually at Shisaka. The recovered zinc is sold to zinc smelters as crude zinc oxide pellets, and the amount of production of the pellets is about 40,000 tons per year. Two types of zinc oxide pellets are produced, and the product for electro-refinery requires high quality, especially in fluorine reduction. Most of fluorine are distributed to the clinker in a reduction roasting process, and others are distributed to the Waelz oxide. Conventionally, high-grade crude zinc oxide pellets were produced by washing the Waelz oxide and by volatilizing and separating fluorine under high temperature operation in a drying rotary kiln. However, formation of accretion in the kiln and damage of the shell and refractories of the kiln often occurred under the influence of high temperature operation, and stable consecutive operation of the kiln was difficult.

To improve this situation, the fluorine input was reduced by compounding management of EAF dust in the reduction roasting process, and the material residence time in the drying rotary kiln was extended by reducing rotary speed. These improvements resulted in stable consecutive operation under low temperature conditions and increased production of high-grade crude zinc oxide pellets. To improve this situation, the fluorine input was reduced by compounding management of EAF dust in the reduction roasting process, and the material residence time in the drying rotary kiln was extended by reducing rotary speed. These improvements resulted in stable consecutive operation under low temperature conditions and increased production of high-grade crude zinc oxide pellets.

1. Introduction

Recently, climate change due to global warming has been recognized and carbon neutrality which reduces greenhouse gas emissions to zero, has been attracting attention to prevent from the global warming. In this situation, steel production of electric arc furnace (EAF) makes it advantageous for carbon neutrality compared to the blast furnace (BF), therefore, shift from BF to EAF is in full swing in Japan. Steel scrap is melted in EAF, and generated dust is called EAF dust, which contains 10 to 30% Zn because steel scrap includes galvanized steel sheets.

EAF dust contains lead and cadmium which recognized as hazardous metal, so its treatment and disposal are strictly regulated. However, many countries and regions still do not have sufficient facilities to treat EAF dust, and the dust is landfilled or accumulated without treatment for zinc recovery and recycling. Therefore, it is necessary to promote stable treatment of EAF dust by developing treatment technologies [1].

EAF dust treatment process is divided into pyrometallurgical and hydrometallurgical processes. The Waelz process using a rotary kiln is the main pyrometallurgical process, and another main process is rotary heath furnace (RHF) process. The ammonium chloride leaching process is used as a hydrometallurgical process [2,3].

In 1977, Shisaka Smelting Co., Ltd. of the Sumitomo Metal Mining (SMM) Group started zinc and lead recycling operations using the Waelz process to recover zinc and lead from dust generated in EAF dust as crude zinc oxide (pellets) for use in an ISP smelter. Figure 1 shows the EAF dust process chart at our plant. This process consists of a reduction, washing and dewatering, and drying process. Figure 2 shows the changes in amount of treated EAF dust at our plant. We started operations in 1977 using a system with one Reduction Rotary Kiln (RRK), washing facilities, and one Drying Rotary Kiln (DRK). At that time, the EAF dust treatment capacity was 50,000 tons per year [4]. And its capacity was increased to 120,000 tons per year by installing another RRK in 1992 [5]. In 2004, we modified the washing facilities to improve crude zinc oxide pellets production capacity to 50,000 tons per year, which enabled EAF dust treatment capacity of 120,000 tons per year. Recently, approximately 90,000 tons of EAF dust is treated per year.

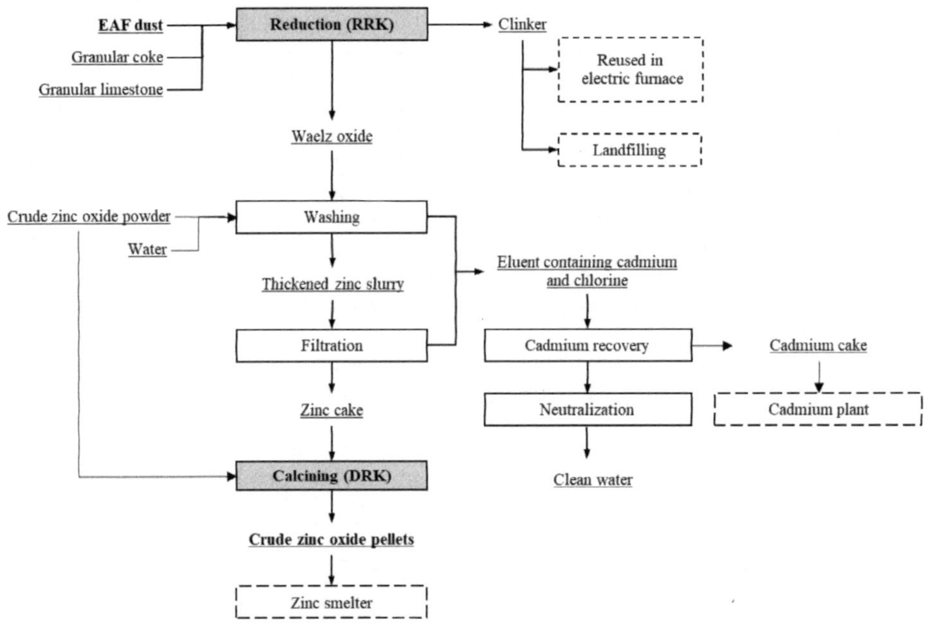

Fig.1 Process chart at Shisaka smelter

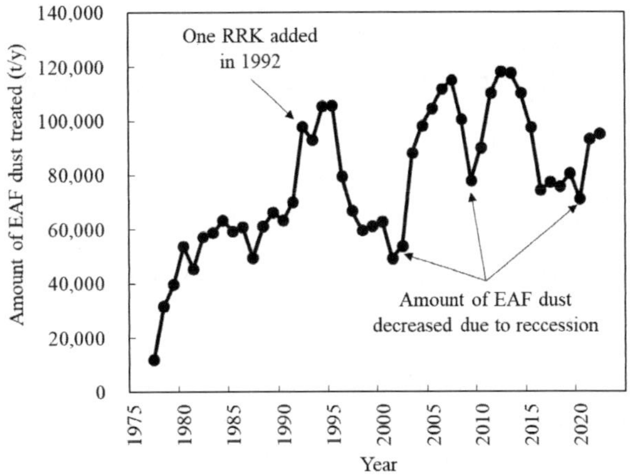

Fig.2 Changes in amount of treated EAF dust

2. Production of high-grade crude zinc oxide pellets

The recovered zinc from EAF dust is sold to zinc smelters as crude zinc oxide pellets, and there are two types of crude zinc oxide pellets marketed: one type is for ISP and the other type is for electrolytic smelting. The latter is called high-grade crude zinc oxide pellets, which are high valued as containing less than one-tenth of fluorine and chlorine compared with crude zinc oxide pellets for ISP. Therefore, increased production of high-grade crude zinc oxide pellets leads to an expansion of sales channels and improvement in profit margins.

In order to reduce the fluorine content of high-grade crude zinc pellets less than one-tenth of crude zinc oxide pellets for ISP, additional operation control is implemented as below.

- Controlling fluorine content by blending EAF dust
- Increasing in volatilization rate of fluorine due to high temperature operation of the DRK

However, there were two problems for production increase. One is that fluorine in the crude zinc oxide pellets frequently exceeds the standard value. Another is that operation rate is decreased due to formation of accretion and damage of refractories in high temperature operation. This report describes the improvement of those problems by reforming the blending method and lowering temperature of the DRK, which were implemented to stably increase the production of high-grade crude zinc oxide pellets.

3. Blending F/Zn control

Three kinds of EAF dust are blended to keep zinc content constant and fed to RRK. The amount of the Waelz oxide produced from the RRK is proportional to the zinc content of blended EAF dust.

During the production campaign of high-grade crude zinc oxide pellets, we focused only on the blended fluorine content of EAF dust to satisfy the fluorine content of the pellets. However, when the blended zinc content of EAF dust was low, the proportion of fluorine content of Waelz oxide is increased. Although fluorine in the Waelz oxide is removed in the DRK, producing high fluorine content of the Waelz oxide leads to lowering product yield rate. Therefore, we set F/Zn ratio as a new parameter to keep the fluorine content of the Waelz oxide below to a fixed value stably.

4. Extension of residence time in the kiln

To prevent damage of the refractories and formation of accretion, which cause decreasing operation rate, we have developed an operation method that lowers the operating temperature of the DRK while keeping the amount of volatilized fluorine.

Low operating temperature causes the reaction rate of fluorine volatilization decreases and the volume of fluorine volatilization becomes low. Komori et al [6] showed that the volatilization behaviour of fluorine depended on the calcining temperature and the volatilization rate of fluorine declined by about from 15 to 30% as the temperature decreases by 100 °C in the range of from 950 to 1,150 °C. Based on these findings, we investigated the residence time of kiln to keep the volatilization rate of fluorine.

5. Results and discussion

5.1. Blending F/Zn control

Figure 3 shows the changes in fluorine content of the Waelz oxide and blending F/Zn ratios. The F/Zn ratio varied in the range of 0.02 to 0.03 before April 2021, because we controlled only fluorine content of EAF dust. Therefore, since June 2021, the F/Zn ratios have been controlled below to 0.013, and fluorine content of the Waelz oxide has been also below 0.2%. Figure 4 shows the fluorine content of the pellets before and after the change of control. After controlling blending F/Zn ratios, the fluorine content of the pellets has been decreased, and product yield rate has been improved.

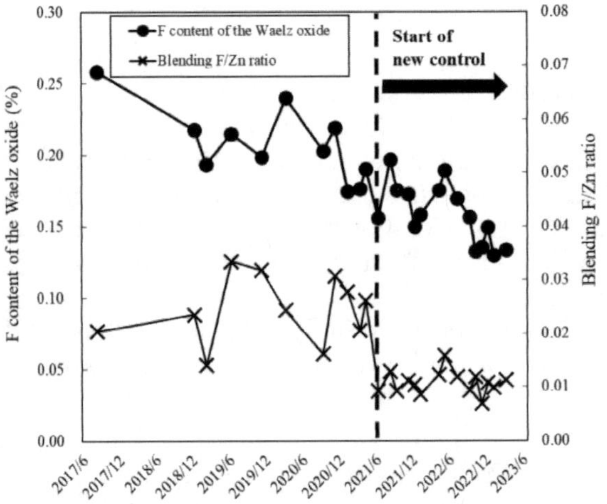

Fig.3 Changes in fluorine content of the Waelz oxide and blending F/Zn ratios

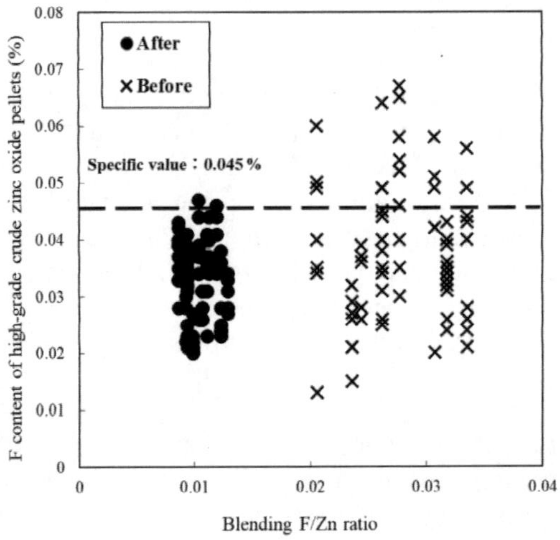

Fig.4 Fluorine content of the pellets before and after the change of control

5.2. Extension of residence time in the kiln

Lowering the operating temperature decreases the volatilization rate of fluorine and increases fluorine content of the pellets as formula (1) and Arrhenius formula (2). Previous study [6] showed that reducing the operating temperature by approximately 100 °C from 900-1,000 °C to 800-900 °C decreased the volatilization rate in the range of approximately 15-30 %.

To avoid the decrease of the volume of fluorine volatilization, the residence time in DRK needs to be extended. The residence time was obtained by Sullivan's formula (3) [7] and is inverse proportion to the rotary speed, so the residence time was extended by reducing the rotation speed. Under low temperature conditions, the volatilization rate of fluorine was reduced by up to 30%, the residence time was extended by 40% to keep a same volatilization rate.

$$V = \frac{dC}{dt} = -kC^{n} \quad (1)$$

V : reaction rate(mol/s), C : concentration(mol), t : time(s), k : reaction rate coefficient, n : the order of reaction.

$$k = \mathrm{A}\exp\left(-\frac{\mathrm{E}}{\mathrm{R}T}\right) \quad (2)$$

k : the reaction rate coefficient, A : the pre-exponential factor, E : the activation energy, R : the gas constant, T : the absolute temperature.

$$\mathrm{T} = \frac{0.03102\,\mathrm{L}\,\beta^{0.5}}{\mathrm{D\,S\,N}} \quad (3)$$

T : the residence time, L : the length, β : the angle of repose, D : the diameter, S : the inclination, N : the rotation speed.

Figure 5 shows the relationship between the calculated residence time and fluorine volatilization rate at each temperature. By extending the residence time from 70-90 minutes (rotation speed: from 0.8 to 1.0 rpm) to 100-130 minutes (rotation speed: from 0.5 to 0.6 rpm), the fluorine volatilization rate could be maintained at 90-95%, which is equivalent to high-temperature operation. The operation campaign was extended from 4 days to 10 days or more without serious formation of accretion or damage of refractories.

Fig.5 DRK residence time vs. fluorine volatilization rate

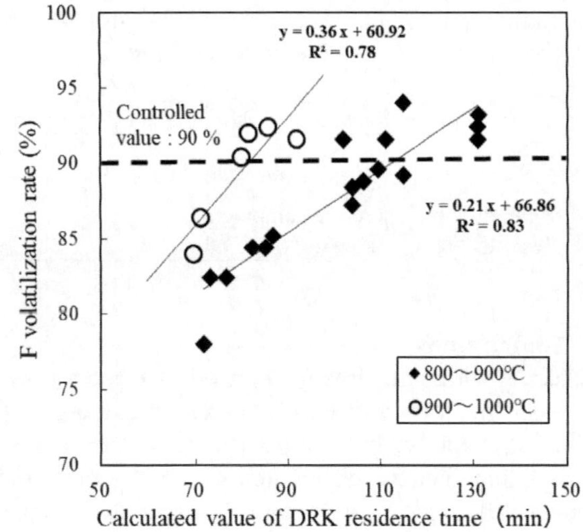

Table 1 shows the volatilization rate of fluorine. Lowering the operating temperature resulted in approximately 35% decrease, which close to expected value.

Table 1 Volatilization rate of fluorine				
Temperature	°C	900 ~ 1,000	800 ~ 900	Δ 100
Reaction rate	mol/s× 10^{-2}	8.92	5.85	Δ 3.07 (Δ 35%)

5.3. *Production increase of high-grade zinc oxide pellets*

Figure 6 shows the amount of high-grade crude zinc oxide pellets production. After 2021, annual production increased 4,000 tons per year from 2,000 to 6,000 tons per year without any formation of accretion and damages to refractories during campaign.

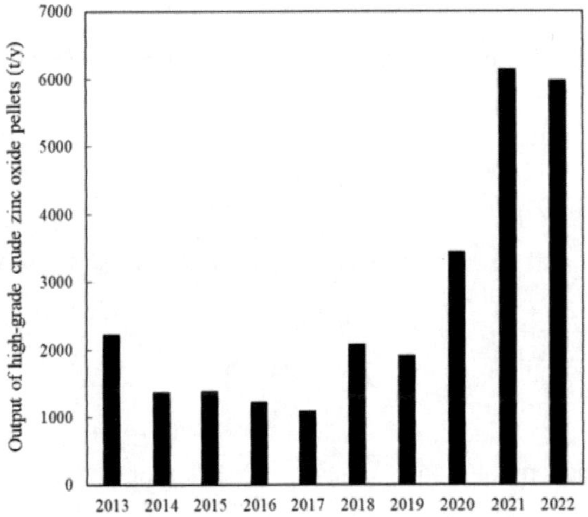

Fig.6 The amount of high-grade crude zinc oxide pellets production

Table 2 shows the production increasing impact of high-grade crude zinc oxide pellets. By controlling the blending F/Zn ratio, the variation of the fluorine content of high-grade crude zinc oxide pellets was reduced and the product yield rate was improved. About 920 tons per year of high-grade crude zinc oxide pellets which was approximately 23 % of the production increase was achieved. On the other hand, extending the operation campaign by lowering operating temperature and extending the residence time in the kiln about 3,080 tons per year of high-grade crude zinc oxide pellets which was approximately 77 % of the production increase was achieved.

Table 2. The production increasing impact of high-grade crude zinc oxide pellets		
Effort	Amount of increase (t)	Contribution ratio (%)
Controlling the blending F/Zn of EAF dust	920	23
Extending the operation campaign	3,080	77
Total	4,000	100

6. Conclusions

Shisaka Smelting has greatly improved the technology used in its zinc recycling over the 45 years since commencement of EAF dust treatment operation. The high-grade crude zinc oxide pellets production technology reported here, in particular, has been indispensable for continuing stable treatment of EAF dust, leading to increased production of high-grade crude zinc oxide pellets as well as expansion of sales channels and improvement in profit margins. In Japan, as moves are made toward carbon neutrality, there is an ongoing shift from blast furnaces to electric furnaces as a means of iron and steel production, and it is expected that the amount of EAF dust generated will increase. We will continue to strive to make various improvements for stable treatment of EAF dust and stable supply of pellets and move forward with even more effective use of resources.

References

[1] S. Chikujou March 2021 *The Global Trend of Zinc Recycling* Metal Economics Research Inst. Japan
[2] L. Southwick January 2015 *Zinc Recovery from Electric Arc Furnace (EAF) Dust Worldwide Survey* International Lead and Zinc Study Group
[3] T. Nakamura, E. Shibata, T. Takasu and H. Itou 2008 *Resources Processing* **55** 144-148
[4] M.Yasukawa and H.Kuwata 1981 *J.MMIJ.* **97** 820-821
[5] M.Takewaki and Y.Ojima 1993 *J.MMIJ.* **109** 1112-1117
[6] S.Komori, H.Fujiyama, H.Watanabe and K.Tomoda 2015 *Proc. Pb-Zn 2015* 831-842
[7] J D.Sullivan, C G.Maier and O C. Ralston 1927 *Bur. Mines Tech. Paper* 384

Development and application of matte/speiss/metal thermodynamic database for optimization of processing of drosses, dusts and reverts from lead, zinc and copper production

Denis Shishin*[,1], Maksym Shevchenko[1], Roman Starykh[1], Svetlana Sineva[1], Viktoria Prostakova[1] and Evgueni Jak[1]

Pyrometallurgy Innovation Laboratory (PYROSEARCH), School of Chemical Engineering, University of Queensland, Brisbane, Queensland, Australia, 4072

d.shishin@uq.edu.au

Abstract. Detrimental elements such as arsenic and antimony tend to accumulate in dusts and drosses of lead-zinc as well as copper smelters. These by-products are commonly treated in dedicated reductive smelting units producing Cu-rich matte, As-Sb-collecting speiss and Pb-rich metal, as well as liquid slag. Such process would have a complex distribution of elements (Cu, Fe, Pb, Zn, Ni, Sn, Sb, As, Ag, Au, S, etc.) among four liquid phases. Thermodynamic calculations can be used for predictions and optimization of such processes. They must rely on accurate models developed in integration with experimental program. Present study reports results of recent progress in experimental and modelling studies of As, Sb-containing speiss systems and demonstration of calculations relevant for industrial conditions. Experimental methodology consisted of equilibration, quenching and electron probe X-ray microanalysis. Calculations were performed using FactSage software and thermodynamic database developed based on the dedicated recent experiments. The main findings of the paper are: a set of binary and ternary diagrams relevant to matte and speiss formation conditions; distribution of Ag and Au among liquid Pb and speiss in key quaternary systems; and distributions of 10 main elements for industrially relevant matte/speiss/metal multicomponent equilibria.

1. Introduction

Modern lead and zinc pyrometallurgical production sites need to adapt to multiple, often conflicting, trends. All countries worldwide implement stricter environmental regulations. In Europe, Australia, and the USA, the perception of production sites as "polluting" has led to public pressure on both industries and governments. Production in these countries stagnated, and no major investment in metallurgical infrastructure was observed in 2010-2020 [1, 2]. In 2022, energy crisis in Europe lead to several shutdowns. As a consequence, metallurgical research and education in these countries have been underfunded, leading to reduced number of students [3]. At the same time, the global shift towards renewable energy, if taken seriously, would require a substantial increase in base metal production, including lead and zinc [4]. Hence, lower quality ores containing higher amounts of impurities such as arsenic and antimony would need to be incorporated into the production streams, posing challenges to traditional lead, zinc and copper processes. The signs of de-globalization have prompted governments to compile lists of critical metals or minerals. These lists are intended to secure manufacturing capability and often include antimony, bismuth, zinc, nickel, indium, and others, but not lead or copper [5]. Metallurgy experts recognize that the recovery of "critical" elements from primary and secondary sources would heavily rely on lead, zinc, and copper pyrometallurgical infrastructure [6]. However, accepting this fact requires a fundamental shift in the mindset of policymakers.

Several solutions addressing these challenges have been proposed. First, an expansion of end-user product recycling into primary metals production, often referred to as closing the gap in the circular economy [6]. Second, a closer integration among lead, zinc, and copper primary and recycling processes with an exchange of "waste streams" [7]. If implemented, this approach will enhance the recovery of all valuable metals and reduce the environmental impact of their disposal. The concept of such polymetallic integrated plant is shown in Figure 1 and was largely inspired by Alvear et al. [8], who recently reviewed existing polymetallic plants. The challenges of logistics and transport costs are evident, but they do not appear insurmountable, considering that nowadays many concentrates with relatively low contents of valuable metals travel thousands of kilometers from miners to smelters. Economics of the process can be evaluated and optimized using digital twins of metallurgical infrastructure [9].

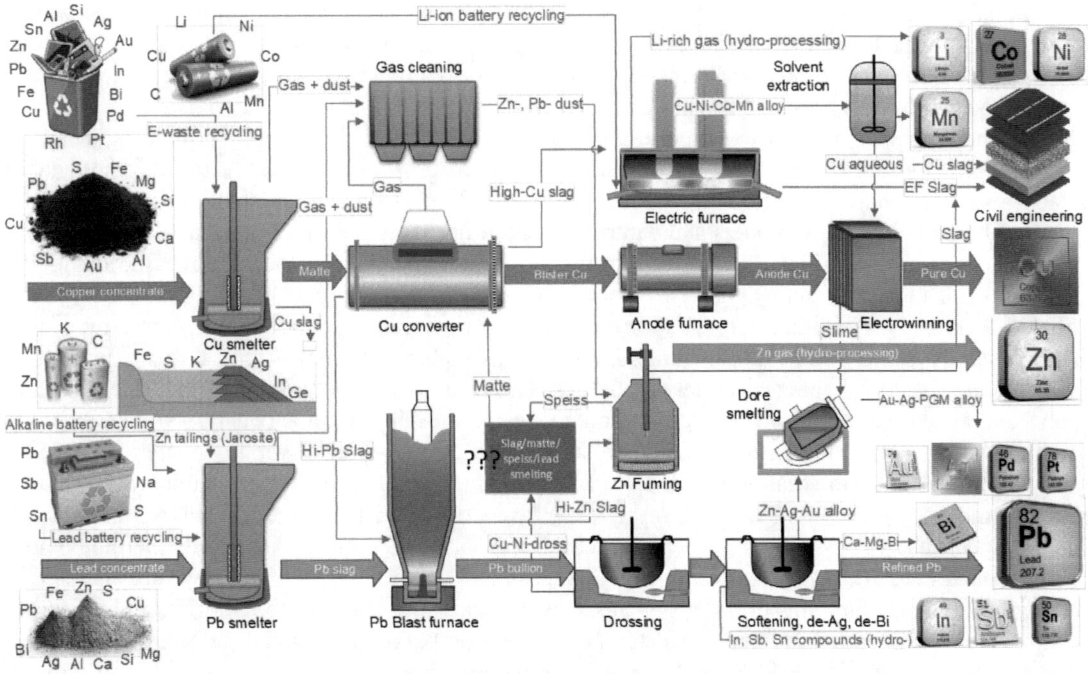

Figure 1. The concept of future integrated polymetallic plant. Flowsheet configuration separating slag/matte/speiss/metal in the center is to be determined in future.

As shown in Figure 1, the combination of process units, smelting and separating the intermediate products among the slag, matte, speiss, metal phases would play an important role in integrated polymetallic plant. The "Zn-, Pb- dust" stream from gas cleaning systems would inevitably contain high concentrations of arsenic and copper [10]. These must be removed from the main "Pb bullion" stream. The "Cu-Ni-dross" can be rich in Sb, Sn and S. Elements of slag/metal, matte/metal or speiss/metal separation exists in many lead smelters, and the chemistry is expected to grow more and more complex in future. The final flowsheet can only be developed by process engineers on-site because the compositions of streams are highly confidential.

The role of PYROSEARCH group is to provide fundamental information on high temperature chemistry, which is being used by the sponsor companies to simulate the individual process units. The research is based on integrated experimental and thermodynamic modeling studies. As a product, we provide computational tools capable of predicting the conditions of formation of phases, the distribution of all elements among these phases, and energy balance for the process. The present study reports some recent modeling and experimental results limited to the chemistry within the Cu-Fe-Pb-Zn-Ni-Sn-Sb-As-Ag-Au-S system.

2. Development of the thermodynamic database

Thermodynamic calculations of the present study are based on Gibbs energy minimization routine included in FactSage software [11]. FactSage software also contains programmed mathematical expressions for the Gibbs energies of complex solution phases. These expressions in the form of $G = f(T, P, x_1..x_n)$ are often referred to as thermodynamic models. In thermodynamic models, T is temperature, P is total pressure, and $x_1...x_n$ are compositional variables. FactSage software was tested to perform reasonably fast and reliable calculations for 20 elements, dozens of potential phases, and thousands of bond fractions within the solution phases. Thermodynamic models contain adjustable parameters, which must be found from experiments, or in rare cases, from the first-principal methods, i.e. statistical thermodynamics or quantum mechanics. Examples of model parameters are enthalpy of formation, entropy, and heat capacity expressions ($\Delta H_f°$, S°, C_P) for solution components, as well as energy parameters for the interactions between solution components. Thermodynamic models for possible phases together with a set of model parameters form a thermodynamic database. A portion of the database used in the present study is shown schematically in Table 1.

The modeling methodology used in the present study is based on CALPHAD principles. They can be formulated as follows: for a given chemical system (a group of elements), the thermodynamic database should contain a single set of model parameters for all potentially stable phases within the system. The combination of models and parameters should provide the best description different types of available experimental data, including information on phase equilibria, activity measurements, the distribution of elements among phases, calorimetry, and crystallography. In PYROSEARCH, these principles have been further expanded to proactively integrate thermodynamic modelling with experiments. In cases when experimental data are not available or contradictory, target experimental series are designed to fix certain model parameters. The results of these experiments are then included in the overall dataset and used to optimize all model parameters simultaneously. To determine the conditions of experiments, thermodynamic calculations with intermediate version of the database are used, which substitutes the trial-and-error approach. Still, the combinatorial nature of interactions among elements means that the number of sub-systems to be investigated using the CALPHAD approach is huge. For instance, the 11-composnent system from Table 1 requires complete thermodynamic assessment of 11!/2!/(11-2)! = 55 binary subsystems and 11!/3!/(11-3)! = 165 ternary subsystems. The number of quaternary and higher order systems is even higher, but they may be prioritized. The models in the present study only use binary and ternary interaction parameters, which reflect the physics of interactions on atomic level. Still, multicomponent experiments are necessary to verify and improve binary and ternary parameters. The following sections contain examples of thermodynamic modelling and experimental results.

The Modified Quasichemical Model (MQM) [12-15] used in the present study, a collaborative effort between CRCT (Montreal, Canada) and GTT (Germany). It has demonstrated its ability to describe the thermodynamics of many chemical systems, binary, ternary and of more components. Over the years, the model has undergone several improvements, as detailed in recent publications [16]. These enhancements now offer greater flexibility in choosing constituents, formulating expressions for excess Gibbs energy, and interpolating binary parameters into ternary and higher-order systems. Notably, other established models, such as the Bragg-Williams ideal solution and the associate solution model can be reproduced within the MQM framework. Notable, the MQM was used not only for the liquid phase, but also for a number of solid solutions, as shown in Table 1.

Table 1. A portion of thermodynamic database used in the present study for the calculations within the Cu-Fe-Pb-Zn-Ni-Sn-Sb-As-Ag-Au-S system.

Phase name	Site occupancy	Model (formalism)
Liquid matte/metal/speiss	(Cu^{I}, Cu^{II}, Fe^{II}, Fe^{III}, Pb^{II}, Zn^{II}, Ni^{II}, Sn^{II}, Sb^{III}, As^{III}, Ag^{I}, Au^{I}, S^{II})	Modified Quasichemical Model in Pair Approximation (MQMPA)
fcc and bcc solids alloys	(Fe, Cu, Ni, S, Pb, Zn, As, Sn, Sb, Ag, Au)	Bragg-Williams (B-W) Formalism
Pyrrhotite	(Fe, Cu, Ni, Vacancy)(As, S)	Compound Energy Formalism (CEF)
Pyrite (MeS2):	(Fe, Cu, Ni)$(S)_2$	CEF
Digenite-bornite	(Cu_2S, FeS, PbS, ZnS, Ni_2S, Ag_2S, $Vacancy_2S$),	B-W
ORTH	[Fe, Ni]$[As, Sb, S]_2$	CEF
GAMM Cu3X	(Cu, Fe, Ni, Sb, Sn, Ag, Au)	MQMQA
Beta-Cu3X	(Cu, Ni, Sb, Sn, Au)	MQMQA
Cu3As	(Cu^{I}, As^{III}, Ag^{I}, Au^{I})	MQMQA
HEX MeX_	(Fe, Ni, As, Sb, Sn, Cu, S, Au)	MQMQA
Beta-Ni3S2	$(Fe, Ni, Cu, Vacancy)_2S$	CEF
Gersdorfite	(Ni)$(As, S)_2$	CEF
Sphalerite	(Zn, Fe)S	B-W
Wurtzite	(Zn, Fe)S	B-W
$(Cu, Ni)_7As_3$	$(Cu, Ni)_7As_3$	B-W
Villamaninite	$(Cu, Ni, Fe)S_2$	B-W
Millerite	(NiS, CuS)	B-W
>100 solid phases	As_2S_3, Fe_2As, CoAs, Cu_3AsS_4, $ZnAs_2$, $FeSn_2$, ZnSb, Sn_4Sb_3, etc.	Stoichiometric compounds
>50 gaseous species	S_2, As_2, AsS, Zn, ZnS, AgS, Pb, PbS, Sn, SnS, Ag, AgS, Sb, Sb_2, SbS, etc.	Ideal gas

2.1. Phase diagrams of key binary systems

Literature search has been conducted for all the binary sub-systems within Table 1. The available experimental results or thermodynamic assessments have been analyzed and used to establish model parameters. Examples of publications describing the optimization of model parameters are available for some systems: Pb-S [17], As-S [18], Cu-As [19], Pb-As [20].

A summary of binary systems important for the matte formation is given in Figure 2 and Figure 3. The term “matte” is typically used for the molten sulfide-based material. In Figure 1, liquid matte is an important product in “Cu smelter” and “Cu converter” units. Copper is a “chalcogenic” element, which means it has strong interaction with sulfur. When a significant amount of copper enters lead production stream, it is separated in a form of liquid matte [21] or solid copper sulfides. In Figure 1, the formation of liquid matte is expected in “Pb blast furnace”, but separation will be conducted in continuous or batch-type “Drossing” units. As can be seen in Figure 2, there a is miscibility gap between metal and sulfide liquids in the Cu-S system. Iron, nickel and lead do not form a miscibility gap with sulfide, which means metal and matte phases generally must be described using a single liquid solution thermodynamic model. Other elements, such as tin, antimony, or silver do form miscibility gaps with sulfides (Figure 3). Arsenic and zinc are very volatile. The metallic phases in the As-S and Zn-S systems can only be observed if high external pressure is applied. Solid zinc sulfide has a remarkably high melting point.

Figure 4, Figure 5 and Figure 6 show important binary systems to understand the “speiss” phases. Speiss can be defined as a series of solid compounds and liquids that are formed due to the strong interaction between d-metals (Fe, Co, Cu, Ni, Zn) and metalloids (As, Sb) [22]. Tin (Sn) is sometimes also included as speiss-forming component because of high-melting solid phases with Cu, and particularly with Ni. No liquid miscibility gaps are observed, except for the Fe-Sn system. Metallic lead

Journal of Physics: Conference Series **2738** (2024) 012032 doi:10.1088/1742-6596/2738/1/012032

does not form high melting phases or miscibility gaps with As, Sb or Sn (Figure 6). Thus, from the modelling point of view, liquid metal and liquid speiss again must be described as a single solution. Historically, the word speiss was first used in Germany to characterize some of the products of the "drossing" process of lead bullion. These phases mainly consist of Cu-As, Cu-Sb, Fe-As, Ni-As, or Ni-Sb compounds [23]. The formation of a solid speiss crust on top of molten lead is a key step in lead fire refining after first copper drossing, but it also may negatively interfere with the separation of copper in a form of liquid matte [24]. Typically, solid speiss compounds are considered as undesirable by-products and very few companies are willing to treat them to extract metals. Other areas of potential speiss formation in Figure 1 are: Fe-As and Zn-As compounds in the products of corrosion of steel jackets in the "Pb blast furnace" [25], droplets of Cu-As, Cu-Sb, and Fe-As in the slag of "Zn fuming [26].

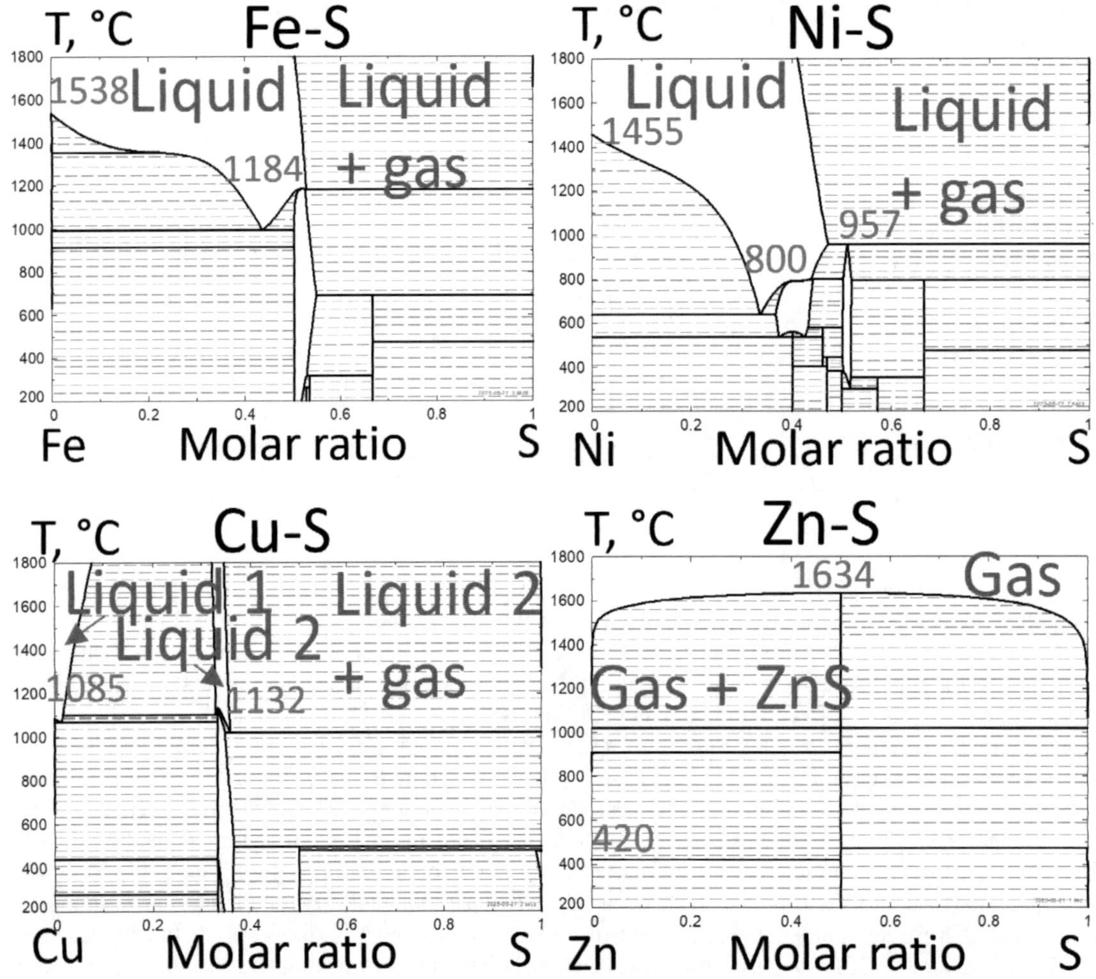

Figure 2. Phase diagrams of matte forming binary systems at 1 atm total pressure: Fe-S, Ni-S, Cu-S, Zn-S. Lines are calculated using thermodynamic database of the present study based on assessment of experimental data.

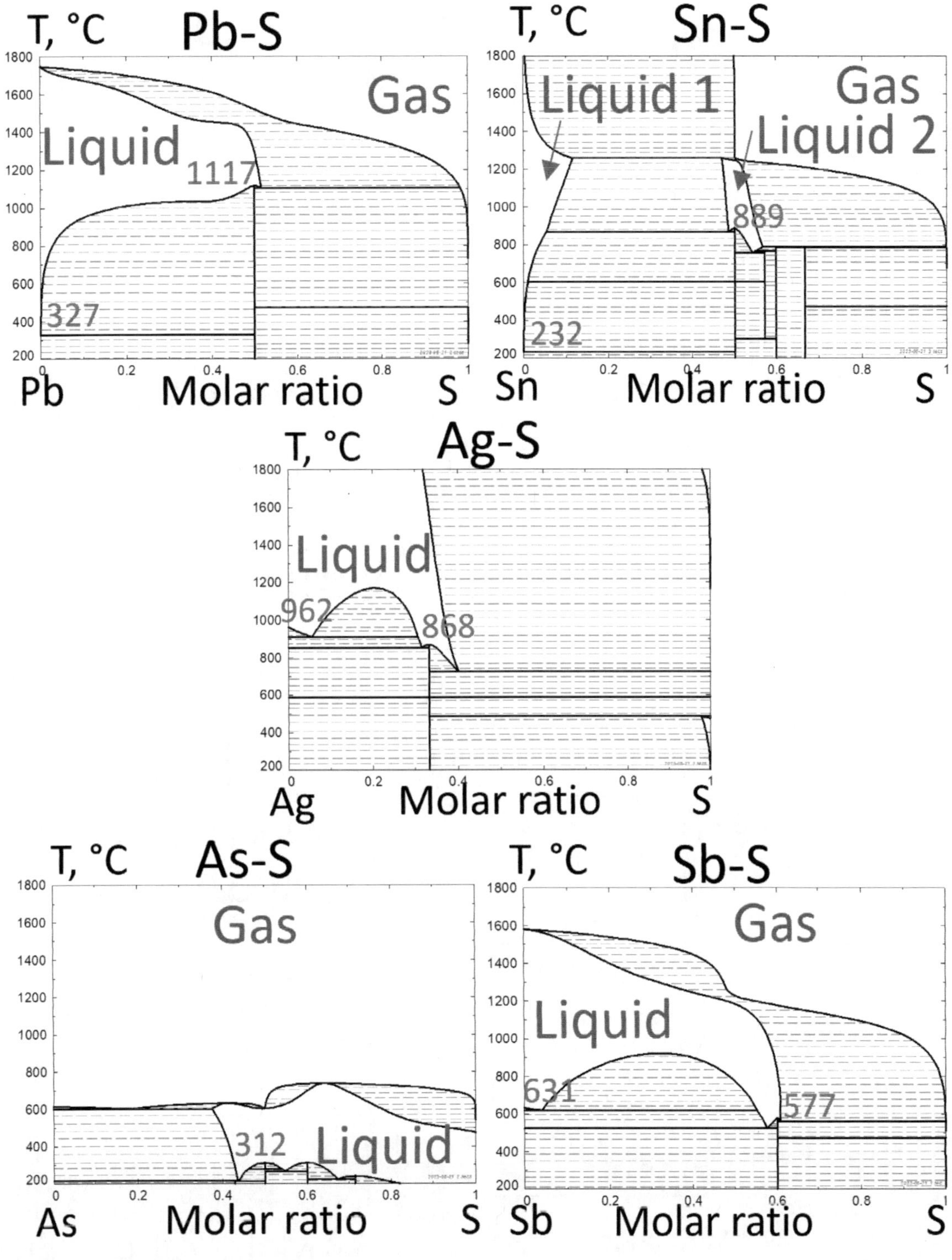

Figure 3. Phase diagrams of matte forming binary systems at 1 atm total pressure: Pb-S, Sn-S, Ag-S, As-S and Sb-S. Lines are calculated using thermodynamic database of the present study based on assessment of experimental data.

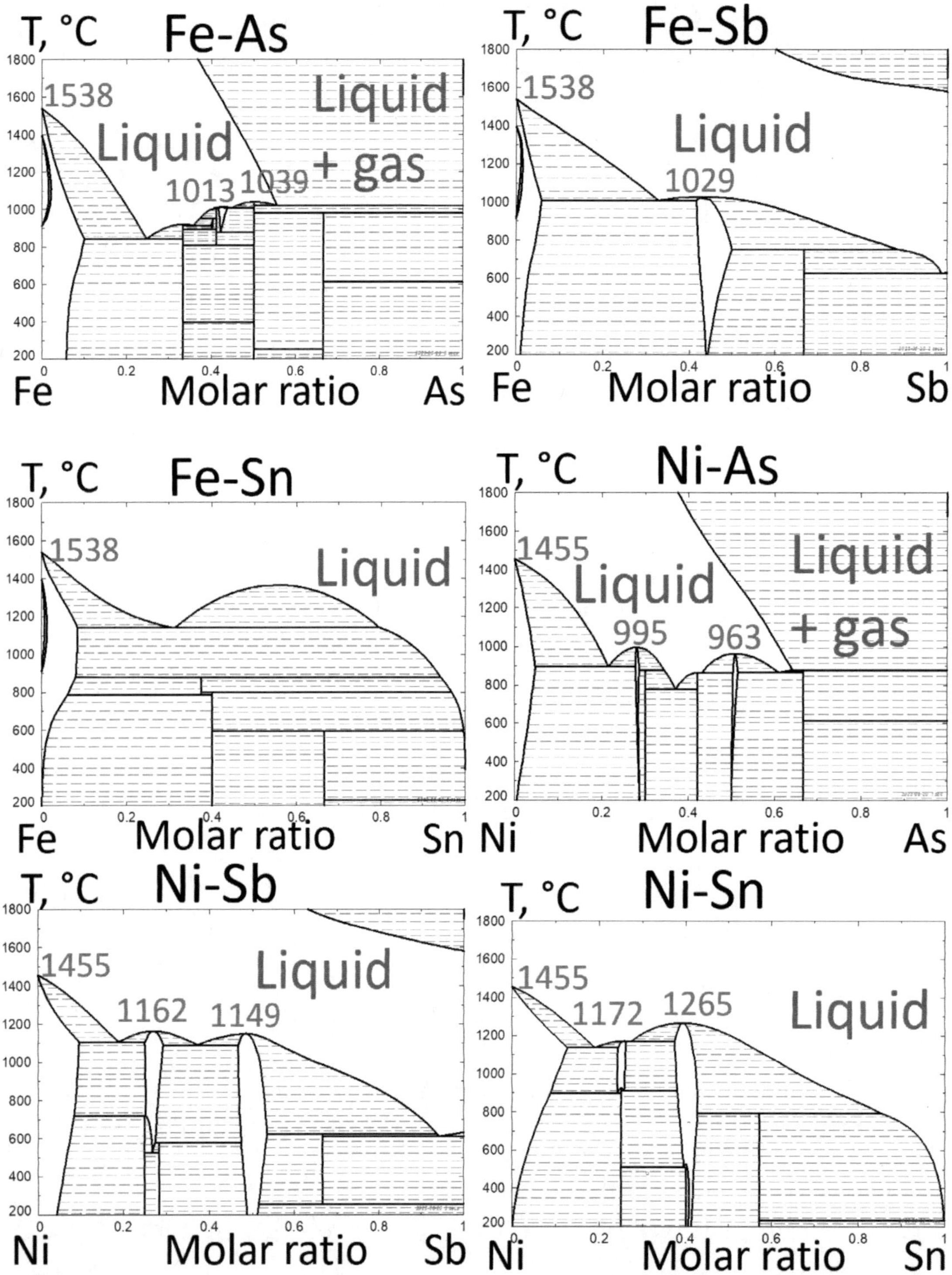

Figure 4. Phase diagrams of speiss forming binary systems at 1 atm total pressure: Fe-As, Fe-Sb, Fe-Sn, Ni-As, Ni-Sb, Ni-Sb. Lines are calculated using thermodynamic database of the present study based on assessment of experimental data.

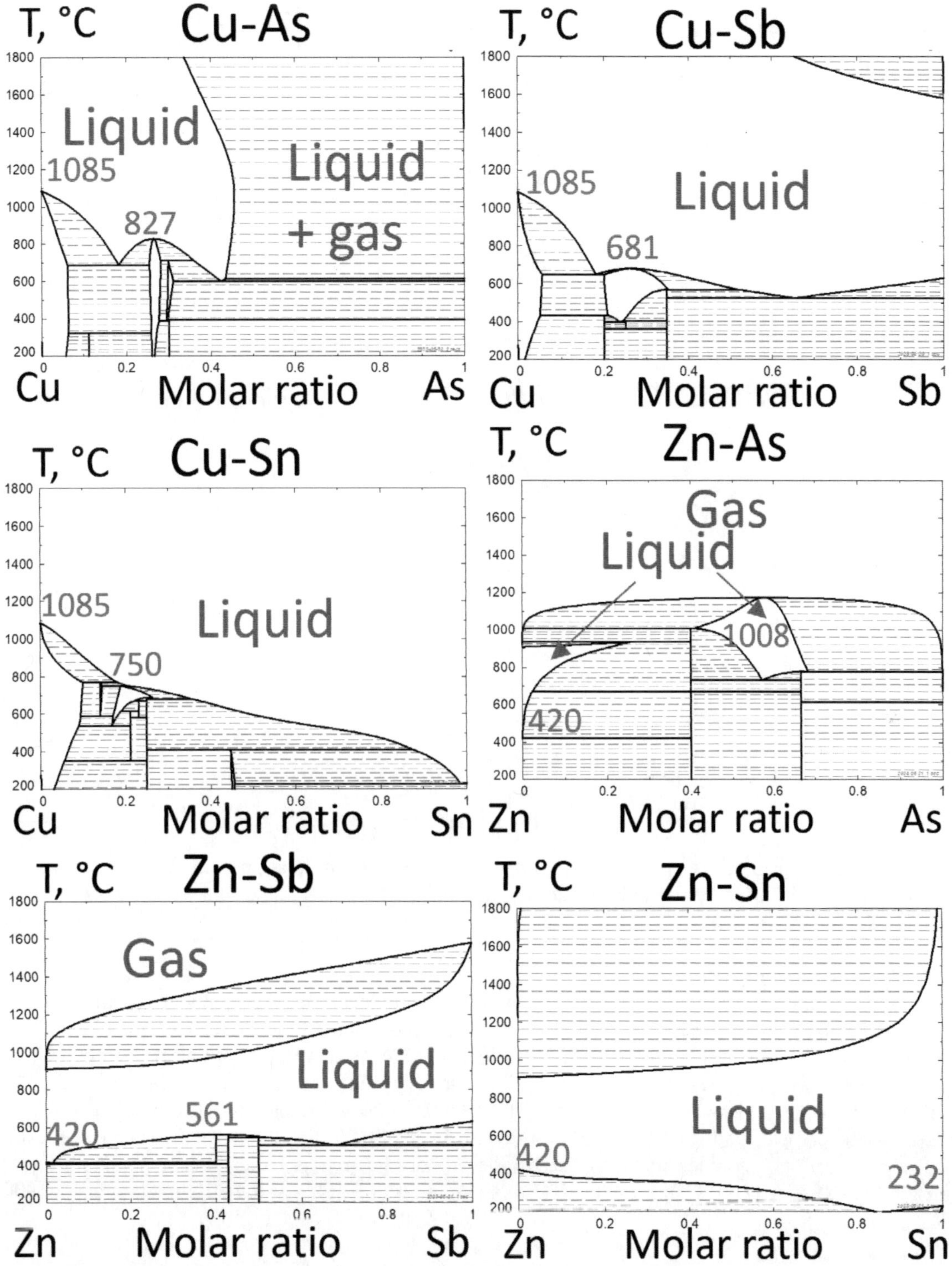

Figure 5. Phase diagrams of speiss forming binary systems at 1 atm total pressure: Cu-As, Cu-Sb, Cu-Sn, Zn-As, Zn-Sb, Zn-Sb. Lines are calculated using thermodynamic database of the present study based on assessment of experimental data.

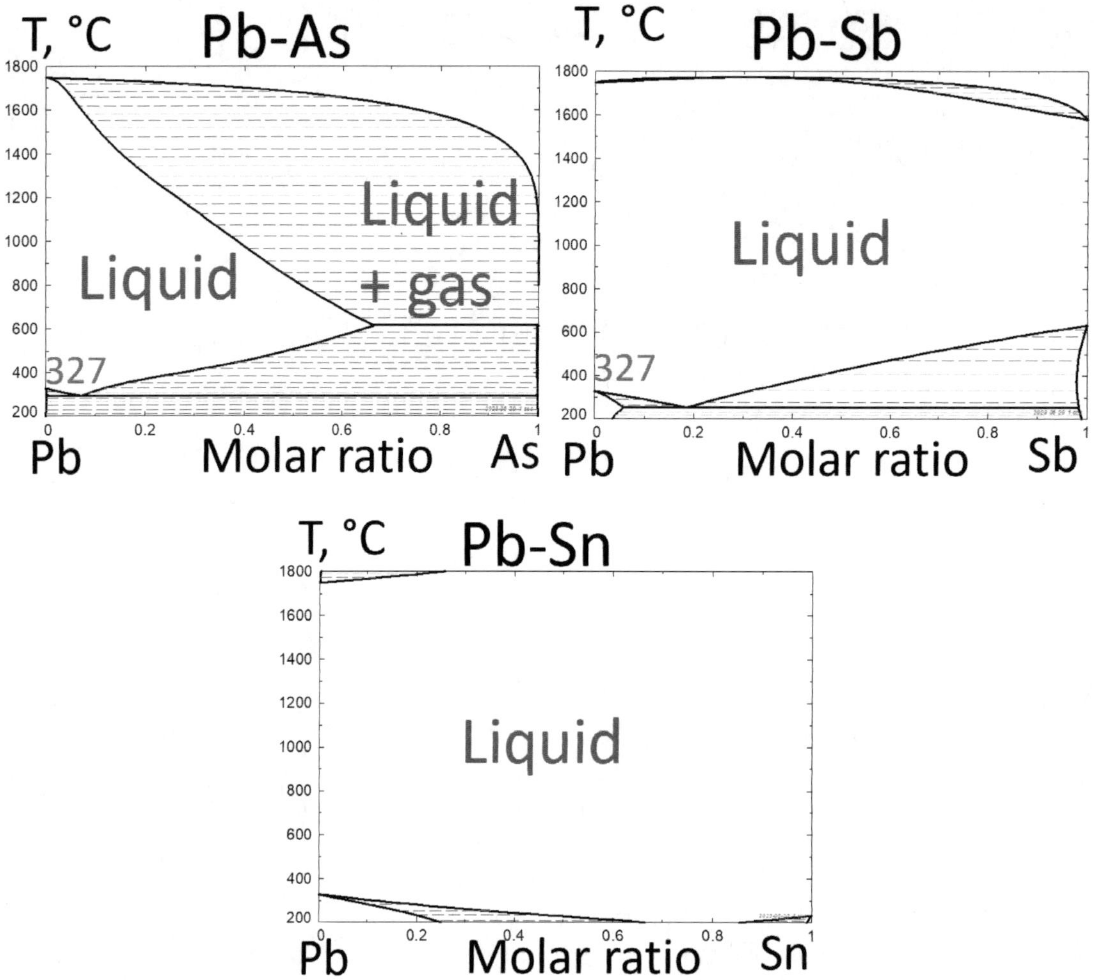

Figure 6. Phase diagrams of speiss forming binary systems at 1 atm total pressure: Pb-As, Pb-Sb, Pb-Sn. Lines are calculated using thermodynamic database of the present study based on assessment of experimental data.

2.2. Phase diagrams of key ternary systems

The liquid-liquid equilibria important for the matte/metal or speiss/metal is summarized in Figure 7. There is a good separation between lead-based metal and matte, but the separation between lead-based metal and speiss exists only under certain conditions: the Cu/As, Cu/Sb or Cu/Sn ratio show be high enough to result in the appearance of the miscibility gap. The extension of the miscibility gap in the Pb-Cu-As, Pb-Cu-Sb and Pb-Cu-Sn is the topic of current experimental research. Earlier publications [27, 28] showed very wide gap at 1200 °C, both for Pb-Cu-As and Pb-Cu-Sb systems, which is now believed to be incorrect, as explained in the paper by the authors [20]. It is believed that no liquid-liquid miscibility exists at 1200 °C in the Pb-Cu-As system, but it does exist at lower temperatures, i.e. 900 °C.

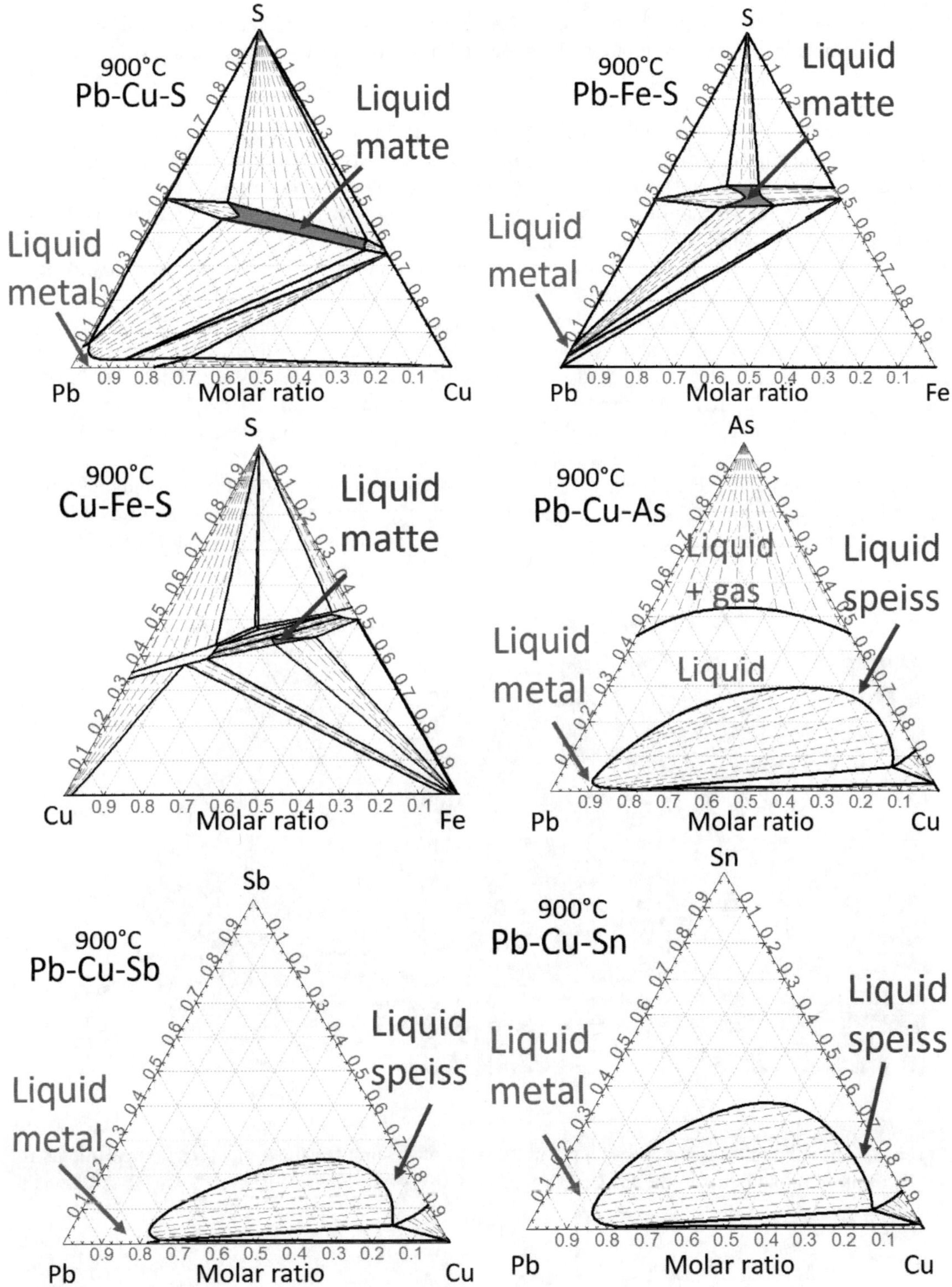

Figure 7. Phase diagrams showing matte/metal and speiss/metal immiscibility: Pb-Cu-S, Pb-Fe-S, Cu-Fe-S, Pb-Cu-As, Pb-Cu-Sb and Pb-Cu-Sn. Lines are calculated using thermodynamic database of the present study based on assessment of experimental data.

2.3. Example of quaternary system experiments and modelling

The partitioning of precious metals, such Ag and Au in solid or liquid speiss an important factor and can motivate companies to further process them before dumping. Thus, it is important to know the distribution of Ag and Au for liquid-liquid equilibria involving speiss [29]. Some works are available for distribution of precious metals between lead and iron-rich speiss [30], between matte and iron-rich speiss [31], and between matte and copper-rich speiss [32], but no study was found for the distribution between lead and Cu-rich speiss. A series of 20 experiments was designed to verify and improve the thermodynamic model. The experimental methodology involved high temperature equilibration of experimental mixtures at controlled conditions, rapid cooling (quenching) of the samples and direct measurement of the compositions of equilibrium phases using Electron Probe X-ray Microanalysis (EPMA) [22]. Figure 8 and Figure 9 show a typical microstructure, schematic location of liquid-liquid two phase zone inside the tetrahedron and compositions of obtained phases. The black tie-line inside the tetrahedron approximately shows the compositions targeted in the experiment. The perfect quenching and polishing are very hard to achieve for these types of systems, but significant progress has been achieved for these extremely fluid liquids over the last years, and the results are believed to have the necessary statistical significance to verify and improve the thermodynamic database. The results of this experimental series, together with available literature data were used to improve the description of the Cu-Ag-As, Cu-Au-As, Cu-Ag-Sb, Cu-Au-Sb, Cu-Ag-Sn ternary sub-systems. The concentrations of Ag and Au in the experiment were quite high. Blue symbols represent experimental results and red symbols are model predictions for the same bulk composition as in experiment. Good agreement was achieved, which provides the confidence of model predictions at lower concentrations of Ag and Au.

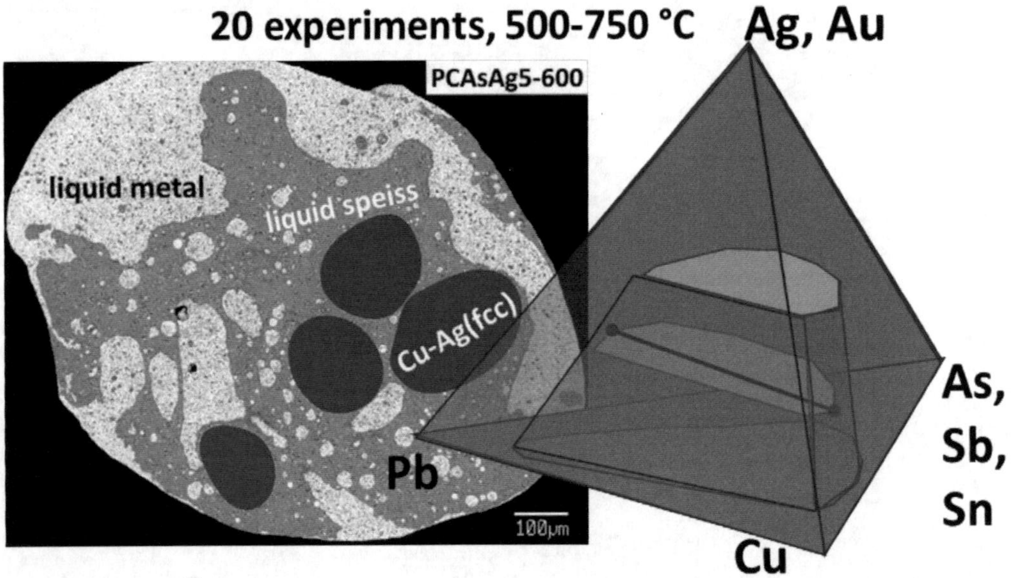

Figure 8. Example micrograph and schematic location of tie-line for the series of 20 experiments for the distribution of elements in the Pb-Cu-As-Ag, Pb-Cu-As-Au, Pb-Cu-Sb-Ag, Pb-Cu-Sb-Au and Pb-Cu-Sn-Ag systems.

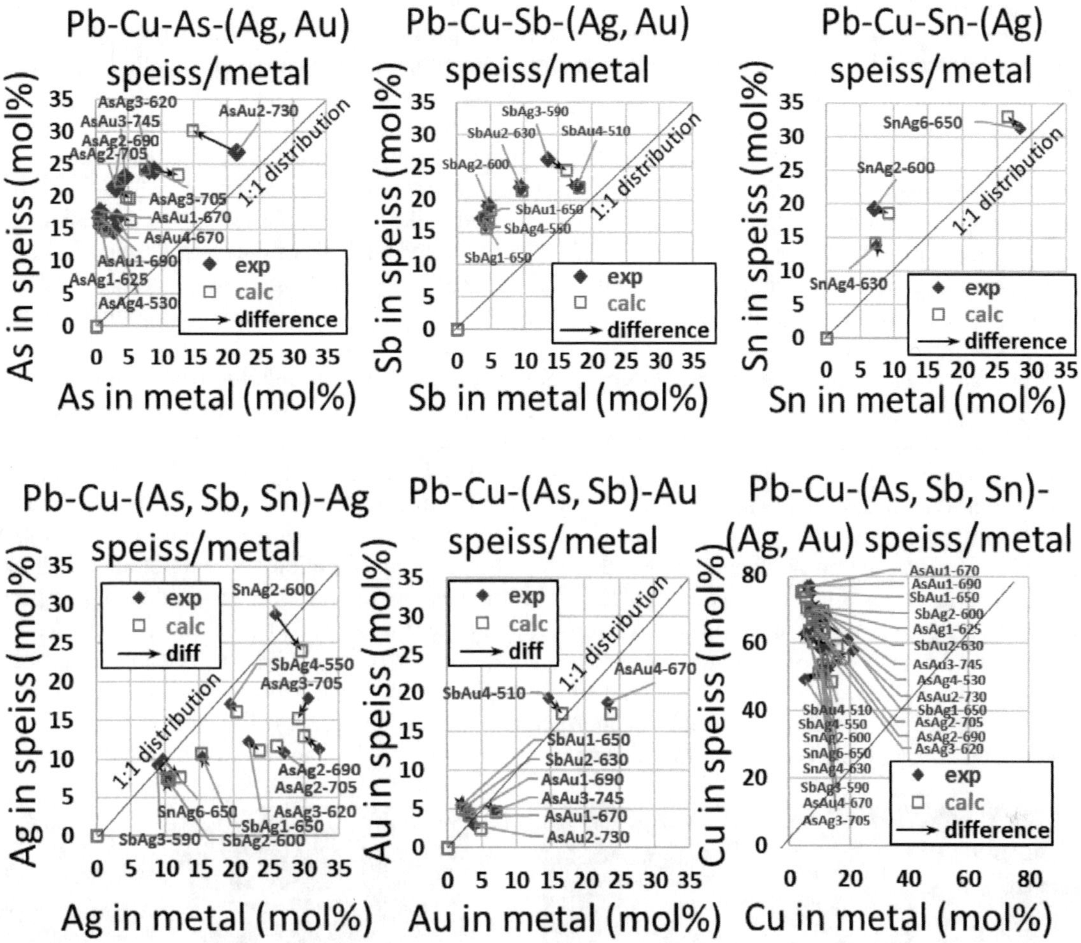

Figure 9. A series of 20 experiments for the distribution of elements in the Pb-Cu-As-Ag, Pb-Cu-As-Au, Pb-Cu-Sb-Ag, Pb-Cu-Sb-Au and Pb-Cu-Sn-Ag systems. Blue symbols are measured compositions, red symbols are model predictions using thermodynamic database of the present study.

3. Experiments and modelling for industrially relevant multicomponent equilibria

Several companies-sponsors of PYROSEARCH were interested in the potential for the conditions of 3-liquids matte/speiss/metal equilibrium which may be realized during the lead reduction process with high input of secondary complex materials. A series of 10 multicomponent experiments has been designed. Figure 11 shows some of the results for this type of equilibria. Due to very low viscosities of liquids, it is virtually impossible to achieve the perfect quenching, but valuable information can be obtained. The overall idea of the measurement is that droplets observed within the phase regions appeared during the quenching and must be included in the measurement by averaging many points over the regions and using relatively large diameter of electron beam in EMPA. The assumption is that no cross contamination existed outside of the established phase regions at the temperature of equilibrations, and that the quenching rate was sufficiently fast to prevent further segregation and growth of phase regions. By itself, these assumptions seem too big, but the accumulated experience in micrographs of hundreds of tests gives certain confidence. Model calculations also allow to cross-check challenging experiments with simpler and well-established experiments.

Different microstructures were obtained for the sample with the same bulk compositions at 1100 and 1000 °C, indicating the transition from 2-liquids to 3-liquids equilibria. Thermodynamic calculations were performed using FactSage software with the J-option on Liquid matte/metal/speiss solutions, allowing 3-phase immiscibility. Essentially, it adds 3 copies of the same multicomponent solution, which is substantially more complicated than a typical equilibrium calculation. The model reproduces the appearance of additional metal/speiss immiscibility between 1100 and 1000 °C, after small corrections were made within earlier model for the Cu-Pb-As system [20]. These corrections were supported by unpublished experimental results within the Cu-Pb-As system. The calculated miscibility gap between metal and matte is somewhat "wider" compared to the experiment, which is mainly observed as lower concentration of Cu, Fe and S in the lead bullion (metal phase), and higher calculated mass of matte for a given virtual bulk composition. The detailed analysis of the distribution of all other elements is outside of the scope of this publication. Overall, 10 multicomponent experiments (Figure 10) have been done, and the difference between the model predictions and measured compositions is shown in Figure 11 and Figure 12.

The challenge of establishing the correct solubility of Cu and S in metallic lead in the range of temperatures between 1000 and 1200 °C for the Cu-Pb-S system have been discussed in 2020 publication [17]. Since then, more data have been produced, resulting in a decision an increase the solubility compared to the model of 2020 [17]. The results in Figure 11 suggest even further increase of solubilities within the Pb-Fe-Cu-S system, which now cannot be reproduced by the model without compromising the data in ternary-subsystems. An argument can be made that some of the droplets observed in metal region were entrained at the temperature of equilibration and did not appear during the quenching. Let's assume the model indeed overestimates the amount of mass produced, compared to reality, and it is used in industrial for the simulation of the process unit. In this case, several kinetic factors would have to be introduced. Suspension indices, reflecting imperfect settling of droplets compared to equilibrium would mask some of model imperfection. If we assume that the model is correct, the suspension indices would still be required and calibrated based on the operation of the specific unit. Thus, the model provides reasonable approximation, which in any case requires correction.

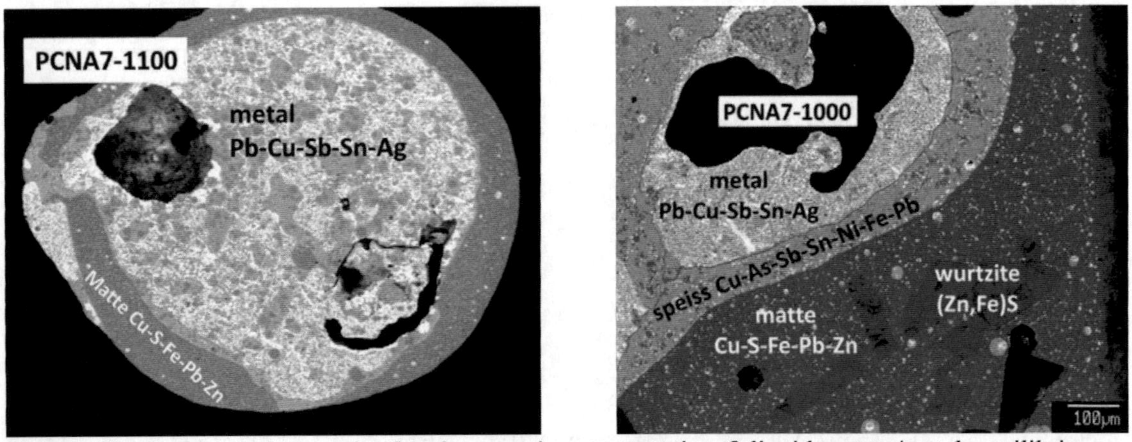

Figure 10. Example micrographs for the experiments targeting 2-liquids matte/metal equilibrium conditions at 1100 °C and 3-liquids matte/speiss/metal equilibrium conditions at 1000 °C.

PCNA7-1100	grams	wt%, Pb	Cu	Ni	As	Sb	Sn	Fe	S	Zn	Ag
Matte	**0.20**	10.74	42.82	0.77	0.46	0.27	0.22	17.19	23.51	3.63	0.40
Bullion	**0.70**	58.89	22.35	1.96	1.73	4.06	3.19	2.69	3.21	0.73	1.20
Speiss	**0.00**	0.00	0.00	0.00	0.00	0.00	0.00	0.00	0.00	0.00	0.00
FactSage calc	grams	wt%, Pb	Cu	Ni	As	Sb	Sn	Fe	S	Zn	Ag
Matte	**0.29**	11.83	41.88	0.52	0.76	0.53	0.55	16.82	22.16	4.47	0.33
Bullion	**0.61**	65.38	19.69	2.25	1.77	4.48	3.43	0.50	0.89	0.26	1.34
Speiss	**0.00**	0.00	0.00	0.00	0.00	0.00	0.00	0.00	0.00	0.00	0.00

PCNA7-1000	grams	wt%, Pb	Cu	Ni	As	Sb	Sn	Fe	S	Zn	Ag
Matte	**0.20**	11.08	42.54	0.62	0.41	0.22	0.20	16.40	23.54	4.58	0.40
Bullion	**0.70**	70.93	16.56	1.21	0.86	3.81	2.81	0.85	1.30	0.35	1.32
Speiss	**0.10**	13.44	43.46	6.98	8.40	8.39	7.06	6.73	2.77	1.59	1.17
FactSage calc	grams	wt%, Pb	Cu	Ni	As	Sb	Sn	Fe	S	Zn	Ag
Matte	**0.25**	10.25	42.53	0.31	0.80	0.42	0.39	16.17	22.64	6.00	0.29
Bullion	**0.66**	73.13	14.72	1.77	1.28	3.89	2.95	0.22	0.51	0.12	1.42
Speiss	**0.10**	26.26	44.09	4.39	5.05	9.08	6.70	1.78	0.63	0.85	1.17

Figure 11. Experiments and modelling results for the matte-speiss-metal 2-liquids matte/metal equilibrium conditions at 1100 °C and 3-liquids matte/speiss/metal equilibrium conditions at 1000 °C.

RMS	wt%, Pb	Cu	Ni	As	Sb	Sn	Fe	S	Zn	Ag	Distance
Matte	1.66	2.02	0.31	0.71	0.29	0.24	1.59	1.08	0.64	0.10	3.42
Bullion	4.02	2.02	0.83	0.95	0.52	0.45	0.96	1.18	0.24	0.09	4.97
Speiss	13.79	24.78	5.90	8.28	5.35	4.43	7.02	3.32	0.63	0.71	31.89

Figure 12. Root mean square difference in the compositions of matte, bullion and speiss between the experiment and model for 10 equilibrium experiments.

Conclusions

The fundamentals of the matte/speiss/metal equilibria, which are important for the increasing integration among Pb, Cu and Zn processes have been reviewed. Present study provides the set of assessed binary and ternary diagrams which can be used to understand the conditions of metal/matte/speiss formation. Also, the distribution of Ag and Au among liquid Pb and speiss in key quaternary systems, Pb-Cu-As-Ag, Pb-Cu-As-Au, Pb-Cu-Sb-Ag, Pb-Cu-Sb-Au and Pb-Cu-Sn-Ag, have been measured and compared with model predictions. The distributions of Pb, Cu, Ni, As, Sb, Sn, Fe, S, Zn and Ag among matte, speiss and metal have been measured for conditions, relevant to recycling of WEEE using the reductive smelting of Pb route. The model reasonably describes these challenging experiments.

Acknowledgments

The authors acknowledge the financial support and technical guidance by the consortium of lead producers: Aurubis (Germany), Kazzinc Glencore (Kazakhstan), Glencore Nordenham (Germany), Umicore (Belgium), Nyrstar (Australia), Peñoles (Mexico) and Boliden (Sweden) through Australian Research Council Linkage program LP180100028 "New copper lead processes for critical metals recovery from complex sources". The present study would not be possible without the facilities and technical assistance of the Australian Microscopy & Microanalysis Research Facility at the Centre for Microscopy and Microanalysis in The University of Queensland.

References

[1] Kania H, Saternus M. Evaluation and Current State of Primary and Secondary Zinc Production—A Review. Applied Sciences. 2023; 13: 2003.

[2] Roberts H. Evolution of Global Secondary Lead Production. In: Cham. Springer International Publishing; 2020. 3-15.

[3] Jak E, Shishin D, Shevchenko M, Hayes PC. University Research, Implementation and Education in support of the Sustainable Metals Production and Recycling. In: EMC 2023, Duesseldorf; Germany. 2023.

[4] Watari T, Nansai K, Nakajima K. Major metals demand, supply, and environmental impacts to 2100: A critical review. Resources, Conservation and Recycling. 2021; 164: 105107, https://doi.org/10.1016/j.resconrec.2020.105107.

[5] Australian Government, Department of Industry Science and Resources Critical Minerals Strategy 2023–2030, industry.gov.au/CriticalMineralsStrategy, 2023

[6] Bartie NJ, Abadias Llamas A, Heibeck M, Froehling M, Volkova O, Reuter MA. The simulation-based analysis of the resource efficiency of the circular economy - the enabling role of metallurgical infrastructure. Miner. Process. Extr. Metall. 2020; 129: 229-249, https://doi.org/10.1080/25726641.2019.1685243.

[7] Shibata E. Contributions of Non-ferrous Smelters to Metal Resource Circulation in Japan. In: PbZn 2020: 9th International Symposium on Lead and Zinc Processing, Cham. Springer International Publishing; 2020. 285-290.

[8] Flores GA, Risopatron C, Pease J. Processing of Complex Materials in the Copper Industry: Challenges and Opportunities Ahead. Jom. 2020: Ahead of Print, https://doi.org/10.1007/s11837-020-04255-9.

[9] Reuter MA. Digitalizing the Circular Economy - Circular Economy Engineering Defined by the Metallurgical Internet of Things. Metall. Mater. Trans. B. 2016; 47: 3194-3220, https://doi.org/10.1007/s11663-016-0735-5.

[10] Shishin D, Prostakova V, Jak E. Review and thermodynamic analysis of As (arsenic) in copper smelting liquid mattes, metals, slags, speiss and solid phases. In: COPPER 2022, Santiago, Chile. 2022.

[11] Bale CW, Belisle E, Chartrand P, Decterov SA, Eriksson G, Gheribi AE, et al. FactSage thermochemical software and databases, 2010-2016. CALPHAD. 2016; 54: 35-53, https://doi.org/10.1016/j.calphad.2016.05.002.

[12] Pelton AD, Decterov SA, Eriksson G, Robelin C, Dessureault Y. The Modified Quasichemical Model. I - Binary Solutions. Metall. Mater. Trans. B. 2000; 31: 651-659, https://doi.org/10.1007/s11663-000-0103-2.

[13] Pelton AD, Chartrand P. The Modified Quasichemical Model. II - Multicomponent Solutions. Metall. Mater. Trans. A. 2001; 32: 1355-1360, https://doi.org/10.1007/s11661-001-0226-3.

[14] Chartrand P, Pelton AD. The Modified Quasichemical Model. III - Two Sublattices. Metall. Mater. Trans. A. 2001; 32: 1397-1407.

[15] Pelton AD, Chartrand P, Eriksson G. The modified Quasichemical Model. IV - Two Sublattice Quadruplet Approximation. Metall. Mater. Trans. A. 2001; 32: 1409-1415.

[16] Jung I-H, Van Ende M-A. Computational Thermodynamic Calculations: FactSage from CALPHAD Thermodynamic Database to Virtual Process Simulation. Metall. Mater. Trans. B. 2020; 51: 1851-1874, https://doi.org/10.1007/s11663-020-01908-7.

[17] Shishin D, Chen J, Jak E. Thermodynamic Modeling of the Pb-S and Cu-Pb-S Systems with Focus on Lead Refining Conditions. J. Phase Equilib. Diffus. 2020; 41: 218-233, https://doi.org/10.1007/s11669-020-00811-7.

[18] Prostakova V, Shishin D, Jak E. Thermodynamic optimization of the As-S system. Calphad. 2021; 72: 102235, https://doi.org/10.1016/j.calphad.2020.102235.

[19] Shishin D, Jak E. Critical assessment and thermodynamic modeling of the Cu-As system. CALPHAD. 2018; 60: 134-143, https://doi.org/10.1016/j.calphad.2017.12.005.

[20] Shishin D, Chen J, Hidayat T, Jak E. Thermodynamic modelling of the Pb-As and Cu-Pb-As systems supported by experimental study. J. Phase Equilib. Diff. 2019; 40: 758–767, https://doi.org/10.1007/s11669-019-00764-6.

[21] Shishin D, Fallah-Mehrjardi A, Shevchenko M, Hidayat T, Jak E. Experimental study, thermodynamic calculations and industrial implications of slag-matte-metal equilibria in the Cu-Pb-Fe-O-S-Si system. Journal of Materials Research and Technology. 2022; 19: 899-912, https://doi.org/10.1016/j.jmrt.2022.05.058.

[22] Starykh R, Sineva S, Shishin D, Jak E. Experimental investigation of phase equilibria in the metal-speiss systems for slag cleaning and recycling applications. In: EMC 2023, Duesseldorf; Germany. 2023.

[23] Ettler V, Johan Z, Bezdicka P, Drabek M, Sebek O. Crystallization sequences in matte and speiss from primary lead metallurgy. Eur. J. MIneral. 2009; 21: 837-854, https://doi.org/10.1127/0935-1221/2009/0021-1942.

[24] Richards G, Curtis C. Integrated Pyrorefining of Lead at Teck's Trail Operations. In: Extraction 2018, Cham. Springer International Publishing; 2018. 623-629, https://doi.org/10.1007/978-3-319-95022-8_48.

[25] Watt W, Hidayat T, Shishin D, Jak E. Advanced Thermochemical Fundamental and Applied Research to Improve Integrity of the Steel Water Jacketed Furnace at Port Pirie. In: Extraction 2018, Ottawa, Canada. 2018.https://doi.org/10.1007/978-3-319-95022-8_18.

[26] Attah-Kyei D, Klemettinen L, Michallik R, Salminen J, Taskinen P, Lindberg D. Hydrogen as Carbon-Free Reducing Agent in Non-ferrous Slag Fuming. Metall. Mater. Trans. B. 2022; 53: 3775-3792, https://doi.org/10.1007/s11663-022-02640-0.

[27] Espeleta AK, Hino M, Azakami T. Phase separation and silver and gold distribution in the lead-iron-antimony and lead-copper-antimony systems at 1200°C. Shigen to Sozai. 1990; 106: 825-830, https://doi.org/10.2473/shigentosozai.106.825.

[28] Azakami T, Hino M, Yazawa A. The liquid miscibility gap and the distribution of silver between speiss and metallic lead in the lead-iron-arsenic, lead-copper-arsenic and lead-iron-copper-arsenic systems at 1200°C. Can. Metall. Q. 1979; 18: 389-394, https://doi.org/10.1179/000844379795317607.

[29] Chaidez-Felix J, Romero-Serrano A, Hernandez-Ramirez A, Peraz-Labra M, Almaguer-Guzman I, Benavides-Perez R, et al. Effect of copper, sulfur, arsenic and antimony on silver distribution in phases of lead blast furnace. Trans. Nonferrous Met. Soc. China. 2014; 24: 1202-1209, https://doi.org/10.1016/s1003-6326(14)63180-0.

[30] Voisin L, Henao HM, Hino M, Itagaki K. Phase relations, activities and minor elements distribution in Fe-Pb-As and Fe-Pb-Sb systems saturated with carbon at 1473 K. Mater. Trans. 2005; 46: 3030-3036, https://doi.org/10.2320/matertrans.46.3030.

[31] Itagaki K, Voisin L, Mendoza D. Phase relations and activities in the Cu-Fe-S-X and Cu-Fe-S-C-X (X = As or Sb) systems and distribution of precious metals relating to reduction smelting of copper. In. Minerals, Metals & Materials Society; 2006. 1: 289-300.

[32] Hino M, Pagador RU, Itagaki K. Phase separation and silver distribution between matte and speiss. In. Minerals, Metals & Materials Society; 1999. 2: 1191-1200.

Journal of Physics: Conference Series **2738** (2024) 012033 doi:10.1088/1742-6596/2738/1/012033

Scorodite synthesis process using solid iron oxides

Ken Adachi [1], Tomoro Karube[2], Takumi Anezaki[2], Atsushi Iizuka[1] and Etsuro Shibata[1]*

[1]Center for Mineral Processing and Metallurgy, Institute of Multidisciplinary Research for Advanced Materials, Tohoku University, Sendai, Miyagi, Japan 980-8577.,
[2]Graduate School of Environmental Studies, Tohoku University, Sendai, Miyagi, Japan 980-0845.
*etsuro.shibata.e3@tohoku.ac.jp

Abstract The immobilization of arsenic in the form of scorodite (FeAsO4·2H2O), which has excellent chemical stability, is attracting attention as a method for treating wastewater containing high concentrations of arsenic generated at nonferrous metal smelting plants. Scorodite, with its low solubility and high arsenic content per volume, is expected to be an environmentally friendly arsenic fixation method suitable for final disposal.Various scorodite synthesis methods have been studied. The first synthesis method proposed is the hydrothermal method using an autoclave to synthesize highly crystalline scorodite. Although the hydrothermal method is capable of synthesizing scorodite with good crystallinity, from an economic point of view, scorodite synthesis under ambient pressure and at low temperatures is more attractive. As a low-temperature scorodite synthesis method under atmospheric pressure, the oxidation of Fe(II) process by O_2 bubbling was proposed. In this method, ferrous sulfate is added to an arsenic-containing solution as a Fe ion source, and in situ oxidation of Fe(II) by O_2 or air bubbling to form scorodite with good crystallinity. In addition to temperature, other conditions, pH, Fe/As ratio, and reaction time have been reported to affect scorodite crystallization. Recently, scorodite synthesis using solid iron oxide as the Fe source for scorodite synthesis, instead of aqueous Fe salt solutions, has attracted much attention. In this presentation, we will report on the investigation of reaction parameters, such as the type of iron oxide and reaction temperature, for the scorodite synthesis using solid iron oxide.

1. Introduction

Arsenic (As) is a highly toxic element that is widely distributed throughout the earth. Non-ferrous metal smelters that process ores containing As emit large amounts of As in the form of slag, sludge, dust and wastewater. Wastewater poses a particular threat to the natural environment and human health due to its high toxicity and mobility. Therefore, efforts have been made to find ways to safely fix As in wastewater in a stable form that can be safely disposed of.

The iron co-precipitation method is commonly used to immobilize arsenic, but it produces large amounts of arsenic-containing sludge, which is harmful to the environment. In addition, adsorption of arsenic on iron compounds raises concerns about potential As leaching. [1 2] One promising method that addresses these issues is the immobilization of arsenic as scorodite. Scorodite is a stable compound with low solubility, making it suitable for arsenic immobilization. This method can significantly reduce the amount of sludge produced. Scorodite synthesis is expected to be introduced especially in areas with high concentrations of arsenic-containing wastewater[3 4].

Several methods for the synthesis of scorodite have been investigated. The hydrothermal method in an autoclave was first proposed and produced highly crystalline scorodite. [5-7] However, the synthesis of scorodite at ambient pressure and low temperature is more cost effective. Fujita *et al.* reported an excellent low-temperature scorodite synthesis method under atmospheric pressure at 95°C using ferrous sulfate as the Fe ion source and in situ oxidation of Fe^{2+} by air bubbling to form scorodite with good crystallinity. [8, 9] Other conditions such as pH, Fe/As ratio and reaction time were found to affect scorodite crystallization. Recently, the use of solid iron oxide instead of aqueous Fe salt solutions as the Fe source for scorodite synthesis has attracted much attention. Hematite [10-12], ferrihydrite [13], magnetite [14] and limonite [15] have been investigated as solid iron oxides. Scorodite synthesis using these stable iron oxides is expected to be an environmentally friendly arsenic removal process because they are inexpensive and can utilize by-products of other industrial processes.

This study investigates the immobilization behavior of arsenic in solid iron oxides, particularly magnetite (Fe_3O_4) and wustite (FeO) in comparison with hematite (Fe_2O_3), and clarifies their characteristics in terms of solution composition, pH, and ORP. The effect of the initial concentration of Fe(II) added on the reaction is also studied in the case of magnetite.

2. Experimental

Aqueous solutions containing As(V) were prepared using arsenic acid reagent (60 mass% H_3AsO_4, FUJIFILM Wako Pure Chemical Corporation) and distilled water, and solid iron oxides, hematite (Fe_2O_3, Alfa Aesar), magnetite (Fe_3O_4, Alfa Aesar) and wustite (FeO, Alfa Aesar). The initial Fe(II) concentration was adjusted by adding ferrous sulfate heptahydrate ($FeSO_4 \cdot 7H_2O$) to study the effect of Fe(II) concentration. A 1 L multi-port flask was used as the reaction vessel, which is common with the previous report[13], and solid iron oxide was added to 700 mL of arsenic acid solution $[H_3AsO_4] = 0.67$ mol L^{-1} (i.e., $[H_3AsO_4] = 50$ g L^{-1}) maintained at 95°C to a pulp concentration of 0.35 mol L^{-1}. The synthesis was carried out under Ar flow (700 mL min^{-1}) to prevent oxidation by atmospheric O_2, and stirring was performed with an impeller at 200 rpm. The pH transition of the solution was monitored with a pH meter and an ORP meter (Pt electrode) with the reference electrode of Ag|AgCl in 3.33M KCl aq., and the pH was adjusted by the addition of dilute sulfuric acid (2.0 mol L^{-1}) prepared with concentrated sulfuric acid (H_2SO_4, 98%, FUJIFILM Wako Pure Chemical Corporation).

The slurry was sampled at 10, 60, and 420 min after the addition of solid iron oxide, and the powder sample was obtained as the precipitate by vacuum filtration, and the solution sample was obtained as the filtrate. The powder sample was vacuum freeze-dried overnight to remove water. The solution samples were analyzed for As and Fe concentrations by inductively coupled plasma atomic emission spectroscopy (ICP-AES, Spectro Arcos, Spectro Analytical Instruments, Kleve, Germany). The powder samples were characterized using X-ray diffraction (XRD; D2PHASER, Bruker Corp., Billerica, MA, USA) with Cu Kα radiation (1.542 Å), scanning electron microscopy (SEM; SU-6600, Hitachi, Tokyo, Japan) and raman spectroscopy (Renishaw InVia Raman microscope, Renishaw plc., Wotton-under-Edge, UK).

3. Results and discussions

3.1. Scorodite synthesis using magnetite The behavior of scorodite formation using magnetite with/without the presence of Fe(II) was observed. Fig. 1 shows the changes in As concentration, pH and potential of the Pt electrode when magnetite was added to an aqueous solution of As(V) with / without containing Fe(II) at different concentrations. The initial As(V) concentration was 0.67 mol L^{-1}, and the initial Fe(II) concentrations varied from 0 to 1.0 mol L^{-1}. The results showed that the As concentration decreased with time and was reduced to less than 0.005 mol L^{-1} (*i.e.* less than 0.4 g L^{-1}) after 420 min. As concentrations at 10 and 60 min decreased with increasing initial Fe(II) concentration, indicating that Fe(II) enhanced the As removal rate. This was similar to a previous study using solid iron oxide with hematite for arsenic immobilization, shown in Fig.1(i-Ref) [11]. In such a system, Fe(II) was converted to scorodite by oxidation with hematite, which increased the As removal rate [16]. Similarly, in the present study, Fe(II) in solution could be oxidized and converted to scorodite with magnetite, which increased the As removal rate. The pH of the solution increased rapidly at the beginning of the reaction and then gradually increased or remained constant. This was due to the consumption of protons during the reaction to form scorodite from magnetite, as shown in the following reaction equation.

$$Fe_3O_4 + 2H_3AsO_4 + 2H^+ \rightarrow FeAsO_4{\cdot}2H_2O + Fe^{2+} \quad (1)$$

The pH increase was less pronounced at high initial Fe(II) concentrations due to the pH buffering effect of ferrous sulfate. This suggests that the equilibrium potential of the potential determination reaction is influenced more by pH than by Fe(II) activity.

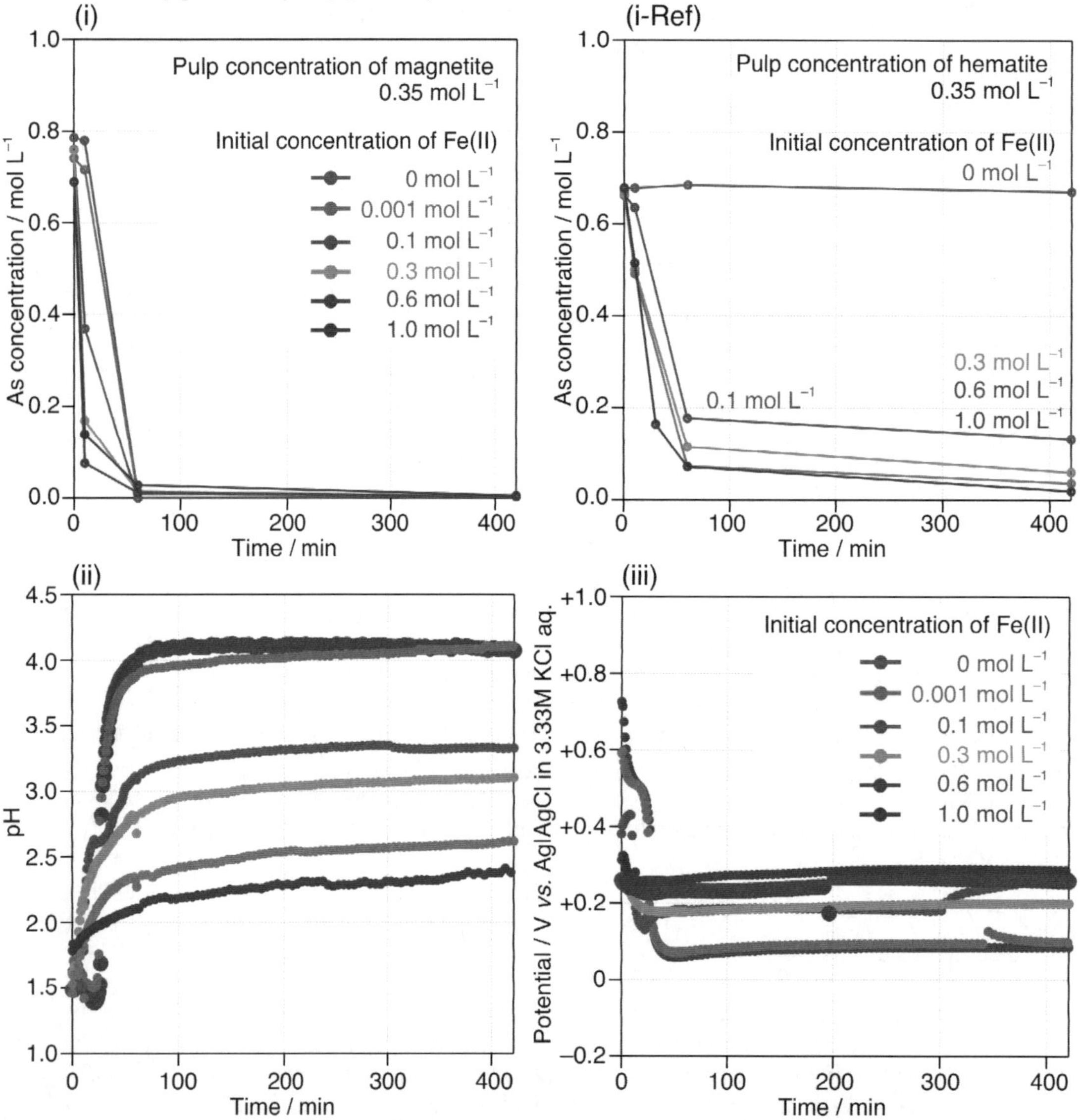

Fig. 1 (i) As concentration, (ii) pH and (iii) potential of Pt electrodes in the reaction solutions with magnetite added during scorodite synthesis for each initial Fe(II) concentration condition. The reported As^{5+} removal behavior by the hematite method is also shown in (i-Ref)[11].

Figure 2 shows SEM images of solid products sampled at 420 min for different initial Fe(II) concentration conditions. Scorodite crystals with facets were formed at high initial Fe(II) concentrations. In contrast, at concentrations below 0.1 mol L^{-1}, only a powder with a cocoon-like appearance covered with an amorphous membrane was observed. This is a gel-like precursor of scorodite, which crystallized under conditions of high initial Fe(II) concentration due to low pH, however, did not crystallize under conditions of low initial Fe(II) concentration due to high pH. In contrast to the hematite method, Fe(II) addition was not necessary for the scorodite formation reaction to proceed when magnetite was used for arsenic fixation. However, the addition of Fe(II) accelerated the reaction rate, and crystalline scorodite was formed.

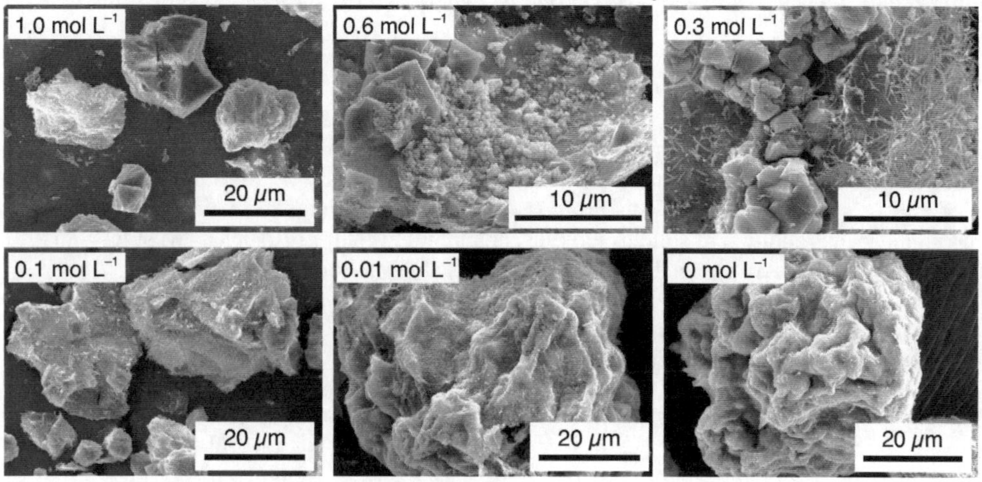

Fig. 2 SEM images of the precipitates in the reaction solutions with magnetite added during scorodite synthesis for each initial Fe(II) concentration condition obtained at 420 min.

Figure 3 shows the microscopic images and the local Raman spectra of solid products sampled at 420 min for Fe(II) concentration of 0.1 mol L^{-1}. The faceted crystals observed in the SEM images are identified as scorodite, while the surrounding film-like material shows spectra different from those of scorodite and iron oxide. Although the small size of the mixture of scorodite crystals and gel-like precursor, typically in the range of tens of microns, makes it difficult to perform local elemental analysis and crystal structure analysis, they can be clearly distinguished through local Raman spectroscopic analysis.

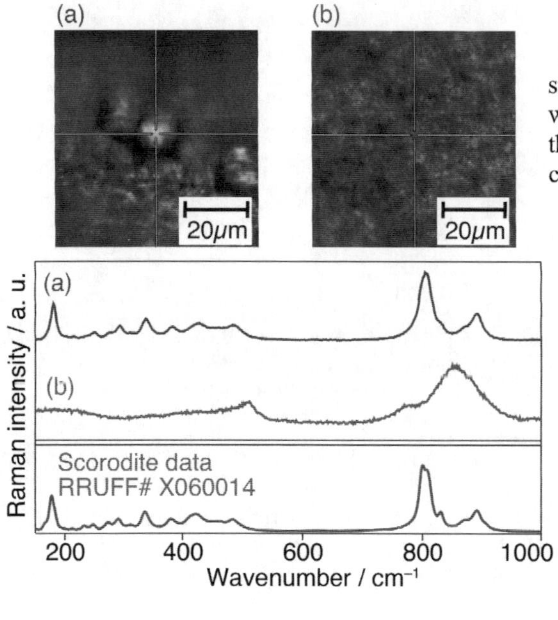

Fig. 3 The microscopic images and the Raman spectra of the precipitates in the reaction solutions with magnetite added during scorodite synthesis for the initial Fe(II) concentration of 0.1 mol L^{-1} condition obtained at 420 min for the parts with (a)

Journal of Physics: Conference Series **2738** (2024) 012033 doi:10.1088/1742-6596/2738/1/012033

3. 2. Scorodite synthesis using wustite Figure 4 shows the behavior of scorodite formation with wustite in As(V) solution without addition of Fe(II). When only wustite was added, no scorodite was formed because wustite is a divalent iron oxide. In contrast, when both wustite and hematite were added simultaneously, the As concentration decreased with time to 0.01 mol L^{-1}. The pH and natural immersion potential of Pt were also measured, and the results were similar to those of magnetite addition under Fe(II)-containing conditions. Considering the rapid decrease of As concentration in the presence of Fe(II) and hematite as shown in Figure 1(i-Ref), this suggests that the decrease of As concentration in the presence of wustite and hematite was caused by the wustite dissolution according to the following equation.

$$FeO + 2H^+ \rightarrow Fe^{2+} + H_2O \qquad (2)$$

This reaction continued because Fe^{2+} produced by dissolution and scorodite was synthesized in the presence of hematite and As(V). SEM images of the product after 420 minutes confirmed the formation of faceted scorodite crystals were formed. This result showed that the scorodite formation reaction could proceed by simultaneous addition of solid iron oxide containing divalent iron, such as wustite, without necessarily adding reagent-derived Fe(II) such as ferrous sulfate in solution.

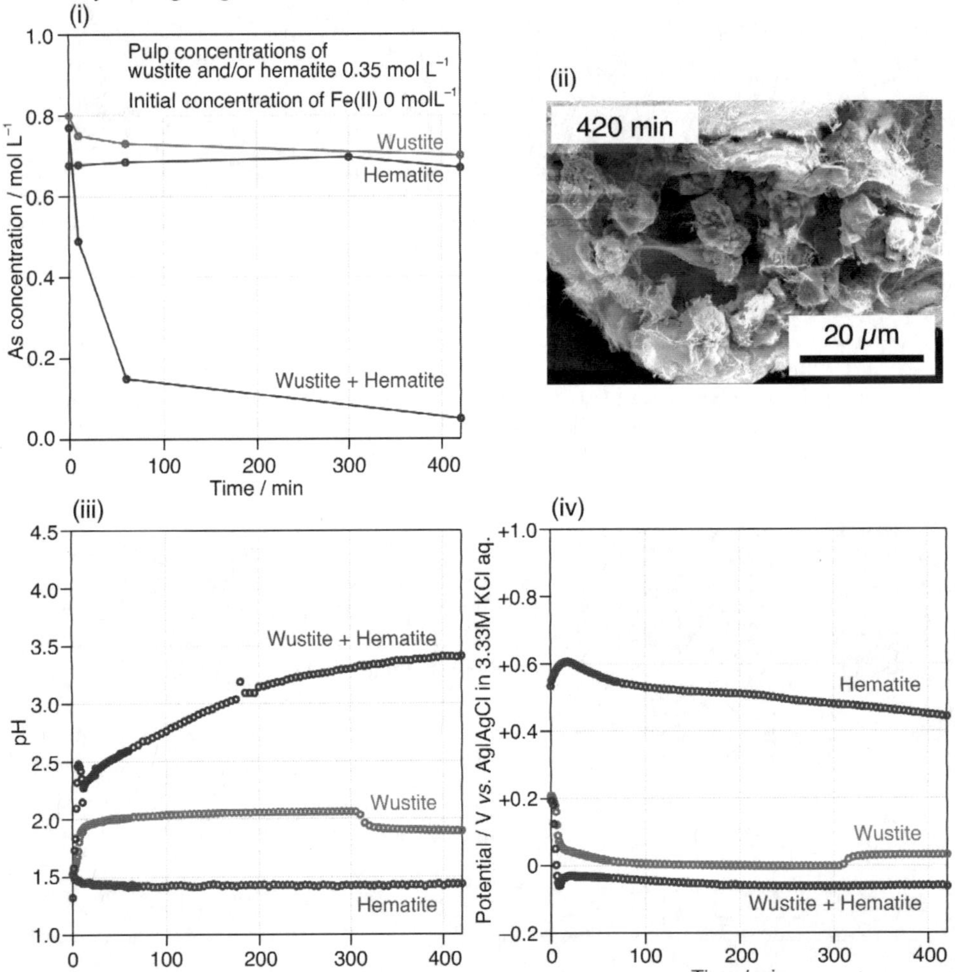

Fig. 4 (i) As concentration, (iii) pH and (iv) potential of Pt electrodes in the reaction solutions with wustite and/or hematite added during scorodite synthesis. (ii) SEM image of the precipitate obtained in the condition of simultaneous addition of wustite and hematite at 420 min.

4. Conclusion

This study investigated the use of magnetite and wustite for the synthesis of scorodite in As-containing aqueous solutions. When magnetite was used, only an amorphous scorodite precursor was produced due to the increased pH. However, the addition of Fe(II) improved the As removal rate and prevented the pH increase, resulting in the formation of crystalline scorodite. When wustite was added alone, the reaction did not proceed, However, when combined with hematite, crystalline scorodite was obtained. Therefore, it is suggested that a suitable process for synthesizing scorodite using solid iron oxide can be achieved without the need for Fe(II) agents in solution by combining different solid iron oxide raw materials.

References

[1] L. G. Twidwell and J. W. McCloskey, *JOM*, 2011, **63**, 94.
[2] M. C. Bluteau, L. Becze and G. P. Demopoulos, *Hydrometallurgy*, 2009, **97**, 221-227.
[3] P. Drahota and M. Filippi, *Environ. Int.*, 2009, **35**, 1243-1255.
[4] K. A. Salzsauler, N. V. Sidenko and B. L. Sherriff, *App. Geochemistry*, 2005, **20**, 2303-2314.
[5] J. E. Dutrizac and J. L. Jambor, *Hydrometallurgy*, 1988, **19**, 377-384.
[6] R. C. M. Mambote, P. Krijgsman and M. A. Reuter, *Miner. Eng.*, 2003, **16**, 429-440.
[7] V. G. Papangelakis and G. P. Demopoulos, *Can. Metall. Quart.*, 1990, **29**, 1-12.
[8] T. Fujita, R. Taguchi, M. Abumiya, M. Matsumoto, E. Shibata and T. Nakamura, *Hydrometallurgy*, 2008, **90**, 92-102.
[9] T. Fujita, R. Taguchi, M. Abumiya, M. Matsumoto, E. Shibata and T. Nakamura, *Hydrometallurgy*, 2008, **90**, 85-91.
[10] A. Iizuka, E. Shibata and S. Ishii, *Mater. Trans.*, 2020, **61**, 402-411.
[11] A. Iizuka, K. Shinoda and E. Shibata, *Mater. Trans.*, 2018, **59**, 843-849.
[12] Z. Rong, X. Tang, L. Wu, X. Chen, W. Dang and Y. Wang, *Journal of Materials Research and Technology*, 2020, **9**, 5848-5857.
[13] K. Adachi, T. Anezaki, T. Karube, A. Iizuka and E. Shibata, *Mater. Trans.*, 2022, **63**, 1287-1293.
[14] X. Z. Li, G. Y. Cai, Y. K. Li, X. Zhu, X. J. Qi, X. Zhang, B. Shu, K. Z. Li, Y. G. Wei and H. Wang, *J. Clean Prod.*, 2021, **278**, 123552.
[15] A. Iizuka, K. Adachi and E. Shibata, *Mater. Trans.*, 2022, **63**, 655-661.

Characteristics and risk assessment of arsenic contamination in a lead and zinc plant area in Yunnan Province

Xiangyang Mei[1], Hua Qing[1*], Qunxing Liu[1], Yuan Yue[1], Hua Zhao[1], Xiaoyan Wa[2] and Qicheng Ma[1]

[1] Yunnan Provincial Appraisal Center of Environmental Engineering, Kunming, Yunnan, China, 650032. Email:21327749@qq.com

[2] Yunnan Scientific Research and Design Institute of Building Materials Co. LTD, Kunming, Yunnan, China, 650032. Email:470709031@qq.com

Abstract: Taking a lead and zinc plant plot in Yunnan as the research object, the land use status and planting types of the region were investigated, the original enterprise distribution area and crop growing area of the plot were defined, and 67 surface soil samples and 14 columnar soil samples were systematically arranged in the plot for monitoring. Combined with the monitoring results and analysis, it is concluded that there is no definite basis that the high arsenic content in soil in the region is caused by the enterprise production, and the high arsenic content in the region may be caused by the background value. Based on the monitoring and investigation results, the risk assessment of the area was carried out. The results showed that the non-carcinogenic risk and carcinogenic hazard index of arsenic exceeded the prescribed acceptable level when the area was developed as the second type of construction land, and soil remediation should be carried out. At the same time, the collaborative distribution monitoring of regional crops showed that the arsenic content of crops was low, which met the requirements of food standards, possibly because the cultivated crops did not have strong adsorption and enrichment effects on the arsenic occurrence form in the regional soil.

1 Introduction

According to relevant investigations and research results, the soil arsenic content in some areas of my country is relatively high, and about 20 million people live in areas with high risk of soil arsenic[1].

* Corresponding author at: Yunnan Provincial Appraisal Center of Environmental Engineering, Kunming 650500, China. E-mail address: 21327749@qq.com.

Due to natural causes and human activities, a large amount of arsenic-containing substances released into the environment will be an important reason for soil arsenic pollution [2]. Arsenic easily enters the human body through crop absorption and contact. When the arsenic in the human body exceeds a certain limit, it will cause acute and chronic arsenic poisoning in the human body, and cause various diseases, which seriously threaten human health. Yunnan Province is rich in non-ferrous metal reserves, and the background value of arsenic in the soil is relatively high. At the same time, due to the development and smelting of non-ferrous metal minerals in the region, a large amount of tailings and metallurgical waste residues are produced. Due to historical reasons, the disposal of some arsenic-containing solid wastes was not standardized, which caused the Yangzonghai arsenic pollution accident, and also caused the excessive arsenic in the Da tun Sea in the Hong he area and the excessive arsenic in the farmland soil in the Bi jiang River Basin [3-5].

According to the nature of pollutants, arsenic has both carcinogenic and non-carcinogenic hazards, and both carcinogenic and non-carcinogenic risks need to be calculated [6]. The heavy metal elements in the soil of Yunnan karst area mainly come from the regional high geological background value and pedogenesis, and the activity of heavy metals is relatively low [7, 8]. Due to human production activities, the proportion of effective states of some heavy metal elements has increased relatively, and the safety of agricultural products has been seriously threatened [9]. Han et al. conducted an investigation and risk assessment of arsenic pollution in an industrial site, and concluded that the arsenic content in the surface soil of the site was high, and the regional groundwater was also polluted [10]. Guo et al. studied the characteristics of heavy metal pollution in the soil of an abandoned sulfuric acid plant, and showed that the excessive levels of heavy metals such as arsenic in different production areas were quite different, exceeding the acceptable risk level [11]. Yao et al. conducted a survey and potential ecological risk assessment of heavy metal pollution in farmland soil in a karst area in eastern Yunnan with high cadmium and arsenic background. They believed that the high background value of heavy metals in the soil in the area seriously threatened the quality of agricultural products [12]. In recent years, there have been many literatures reporting on the risk assessment of arsenic pollution in some plots or related studies on plant uptake and restoration of arsenic-contaminated plots [13-16]. However, there are few reports on the coordinated assessment of arsenic pollution investigation, risk assessment and crop monitoring in plots. In this paper, a preliminary investigation of a plot with high arsenic content in Yunnan was taken as the research object, and the detailed monitoring and investigation, risk assessment and soil collaborative monitoring methods were used to conduct detailed and in-depth research on the plot, so as to provide a scientific basis for the subsequent development of the plot by the government.

2 Monitoring program and risk assessment method in the study area

2.1 Basic situation of the research area

The study area is a 4 km × 4 km area with high arsenic in Yunnan Province, and the latitude and longitude of the site is 23.512035° north latitude and 102.818430° east longitude, the dominant wind direction is southwest wind, the soil type is common red soil, and the land use types are dry land, forest grassland, and bare land, accounting for more than 80%, and a small part is industrial and mining enterprises land. There are no soil-sensitive targets such as hospitals, sanatoriums, and

Journal of Physics: Conference Series **2738** (2024) 012034 doi:10.1088/1742-6596/2738/1/012034

nursing homes in the area, and the soil-environmentally sensitive targets are mainly cultivated land within the survey area. The enterprises in the area are mainly distributed in the southeast of the area. The main enterprises include 10 enterprises including manganese smelting, manganese powder processing, copper ore dressing, and lead-zinc ore dressing plants. The same soil pollutants are all heavy metals, and all of them are currently in suspension. In addition, a domestic garbage sanitary landfill is distributed on the south side. The area is mainly dry land, and the main types of crops are corn, pepper, sugarcane, orange, peach, taro and so on.

2.2 Agricultural land soil and groundwater monitoring program

According to the "Technical Regulations on the Layout of Points for Detailed Investigation of Soil Pollution in Agricultural Land" (Environment Office Soil Letter [2017] No. 1021), the monitoring points are laid out in a grid with a minimum precision of 500m × 500m for various types of soil impact survey units. Combined with the fact that the arsenic content in the preliminary monitoring sampling is greater than the control value in the agricultural land risk management and control standards and the distribution of existing enterprises on the southeast side, the upwind direction control points are arranged in radial directions at different distances, and the upwind direction of the existing enterprise distribution area on the southeast side The point distribution method is densely distributed, and the two sides are densely distributed in the form of umbrella type point distribution. A total of 67 surface soil monitoring points were set up in the survey area, at 3$^{\#}$, 7$^{\#}$, 11$^{\#}$, 14$^{\#}$, 21$^{\#}$, 25$^{\#}$, 28$^{\#}$, 36$^{\#}$, 45$^{\#}$, 51$^{\#}$, 52$^{\#}$, 55$^{\#}$, 62$^{\#}$, 67$^{\#}$ part of the point set up columnar samples, each point in the range of 20 cm – 300 cm to take three samples of surface layer, middle layer and deep layer, the monitoring factor is arsenic. See Figure 1 for site layout. In addition, in order to understand the quality of groundwater in the region, three groundwater monitoring points were arranged upstream and downstream of the landfill.

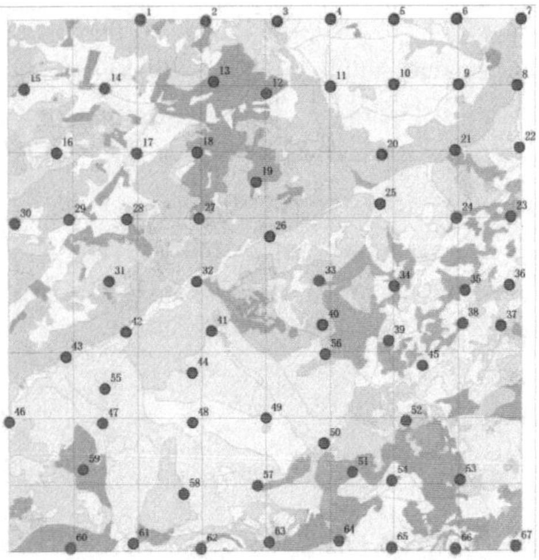

Figure 1. Soil sampling distribution map.

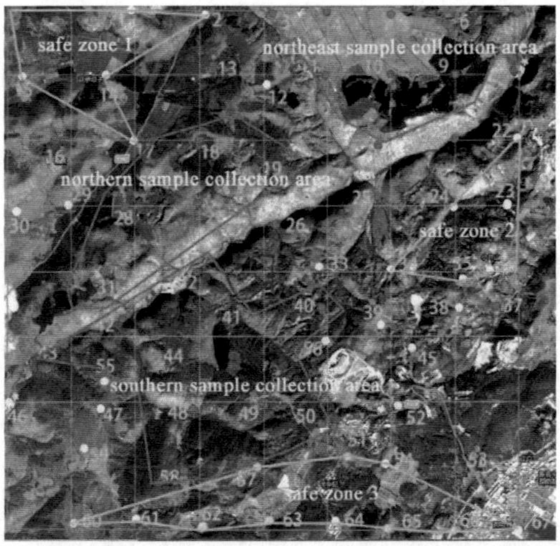

Figure 2. Crop sampling distribution map.

2.3 Collaborative monitoring plan for crops

Among the monitoring points set within the scope of agricultural land soil quality monitoring and investigation, according to the requirements of the "Soil Environmental Quality Agricultural Land Soil Pollution Risk Management and Control Standards (Trial)" (GB15618 - 2018), the soil environmental quality whose arsenic monitoring value exceeds the control value is classified as a control point, and the soil environmental quality that exceeds the screening value and is less than the control value is classified as a super-screening value point. When selecting the cooperative monitoring area for agricultural products, samples were taken from three areas beyond control, namely the southern monitoring area (for the main planted peppers, corn, and taro), and the northeastern monitoring area (for the main planted oranges, peaches, and sugarcane), and the northern monitoring area (mainly planted peppers, corn, and taro for use); The super-screening value distinguishes three areas for sampling, namely the eastern area (for the main planted corn), and the northwestern area (for the main planted peppers) Adoption), Southern District (adoption of the main planted corn). The specific collection method is: not belonging to the planting period, the corn samples that have been harvested are collected at the homes of the growers in the cultivated land in the collection area; peppers, peaches, oranges, green beans, and vegetables can be planted in season to obtain sample varieties, and refer to According to the "Technical Regulations on Preparation and Preservation of Agricultural Product Sample Collection and Transfer" (Environmental Office Soil [2017] No. 59), the collection area is divided according to 20 m × 20 m samples, and samples from 5 points are collected according to the serpentine method, and then mixed in equal amounts to form A mixed sample, the monitoring factor is arsenic. See Figure 2 for crop sampling layout.

2.4 Risk assessment methods for arsenic in soil

Arsenic is not volatile or semi volatile, and the exposure pathways for soil with arsenic monitoring values greater than the control values of construction land are considered to be oral intake of food grown in the soil, skin contact with the soil, and inhalation of soil particles. Using the health risk assessment method of the United States Environmental Protection Agency of the United States, and in combination with HJ25.3-2019 Technical Guidelines for Soil Pollution Risk Assessment on Construction Land[17], we analyzed the exposure of two types of typical land conditions and three exposure routes of the second type of land represented by industrial land in the study area, and selected carcinogenic and non-carcinogenic risk models to assess the health risks of contaminated soil on human beings, The selection of fixed parameters required for its calculation refers to the relevant provisions of HJ25.3 2019 and Regional screening levels (RSLs) general tables [18].

3 Results and discussion

3.1 Distribution characteristics of arsenic in soil

The groundwater quality of the three monitoring points in the monitoring area, 30 monitoring indicators including arsenic, can all meet the requirements of the Class III standard of the "Groundwater Quality Standard" (GB/T 14848 - 2017). Compared with the "Soil Environmental Quality Agricultural Land Soil Pollution Risk Control Standard (Trial)" (GB 15618 - 2018), the distribution of various value points is shown in Figure 3. According to statistics, among the 67 monitoring points, the arsenic content in the soil of 29 points exceeded the screening value, the arsenic content in the soil of 32 points exceeded the control value (up to 533 mg/kg), and only 6 points contained arsenic in the soil was lower than the screening value, and the results show that the regional soil arsenic content is high. The monitoring results show that the monitoring points for arsenic greater than the control value are mainly distributed from southwest to west to north to northeast of the region, mainly along the F1 fault zone and both sides of the region, and the monitoring points greater than the screening value are mainly concentrated in the southeast to east. The southeast side is the distribution area of the surrounding existing enterprises.

The monitoring value of arsenic in the soil around the distribution area of existing enterprises is generally higher than the monitoring value of arsenic in the soil of the distribution area of existing enterprises in the southeast. According to the monitoring results of arsenic in the columnar sample monitoring points shown in Figure 4, the surface monitoring value of the 45# columnar sample in the surrounding enterprise distribution area is greater than the monitoring value at the other two points, while the monitoring value of the 52# columnar sample at the deep layer is greater than the other two points monitoring value. Outside the distribution area of surrounding enterprises, the surface monitoring value of 51# columnar sample in the upwind direction is greater than the monitoring value at the other two points, and the deep monitoring value of 36# columnar sample in the downwind direction is greater than the monitoring value at the other two points. The monitoring data of other columnar samples in the surface layer, middle layer and deep layer also did not show a stable concentration gradient change. From the analysis of the monitoring results of the surface layer and columnar samples of the regional system, the high concentration of arsenic in the regional soil is not caused by industrial pollution, but may be caused by the high natural background value of arsenic in the regional soil layer.

The arsenic monitoring value at the surface layer of the columnar sample monitoring point 21# at the bottom of the gully in the F1 fault zone has an obvious upward trend compared with the monitoring value of arsenic monitoring at the monitoring point 26#, 32#, 42# and 43# on the gully bank of the fault zone , indicating that the arsenic in the soil is accumulated in the surface soil of the gully area at the bottom of the terrain under the action of soil erosion such as rainfall erosion.

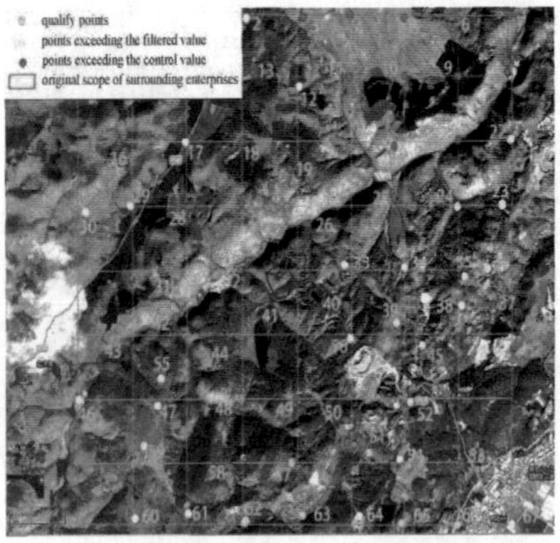

Figure 3. Soil arsenic monitoring results chart.

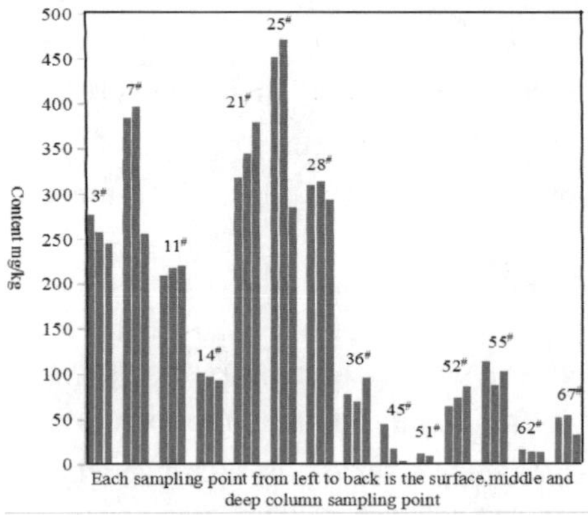

Figure 4. Columnar sample monitoring result chart.

3.2 Crop collaborative monitoring program

The crop monitoring results are shown in Table 1. Compared with the "National Food Safety Standard Food Contaminant Limits" (GB 2762 - 2017), the arsenic content in regional crops can reach the standard and can be eaten safely. The content of arsenic in the regional soil is relatively high, but the arsenic content of crops such as pumpkins, cabbage, oranges, and peppers planted is low, indicating that the arsenic absorption and enrichment effect of the crops grown in the region is weak.

Table 1. Agricultural product monitoring results table Unit: mg/kg

Food name	Division	Test results	Food name	Division	Test results
Fresh pumpkin	Northeast over-control value area	Not detected	Grain corn	Southern over-control value area	0.015
Fresh cabbage	Northeast over-control value area	Not detected	Fresh chili	Southern over-control value area	0.013
Fresh oranges	Northeast over-control value area	Not detected	Green beans	Southern over-control value area	Not detected
Fresh peaches	Northeast over-control value area	0.096	Corn	Eastern super screen value area (Safe Zone 1)	0.018
Fresh chili	Northeast over-control value area	Not detected	Fresh chili	Northwest super screen value area (Safe Zone 2)	0.01
Fresh chili	Northern over-control value area	Not detected	Corn	Southern super screen value area (Safe Zone 3)	0.023
Grain corn	Northern over-control value area	0.018			

Note: The detection index is total arsenic (calculated as As), and the standard value of "National Food Safety Standard Limits of Contaminants in Food" (GB 2762 - 2022) is 0.5 mg/kg.

3.3 Health risk assessment of arsenic in soil

According to the plan of the local government, the area will be mainly used as the second type of construction land. The average concentration of arsenic in the surface soil is calculated to be 185.8 mg/kg, which is used as the concentration of arsenic in the regional soil. The calculation and evaluation results are shown in Table 2. After calculation and evaluation, under the second type of land use conditions, the total non-carcinogenic hazard index of arsenic in the regional surface soil is 2.72, which is greater than the acceptable risk level of the acceptable total hazard index 1 of a single pollutant, and the carcinogenic risk value of arsenic in regional surface soil is 1.32E-04, which is greater than the acceptable risk level of 10^{-6} for a single pollutant. If the next step is to develop the area into a second-class construction land, its risk control value needs to be clarified for restoration and treatment or take measures for risk management and control.

Table 2. Table of calculation results of risk characterization under the second type of land use conditions Unit: dimensionless

Type	Route of exposure	Surface soil depth	0.2m
		Average monitoring value (mg/kg)	185.8
Cancer risk in soil	Hazard quotient for oral ingestion into soil	CR_{ois}	1.02E-04
	Hazard quotient for skin contact with soil	CR_{dcs}	1.84E-05
	Hazard quotient by inhalation route of soil particulate matter	CR_{pis}	1.16E-05
	Hazard index of a single pollutant via all exposure routes	CR_n	1.32E-04
Non-carcinogenic hazard quotient in soil	Hazard quotient for oral ingestion into soil	HQ_{ois}	1.37
	Hazard quotient for skin contact with soil	HQ_{dcs}	0.25
	Hazard quotient by inhalation route of soil particulate matter	HQ_{pis}	1.10
	Hazard index of a single pollutant via all exposure routes	HI_n	2.72

4 Conclusion and suggestions

The concentration of arsenic in the regional soil is high, and most of the monitoring points exceed the control value or screening value. Sampling, monitoring and comparative analysis of the surface and deep soil are carried out through grid distribution, especially the arsenic content in the soil of the enterprise distribution area and the general crop planting area. According to the comparative analysis, the high concentration of arsenic in the regional soil has no definite basis for industrial pollution, which may be caused by the high natural background value of arsenic in the regional soil layer. Through the systematic sampling and analysis of the crops in the area, the results show that the arsenic content in the crops is low, which can meet the standard requirements, and the risk of eating the crops is low.

According to the planning of the local government, the area will be mainly used as the second type of construction land. After calculation and evaluation, under the second type of land use conditions, the non-carcinogenic risk and total carcinogenic hazard of arsenic in the regional surface soil exceed the acceptable risk level. If the area is developed into the second type of construction land in the next step, it is necessary to clarify its risk control value for restoration and treatment or take measures for risk management and control.

Reference:

[1] Rodriguez-Lado L, Sun G, Berg M, et al. Ground water arsenic contamination throughout China. *Science*, 2013, **341** (6148): 866 - 868.

[2] Weng H X, Zhang X Y, Zou L J, et al. The natural existence and cause analysis of arsenic in Chinese soils. *Journal of Zhejiang University: Engineering Edition*, 2000, **34** (1): 88 - 92.

[3] Deng C N, Xu L Q, Li H X, Zhang H. Occurrence characteristics of arsenic in water and sediments after arsenic pollution control in Yangzong-Lake, Yunnan. *Environmental Science Guide*, 2021, **40** (05): 27 - 33.

[4] Zhao X Q, Li L J, Yang H H, et al. The geochemical characteristics of heavy metals in agricultural soils of the Bijiang Watershed in Yunnan Province. *Acta Geosciences Sinica*, 2012, **33** (3): 331 - 340.

[5] Wang J. Geochemical study on arsenic pollution in Gejiu tin polymetallic mining srea, Yunnan. *Yunnan: Kunming University of Science and Technology*, 2011.

[6] Zhou H Y, Cao M H, Wang L L, Lu X H. Health risk assessment of potential heavy metal contaminated site from a sulphuric acid plant. *Environmental Engineering*, 2014, **32** (08): 127 - 130.

[7] Luo H, Liu X M, Wanu S J, et al. Pollution characteristics and sources of cadmium in soils of the karst area in South China. *Chinese Journal of Ecology*, 2018, **37**(5): 1538-1544.

[8] Zhang F G, Feng M, Wang H Y, et al. Ecological Risk Assessment of Heavy Metals of Heavy Metals at Township Scale in the High Background of Heavy Metals, Southwestern, China. *Environmental Science*, 2020, **41**(9): 4197-4209.

[9] Tu C B, He T B, Liu C Q, et al. Effects of land use and parent materials on trace elements accumulation in topsoil. *Journal of Environmental Quality*, 2013, **42**(1): 103-110.

[10] Han X B, Xiong C, Luo Z K, et al. Arsenic Pollution Characteristics and Risk Assessment of an Industrial Site. *Journal of Hebei University of Environmental Engineering*, 2022, **32**(1): 54-58.

[11] Guo D K, Zhu G H, Chen J, et al. Pollution characteristics and health risk assessment of heavy metal in soil of abandoned sulfuric acid plant site. *Environmental Pollution and Prevention*, 2020, **42**(9): 1152-1157.

[12] Yao H S, Yang T M, He L P, et al. Assessment of Heavy Metal Pollution and Potential Ecological Risks of the Soils in Karst Farmland with High Background Levels of Cadmium and Arsenic in Eastern Yunnan. *Journal of Northwest Forestry University*, 2022, **37**(4): 29-26.

[13] Chen Q, Lan J H, Ma Z P et al. Investigation and risk assessment of heavy metal pollution sites in Guangdong Province. *Yunnan Chemical Technology*, 2022, **49** (09): 74 - 76.

[14] Qin X, Zhang C L, Li C B, et al. Characteristics and risk assessment of soil arsenic pollution in an arsenic plant site. *Journal of Guangxi University (Natural Science Edition)*, 2021, **46** (04): 1101 - 1109.

[15] Huang J H. Study on risk assessment of soil arsenic pollution in a plot. *Guangdong Chemical Industry*, 2021, **48** (01): 98 -101.

[16] Song B, Fu F Y, Zhang X H, et al. A survey of arsenic concentrations in vegetables and soils in Guilin and the human health risks assessment. *Acta Scientiae Circumstantiae*, 2014, **34** (03): 728 - 735.

[17] Ministry of Ecology and Environment. *Technical Guidelines for environmental Impact Assessment - Soil Environment (Trial)*: HJ964-2018.

[18] US Environmental Protection Agency. *Regional screening levels (RSLs)-generic tables[A/OL]*. (2017-01-19), [2020-09-20].

Study on strengthening of sulfide tailings high-efficiency solid-liquid separation using water reducing admixtures

Yingdi Dong[1,2], Yong Zeng[1,2], Zhongbao Hua[1,2], Li Wang[1,2], Honghu Tang[1,2,*], Wei Sun[1,2]

[1] School of Minerals Processing and Bioengineering, Central South University, Changsha 410083, China
[2] Engineering Research Center of Carbon Emission Reduction in Development and Utilization of Metal Resources, Ministry of Education, Central South University, Changsha 410083, China

E-mail: honghu.tang@csu.edu.cn

Abstract. The global focus on refined mineral processing and environmental protection has led to stricter requirements for tailings disposal processes. The complex composition and chemical environment of tailings pose challenges in sedimentation and pipeline transportation. This research developed a new additive that is suitable for disposing of copper-lead-zinc sulfide tailings. This additive not only improves sedimentation efficiency and concentration, but also enhances the flowability of high-concentration slurry. This study investigated the effects of four typical water reducing admixtures on the settling and rheological properties of tailings. However, the combination of lignin/PCE and NPAM had a synergistic effect, further improving settling and rheological performance. The addition of PCE increased settling rate by over 2 times, increased sediment concentration by 6.71%, and significantly reduced slurry viscosity and yield stress. The study also used FBRM to analyze the changes in particle/floc size distribution over time. The results showed that the addition of PCE captured more fine particles during flocculation and formed larger flocs. These findings can be used as a reference for the use and optimization of additives in tailings disposal engineering.

Keywords: Tailings; sedimentation; flocculation; rheology; water reducing.

1. Introduction

Copper, lead, and zinc ore are a very important and highly consumed commodity. The processing of copper-lead-zinc ore typically involves wet processing methods to separate them from tailings, resulting in a large amount of sulfide tailings. Based on statistical data [1], China's annual production of tailings has surpassed 1.2 billion tons, exhibiting a consistent upward trend, as shown in Figure 1. In 2020, the production of iron tailings and copper tailings constituted 42 % and 26 % of the total tailings production in China, respectively. The disposal of tailings poses significant challenges due to the complex mineral composition and solution chemistry [2]. Firstly, the fine particle size and high clay content of tailings can affect the efficiency of settling. Secondly, particle interactions, chemical additives, and rheological behavior in tailings slurry can also pose significant obstacles to efficient pumping and pipeline transportation. Besides, the addition of flocculants can lead to the free water be wrapped into flocs and increase the volume of sediment, occupying more tailings storage capacity.

Currently, there has a considerable amount of research focusing on particle surface modification and the development of new tailings disposal reagents for flocculation and settling [3–5]. However, research on using chemical reagents to control tailings rheology is limited. Therefore, it is necessary to explore new reagents to overcome this issue and improve the efficiency and sustainability of tailings disposal.

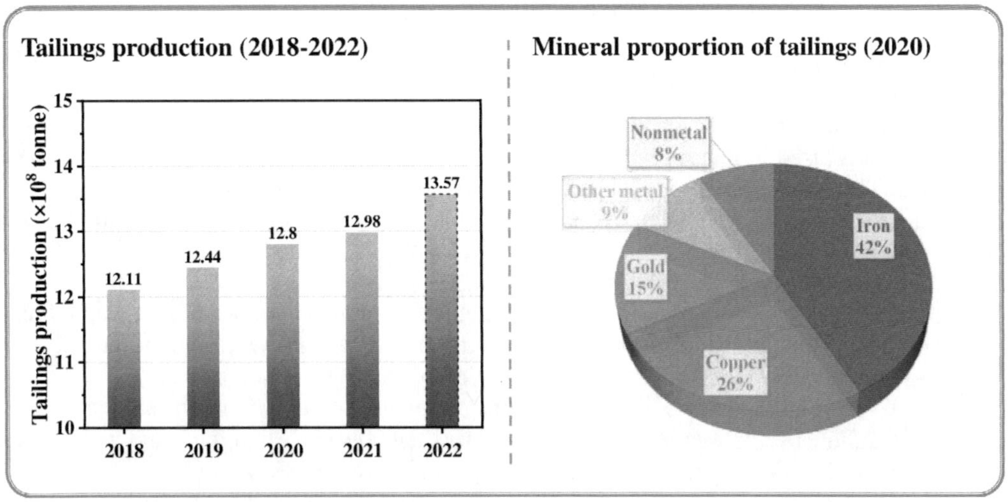

Figure 1. China's tailings production volume and the proportion of different types of tailings production.

The rheological and chemical/interface properties of tailings slurry are crucial parameters for thickened tailings transportation. The rheological properties of the slurry are influenced by various factors, including (but not limited to) slurry solids content, particle shape and size distribution, particle surface properties, additives, pH, and ionic strength [6,7]. A proper particle size distribution can allow more fine particles to fill the void spaces between large particles, achieving the minimum viscosity at high solid concentrations [8,9]. The surface properties of particles are closely related to the rheological behavior of the slurry, such as the anisotropic structure in clay minerals is accompanied by an increase in viscosity and yield stress [10,11]. Ionic strength, pH, and additives are the main methods for controlling the rheological behavior of mineral slurry, applied in engineering and industrial processes such as cement and concrete manufacturing, ceramic production, mineral flotation, tailings slurry, and coal-water slurry transportation. The essence of these methods is to control the rheological behavior of the slurry by adjusting the interparticle interaction forces. Ionic strength and pH can control the double-layer electrostatic repulsion, while suitable surfactants and high molecular weight adsorbing polymers/electrolytes can manipulate hydrophobic forces, spatial hindrance, and bridging forces [7,12].

Water reducing admixtures (WRAs), also known as plasticizers and high-efficiency water reducers, are commonly used additives in the production of high-strength concrete. Their purpose is to reduce the required water content and void volume in the concrete, while maintaining its compressive strength [13]. At the same time, they increase the fluidity of the concrete, making it easier for construction and pouring [14]. Improving the compactness and fluidity of the slurry is also an urgent requirement for tailings slurry pipeline transportation. Therefore, the compatibility of tailings with water reducing admixtures is also one of the focuses of this study. Lignosulfonates, naphthalene sulfonates, and polycarboxylate-based admixtures are well-known reagents that can reduce water content by 5-30% while considering slump and rheology [15,16].

This study aims to investigate the effects of flocculants and WRAs on the settling and rheological behavior of copper-lead-zinc tailings, and to find a combination of additives suitable for tailings disposal, to improve the settling efficiency and sediment concentration of tailings and improve the rheological performance of the slurry. The present study investigated the impact of flocculants and water reducing admixtures (WRAs) on the settling performance of tailings through the measurement of settling rate and sediment concentration of the slurry. Additionally, the rheological properties of the sediment were examined by measuring the shear stress of tailings sediment at various shear rates, in order to assess the effects of flocculants and WRAs. Furthermore, the growth behavior of flocs in tailings was analyzed by monitoring the chord length distribution of particles/flocs over time using FBRM technology, with a specific focus on the influence of WRAs.

2. Materials and methods

2.1. Tailings and reagents

The tailings samples used in this study were obtained from a copper-lead-zinc sulfide mine located in Jiangxi, China. The main mineral phases in the tailings are quartz, illite, mica, chlorite, and feldspar, with small amounts of pyrite and calcite. The particle size distribution of the tailings was tested using a laser diffraction particle size analyzer (Malvern Instruments Ltd., Worcestershire, UK), with a d_{50} of 18.75 μm. The d_{10} and d_{90} were 3.92 μm and 69.96 μm, respectively.

The tailings disposal reagents used in this study include flocculants and water reducing admixtures (WRAs). The flocculants are divided into cationic polyacrylamide (CPAM), anionic polyacrylamide (APAM), and non-ionic polyacrylamide (NPAM), with molecular weights around 10-11 million. Four different WRAs were used in the study, including sulphonated melamine formaldehyde (SMF), polynapthalene sulfonate/sulphonated naphthalene formaldehyde (SNF), lignin, and polycarboxylic ether (PCE). The molecular structure of water reducing admixtures are shown in Table 1.

Table 1. Molecular structure of water reducing admixtures

WRAs	SMF	SNF
Molecular structure	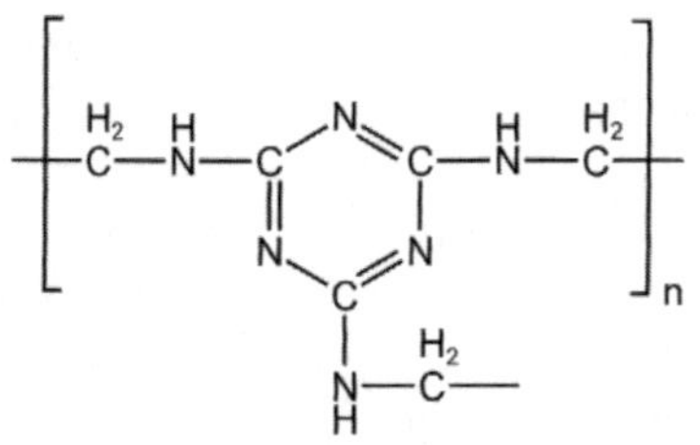	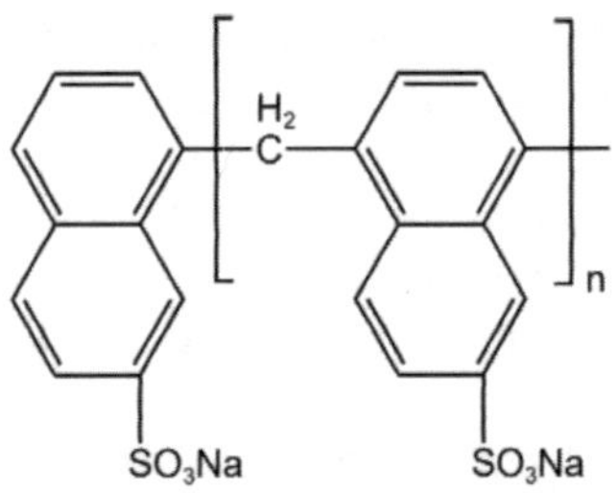
WRAs	Lignin	PCE
Molecular structure	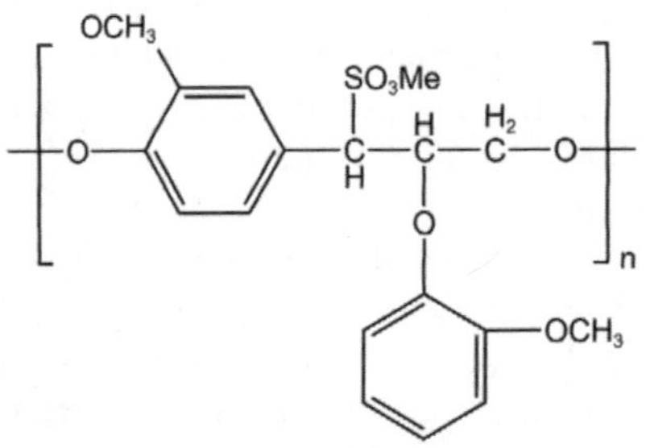	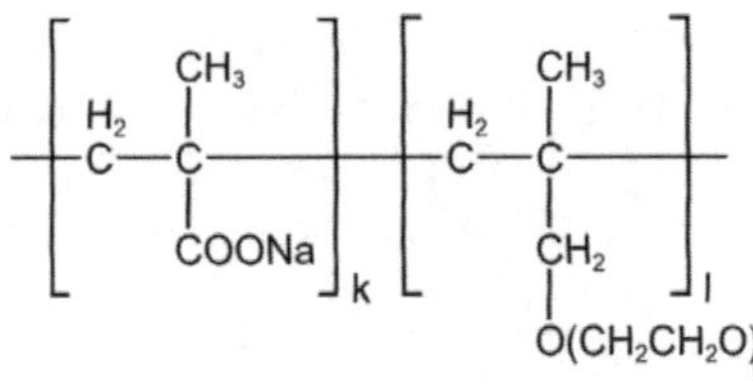

2.2. Sedimentation experiments

The sample prepared for the settling test was a copper-lead-zinc tailings slurry with a solid concentration of 25 wt%. First, 100 mL of the slurry was dispersed by vigorous stirring. The required amount of water reducing agent and flocculant was added at 200 rpm, with a dosing interval of 2 min. Then, the slurry was transferred to a 100 mL graduated cylinder and inverted 10 times. After completion, the mudline height was recorded at regular time intervals. The settling velocity can be determined by the slope of the mudline height versus settling time. The sediment concentration can be calculated based on the mudline height after 30 min of settling.

2.3. Rheological measurements
The rheological properties of the copper-lead-zinc tailings sediment were determined using a rheometer (Anton Paar MCR 102, Austria). The sample used for the rheological test was the sediment obtained after 1 h of settling. 40 mL of the sediment from the bottom was transferred to the rheometer cup for measurement. The testing procedure consists of three stages as follows:

(1) Pre-shearing stage. To homogenize the sample, the slurry was stirred at 200 s^{-1} for 60 s, followed by a 10 s rest.

(2) Up shear rate stage. Rheological data was measured starting from a shear rate of 1 s^{-1} and linearly increasing to 500 s^{-1}.

(3) Down shear rate stage. The shear rate was linearly decreased from 500 s^{-1} to 1 s^{-1}, and the test was ended.

The calculation of yield stress was obtained based on the Herschel-Bulkley model [17,18], with the following equation:

$$\tau = \tau_0 + K\dot{\gamma}^n \tag{1}$$

where τ is the shear stress (Pa), τ_0 is the yield stress (Pa), K is the consistency parameter (Pa·s^n), $\dot{\gamma}$ is the shear rate (s^{-1}), and n is the flow behavior index.

The shear stress forms a hysteresis loop with up shear rate and down shear rate. The area of the thixotropic loop, also known as the hysteresis area, represents the change in rheological properties of the fluid [19]. The hysteresis area can be expressed by the following equation [20,21]:

$$A = \int_{\gamma min}^{\gamma max} \eta_1 \dot{\gamma}\, d\dot{\gamma} - \int_{\gamma min}^{\gamma max} \eta_2 \dot{\gamma}\, d\dot{\gamma} \tag{2}$$

where A is the area of thixotropic loop (Pa/s), η_1 is the up apparent viscosity (Pa·s), η_2 is the down apparent viscosity (Pa·s), and $\dot{\gamma}$ is the shear rate (s^{-1}).

2.4. FBRM measurements
FBRM were conducted using a focused beam reflectance measurement probe (G400 with PI-14/206 probe, Mettler-Toledo LLC, USA) to monitor in real-time the changes in particle chord length and counts in the copper-lead-zinc tailings slurry after the addition of admixtures. The changes in chord length distribution obtained reflect the growth and evolution of flocs [22].

The sample prepared for FBRM was a 5 wt% tailings slurry. Firstly, 250 mL of the slurry was pre-dispersed by vigorous stirring for 5 min. The shear rate was set at 200 rpm, and FBRM started collecting data at a scanning rate of 2 s. During the test, the required amount of water reducing admixtures and flocculant were added at relative times of 3 min and 4 min, respectively. The test ended at a relative time of 15 min.

3. Results and discussion

3.1. Settling performance of tailings

3.1.1. Effects of ionic flocculants on settleability of tailings. Due to the presence of charges on the surface of tailings particles and also ions in the slurry solution, the addition of different ionic flocculants can change the interaction between particles and alter the stability of the slurry. Therefore, the settling behavior of copper-lead-zinc tailings with different ionic flocculants was investigated. The molecular weight of the flocculants was around 10-11 million, and the dosage was fixed at 60 g/t.

Figure 2 shows the effect of different ionic flocculants on the settling rate and sediment concentration of tailings. The mudline height vs. settling time curve indicated that the addition of CPAM had the slowest settling rate on tailings slurry, while adding NPAM had the most significant settling rate. During the mudline height decreased from 100% to 60%, it tended to decrease linearly with settling time for all three flocculants. Below 60%, it entered a slow settling stage. As shown in Figure 2(b), the sediment concentration after 30 min of settling was calculated, and the sediment concentrations of copper-lead-zinc tailings for adding CPAM, APAM, and NPAM were 43.34 %, 44.59 %, and 46.11 %, respectively. In the slurry, at the same solid mass, the higher the sediment concentration, the smaller the sediment volume. Therefore, the sediment formed after flocculation with NPAM has the smallest sediment volume. Thus, NPAM in the flocculant significantly improved the settling performance of copper tailings slurry.

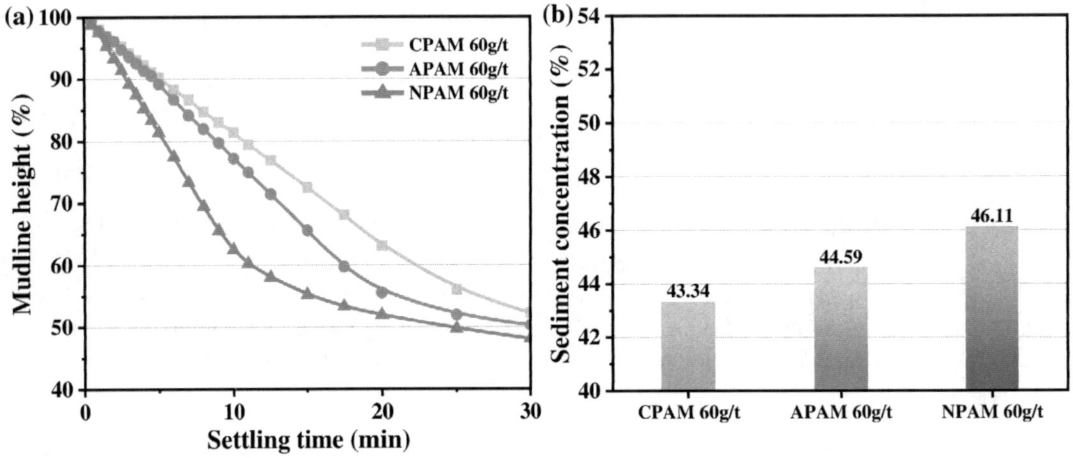

Figure 2. Effects of ionic flocculants on settleability of tailings: (a) mudline height; (b)sediment concentration.

3.1.2. Effects of WRAs on settleability of tailings. In order to further improve the settling rate and sediment concentration of copper tailings, water reducing admixtures were used as surfactants in tailings disposal. The wettability or/and electrical property of particle surfaces could be regulated through the functional groups of WRAs. Therefore, the effects of different types of WRAs on the settling performance of tailings were investigated.

As shown in Figure 3(a), compared to the reference curve with only NPAM added, the addition of SMF and SNF hindered the flocculation and settling of tailings, while lignin and PCE further improved the settling rate, with PCE showing the highest settling rate. Although lignin improved the settling rate of tailings, it did not further concentrate the sediment concentration, with sediment concentration similar to that with only NPAM added, at 46.51% and 46.15% respectively. However, the synergistic effect of PCE and NPAM increased the sediment concentration to 50.68%, which was 4.53% higher than that with only NPAM added. Therefore, PCE has a dual effect of improving the settling rate and sediment concentration of tailings and is an ideal agent for solid-liquid separation of tailings among the water reducing admixtures.

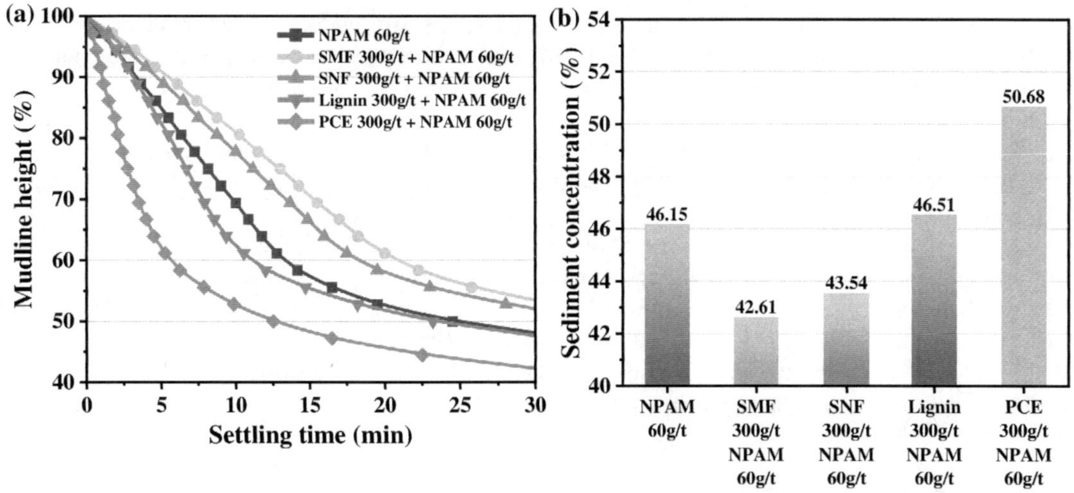

Figure 3. Effects of WRAs on settleability of tailings: (a) mudline height; (b)sediment concentration.

3.1.3. Effects of WRAs dosages on settleability of tailings. To investigate the effect of dosage of WRAs on the settling rate of tailings, settling experiments were conducted in 25 wt% tailings slurry with different PCE dosages (0, 150, 300, 450 g/t) and a fixed NPAM dosage of 60 g/t.

Figure 4(a) shows that the settling rate of copper-lead-zinc tailings slurry significantly increased with the PCE dosage. When the mudline height decreased to 60%, the settling times for only adding NPAM, and for adding PCE 450 g/t and NPAM were 10.5 min and 4.7 min, respectively, indicating a more than 2 times increase in settling rate. Figure 4(b) shows the sediment concentration after 30 min of slurry settling for different PCE dosages. As the PCE dosage increased from 0 to 450 g/t, the sediment concentration increased from 45.92% to 52.63%, representing a 6.71% increase. These results indicate that increasing the dosage of PCE was beneficial for improving sediment concentration and reducing sediment volume. Therefore, both the settling rate and sediment concentration of tailings significantly increased with the increase in PCE dosage.

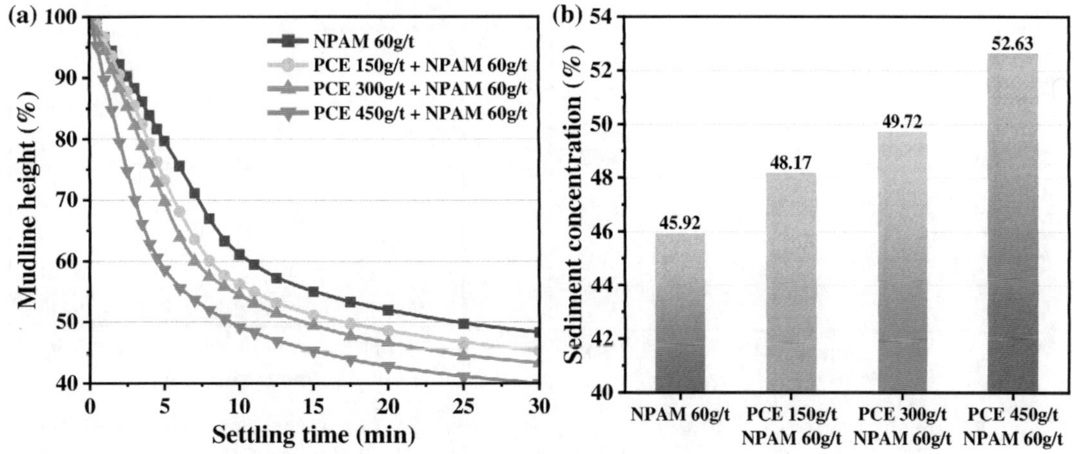

Figure 4. Effects of WRAs dosages on settleability of tailings: (a) mudline height; (b)sediment concentration.

3.2. Rheological performance of tailings

To investigate the differences in rheological behavior of the slurry under different reagent conditions during pipeline transportation, rheological performance tests were conducted using a rheometer, as shown in Figure 5. As the shear rate increased linearly from 1 s^{-1} to 500 s^{-1}, the shear stress of the slurry was recorded, and an ascending curve was formed. Subsequently, the test results formed a descending curve as the shear rate decreased linearly from 500 s^{-1} to 1 s^{-1}. Finally, these two curves formed a thixotropic loop, and the area between the curves was taken as the hysteresis area. Based on the functional relationship between shear rate and shear stress, it can be observed that the samples were all non-Newtonian fluids. According to the Herschel-Bulkley model, the shear stress at shear rate approaching 0 approximates the yield stress, which is the critical stress value for the fluid to start flowing. The results showed that the fluid with only NPAM added had the highest yield stress value, which was 38.21 Pa. The yield stress values for the samples with no additive and with lignin and NPAM added were close, at 13.54 Pa and 11.76 Pa, respectively. The slurry with PCE and NPAM added had the lowest yield stress value, which was 3.15 Pa. Therefore, the addition of PCE facilitated the initiation of slurry flow.

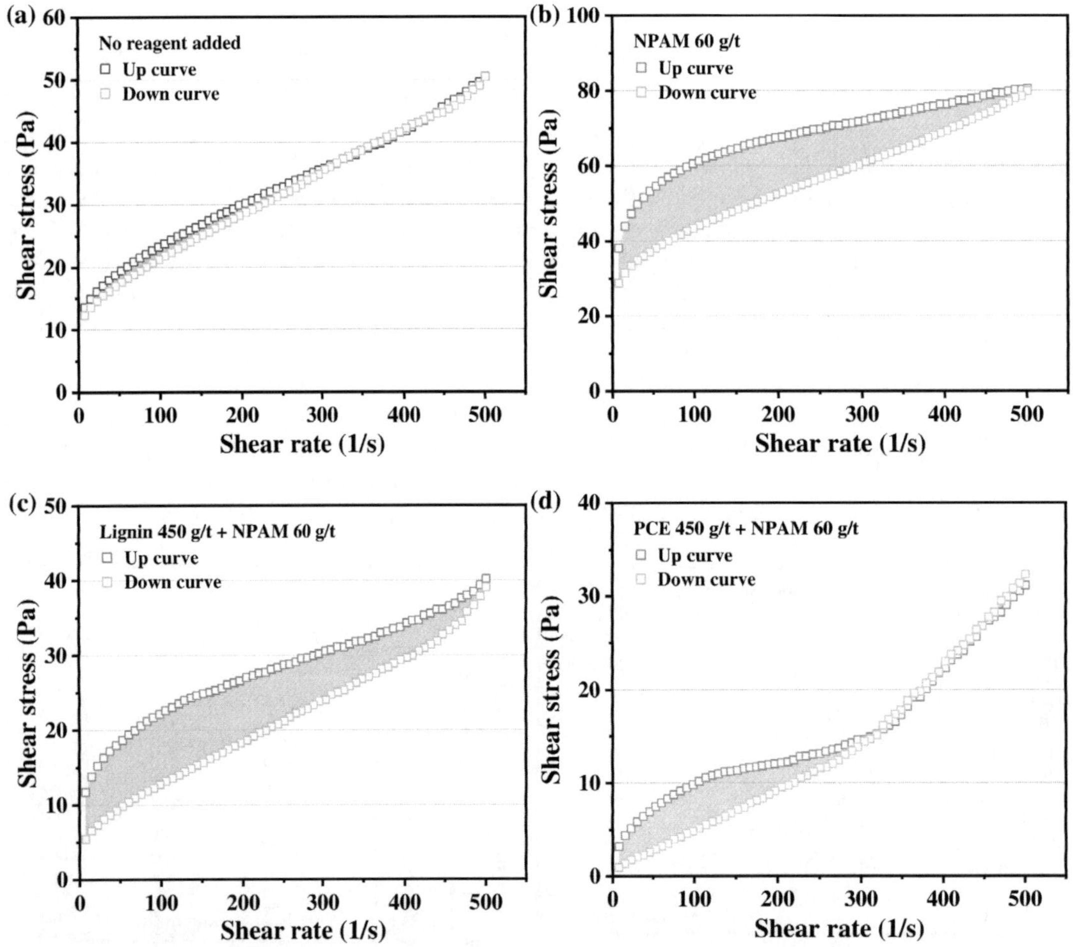

Figure 5. Effects of WRAs on shear rate of tailings slurry: (a) no reagent added; (b) adding 60 g/t NPAM; (c) adding 450 g/t lignin and 60 g/t NPAM; and (d) adding 450 g/t PCE and 60 g/t NPAM.

Figure 6(a) shows the hysteresis area of tailings slurry under different reagent addition. Flocculants are commonly used additives for solid-liquid separation of fine tailings, which significantly increased the thixotropy of the slurry and cause changes in slurry properties during pipeline transportation due to shear. The addition of lignin and PCE both reduced the high thixotropy caused by flocculants. The lowest hysteresis area (887.2 Pa/s) was achieved when 450 g/t PCE and 60 g/t NPAM are added, which was approximately 6.5 times lower than the hysteresis area (5782.5 Pa/s) when only NPAM was added. Thereby, the addition of PCE enhanced the stability of the slurry during pipeline transportation. Figure 6(b) shows the variation of apparent viscosity of tailings slurry with shear rate under different reagents addition. The results indicated that the addition of flocculants significantly increased the apparent viscosity of the slurry and reduced its flowability. However, the addition of WRAs such as lignin and PCE can reduce the apparent viscosity of the slurry, even lower than that of the original slurry. The synergistic effect of PCE and NPAM reduced the apparent viscosity of the slurry to the lowest level, making the slurry highly flowability, which is beneficial for improving the efficiency of slurry transportation and reducing transportation costs.

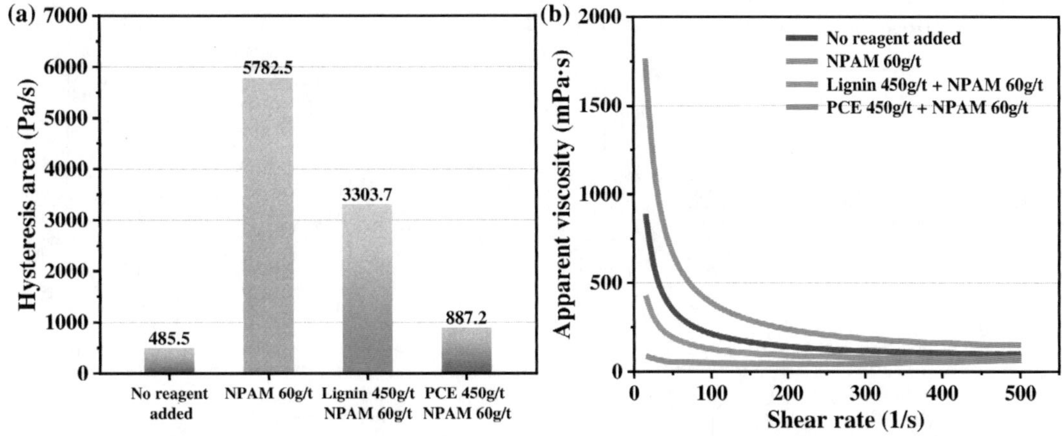

Figure 6. Effects of WRAs on rheological properties of tailings slurry: (a) hysteresis area; (b) apparent viscosity.

3.3. *Floc formation of tailings*

The growth behavior of flocs in copper-lead-zinc tailings particles with various dosages of PCE was investigated by monitoring the changes in particle/floc size (chord length) and count in real-time using FBRM. Firstly, 250 mL of 5 wt% tailings slurry was pre-dispersed by vigorous stirring at a shear rate of 500 rpm for 5 min. Then, the shear rate was reduced to 200 rpm and FBRM data recording was started. Various dosages of PCE and a fixed amount of NPAM were added at relative times of 3 and 4 minutes, respectively. And the shear rate was maintained at 200 rpm until the end of the test at relative time 15 min. Figure 7 shows the chord length distribution of tailings particles/flocs at different time points with various dosages of PCE.

By analyzing the changes in chord length distribution at different times, the growth behavior of flocs formed by particles in the slurry was studied. The chord length distribution of tailings particles/flocs at different time points with different dosages of PCE showed some similarities. Within the relative time of 0 – 5 min, the count of particles in the range of 10 – 50 μm decreased, and > 50 μm flocs began to form gradually. Within the relative time of 5 – 10 min, the count of < 10 μm particles decreased significantly, while the count of flocs > 50 μm flocs increased, and the flocs were mainly concentrated in the range of 100 – 110 μm. Within the relative time of 10 – 15 min, the count of fine particles (< 10 μm) decreased, while the count of > 50 μm flocs increased significantly and the distribution peak shifted to the right. Therefore, the growth behavior of flocs may involve the bridging of relatively larger particles (10 – 50 μm) followed by the gradual capture of fine particles (< 10 μm).

Comparison of the differences in particle/floc chord length distribution at different PCE dosages at the same relative time, the effect of PCE dosage on floc growth behavior was studied. At relative time of 0 min, the chord length distribution of particles at different PCE dosages was basically consistent, indicating that the samples were well dispersed and had similar particle size and number. At relative time of 5 min, with the condition of only adding NPAM, the count of > 50 μm flocs was the highest and decreased with increasing PCE dosage. However, as the PCE dosage increases, the peak of chord length distribution for > 50 μm flocs shifted to the right, indicating the particles formed to larger flocs. At relative time of 10 min, with the increase of PCE dosage, < 10 μm particles showed a decreasing trend, while the number and size of > 10 μm particles increased. At relative time of 15 min, there were significant differences in the chord length distribution results. With the increase of PCE dosage, the number and size of fine particles (< 10 μm) decreased significantly, indicating that more fine particles were captured by the flocs. At the same time, the number and size of > 50 μm flocs increased significantly. Therefore, the PCE additive has the ability to capture more fine particles in tailings and can efficiently flocculate, forming more and larger flocs.

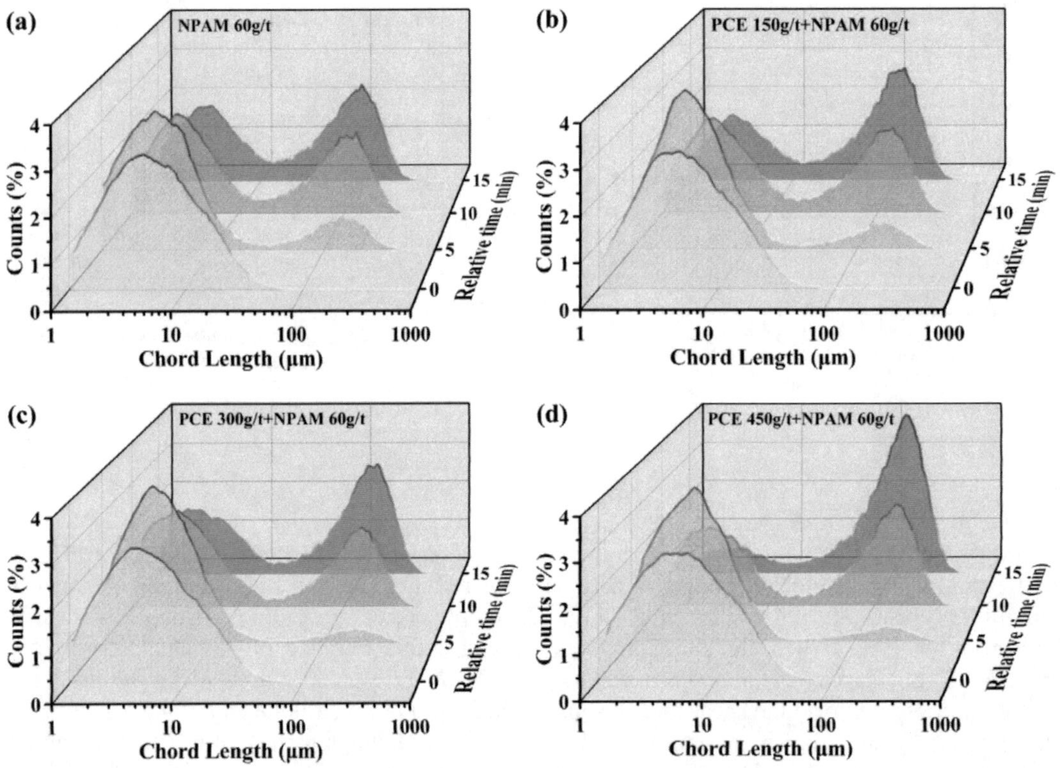

Figure 7. Effects of WRAs dosages on floc formation of tailings slurry: (a) adding 60 g/t NPAM; (b) adding 150 g/t PCE and 60 g/t NPAM; (c) adding 300 g/t PCE and 60 g/t NPAM; and (d) adding 450 g/t PCE and 60 g/t NPAM.

In terms of tailings settling, due to the effect of PCE, more fine particles can be captured and formed larger and denser flocs during flocculation, which was beneficial for the rapid settling of tailings and the reduction of sediment volume. In terms of tailings slurry pipeline transportation, the bridging effect of flocculants increased the interaction between particles, resulting in increased internal friction of the slurry during flow. The PCE additive could reduce the number of fine particles in the slurry, while forming larger flocs to reduce the specific surface area, thereby reducing the internal friction of the slurry, and improving its flowability.

3.4. The potential application of WRAs on tailings disposal process

The solid-liquid separation of tailings not only needs to achieve rapid settling, but also needs to consider the subsequent process. In the industrial processes of tailings slurry pumping and pipeline transportation, there has always been an unresolved contradiction. Efficient tailings transportation requires high-concentration slurry as a prerequisite. However, increasing the slurry concentration leads to decreased slurry flowability, increased additional energy consumption costs, and even the possibility of pipeline blockage. On the other hand, transporting low-concentration slurry not only has low efficiency but also increases the cost of wastewater treatment. Therefore, to achieve efficient pipeline transportation of high-concentration slurry, external forces need to be added. In this study, two WRAs (PCE and lignin) were introduced into tailings disposal, which adsorbed onto the surface of mineral particles and changed the interaction forces between particles. Firstly, this method promoted the capture of more fine particles by flocs, reducing the void volume in flocs, and forming high-concentration sediment. Secondly, this method also promoted the improvement of slurry flowability, solving the problem of balancing high-concentration slurry and easy flow, and achieving efficient pipeline transportation of slurry. As shown in Figure 8, the potential application of WRAs on tailings disposal process is depicted in the schematic.

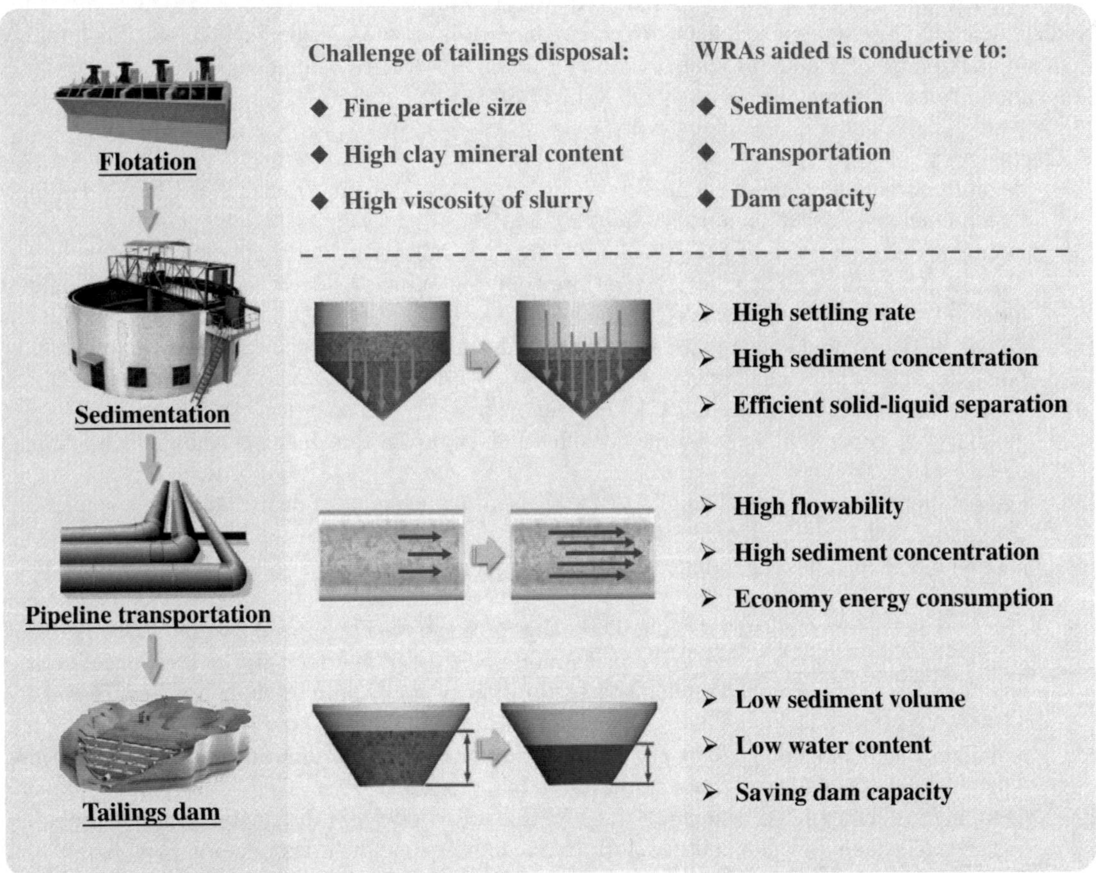

Figure 8. The schematic of the potential application of WRAs on tailings disposal process.

4. Conclusion

Efficient solid-liquid separation of tailings not only requires rapid settling and higher sediment concentration, but also needs to maintain well flowability to save energy consumption costs for pipeline transportation of tailings slurry. In this study, the effects of various flocculants and WRAs on the settling performance and rheological properties of copper-lead-zinc tailings was investigated, and the suitable WRAs for tailings disposal was determined. The results indicated that SMF and SNF were not conducive to the settling and concentration of tailings. However, both lignin and PCE, as water reducing admixtures, contributed to the increase in settling rate and sediment concentration. The addition of PCE had the highest settling efficiency, increasing the settling rate by more than 2 times and the sediment concentration by 6.71 %. In terms of rheological properties, PCE was more effective in regulating the flowability of tailings slurry. After adding 450 g/t PCE, the thixotropic loop area decreased by 6.5 times and the slurry viscosity and yield stress also decreased significantly. The effect of PCE on floc growth behavior in the slurry was studied through FBRM test. The results showed that the assistance of PCE enabled more fine particles to be captured during the flocculation process and formed more and larger flocs. This study confirmed that PCE was a tailings disposal reagent beneficial to settling and pipeline transportation processes, providing potential application prospects for efficient solid-liquid separation of tailings.

Acknowledgments
This research was supported by the National Key R&D Program of China (2022YFC2904501), the Natural Science Foundation of Hunan Province, China (2023JJ20071), the Science and Technology Innovation Program of Hunan Province, China (2023RC3067) and Postgraduate Scientific Research Innovation Project of Hunan Province (No. CX20220197).

References

[1] Anon Report on the Current Competitive Status and Development Prospects of China's Tailings Comprehensive Utilization Industry from 2023 to 2029
[2] Leong Y-K 2021 Controlling the rheology of iron ore slurries and tailings with surface chemistry for enhanced beneficiation performance and output, reduced pumping cost and safer tailings storage in dam *Miner. Eng.* **166** 106874
[3] Zhang Y, Gao W and Fatehi P 2019 Structure and settling performance of aluminum oxide and poly(acrylic acid) flocs in suspension systems *Sep. Purif. Technol.* **215** 115–24
[4] Wu W, Ma J, Xu J and Wang Z 2021 Mechanistic insights into chemical conditioning by polyacrylamide with different charge densities and its impacts on sludge dewaterability *Chem. Eng. J.* **410** 128425
[5] Feng B, Peng J, Zhu X and Huang W 2017 The settling behavior of quartz using chitosan as flocculant *J. Mater. Res. Technol.* **6** 71–6
[6] Farrokhpay S 2012 The importance of rheology in mineral flotation: A review *Miner. Eng.* **36–38** 272–8
[7] Zhu L, Lyu W, Mao X, Zhao Z, Yang D, Zhang H, Wang K, Yang P and Zeng H 2023 Effect of solution pH and polyethylene oxide concentration on surface/interface properties, flocculation and rheology of concentrated monodisperse ultrafine synthetic tailings slurry *Powder Technol.* 119002
[8] Adiguzel D and Bascetin A 2019 The investigation of effect of particle size distribution on flow behavior of paste tailings *J. Environ. Manage.* **243** 393–401
[9] Senapati S, Pothal J K and Mohanty A 2019 Effect of particle size distribution on rheology of high concentration limestone–water slurry for economic pipeline transportation *Part. Sci. Technol.* **37** 707–15
[10] Blachier C, Jacquet A, Mosquet M, Michot L and Baravian C 2014 Impact of clay mineral particle morphology on the rheological properties of dispersions: A combined X-ray scattering, transmission electronic microscopy and flow rheology study *Appl. Clay Sci.* **87** 87–96
[11] Zhang M and Peng Y 2015 Effect of clay minerals on pulp rheology and the flotation of copper and gold minerals *Miner. Eng.* **70** 8–13

[12] Johnson S B, Franks G V, Scales P J, Boger D V and Healy T W 2000 Surface chemistry–rheology relationships in concentrated mineral suspensions *Int. J. Miner. Process.* **58** 267–304

[13] De Schutter G, Ezzat M, Lesage K and Hoogenboom R 2023 Responsive superplasticizers for active rheology control of cementitious materials *Cem. Concr. Res.* **165** 107084

[14] Lokesh G, Nantha Kumar P, Aishwarya N, Parthiban P and Ponshanmugakumar A 2023 Experimental study of the effect of water reducing admixtures on concrete *Mater. Today Proc.*

[15] Shrivastava A K and Kumar M 2016 Compatibility issues of cement with water reducing admixture in concrete *Perspect. Sci.* **8** 290–2

[16] Ercikdi B, Cihangir F, Kesimal A, Deveci H and Alp İ 2010 Utilization of water-reducing admixtures in cemented paste backfill of sulphide-rich mill tailings *J. Hazard. Mater.* **179** 940–6

[17] Hu S, Chen P, Wang Y, Wei C and Li P 2022 Ionic strength and temperature dependence of rheological behavior of deep-water sedimental slurries *Ocean Eng.* **265** 112655

[18] Du M, Liu J, Clode P and Leong Y-K 2019 Microstructure and rheology of bentonite slurries containing multiple-charge phosphate-based additives *Appl. Clay Sci.* **169** 120–8

[19] Chen M, Liu B, Li L, Cao L, Huang Y, Wang S, Zhao P, Lu L and Cheng X 2020 Rheological parameters, thixotropy and creep of 3D-printed calcium sulfoaluminate cement composites modified by bentonite *Compos. Part B Eng.* **186** 107821

[20] Chen M, Yang L, Zheng Y, Huang Y, Li L, Zhao P, Wang S, Lu L and Cheng X 2020 Yield stress and thixotropy control of 3D-printed calcium sulfoaluminate cement composites with metakaolin related to structural build-up *Constr. Build. Mater.* **252** 119090

[21] Kolawole J T, Combrinck R and Boshoff W P 2019 Measuring the thixotropy of conventional concrete: The influence of viscosity modifying agent, superplasticiser and water *Constr. Build. Mater.* **225** 853–67

[22] Wang C, Sun C and Liu Q 2020 Formation, breakage, and re-growth of quartz flocs generated by non-ionic high molecular weight polyacrylamide *Miner. Eng.* **157** 106546

Research on resource comprehensive utilization of granite-type sulfide mine tailings

Boai Zhang[a,b], Hong Zhang[a,b], Rui Xu[a,b], Li Wang[a,b*], Honghu Tang[a,b,*]

[a] School of Minerals Processing and Bioengineering, Central South University, Changsha 410083, Hunan, PR China

[b] Engineering Research Center of Ministry of Education for Carbon Emission Reduction in Metal Resource Exploitation and Utilization, School of Mineral Processing and Bio Engineering, Central South University, Changsha 410083, Hunan, PR China

* Corresponding author. E-mail address: li_wang@csu.edu.cn (Li Wang) honghu.tang@csu.edu.cn (Honghu Tang)

Abstract

In order to recover and utilize mica and quartz from the tailings of a granite-type sulfide ore in Jiangxi Province, the technique of separating mica and quartz from the tailings by flotation without acid was investigated. Pyrite flotation was performed on a 500 g tailing sample to remove pyrite. And the middlings were acquired by desliming of the unfloated products. Finally mica and quartz was separated from the middlings in the mechanical flotation cell. Under conditions of pH=10.3 with a pulp concentration of 35% and a collector concentration of 180 g/t, a yield of 12.89% mica and 42.73% quartz was obtained, with silica grades as high as 71.50% in the quartz. The mica and quartz products can be used in construction ceramics and construction sand, respectively. It is believed that the outcomes of this work could provide ideas for comprehensive utilization of sulfide tailings.

Keywords: granite-type sulfide mine tailings; mica; quartz; pyrite

1. Introduction

Mineral resources are an important material foundation for the country's economic construction and social development. With mineral processing developing, billions of tons of tailings are rejected as a major waste stream, which results in dramatical environment pollution and waste of resources[1-3]. China has a cumulative tailings stock of more than 20 billion tons, and annual production is growing at a rate of 1.2 billion tons. The direct soil pollution caused by tailings covers an area of over a million acres, while the indirect soil pollution covers an area of 667,000 hectares in China. For this reason, the implementation of waste-free discharge of tailings is required to maximize the utilization of mineral resources and protect the ecological environment[4, 5]. Non-ferrous tailings are an important component of tailings, and most non-ferrous tailings are composed of mica, quartz, feldspar and other non-metallic minerals[6-9]. These components cannot be used effectively, resulting in low tailings consumption and a huge waste of resources.

Since 2015, a series of major decisions and deployments regarding the construction of green mines have been introduced in China, which vigorously promoted the full utilization of tailings, and strictly controlled the construction of new tailings ponds. Therefore, a comprehensive utilization of bulk nonmetallic resources in the tailings is imperative. At present, the comprehensive utilization of tailings consists of two main aspects: the recovery of useful minerals as secondary resources and the direct utilization of tailings by directly classifying them into one or several types of non-metallic ores for utilization without re-selection.

The main purpose of the work was to develop the core technology of separating mica and quartz in granite-type sulfide mine tailings under acid-free conditions in the laboratory, in order to achieve the purpose of tailings emission reduction

2. Experimental

2.1 Experimental materials

A run of granite-type sulfide mine tailings was obtained from a processing plant in Jiangxi province, China. Table. 1 displayed the particle size composition of tailings. The XRD spectrum and phase compositions of the tailings were shown in Fig. 1 and Table. 2. Table. 1 illustrated that the proportion of -0.074 mm particles was 69.64%, of which -0.038 mm particles was 51.17%, indicating that the particle size distribution of the tailings was fine and the samples should be desliming. Fig.1 displayed the samples contained principally quartz and mica, followed by small quantities of siderite, pyrite, chlorite, etc. The content of SiO_2 in the samples was low, while the content of Al and K elements were high.

Table 1. Particle size composition of sulfide tailings

Particle class /mm	Yield /%	Accumulative total/%
0.15	5.64	5.64
-0.15+0.074	24.72	30.36
-0.074+0.038	18.47	48.83
-0.038	51.17	100.00
total	100.00	

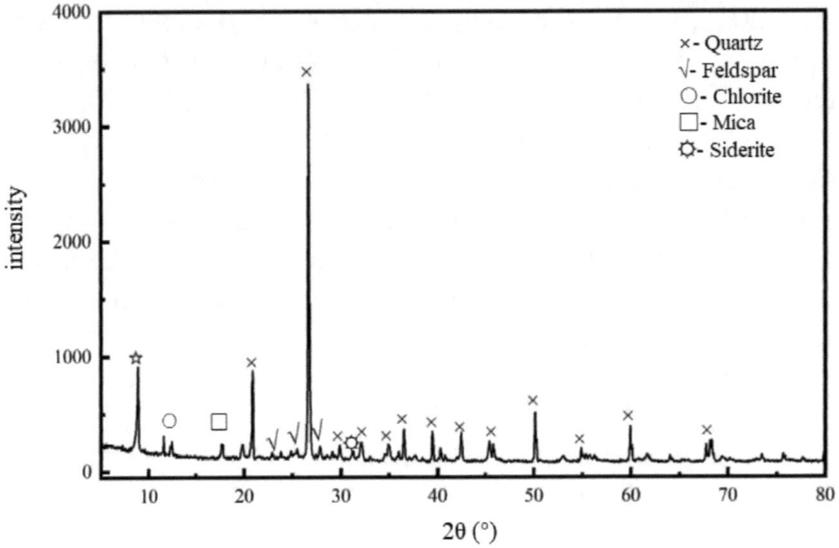

Fig.1 XRD spectrum of tailings particles

Table.2 Phase analysis of tailings

Phase	SiO_2	Al_2O_3	Fe_2O_3	K_2O	MgO
content /%	59.63	18.69	6.47	4.96	1.01
Phase	SO_3	TiO_2	CaO	P_2O_5	Na_2O
content /%	2.94	0.24	0.87	0.11	0.21

The chemical reagents, such as NaOH, HCl, xanthate, terpenic oil used in tests were obtained from Sinopharm Chemical Pharmaceutical Co. Ltd. (Shanghai, China). The BT-3 as a collector was synthetic. The water used for the tests was tap water.

2.2 Flotation tests

2.2.1 Pretreatment of granite-type sulfide tailings

The tailings samples contained some pyrite and the proportion of –0.0385 mm particles were high. Therefore, the samples were supposed to be pre-treated. Flotation flowchart was shown in Fig. 2. Flotation experiments were performed in an XDF-1000 machine at 1900 rpm to recover the valuable mineral from the tailings. First, 500 g of tailings samples and 900 mL of water were added to the cell and stirred for 3 min. The xanthate and terpenic oil were then added and stirred for 3 min. Flotation was performed for 6 min after pre-conditioning, and the floated (pyrite) products were received, filtered, and dried, while the unfloated products were deslimed. And the acquired slime was rejected and the middling was treated further.

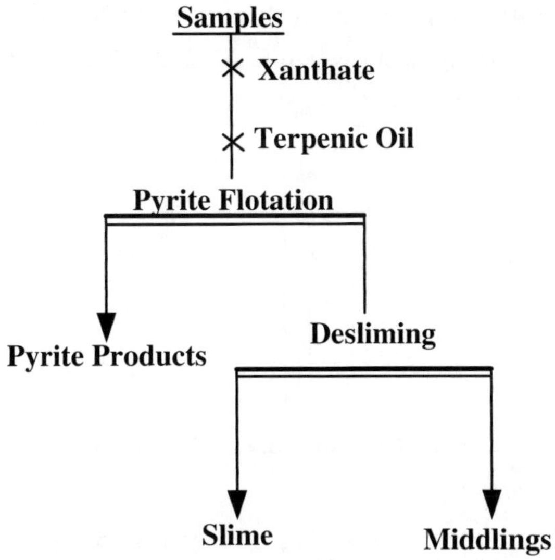

Fig. 2 Pre-processing flowchart of granite-type sulfide mine tailings

2.2.2 Flotation separation of mica and quartz

Fig.3 displayed the flowchart of flotation separation of mica and quartz. After desliming, the collector was added, and the pH of the pulp was adjusted to the experimental demand with NaOH or HCl and stirred for 4 min. Finally, the mica flotation was performed for 5 min, and the mica products and quartz products were recovered, filtered and dried.

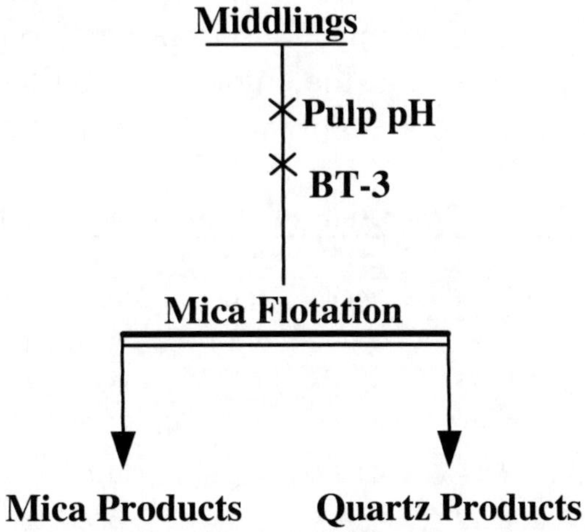

Fig.3 Flowchart of flotation separation of mica and quartz

3 Results and discussion

3.1 Pyrite flotation

Fig.4 showed the influence of dosage of xanthate on the yield of pyrite (a) and the XRD spectrum of pyrite products (b). With the increase of xanthate dosage, the yield of pyrite increased, while the yield of quartz in the foam also increased. Due to entrainment, a large amount of quartz floated up with the increase of xanthate dosage, which led to a massive loss of quartz. Therefore, the optimal dosage of xanthate was 50 g/t.

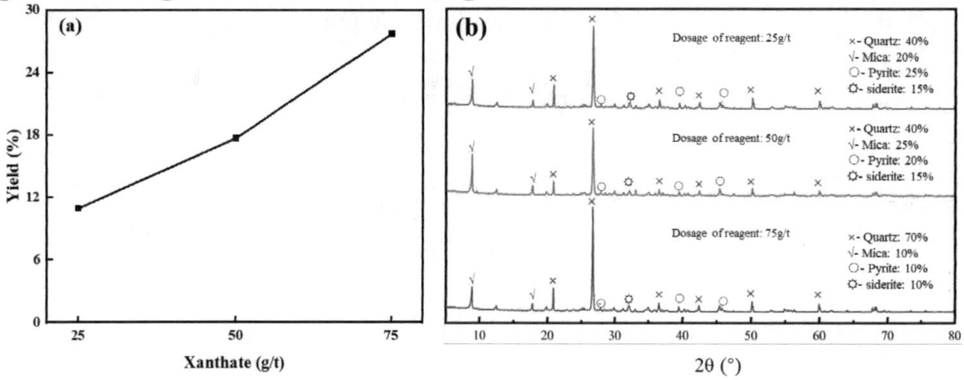

Fig.4 Pyrite flotation yield (a) and XRD spectrum of pyrite products (b) at different dosage of xanthate.

3.2 Mica flotation

3.2.1 Influence of pH on the flotation results

The influence of pH on the flotation results of mica with BT-3 (240 g/t) as a collector was displayed in Table. 3. As shown in Table. 3, as the pH of the pulp increased from 9.0 to 10.8, the yield of mica continuously increased from 4.54% to 22.07%. And the grade of SiO_2 in the quartz product initially increased to 71.60% and then decreased to 69.48%. When the pH was increased to 10.3, the froth flotation phenomenon of mica was improved. What's more, the grade of SiO_2 in the quartz product was the highest, 71.60%. Taking all factors into consideration, the optimal flotation pH was determined to be 10.3.

Table. 3 Test results of mica flotation at different pH

pH	Product	Yield/%	Grade/%			Recovery/%		
			K_2O	Na_2O	SiO_2	K_2O	Na_2O	SiO_2
9	Mica	4.54	3.14	0.10	56.48	6.35	6.15	7.06
	Quartz	50.65	4.15	0.14	66.61	93.65	93.85	92.94
	Tailings	55.19	4.07	0.13	65.78	100.00	100.00	100.00
9.7	Mica	7.02	4.42	0.12	59.78	13.80	11.40	11.51
	Quartz	48.22	4.02	0.14	66.93	86.20	88.60	88.49
	Tailings	55.24	4.07	0.13	66.02	100.00	100.00	100.00
10.3	**Mica**	**16.32**	**4.98**	**0.18**	**53.84**	**36.17**	**38.73**	**32.48**
	Quartz	**38.86**	**3.69**	**0.12**	**71.60**	**63.83**	**61.27**	**67.52**
	Tailings	**55.18**	**4.07**	**0.13**	**66.35**	**100.00**	**100.00**	**100.00**
10.8	Mica	22.07	4.78	0.16	61.84	47.01	47.86	37.27
	Quartz	33.07	3.60	0.12	69.48	52.99	52.14	62.73
	Tailings	55.14	4.07	0.13	66.42	100.00	100.00	100.00

3.2.2 Influence of pulp concentration on the flotation results

At pH 10.3, the effect of pulp concentration on the flotation yield of mica with BT-3 (240 g/t) as a collector was shown in Table. 4. Table. 4 showed an increase from 20% to 40% in the pulp concentration led to an increase from 9.81% to 17.41% in the yield of mica. As the flotation concentration exceeded 35%, the yield of mica remained relatively constant, but the amount of impurities in the mica products significantly increased. All things considered, the optimal pulp concentration for flotation was 35%.

Table 4 Yield of mica at different pulp concentration

Pulp concentration/%	Yield of mica/%
20	9.81
30	10.25
35	**16.32**
40	17.41

3.2.3 Influence of collector dosage on the flotation results

At pH 10.3 and pulp concentration 35%, the influence of collector dosage on the flotation results was presented in Table. 5. Table. 5 illustrated the dosage of the collector had a significant impact on mica flotation. As the dosage of the collector increased from 60 g/t to 300 g/t, the yield of mica continuously increased from 5.89% to 17.78%, while the yield of quartz decreased from 49.41% to 36.91%, and the grade of quartz initially increased to 71.50% and then decreased to 70.86%. A little of collector dosage resulted in incomplete recovery of mica, leading to a higher content of impurities such as potassium and sodium in the subsequent quartz products. Excessive dosage of the collector not only affected the selectivity of flotation and increased costs but also led to the loss of quartz products. Based on this, the optimal dosage of the collector was determined to be 180 g/t, with a mica yield of 12.89% and a SiO_2 content of 71.50% in the quartz product.

Through the above tests, the optimal flotation flowchart of granite-type sulfide mine tailings was obtained in Fig. 5, by which pyrite products, mica products and quartz products with SiO_2 grade 71.50% could be acquired. Furthermore, these products could be used in their respective fields, as shown in Fig. 6.

Table 5 Test results of flotation separation at different collector dosage

Dosage of collector/g/t	Product	Yield/%	Grade/%	Recovery/%
			SiO_2	SiO_2
60	Mica	5.89	48.76	6.11
	Quartz	49.41	67.67	93.89
	Samples	55.3	65.66	100.00
120	Mica	9.94	49.02	9.44
	Quartz	45.27	69.75	90.56
	Samples	55.21	66.02	100.00
180	**Mica**	**12.89**	**49.85**	**22.18**
	Quartz	**42.73**	**71.50**	**77.82**
	Samples	**55.62**	**66.48**	**100.00**
240	Mica	16.32	53.84	32.48
	Quartz	38.86	71.60	67.52
	Samples	55.18	66.35	100.00
300	Mica	17.78	55.84	33.87
	Quartz	36.97	70.86	66.13
	Samples	54.75	65.98	100.00

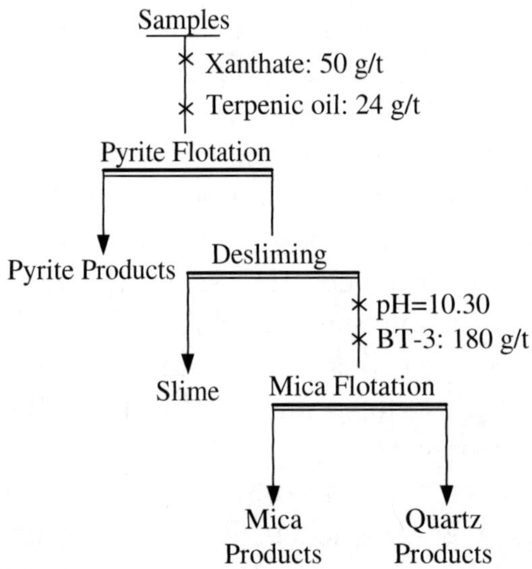

Fig. 5 Flotation flowchart of granite-type sulfide mine tailings

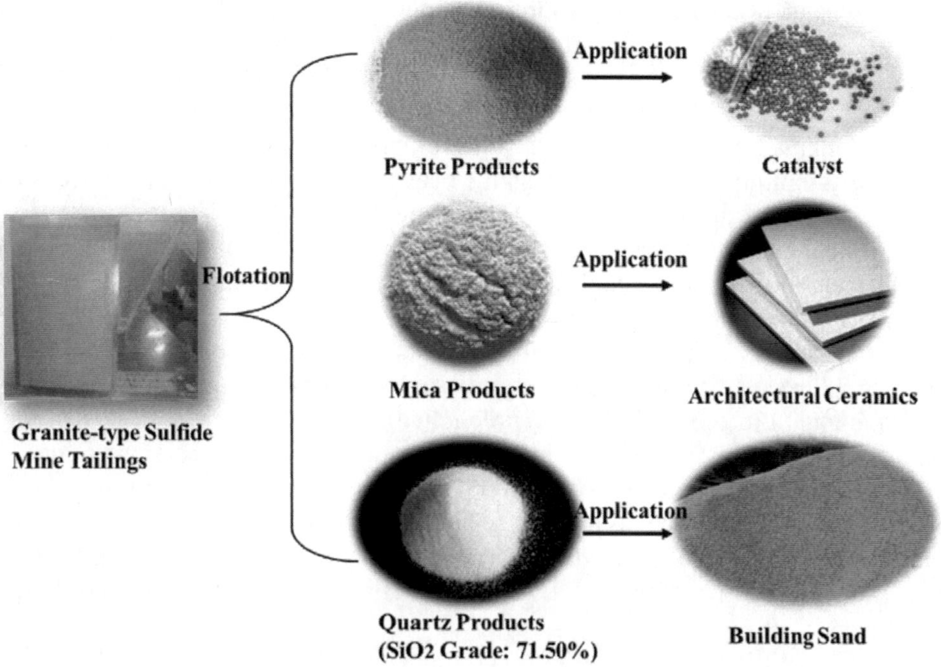

Fig. 6 Utilization of granite-type sulfide mine tailings

Conclusion

1.The pH, pulp concentration, and collector concentration all have an impact on the flotation separation of mica and quartz, with the collector having the greatest influence.

2.At xanthate concentration of 50 g/t, pH of 10.30, pulp concentration of 35%, and BT-3 concentration of 180 g/t, the tailings of granite sulfide ore can be more effectively recovered and utilized through flotation. Through flotation, a yield of 17.66% of pyrite products, a yield of 12.89% of mica products, and a yield of 42.73% of quartz products can be obtained. The quartz product has the silica content of 71.50%.

3.The three products obtained from the experiment can all be applied in their respective fields. The research results provide ideas for comprehensive utilization and tailings reduction of granite-type sulfide tailings

Acknowledgements

We gratefully appreciate the financial support from the National Natural Science Foundation of China (52274285).

References

[1] Qingjun G, Ying S, Weijian Y, et al. Deep removal of phosphorus and synchronous

Journal of Physics: Conference Series **2738** (2024) 012036 doi:10.1088/1742-6596/2738/1/012036

preparation of high-strength gypsum from phosphogypsum by crystal modification in NaCl-HCl solutions[J]. Separation and Purification Technology, 2022, 298.

[2] Uugwanga M N, Kgabi N A. Assessment of Metals Pollution in Sediments and Tailings of Klein Aub and Oamites Mine Sites, Namibia[J]. Environmental Advances, 2020, 2.

[3] Lei J, Hongjuan S, Tongjiang P, et al. Comprehensive evaluation of environmental availability, pollution level and leaching heavy metals behavior in non-ferrous metal tailings[J]. Journal of Environmental Management, 2021, 290.

[4] Wei C, Shenghua Y, Genmao Z, et al. Copper recovery from tailings through bioleaching and further application of bioleached tailings residue: Comprehensive utilization of tailings[J]. Journal of Cleaner Production, 2022, 332.

[5] Georgy L, Anton K, Faiz S, et al. Utilization potential of mine tailings in geopolymers: Part 1. Physicochemical and environmental aspects[J]. Process Safety and Environmental Protection, 2021, 147(prepublish).

[6] Yunlong L, Fu W, Qilong L, et al. Effect of TiO2 on crystallization kinetics, microstructure and properties of building glass-ceramics based on granite tailings[J]. Journal of Non-Crystalline Solids, 2021, 572.

[7] Lingfeng S, Ning S, Rui X, et al. Adsorption mechanisms of activated surface of quartz and feldspar with mixed NaOL/DDA[J]. Separation and Purification Technology, 2023, 314.

[8] Silva A C, Carolina S D, Sousa D N, et al. Feldspar production from dimension stone tailings for application in the ceramic industry[J]. Journal of Materials Research and Technology, 2019, 8(1).

[9] Ning S, Guodong W, Peng G, et al. Selective flotation of quartz from feldspar using hydroxypropyl starch as depressant[J]. Minerals Engineering, 2023, 195.

Journal of Physics: Conference Series **2738** (2024) 012037 doi:10.1088/1742-6596/2738/1/012037

Recycling and detoxification process of waste tire ash via acid leaching and neutralization

T Wajima

Department of Materials Science, Graduate School of Engineering, Chiba University, 1-33, Yayoi-cho, Inage-ku, Chiba 263-8522, Japan

wajima@chiba-u.jp

Abstract. The possibility of a novel process of recycling and detoxification for waste tire ash, which contains hazardous Zn and Al, using acid leaching and neutralization was investigated. The waste tire ash mainly contains Zn, Al, Fe, Ca and Si. The leached contents of Zn, Al, and Fe with 1 mol/L of sulphuric acid (H_2SO_4) and hydrochloric acid (HCl) are almost same and the highest, while those of Ca with HCl is high and with H_2SO_4 is almost zero due to the formation of gypsum ($CaSO_4 \cdot 2H_2O$). The optimum leaching condition is that waste tire ash was treated with 1 M H_2SO_4 for 1 h at 60 °C (solid-liquid ratio is 125 g/L). The elution of Zn and Al from the residue via H_2SO_4 leaching are 0.7 mg/L and 0.15 mg/L, respectively, which are lower than Japanese elution standard (Zn: 1 mg/L, Al: 0.2 mg/L), while those from raw ash are 3.3 mg/L and 1.4 mg/L. Zn-type layered double hydroxide (Zn-LDH) can be synthesized from leached liquor at 40 °C for 1 h by neutralization at pH 7. These results suggested that detoxified ash and environmental friendly material can be obtained from waste tire ash using H_2SO_4 acid leaching and neutralization.

1. Introduction

In 21st century, the efficient utilization of resources and energy is becoming more essential. Among various industrial wastes, much attention has been paid to combustion ashes containing heavy metals, and research on their treatment is becoming an important field from both viewpoints of environmental protection and resource recovery.

Waste tires have a disposal problem in many countries for decades. Many waste tires are currently stockpiled in various countries all over the world, and are dangerous because they pose potential environmental problems for fire hazardous and providing mosquitoes breeding grounds. The landfill disposal of waste tires is becoming unacceptable because of the shortage of available landfill sites.

As illegal dumping of waste tires is becoming an public concern all over the world, their disposal and treatment come under the spotlight. In Japan, nearly 1 million tons of waste tires have been discharged annually in the past decade [1], more than 4.4 million tons of spent tires are stockpiled in United States [2] and over 3 million tons of passenger tires are generated in Europe [3]. The most popular utilization of waste tires is thermal recovery, given that the combustion heat of tires (7.2-8.5 kcal/kg) is much higher than that of other solid residues such as plastic-derived fuel or municipal solid wastes and is comparable to that of C-grade heavy oil (9.2 kcal/kg). In the United States, Canada, Germany, the United Kingdom, and Japan, waste tires have been used as a supplemental fuel for cement kilns and in paper mills. As automobile tires contain acceptable amounts of heavy metals such as zinc, iron, and aluminium, these metals are eventually concentrated in the combustion ashes, especially in fly ash. In view of resource recycling, waste tire fly ash would be a promising secondary source for these valuable metals.

In general, automobile tires consist of rubbers, carbon black, metal codes, additives compounds and bead wires [4]. The codes and additives contain iron, copper, zinc and aluminium [5]. Zinc is added as zinc oxide (ZnO) to the rubber compound for tires as a vulcanizing agent. The average value of ZnO is 1.2 % for car tire (0.4 to 9 %) and 2.1 % (1.2 to 3 %) for truck tires [6, 7]. On combustion, the ZnO significantly concentrate within the ash. Therefore, variable Zn in the ash is desired to extract to be used for some applications, especially using acid leaching as conventional method [5, 8]. However, these process discharged acid hazardous wastewater or hazardous residue, meaning that not eco-friendly.

Layered double hydroxide (LDH) is a nonstoichiometric compound represented by the general formula $[M^{2+}_{1-x}M^{3+}_{x}(OH)_2][(A_{x/n})^{n-}\cdot mH_2O]$. Here, M^{2+} is a divalent metal ion, such as Ca^{2+}, Mg^{2+} or Zn^{2+}, and M^{3+} is a trivalent ion, such as Fe^{3+}, Al^{3+} or Cr^{3+}. By replacing a part of the divalent metal ions with trivalent metal ions, a positively charged hydroxide basic layer and a laminated structure having an anion in the intermediate layer to compensate for the charge, are adopted. The amount of positive charges in the basic layer results in an anion exchange capacity and the anion in the intermediate layer is an inorganic anion, such as Cl^-, NO_3^-, OH^-, CO_3^{2-} or SO_4^{2-}.

In this study, we attempted to extract hazardous contents of Zn and Al from waste tire ash using acid leaching to synthesize a material including Zn layered double hydroxide (Zn-LDH) by neutralizing extracted solution. The novelty of this study is eco-friendly recycling process to discharge no hazardous wastes by converting extracted element into functional material, Zn-LDH.

2. Experimental

2.1. Samples

In this study, waste tire ash was received from one of the companies in Japan. The chemical and mineralogical compositions of raw ash were determined by X-ray fluorescence spectrometry (XRF) (Epsilon1, Malvern Pananalytical, Japan) and X-ray diffraction (XRD) (MiniFlex600, Rigaku, Japan), respectively (Table 1 and Fig. 1). The raw ash was mainly composed of ZnO (5.7 %), Al_2O_3 (4.9 %), Fe_2O_3 (5.0 %), CaO (53.0 %), SiO_2 (5.4 %) and SO_3 (11.1 %), found as zincite (ZnO), magnetite (Fe_3O_4), quartz (SiO_2), ettringite ($Ca_6Al_2(SO_4)_3(OH)_{12}\cdot 36H_2O$) and gypsum ($CaSO_4\cdot 2H_2O$), respectively.

Table 1. Chemical compositions of waste tire ash, H_2SO_4-leached ash and HCl leached ash (wt. %).

Sample	ZnO	Al_2O_3	Fe_2O_3	CaO	SiO_2	SO_3
Raw ash	5.7	4.9	5.0	53.0	16.6	11.1
HCl-leached ash	3.3	6.8	6.8	19.3	41.4	19.3
H_2SO_4-leached ash	1.7	1.4	16.2	26.8	26.2	27.3

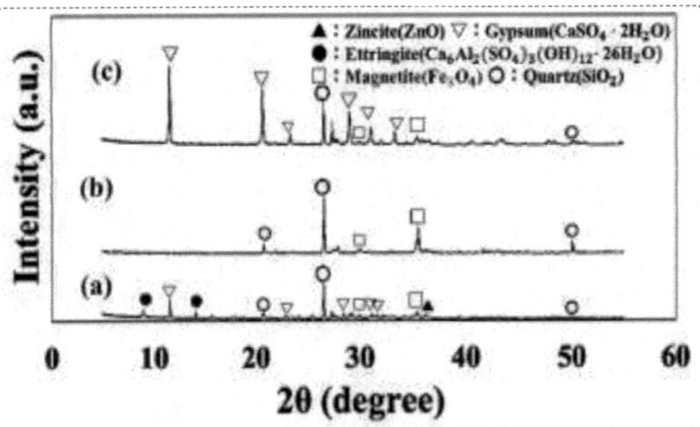

Figure 1. XRD patterns of (a) raw ash, (b) H_2SO_4-leached ash and (c) HCl-leached ash.

2.2. Extraction using acid leaching

To obtain non-hazardous ash by the extraction of Zn and Al from the ash, the ash (0 - 20.0 g) was added to 40 mL of HCl solution with 0 - 18 M, and shaken for 60 min at room temperature. After shaking, filtration was conducted, and then the extraction amounts of Zn, Al, Fe and Ca (the four major elements in the ash) in the solution from the ash were investigated using an atomic absorption spectrometer (AAS, AAnalyst200, Perkin Elmer). The residue after filtration was dried, and analyzed using a powder X-ray diffraction device (XRD, MiniFlex600, Rigaku). The surface structures of the ash and the residue following acid extraction were analyzed by a scanning electron microscope (SEM, JEOL, JSM-6510A).

To examine the effect of temperature and reaction time, the ash (25 g) was added into 1 M H_2SO_4 solution (200 mL), and stirred for 120 min with a hot magnetic stirrer at 20, 40 and 60 °C. During stirring, a part of slurry was collected and filtrated, and the concentrations of extracted Zn, Al, Fe and Ca in the filtrate were determined using AAS.

Elution test was done for the ash and the residues after acid leaching with 1 M H_2SO_4 or 1 M HCl. 2 g of the raw ash or the residues was added into 20 mL of distilled water in 50 mL tube, and the tube was shaken for 6 h at room temperature. After shaking, filtration was conducted, and then the eluted amounts of Zn and Al in the solution from the ash was investigated using AAS.

2.3. Synthesis

The obtained mixed acid solution (50 mL) after leaching with 1 M H_2SO_4 for 60 min at 60 °C (125 g/L) was neutralized to pH 7 by adding 1 M NaOH solution, and stirred with a hot stirrer at 40 °C. After 60 min stirring, the slurry was filtered, dried, and washed with distilled water to obtain a product. The mineralogical phases of the product were identified using XRD, and the surface morphologies of the product were analyzed using an SEM.

3. Results and discussion

Figure 2 shows the extraction amounts of Zn, Al, Ca and Fe from the ash using 0 - 18 M acid solution of HCl and H_2SO_4. In the case of HCl extraction, Zn, Al and Ca increased up to 1 M HCl and remained constant thereafter, while the concentration of Fe increased above 8 M HCl. In the case of H_2SO_4, the concentrations of Zn and Al increased up to 1 M H_2SO_4 and then decreased. Ca and Fe were hardly eluted regardless of concentrations of acid solution. It may be considered that the surface of the ash was more covered by $CaSO_4$ formed from Ca content in the ash and SO_4^{2-} in H_2SO_4 solution with increasing H_2SO_4 concentration above 1 M.

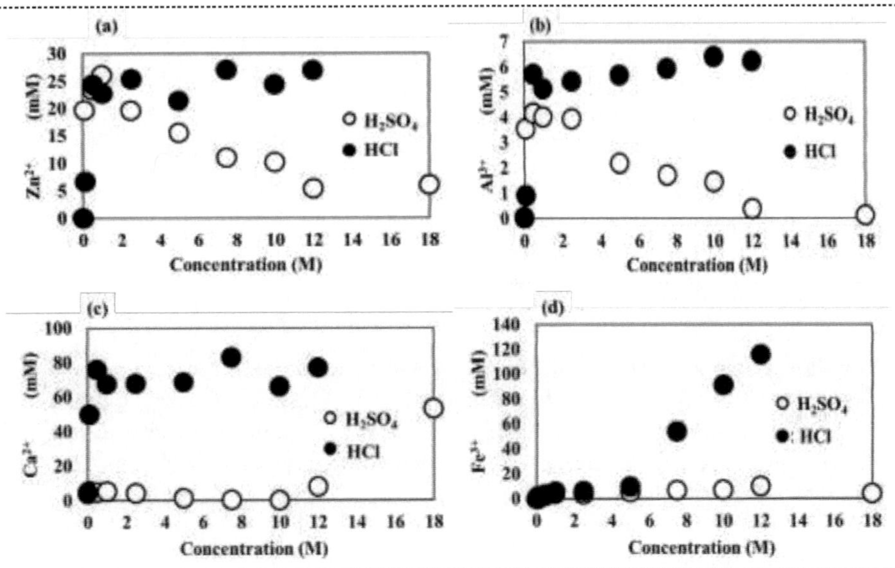

Figure 2. Extraction of (a) Zn, (b) Al, (c) Ca and (d) Fe from the ash using HCl and H_2SO_4.

Table 1 and Figure 1 show the chemical and mineralogical compositions of raw ash and the residues after leaching with 1 M HCl and 1 M H_2SO_4. The H_2SO_4-treated ash contained less ZnO and Al_2O_3 than the raw ash, and the contents of ZnO and Al_2O_3 were lower and CaO content was higher than HCl-treated ash. After leaching with HCl and H_2SO_4, peaks of zincite (ZnO) containing Zn and ettringite ($Ca_6Al_2(SO_4)_3(OH)_{12} \cdot 36H_2O$) containing Al in the raw ash were eliminated. The peaks of gypsum ($CaSO_4 \cdot 2H_2O$) in the H_2SO_4-leached ash, which were not observed in the HCl-leached ash, are confirmed, due to the reaction between Ca^{2+} extracted from the ash and SO_4^{2-} in H_2SO_4 solution.

Figure 3 shows SEM photographs of (a) the raw ash and the residue leached with (b) HCl and (c) H_2SO_4. In the raw ash, agglomerates of some particles on the surface of the ash were detected (Fig. 3 (a)). The smooth surface of the residue after leaching with HCl was observed (Fig. 3 (b)), while needle-like crystals, which appear to be gypsum crystals, were observed on the surface of the ash leached with H_2SO_4 (Fig. 3 (c)).

From these results, Zn and Al (hazardous elements) can be extracted from the ash using 1 M H_2SO_4 solution without dissolving Ca content (Non-hazardous elements).

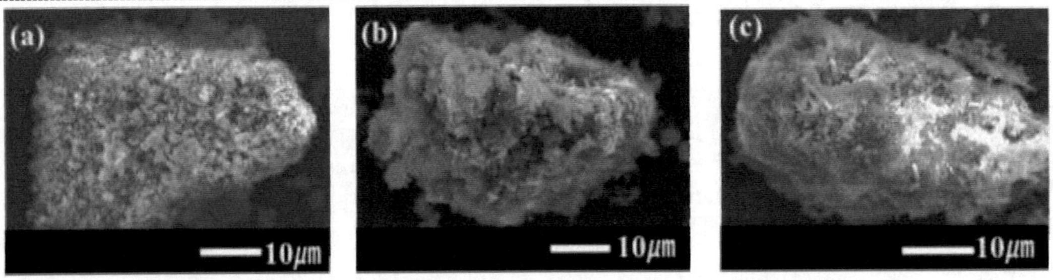

Figure 3. SEM photographs of (a) the raw ash, the residue leached with (b) HCl and (c) H_2SO_4.

Table 2 shows the elution of Zn and Al from the ash and the residue via acid leaching with HCl and H_2SO_4 in elution test. The elution of Zn and Al from the residue via H_2SO_4 leaching are 0.7 mg/L and 0.15 mg/L, respectively, which are lower than Japanese elution standard (Zn: 1 mg/L, Al: 0.2 mg/L), while those from raw ash are 3.3 mg/L and 1.4 mg/L. It is noted that the elution of Zn and Al from the residue via HCl leaching are 0.5 mg/L and 0.17 mg/L, respectively, which are also lower than Japanese elution standard.

Table 2. Elution of Zn and Al from waste tire ash, H_2SO_4-leached ash and HCl leached ash in elution test (mg/L).

Sample	Zn	Al
Raw ash	3.3	1.4
HCl-leached ash	0.3	0.17
H_2SO_4-leached ash	0.7	0.15

Figure 4 shows the extraction amounts of Zn, Al, Ca and Fe from the ash using 1 M H_2SO_4 by varying dosage from 0 to 500 g/L at 20 °C for 1 h. At low addition, Zn, Al, Fe and Ca were extracted, as addition increased to 125 g/L, Zn and Ca decreased while Al and Fe were almost constant. At 125 g/L, Ca was almost completely un-extracted, and above 125 g/L, the extraction of all metal ions decreased. Zn and Al were most extracted at 125 g/L addition.

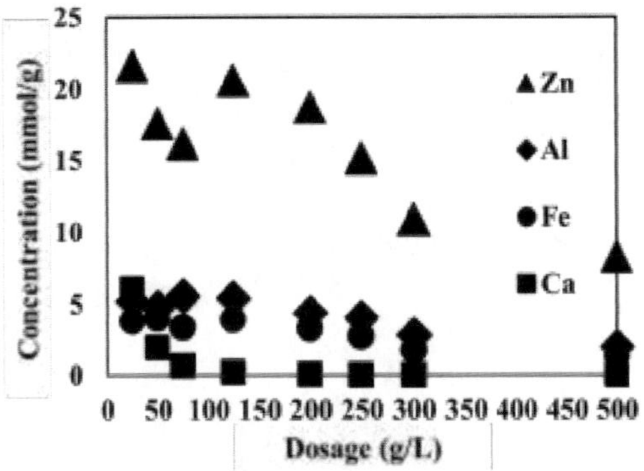

Figure 4. Extraction of Zn, Al, Fe and Ca from the ash using 1 M H_2SO_4 solution by varying dosage from 0 to 500 g/L at 20 °C for 1 h.

Figure 5 shows the concentrations of (a) Ca^{2+} and Zn^{2+}, and (b) Al^{3+} and Fe^{3+} extracted from the ash in 1 M H_2SO_4 solution (125 g/L) during stirring at 20, 40 and 60 °C. While Ca was hardly extracted at any temperature, extracted Zn increased with increasing temperature and remained constant after 60 minutes. Extracted Al was almost constant regardless of temperatures, and extracted Fe increased with increasing temperature.

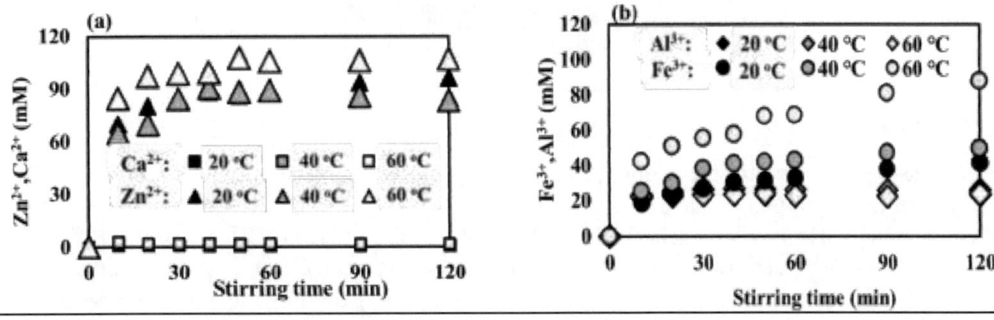

Figure 5. Concentrations of (a) Ca^{2+} and Zn^{2+}, and (b) Al^{3+} and Fe^{3+} extracted from the ash in 1 M H_2SO_4 solution (125 g/L) during stirring at 20, 40 and 60 °C..

From these results, the optimum leaching condition is that waste tire ash was treated with 1 M H_2SO_4 for 1 h at 60 °C (solid-liquid ratio is 125 g/L). Contents of Zn, Al, Fe and Ca in the extracted solution and pH of the solution are shown in Table 3.

Table 3. Zn, Al, Fe and Ca contents in extracted solution and pH of the solution.

Concentration (mM)				M^{2+}/M^{3+}	pH
Zn	Al	Fe	Ca		
103	21	63	2	1.2	3.1

Figure 6 and 7 show the XRD patterns and SEM photo of the product from the extracted solution via neutralization. Zn-type layered double hydroxide (Zn-LDH), which is one of the environmental purification material, can be synthesized from extracted solution by neutralization at pH 7, and hexagonal plane crystals, which seems to be Zn-LDH, can be observed.

Journal of Physics: Conference Series **2738** (2024) 012037 doi:10.1088/1742-6596/2738/1/012037

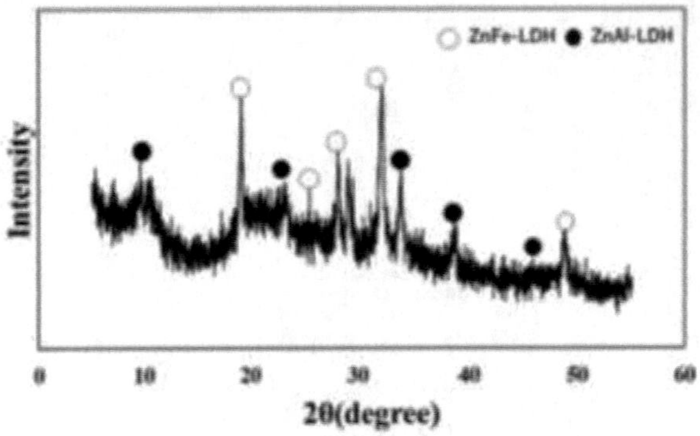

Figure 6. XRD patterns of the product from the extracted solution via neutralization.

Figure 7. SEM photo of the product from the extracted solution via neutralization.

Figure 8 shows the summery for the proposed process of this study. Waste tire ash, including zincite (ZnO), magnetite (Fe_3O_4), quartz (SiO_2), gypsum ($CaSO_4{\cdot}2H_2O$) and ettringite ($Ca_6Al_2(SO_4)_3(OH)_{12}{\cdot}36H_2O$), was leached with 1 M H_2SO_4 solution at 60 °C for 1 hour (125 g/L) to extract Zn^{2+}, Al^{3+} and Fe^{3+} in the solution. The residue contains SiO_2 and Fe_3O_4 remained in the ash and $CaSO_4{\cdot}2H_2O$ formed from Ca content in the ash and SO_4^{2-} in the solution, which is non-hazardous residue with lower elution level of Zn and Al. Extracted elements in the solution can be converted into the product including Zn-LDH as functional material by neutralizing to pH 7 with NaOH.

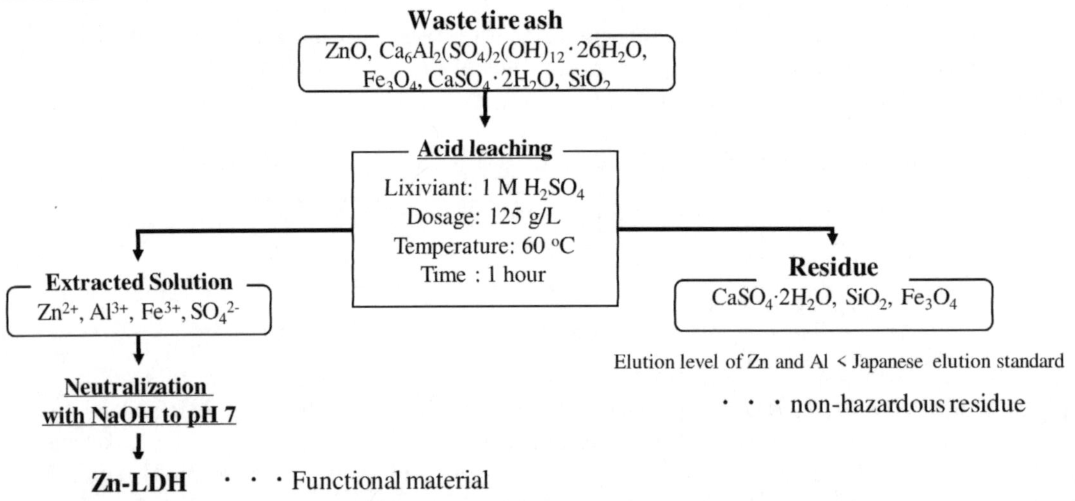

Figure 8. Proposed process of this study.

Journal of Physics: Conference Series **2738** (2024) 012037 doi:10.1088/1742-6596/2738/1/012037

These results suggested that detoxified ash and environmental friendly material can be obtained from waste tire ash using H_2SO_4 acid leaching and neutralization.

4. Conclusion

In this study, hazardous contents of Zn and Al was extracted from waste tire ash using acid leaching to synthesize a material including Zn-LDH by neutralizing the extracted solution was investigated. As a result, it was possible to extract Zn and Al from the ash using H_2SO_4 without extracting Ca, and the optimum leaching condition is that waste tire ash was treated with 1 M H_2SO_4 for 1 h at 60 °C (solid-liquid ratio is 125 g/L). The elution of Zn and Al from the residue via H_2SO_4 leaching are lower than Japanese elution standard. Zn-LDH can be synthesized from the extracted solution by neutralizing at pH 7. These results suggested that detoxified ash and environmental friendly material can be obtained from waste tire ash using H_2SO_4 acid leaching and neutralization.

5. References

[1] JATMA (The Japan Automobile Tyre Manufactures Association, Inc.) *Tire Industry of Japan*, http://www.jatma.or.jp.
[2] RMA (Rubber Manufactures Association) *US Scrap Tire Markets*, http://www.rma.org.
[3] ETRMA (European Tire and Rubber Manufactures Association), http://www.etrma.org.
[4] Smolders E and Degryse F 2002 *Environ. Sci. Technol.* **36** 3706-3710.
[5] Kinosita T, Yamaguchi K, Akita S, Nii S F, Kawaizumi S and Takashi K 2005 *Chemosphere* **59** 1105-1111.
[6] Adachi K and Tainosho Y 2004 *Environ. Int.* **30** 1009-1017.
[7] Kotani M, Tamura I and Ishii K 1993 *Soc. Rubber Ind. Jpn.* **60** 883-889.
[8] Khoshgoftarmanesh A H, Taheri S, Shariatmadari H, Ghaziaskar S H and Chaney R L 2010 *J. Residuals Sci. Technol.* **7** 69-72.

Journal of Physics: Conference Series **2738** (2024) 012038 doi:10.1088/1742-6596/2738/1/012038

Life cycle assessment and product carbon footprint for SHG zinc production

Sabina C. Grund[1] and Eric van Genderen[2]

1. International Zinc Association, Avenue de Tervueren 168 1000 Park Forty Plaza, Suite 130, 1150 Brussels, Belgium
2. International Zinc Association, 1000 Park Forty Plaza, Suite 130 Durham, NC 27713, USA

sgrund@zinc.org

Abstract. The industry's environmental impact has moved into focus at global level. Regulators, supply chains, and the financial sector need credible, standardized, and transparent information for multiple reasons. These include regulatory targets for limiting emissions, downstream user optimizations, and developing science-based emission reduction strategies. By characterizing environmental impacts, companies can satisfy stakeholder demands and measure progress towards goals. In this paper, an introduction to life cycle assessment (LCA) for reporting on environmental impacts, including the product carbon footprint as global warming potential is given. Average results from the most recent global LCA update for special high-grade zinc production are summarized. Resulting carbon footprint information is then used to derive the zinc industry's baseline carbon footprint as the starting point for the decarbonization roadmap for the global zinc industry.

1. Introduction

Modern life and increasing living standards have unintended consequences, among these emissions to the environment. Life Cycle Assessment (LCA) over the past two decades has developed into the widely used tool to quantify the environmental footprint of a process, a product, or a service. Specifiers and designers use environmental footprint information for material selection to minimize impact of future products. Regulators increasingly set targets based on LCA information. For life cycle impact assessment (LCIA) various impact categories are used. The focus today lies on the product carbon footprint expressed as global warming potential (GWP).

This paper will provide an overview of the most recent global average LCA for primary, special high-grade zinc production based on 2020 and 2021 industry data. It will then be demonstrated how results feed into the derivation of the baseline carbon footprint of the zinc industry and of carbon abatement scenarios considering both Scope 1 (direct) and Scope 2 (indirect from purchased electricity) and selected Scope 3 emissions.

2. Life cycle assessment

LCA quantifies the environmental impact caused by a material, product, process, or service over its life cycle from cradle to gate (typical for basic raw materials and commodities) or cradle to grave (typical for products and services). LCA methodologies are by the International Organization for

Standardization (ISO 14040 [1] and 14044 [2]). More specifically, for the carbon footprint the Green House Gas (GHG) Protocol "Product Life Cycle Accounting and Reporting Standard" [3] is available. It is based on International Organization of Standardization (ISO 14040 and 14044) but limited to carbon footprint calculation only.

There are four components to a typical LCA study (Figure 1): 1) Goal and scope, 2) Life cycle inventory, 3) Life cycle impact assessment, 4) Interpretation of results.

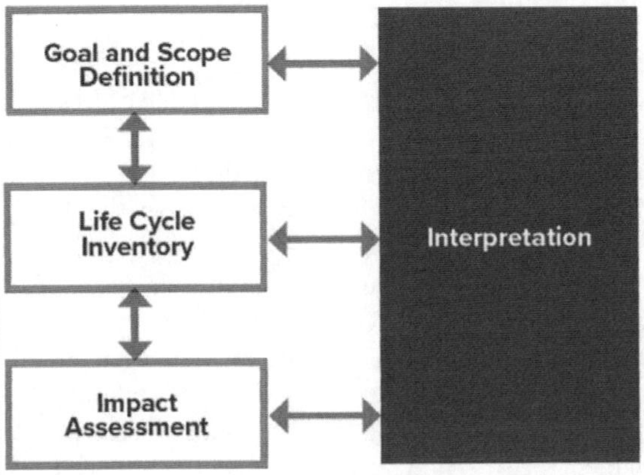

Figure 1. Life cycle assessment framework [5]

2.1. *Life cycle assessment for zinc*

Since 2009, the International Zinc Association (IZA) provides global average LCA information for primary zinc production [4] (cradle-to-gate, Figure 2), which established an environmental profile [5] for zinc that represents geographic differences in mining, smelting, energy use, and transportation. This Cradle-to-Gate LCA model is based on anonymized industry data from zinc mines and smelters worldwide.

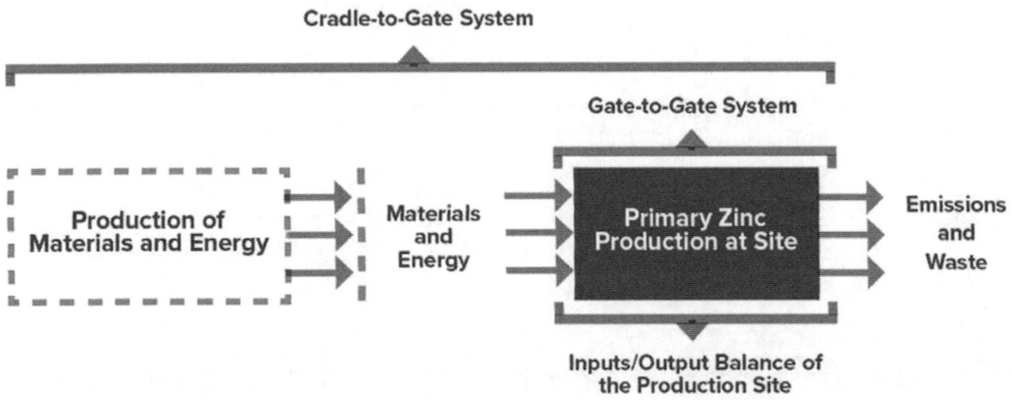

Figure 2. Schematic illustration of the cradle-to-gate and the gate-to-gate system of primary

Frequently, this environmental profile for special high-grade zinc (SHG; 99.995 % zinc) is updated. With each update, the number of participating companies increases. As a result, life cycle inventory data for primary (SHG) zinc production today are representativity for the sector. IZA submits global

Journal of Physics: Conference Series **2738** (2024) 012038 doi:10.1088/1742-6596/2738/1/012038

and in some cases regional average life cycle inventory data to relevant databases such as GaBi [6], Ecoinvent [7], or PEF [8].

2.2. System boundaries of zinc production

Natural zinc ores usually contain 5-15% of zinc as zinc sulfide. Accompanying elements are lead, copper, silver, cadmium, and sometimes indium, and germanium. Already at the mine site, zinc ores are crushed, ground, and concentrated by flotation. Zinc ore concentrates have a zinc concentration of about 55%.

To produce primary zinc of special high-grade quality (SHG; 99.995%), the combination of roasting, leaching and electrowinning (RLE) is the widely used methodology. Accompanying elements are concentrated in co-products such as sulfuric acid or metals concentrates for recovery in other metals smelters. More than 90% of all zinc worldwide is produced using RLE technology varying in details and single process steps. Pyrometallurgical process types such as the Imperial Smelting furnace are used in other cases.

In primary zinc production, secondary raw materials e.g. steel mill dusts with an enriched zinc concentration (Waelz or crude zinc oxide) can be processed together with zinc concentrates.

Figure 3 shows the system boundaries for the most recent LCA update for SHG zinc based on industry data from 2020 and 2021 [9]. Participating companies reported relevant input and output data for the processes as included in the system boundaries (Figure 3). As a result, global average LCIA and LCIA information is available, and companies receive their company-specific environmental footprint information benchmarked against the average.

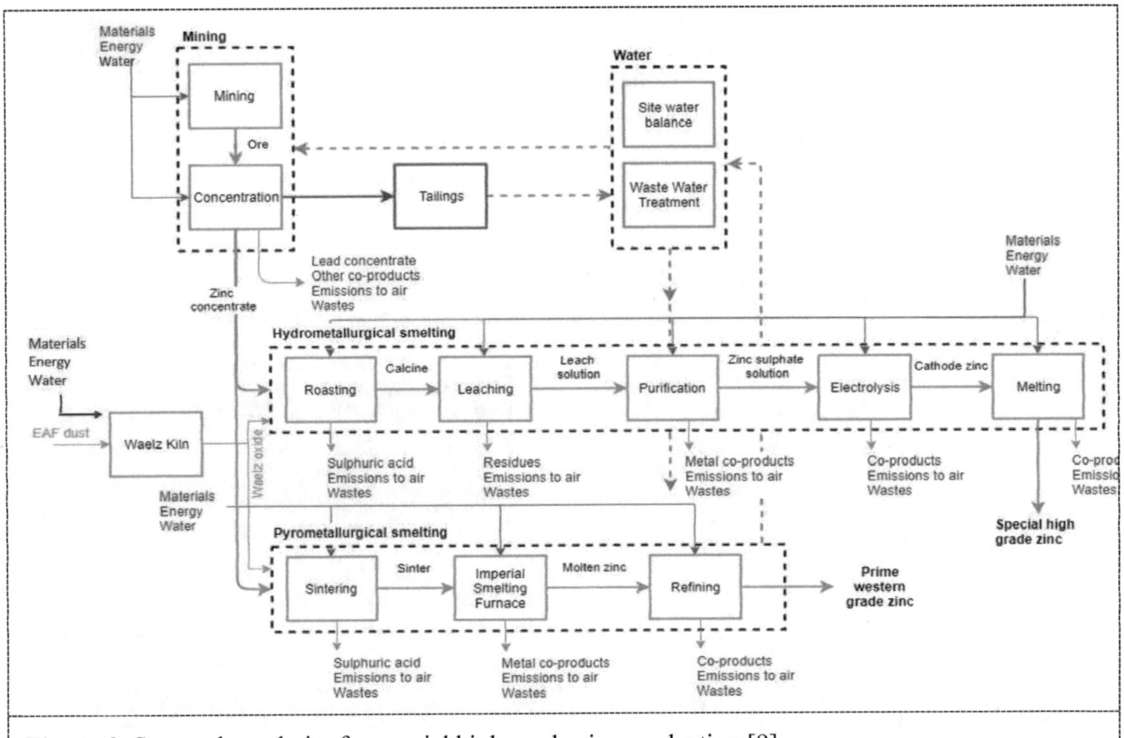

Figure 3. System boundaries for special high-grade zinc production [9].

2.3. Results

Overall results for SHG zinc, zinc ore concentrate, and Waelz oxide are presented in Table 1. Based on zinc content, 86% of SHG production is sourced from concentrate and 14% from Waelz oxide. [9]

Table 1. Life cycle impact analysis (LCIA) results per metric ton [9].

Environmental Indicator	Unit	*SHG zinc*	Zinc conc. (52 % Zn)	*Zinc conc. [per t Zn contained]*	Waelz oxide (56% Zn)	*Waelz oxide [per ton Zn contained]*
Global Warming Potential (GWP)	kg CO_2-eq	3800	580	1100	2900	5200
Acidification Potential (AP)	kg SO_2-eq	33.0	4.50	8.7	30.00	54.00
Eutrophication Potential (EP)	kg PO_4^{-3-}eq	1.30	0.37	0.72	0.23	0.41
Photochemical Ozone Creation Potential (POCP)	kg C_2H_4-eq	1.50	0.24	0.47	1.50	2.60
Abiotic Resource Depletion Potential, minerals and metals (ADPe)	kg Sb-eq	0.78	2.20	4.20	6.7E-5	1.2E-4
Primary Energy Demand, non-renewable (PEDnr)	MJ	48000	8400	16000	39000	70000
Blue Water Consumption	L water	74000	27000	51000	2400	4200

Figure 4 shows the results for SHG zinc production, broken down by process steps. It can be seen that smelting often accounts for most of the burden (47% and 61%; excluding ADPe). Concentrate burdens range from 24% to 52%, while Waelz oxide burdens range from 1% to 15%. Finally, intermediate transport of concentrate and Waelz oxide can range from 0% to 7%, depending on the impact category. Based on zinc content, 86% of the zinc is coming from concentrate and 14% from Waelz oxide. [9]

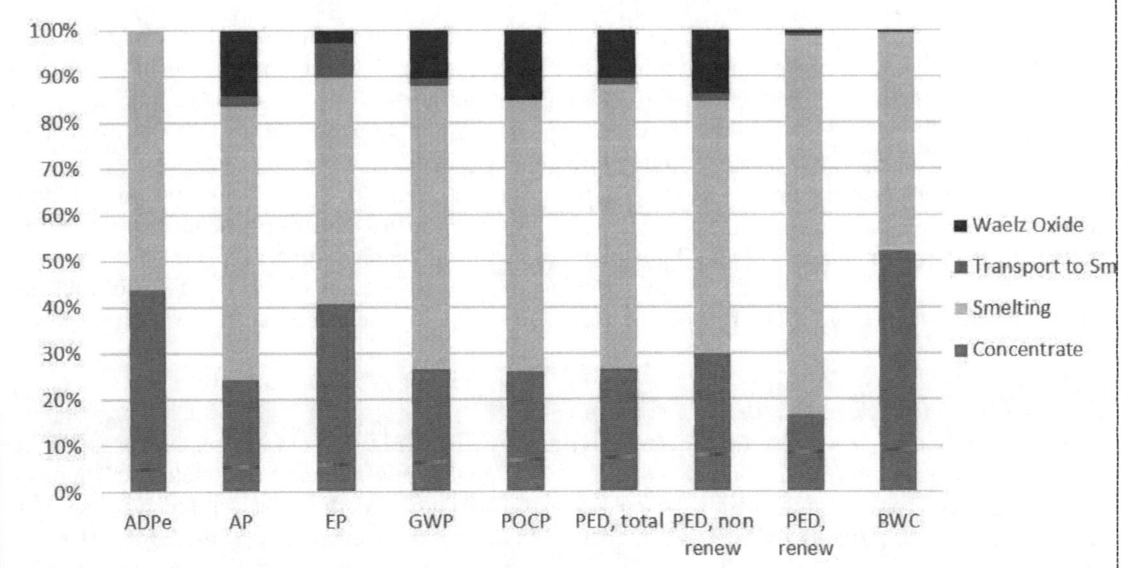

Figure 4. Relative results of the life cycle impact assessment for SHG zinc production [9].

Detailed results for the smelting process are shown in Figure 5. All results, except from ADPe, and Blue Water Consumption (BWC) are driven by electricity consumption. Direct emissions and credit from sulfuric acid production are the second highest impact contributors. Smelting contributes 56% to total ADPe, of which the main impact is coming from the production of raw materials (mainly Zinc oxide). The consumption of electricity impacts the Blue Water Consumption impact category higher than the water consumption itself, mainly because of the hydro power sources. Direct emissions of SO_2 from the smelting process are also a noticeable contributor to overall AP and POCP impacts. [9]

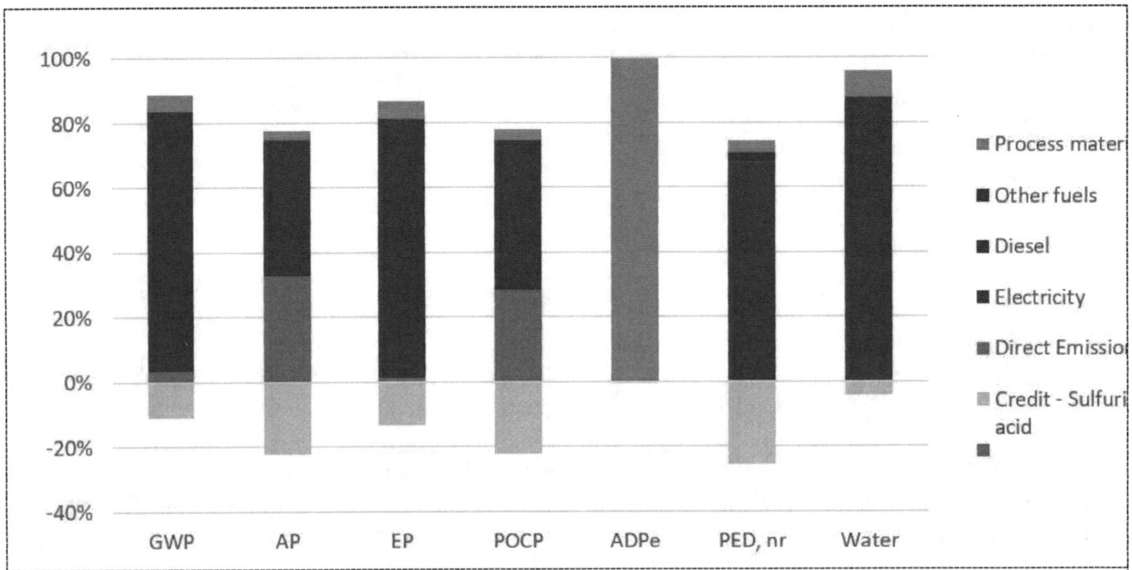

Figure 5. Relative life cycle impact results for smelting and refining in SHG zinc production [9].

2.4. *Representativity*

The most recent LCA update for SHG zinc production provided by IZA [9] includes primary data from 25 mines and 24 smelters from 10 companies, which cover 2.5 million metric tons of mined zinc (30% of world mined zinc production) and 3.97 million metric tons of SHG zinc (31% of the world zinc production). The best covered region was Europe (including Norway) with up to 94% coverage of the regional production while Asia was underrepresented (13% of Asian zinc production).

Exemplarily, the variation of environmental footprints between zinc smelters is depicted in Figure 6. The anonymized carbon footprints for SHG zinc produced in participating smelters are shown as global warming potential (GWP) in kg carbon dioxide equivalents (CO2e) per t of SHG zinc produced. Huge variations become visible spanning over a range from below 1000 up to 9000 kg CO2e/kg, depending largely on the type of electricity supply (renewables versus coal-fired power plants (Figure 6).

2.5. *Chinese zinc production in the LCA subsection*

Based on the recent IZA LCA update for SHG zinc (2020/2021), three Chinese mines and three Chinese smelters participated. Due to conditions for confidentiality not being satisfied (number participating companies), information on an average environmental footprint for the Chinese zinc industry is not yet viable. However, the global average can be compared with the result of study published by Qi et al. (2017; Table 2). The global carbon intensity for SHG zinc production is currently estimated to be 3.8 kg CO2e per kg SHG zinc, while the Chinese average published by Qi et al. (2017) was 6.12 kg CO2e per kg SHG zinc.

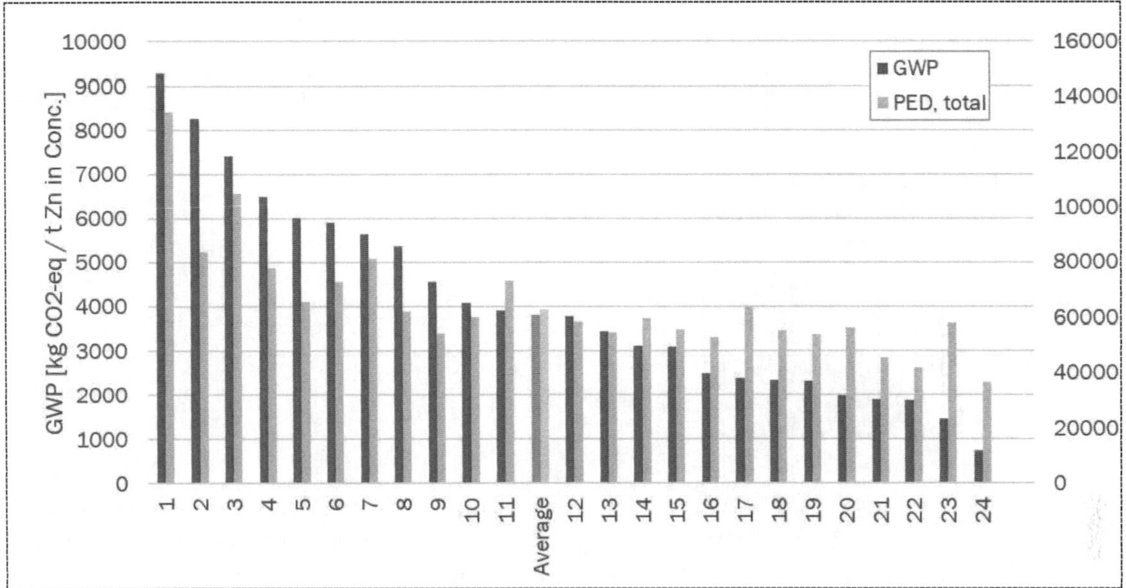

Figure 6. Global average and anonymized Global Warming Potential (GWP) for special high-grade zinc production sites participating in the LCA update based on 2020 and 2021 industry data [9] Includes zinc mining, concentration, crude zinc oxide production, and smelting/refing.

Table 2. Comparison of the carbon footprint of SHG zinc as reported in literature and as a result of the IZA update of the global average carbon footprint of SHG zinc production based on 2021 industry data [9].

Author	Global Warming Potential in kg CO_2e per kg SGH zinc	*Year of publication*	Reference
Norgate et al.	4.6	2007	[12]
Nuss and Eckemann	3.1	2014	[11]
Van Genderen et al.	2.66	2016	[4]
Qi et al.	6.12	2017	[10]
IZA	4.01	2022	[5]
Blano Schueller and Gediga	3.8	2023	[9]

3. Company-specific product carbon footprint

Companies are asked for the specific product carbon footprint of zinc produced at their individual sites. For reference, companies and customers may compare their site-specific product-specific carbon footprint to the global average for marketing purposes. For this purpose, IZA launched the Carbon Footprint Guidance [13] to harmonize calculation and reporting.

This guidance is an excerpt of the LCA methodology and model used for calculating the full environmental footprint in IZA's LCA update projects with the product carbon footprint for special high-grade zinc. Version 1.0 of the ISO-compliant reporting guidance for GWP reporting has been recognized by the London Metals Exchange (LME) as the single methodology to be used when submitting carbon footprints for registered zinc brands to the LMEpassport (Figure 7).

Figurure 7. Carbon footprint calculation and reporting guidance for special high-grade zinc [13].

Anticipated update (version 2.0, 2023) will account for carbon footprint allocation to secondary raw materials as a substitute for zinc ore concentrates [14] and seek alignment with end-use product sectors requirements such as e.g., the Pathfinder Framework [21] by the Partnership for Carbon Transparency (PACT). PACT represents a large group of companies from the automotive and various companies.

4. Decarbonization Roadmap for the global zinc industry

The zinc sector decarbonization roadmap has characterized the baseline carbon footprint and identified goals for net-zero. It builds on zinc stock and flow analysis, future demand and availability scenarios, and carbon footprint information from IZA LCA updates and other sources [5][15][16]. As a result, the baseline carbon footprint of the global zinc industry (2019), was estimated to be 59 million tons (Mt) CO2e (0,12% global GHG emissions; [17]). The total GHG emissions were calculated relative to applying global mining, recycling, and smelting emission factors for production amounts – 1030 kg CO2e / t Zn concentrate production, 2500 kg CO2e / t Zn smelting, 4030 kg CO2e / t Zn Waelzing, [22] and 78 kg CO2e / t Zn from remelting [18], respectively (Figure 8). Plausibility was checked with the zinc database by Skarn Associates [19].

For zinc mining, 73% of all GHG emissions are associated with electricity consumption. Further 23% of emissions come from diesel fuel consumed by mine trucks. About 90% of all GHG emissions from zinc smelting come from the use of electricity due to a majority of zinc smelters using the roast-leach electrowinning process. More than 50% of all zinc today is produced in China. Due to the importance of purchased energy sourcing, governmental decarbonization strategies for the power impact the success of industry initiatives.

Looking at zinc production technologies used today, Figure 9 shows the broad range of site-specific carbon footprints including information on tonnage and production technology. For sites producing zinc pyrometallurgically, via the Imperial Smelting Furnace or vertical retorts, the carbon footprint is related to the use of carbon as a reductant. Carbon abatement strategies for these technologies will face their own challenges.

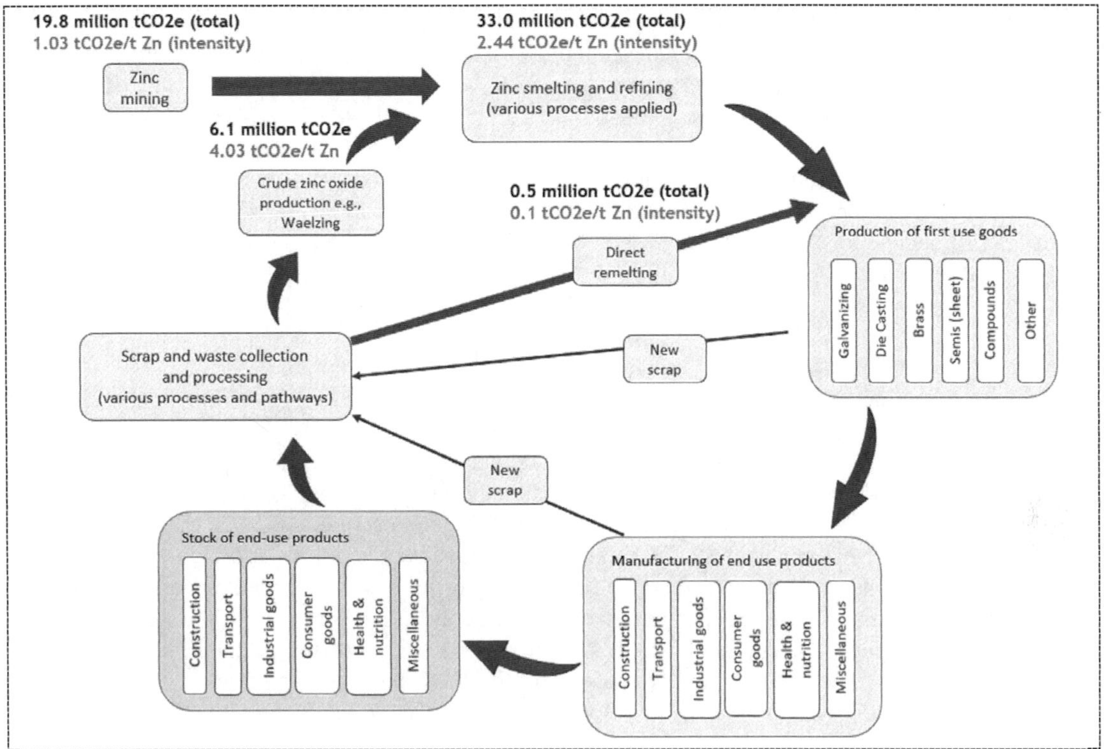

Figure 8. Baseline carbon footprint and carbon intensity for the global zinc industry, reference year 2019 [14].

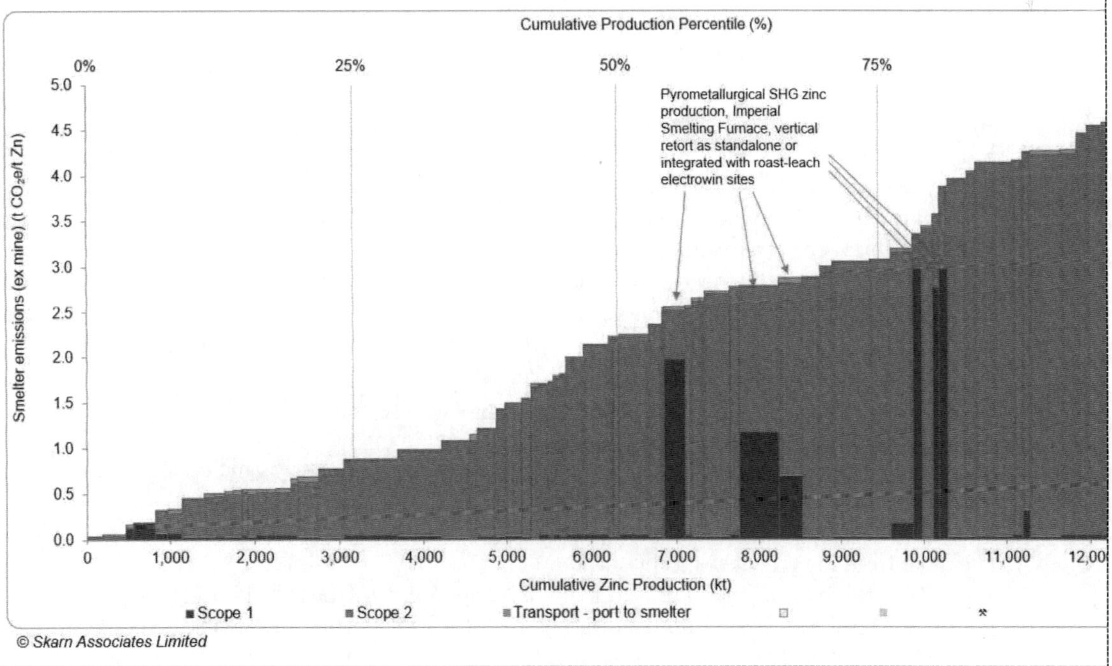

Figure 9. Zinc Smelter GHG intensity curve for 2021 based on [19].

It is estimated from 1.2% global domestic product growth that the baseline carbon footprint of the zinc industry of approximately 59 Mt CO2e annual emissions (~%0.12 global artificial emissions) continuing with a "business-as-usual" scenario would increase to approximately 95 Mt by 2050 (17.1 Mt associated with Scope 1 emissions). Based on current electrical utility decarbonization trends reported by the International Energy Agency [20], the impact of purchased energy will decrease by 13.4 Mt CO2e by 2030. Assuming that country ambitions are met between 2030 and 2050, the impact of purchased energy is expected to decrease by an additional 58.1 Mt CO2e. Consequently, the zinc industry is responsible for decreasing its Scope 1 footprint by 15.4 Mt, thus achieving 90% total Scope 1 and 2 reductions by 2050.

5. Summary and conclusions

Lifecycle assessment is the established tool for regulators, specifiers, and designers used for calculating environmental impacts of products, processes, and services. Providing credible and transparent life cycle information has become key for market access.

The International Zinc Association regularly updates the global average environmental profile for special high-grade zinc and submits the associated data sets to relevant LCA data bases.

Increasingly, end use product sectors demand company specific information on environmental footprints including the carbon footprint as global warming potential. Companies can use the IZA carbon footprint calculation and reporting guidance when reporting information to the London Metals Exchange LMEpassport. With this guidance, IZA contributes to harmonizing calculation and reporting thus enhancing transparency, comparability, and credibility for the whole zinc sector.

Carbon footprint and mass flow analysis information was used for developing the Decarbonization Roadmap for the global zinc industry. Scenario analysis shows that impact opportunities for decarbonization lie in decarbonizing electricity production and coal-based reduction processes used in zinc production and recycling such as vertical retorts, Imperial Smelting, and Waelzing.

References

[1] ISO 14040: "Environmental management — life cycle assessment — principles and framework", 2006, https://www.iso.org/standard/37456.html
[2] ISO 14044: "Environmental management — life cycle assessment — requirements and guidelines", 2006, https://www.iso.org/standard/38498.html
[3] GHG Protocol "Product life cycle accounting and reporting standard", https://ghgprotocol.org/sites/default/files/standards/Product-Life-Cycle-Accounting-Reporting-Standard_041613.pdfreference
[4] Eric van Genderen, Maggie Wildnauer, Nick Santero, Nadir Sidi: „A global life cycle assessment for primary zinc production", Int J Life Cycle Assess (2016) 21:1580–1593
[5] "Environmental Profile of SHG zinc", IZA 2022, available from the IZA website www.zinc.org
[6] GaBi Database: https://sphera.com/life-cycle-assessment-lca-database/
[7] Ecoinvent Database: www.ecoinvent.oeg
[8] Product Environmental Footprint data base: https://eplca.jrc.ec.europa.eu/EnvironmentalFootprint.html
[9] Sphera: "Special high - grade zinc life cycle assessment update for 2021", project report available from the IZA website (www.zinc.org)
[10] Congcong Qi, Liping Ye, Xiaotian Ma, Donglu Yang, Jinglan Hong: "Life cycle assessment of the hydrometallurgical zinc production chain in China", Journal of Cleaner Production 156 (2017) 451e458
[11] Nuss, P., Eckelman, M.J.: "Life cycle assessment of metals: a scientific synthesis", 2014, Plos

One. http://dx.doi.org/10.1371/journal.pone.0101298.
[12] Norgate, T.E., Jahanshahi, S., Rankin,W.J., "Assessing the environmental impact of metal production processes." 2007, J. Clean. Prod. 15, 838e848.
[13] "SHG zinc carbon footprint – technical guidance on carbon footprint calculation for special high-grade zinc" version 1.0, 2022, available from the IZA website www.zinc.org
[14] Sabina Grund, Eric van Genderen, Martin van Leeuwen: „Decarbonizing the zinc industry while maximizing zinc circularity", proceedings of the European metallurgical Conference (EMC) 2023
[15] Rostek, L., et al (2022): A dynamic material flow analysis of the global anthropogenic zinc cycle: Providing a quantitative basis for circularity discussions, Resources, Conservation & Recycling 180 (2022) 106154, available at https://www.sciencedirect.com/science/article/abs/pii/S0921344922000027?via%3Dihub
[16] ILZSG (2022): "Monthly bulletin of the International Lead and Zinc Study Group", volume 62, No 11, November 2022
[17] World Resources Institute (2019): "World Greenhouse Gas Emissions: 2019", available at https://www.wri.org/data/world-greenhouse-gas-emissions-2019
[18] Reuter M. and Neumann F. (2021), "Simulation-based calculation of the environmental Impact of zinc die casting when using primary and secondary raw materials for alloy production, World of Metallurgy - Erzmetall 74(6):325-330
[19] Skarn Associates (2022): "Greenhouse gas intensity curves for zinc", version 1.0
[20] International Energy Agency: "Data and statistics", 2023, available online via www.iea.org
[21] WBCSD/PACT: "Pathfinder Framework – guidance for the accounting and exchange of product life cycle emissions", 2023, version 2.0, available at https://www.wbcsd.org/Programs/Climate-and-Energy/Climate/SOS-1.5/Resources/Pathfinder-Framework-Version-2.0
[22] Sphera: "Special high - grade zinc life cycle assessment update for 2018", project report available from the IZA

Journal of Physics: Conference Series **2738** (2024) 012039 doi:10.1088/1742-6596/2738/1/012039

Fundamental experimental and thermodynamic modelling study in support of the lead battery recycling slag

Maksym Shevchenko*[1], Denis Shishin#[1], Daniel Graf[2], Joseph Grogan[2] and Evgueni Jak[1]

1. Pyrometallurgy Innovation Centre (PYROSEARCH), School of Chemical Engineering, The University of Queensland, Brisbane, Queensland, 4072, Australia.
2. Gopher Resource, 685 Yankee Doodle Road, Eagan, MN 55121, United States of America.

* Corresponding author: Maksym Shevchenko, m.shevchenko@uq.edu.au

Presenting author: Denis Shishin, d.shishin@uq.edu.au

Abstract. Antimony, tin, and other higher value metals, as well as elements such as arsenic and lead, can be found in the slags produced during the lead battery recycling process. Gopher Resources operates lead battery recycling furnaces that use sodium slag systems for which there has been less thermodynamic research than more common slag systems. Better thermodynamic information could help improve the process efficiency and control of these furnaces, as well as the recovery of higher value metals. The present study focuses on experimental research and thermodynamic modeling of slags belonging to the Na-Si-Fe-O system, with minor elements including S, Sb, Sn, Pb, and As. Examples of important systems studied extensively for the first time include Fe-Sb-Si-O and Na-Sn-Si-O. Phase equilibria methods are used to determine all the interaction parameters between the impurity metals and the main components of the slag. The experimental methodology involves equilibration, quenching, and electron-probe X-ray microanalysis of the samples. The modified Quasichemical model is used to describe the thermodynamics of the slags. The model also takes into account possible formation of matte/metal/speiss liquids, and numerous solid phases, which is important for understanding of fundamentals operation of various process units.

1. Introduction

Pyrometallurgical recycling of lead acid batteries can be performed using at least three different routes, depending on the sulfur removal technology: 1) hydrometallurgical separation of sulfur in a form of Na_2SO_4 through precipitation of lead basic carbonate followed by pyrometallurgical processing [1]; 2) pyrometallurgical removal of sulfur in a form of SO_2 using oxidative smelting, often conducted in top submerged lance furnace [2]; 3) pyrometallurgical removal of sulfur in a form of Na_2S-FeS matte during reductive smelting [3]. Gopher Resource operates using the first route. Process steps in terms of main chemical reactions are described below (**Figure 1**).

Step 1: Hydro de-sulfurization:

- $3PbSO_4(aq) + 3Na_2CO_3(aq) + H_2O = Pb_3(CO_3)_2(OH)_2(solid)\downarrow + 3Na_2SO_4(aq) + CO_2$
- Lead basic carbonate is contaminated with Na_2SO_4 and Na_2CO_3

Step 2. Oxidizing smelting (Reverberatory furnace) of desulfurized paste and grids:

- $Pb_3(CO_3)_2(OH)_2$(solid) + heat = 3PbO(slag) + H_2O + $2CO_2$
- Pb(solid) + heat = Pb(liq)
- Pb(liq) + PbO_2 = 2PbO (slag)
- Fluxing (and fiber impurities dissolution): SiO_2(solid) + PbO (slag) = (PbO), (SiO_2) in slag
- Lead softening: [Sb, As, Sn] in Pb(liq) + PbO_2 = ($SbO_{1.5}$, $AsO_{1.5}$, SnO_2, PbO) in slag
- Dissolution of Na_2SO_4, Na_2CO_3 into slag: (PbO, SiO_2, $SbO_{1.5}$, $AsO_{1.5}$, $NaO_{0.5}$, Na_2SO_4 (50-60 wt% Pb) – form of sulfur in this slag is not well understood.

Step 3. Reduction smelting (Blast furnace), <u>Focus of present study</u>:
- 2C + O_2 = 2CO
- Energy burn: CO + $0.5O_2$ = CO_2
- PbO(slag) + CO = Pb(liq) + CO_2
- Possible: Na_2SO_4(slag) + 4Fe = Na_2O + 3FeO + FeS (matte)
- Possible: Na_2SO_4(slag) + 4CO = Na_2S + $4CO_2$ (matte)
- ($SbO_{1.5}$, $AsO_{1.5}$, SnO_2) (slag) + CO = **[Sb, As, Sn] in Pb(liq)** + CO_2
- **Products: slag Na_2O-FeO-SiO_2 (small wt% Pb, Sb, Sn, As, S), metal Pb (with Sb, As, Sn), possibly matte FeS-(Na_2S, PbS, ...)**

Thus, the thermodynamics of slag, metal and matte under reducing conditions is fundamental for improving the overall efficiency of the process and developing possible options for slag cleaning. The purpose of the present project is to develop a computational tool based on thermodynamics. The tool should be able to predict conditions when the phases are liquid or solid. It can estimate the energy required for the process, and optimal amount of fuel. It uses non-ideal solutions to predict the distribution of every element among the phases in the system. The range of conditions targeted in the present study represent the reduction smelting of materials formed during lead acid battery recycling.

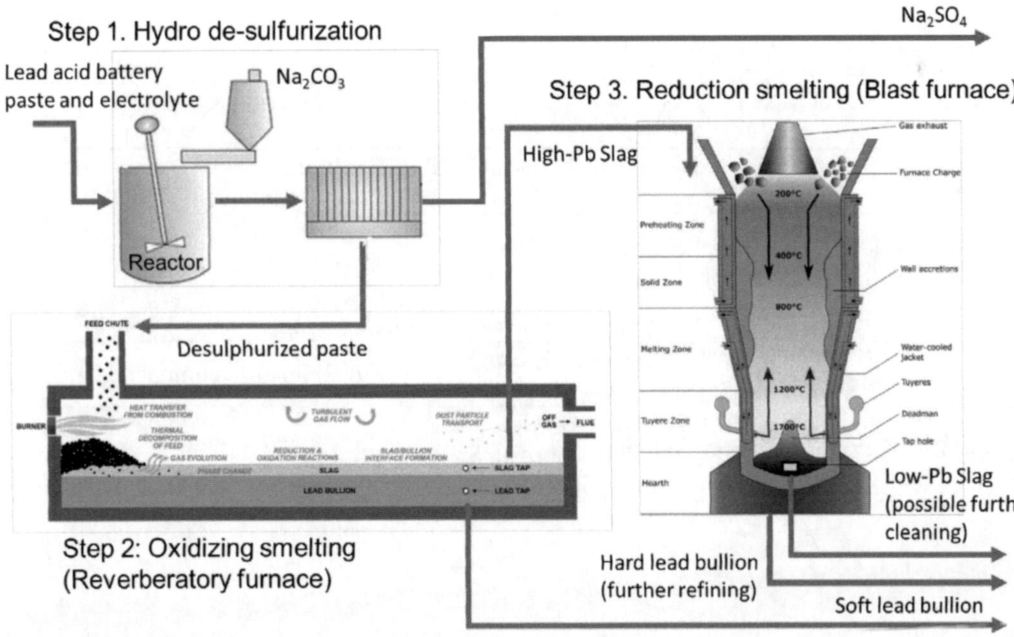

Figure 1. Schematic of the lead acid battery recycling using the lead carbonate precipitation and pyrometallurgical processing route. Furnace pictures are from [4, 5]

2. Methods and scope of the project

The methodology is based on the integrated experimental and thermodynamic modelling work. The principles are described in the parallel PbZn'2023 paper [6]. In brief, thermodynamic models are developed for the potential phases in the system, which are shown in **Table 1**. Thermodynamic models contain adjustable parameters, which need to be optimized using experimental information. These parameters represent thermodynamic properties of end-members of solution phases, as well as binary and ternary interaction parameters between solution constituents. Many of model parameters for the selected systems have been determined earlier, through consortia programs conducted in PYROSEARCH since about 2013. The remaining model parameters are determined through systematic assessment of chemical systems, selected for the priority studies. They are shown in Figure 2. When available, experimental literature data are used. When data are not available, experiments are designed to fix the particular model parameters.

The model for the slag phase has been developed within the modified Quasichemical framework [7-11]. For the geometric interpolation of binary parameters into ternary space [12, 13], the acidic component SiO_2 was treated as an asymmetric one with respect to basic FeO, PbO, Na_2O, SnO, etc. oxides. The decisions on geometric grouping of amphoteric components Fe_2O_3, Sb_2O_3, SnO_2 were made on case-by-case basis, selecting the one that ensures the best approximation of the experimental data with the least number and values of additional ternary parameters. For slags containing sulfur, quadruplet approximation formalism was used, which considers mixing of pure oxide and sulfide endmembers (e.g. Na_2O-FeO-Fe_2O_3-SiO_2-Na_2S-FeS-Fe_2S_3-SiS_2-...) with possible additional interaction parameters such as Na-Fe-O-S.

While two sublattices (cationic for metals and anionic for O^{-2} and S^{-2}) were used in slag model, the matte/metal/speiss model treats all species equally on one sublattice by assigning binary interaction parameters like Fe-O, Fe-S, Na-S, and ternary Fe-O-S, Na-Fe-O, etc. This provides additional freedom in describing non-stoichiometric mattes (with deficiency and excess of sulfur), extending to sulfur-free metal within one model, but has a limitation on lacking accuracy in ternary oxide or sulfide systems, e.g. FeO-Sb_2O_3-SiO_2, for which therefore the slag model is preferred.

Table 1. Main phases and thermodynamic models used in the present study

Phase name	Site occupancy	Model (formalism)
Liquid matte/metal/speiss	(Na^{I}, Fe^{II}, Fe^{III}, Pb^{II}, Sn^{II}, Sb^{III}, As^{III}, O^{II}, S^{II})	Modified Quasichemical Model in Pair Approximation (MQMPA)
Slag	(Na^{+1}, $[NaFe]^{+4}$, Fe^{+2}, Fe^{+3}, Si^{+4}, Pb^{+2}, Sn^{+2}, Sn^{+4}, Sb^{+3}, As^{+3})(O^{-2}, S^{-2})	Modified Quasichemical Model in Quadruplet Approximation (MQMQA)
Spinel	$[Fe^{+2}, Fe^{+3}]^{tetr}[Fe^{+2}, Fe^{+3}, Sn^{4+}, Vacancy^{0}]^{oct}{}_2O_4$	Compound Energy Formalism (CEF)
Solid phases	SiO_2, Fe_2SiO_4, $NaFeO_2$, SnO_2, $FeSb_2O_4$, Na_2SnO_3, $Na_4Sn_2Si_3O_{12}$, etc.	Stoichiometric compounds
Gaseous species	N_2, O_2, H_2O, CO, CO_2, S_2, SO_2, SO, As_4, As_2, AsS, AsO, As_4O_6, Pb, PbS, PbO, Sn, SnO, SnS, Sb, Sb_2, SbO, Sb_4O_6, SbS, Na, and more.	Ideal gas

Journal of Physics: Conference Series **2738** (2024) 012039 doi:10.1088/1742-6596/2738/1/012039

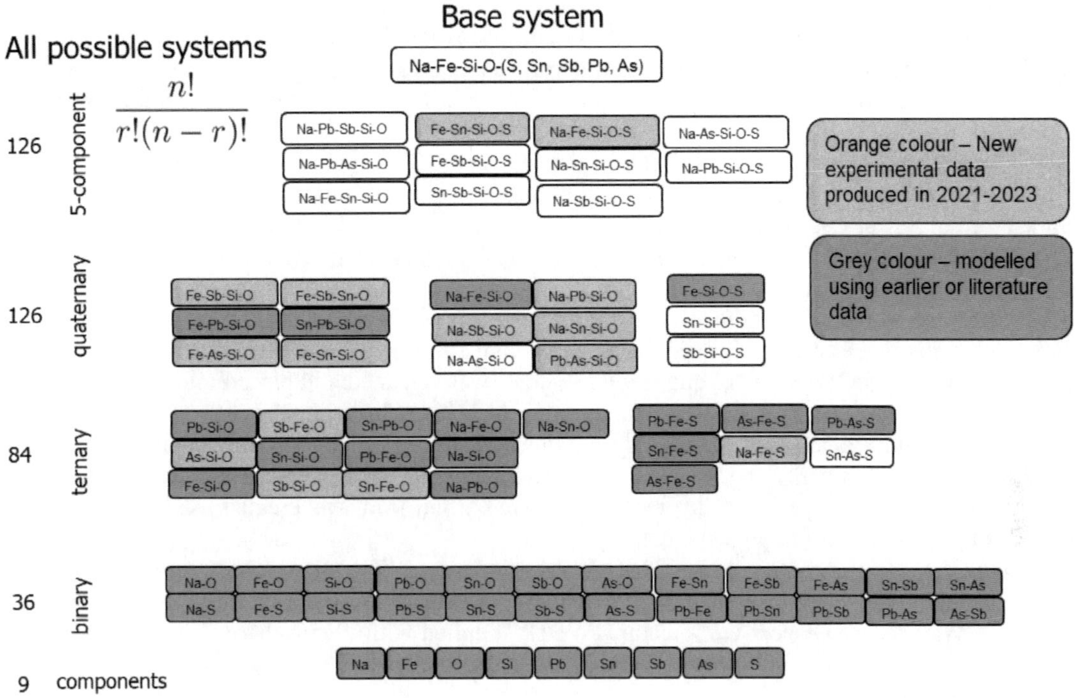

Figure 2. A summary of chemical systems selected for thermodynamic modelling and experimental study.

Experiments of the present study are conducted using the equilibration, quenching, EPMA analysis methodology. The steps required to produce one experimental point are shown in **Figure 3**. They were described in earlier publications [14, 15].

Powder mixtures were prepared with an agate mortar and pestle from synthetic reagents (SiO_2, Fe_2O_3, Fe, Sb_2O_3, Sb, Na_2CO_3, SnO_2, Sn, As_2O_3, As, PbO, Pb, FeS – all at least 99.8% purity) to obtain the desired bulk composition. To prevent explosion of sealed ampoules by released gases, and minimize loss of volatile components on open substrates, some of these reagents were pre-reacted to form master-slags or master-compounds, such as:

- $Na_2CO_3 + SiO_2 = Na_2SiO_3 + CO_2(g)$. Na_2SiO_3 master-compound can be used in closed ampoule.
- $PbO + SiO_2$ = "$Pb_2Si_3O_8$" master-slag (900°C). This master-slag prevents the loss of volatile PbO in during high-temperature experiments;
- $Na_2CO_3 + Sb_2O_3 + O_2 = 2NaSbO_3 + CO_2(g)$. This reaction binds volatile Sb_2O_3 together with Na_2CO_3 into more stable compound;
- Fe + As = FeAs (600-700°C). This compound can be heated up to 1200°C, while pure unreacted As sublimates and causes explosion of sealed ampoules.

The mixtures were pressed into 0.3-0.5 g pellets and placed into a substrate. Several types of holding materials were used: sealed SiO_2 ampoules, open SiO_2 crucible, Pt-25%Ir foil, or Ir wire in a shape of spiral. The innovation of the present study was the use of rhenium foil as a new type of holding material. It is less stable than other PGMs at oxidizing conditions, but has a high melting temperature (second only to W), less expensive, relatively ductile (compared to Ru, Rh, Os, Ir), and more stable against corrosion by liquid metals (Pb, Sn, Sb) at low oxygen partial pressure. The use of rhenium foil allowed successful experiments in the Fe-Sb-Si-O, Na-Sn-Si-O systems equilibrated with Sb or Sn

Journal of Physics: Conference Series **2738** (2024) 012039 doi:10.1088/1742-6596/2738/1/012039

metals. It was even possible to re-use rhenium foil for several experiments within the same chemical system. Fun fact: Re was the last stable element to be discovered in 1925.

Equilibration experiments were conducted in a vertical tube furnace. Thermocouple based on Pt-Rh 6%/Pt-Rh 30% alloy protected by the alumina shield allowed the temperature control within the furnace of ±1 K. The location of the thermocouple was 2-3 mm from the sample. The working thermocouple was calibrated against a standard one every 1-2 months. The combined uncertainty in calibration, thermal contact, and other factors provide the overall uncertainty of equilibration temperature of 5 K or less. The equilibration time was pre-determined in trial experiments, specific for each chemical system and the range of compositions. It ranged from 10-15 min for highly volatile materials to several weeks for low-temperature viscous materials. To preserve the composition of phases existing at the temperature of equilibration, samples were quenched into the solution of calcium chloride in water. High concentration of calcium chloride prevents freezing and can provide a quenching medium at 253 K. After quenching, samples were cleaned and placed into epoxy resin. The Epoxy resin provides a stable substrate for the polishing and electron microscopy analysis. Silicon carbide paper and diamond paste used during the polishing, which was rub by the TegraPol-31 machine (Struers, Denmark). To prevent the accumulation of charge on the surface of the sample during the electron microscopy, a layer of conductive carbon film was created used the QT150TES apparatus by Quorum Technologies, UK.

Electron probe X-ray microanalysis (EPMA) was run using JEOL JXA 8200L machine by Japan Electron Optics Ltd., Tokyo. Elemental compositions of each phase were measured. For quantitative analysis, a Wavelength Dispersive Detectors (WDD) supplied with the machine was used. During the X-ray count, the acceleration voltage of 15 kV and a probe current of 20 nA were used. Collected characteristic X-ray counts were converted to concentration using the appropriate standards: hematite Fe_2O_3 for Fe, wollastonite $CaSiO_3$ for Si and Ca, albite $NaAlSi_3O_8$ for Na, InAs for As, metallic Sb for Sb, chalcopyrite $CuFeS_2$ or pyrite FeS_2 for S, cassiterite SnO_2 for Sn. The standards were supplied by Charles M. Taylor Co., Stanford, CA, USA, and 71.4 wt.% PbO-SiO_2 K456 glass from NIST for Pb. Electron interactions, X-ray absorption, effects of fluorescence and other factors, that differ between the sample and the standard were taken into account by the Duncumb-Philibert ZAF correction. The correction calculations were performed by the software supplied with the machine. The overall accuracy of measurement of compositions for elements in high concentrations was estimated to be at least 1 wt %. The information on the oxidation states, e.g. Fe^{2+}/Fe^{3+}, cannot be obtained using the current analytical methodology. The following standards were used for EPMA: For the slag phase, non-zero probe diameters of 20-50 micron may be used to decrease the scatter due to dendrite formation on quenching and limit the loss of volatile species (Na) under electron beam.

Journal of Physics: Conference Series **2738** (2024) 012039 doi:10.1088/1742-6596/2738/1/012039

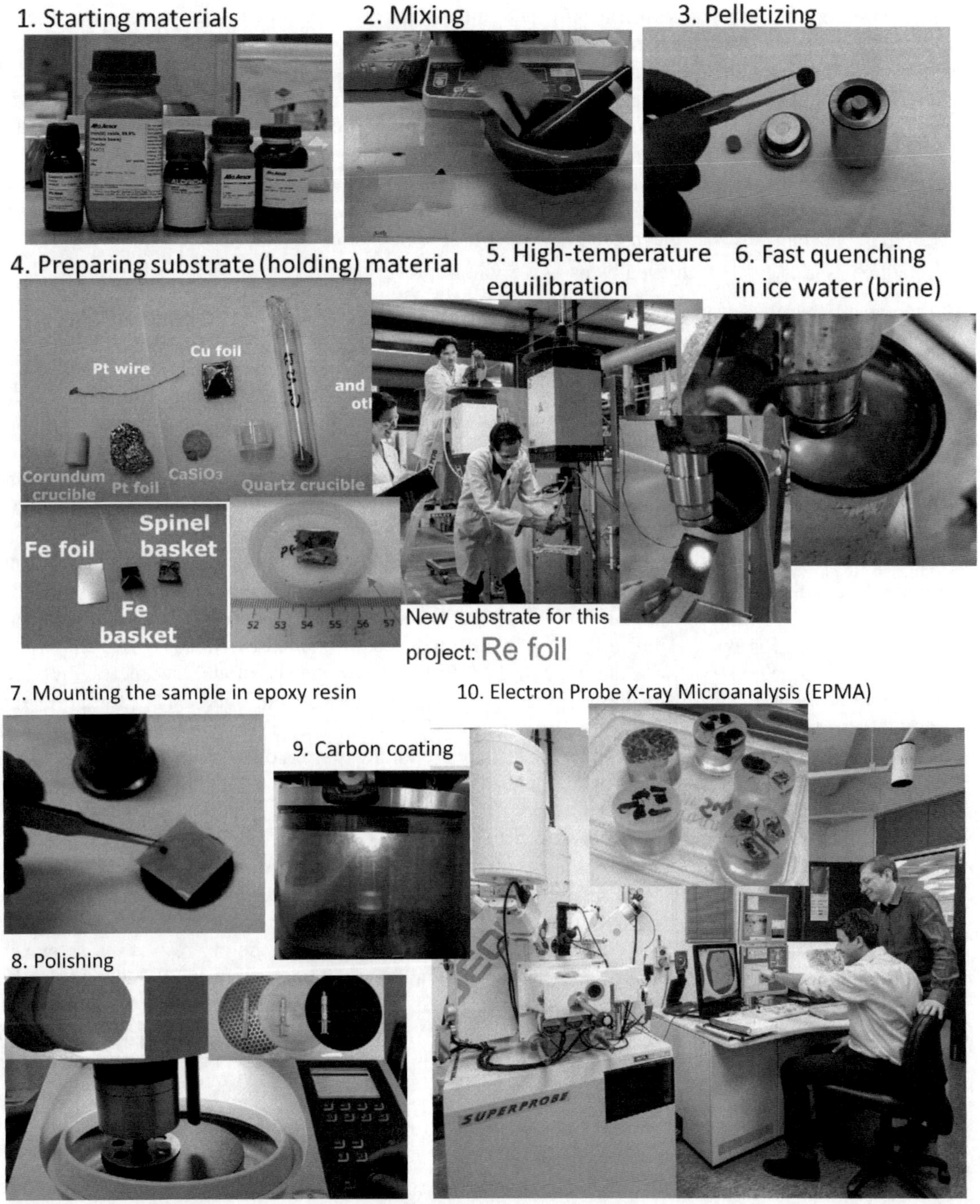

Figure 3. Steps required to produce an experimental point.

3. Selected results

3.1. Na-Fe-Si-O thermodynamic model

Thermodynamic assessment of the Na-Fe-Si-O and its sub-systems has been conducted using experimental data from literature or experimental results produced earlier. The Fe-Si-O system is largely based on the earlier assessment [16], with modifications. The same modelling framework was used, but parameters have been re-optimized to achieve equal or better agreement with collected experimental data [16], including more recent experimental results [17]. Thermodynamic properties of liquid SiO_2 in slag have been changed in 2022. The modelling framework for the Na-Si-O system was described by Wu et. [18], but model parameters were modified to describe experimental phase diagram data collected in later work by Nekhoroshev [19]. Phase diagram is shown in **Figure 4**. The modelling framework of Nekhoroshev [19] used more complex slag structure using $(Na_2)^{2+}$ species to describe the cristobalite liquidus at high temperatures, and it is harder to expand in large thermodynamic database. In the present study it was demonstrated that reasonable description of cristobalite liquidus can be achieved by using very high-power parameters on Si^{4+}, such as $g^{0,15}_{Na^+,Si^{4+}//O}$. The model of the present study provides inferior results for metastable low-temperature miscibility gap in high-SiO_2 glasses compared to Nekhoroshev [19], and for very low-SiO_2 liquids (< 0.3 mole fraction of SiO_2), but these conditions are of low priority for reductive smelting. The Na-Fe-O system has been modelled by Moosavi-Khoonsari and Jung [20]. In their modelling framework, the $NaFe^{4+}$ associate was included in for the slag solution. The formation of $NaFe^{4+}$ in slag is expected at higher oxygen partial pressures (higher oxidizing potentials), compared to those observed in Step 3 of **Figure 1**, but for model consistency and possible other uses of the database, the same modelling framework has been adopted in the present study. Model parameters have been re-optimized to achieve consistency with updated Fe-O systems. Models for the Fe-O system have been re-evaluated by PYROSEARCH in 2015-2016 [21, 22] to correct the properties of wüstite, but they have not yet been corrected in FactSage FToxid database, including the 8.2 version. Moosavi-Khoonsari and Jung [20] used the Fe-O system from FactSage FToxid database. Selected results of the modelling of the present study are shown in **Figure 4**. The Na-Fe-O-Si system has been modelled by Moosavi-Khoonsari and Jung [23], who collected existing experimental data. In our investigation, model parameters were re-evaluated to reproduce the collected experimental data and the above mentioned changes in Fe-O and Fe-O-Si. Example is shown in **Figure 4**. Results at higher $P(O_2)$ were also included in the model of the present study, but they are not shown because they are far away from operational conditions.

3.2. *Introduction of Na in matte-metal thermodynamic model*

During the course of the project, it became evident that the formation of liquid matte (sulfide-rich liquid) phase is possible in the blast furnace, or during the pyrometallurgical cleaning of lead blast furnace slags. Liquid matte is expected to be close to Na_2S-FeS-FeO stoichiometry. Existing literature data were used to model the Na-S and Na-Fe-S system. Examples of modeling results are shown in **Figure 5**. The model for the Fe-S is based on the thermodynamic assessment of Waldner [24]. The As-S [25] and Pb-S [26] were modelled recently. The Sb-S, Sn-S were modelled as well, based on literature data, but the assessments are not published yet. Preliminary model parameters are introduced for the Na-As-S, Na-Sb-S, Na-Sn-S and Na-Pb-S systems to provide reasonable description of the corresponding ternary phase diagrams.

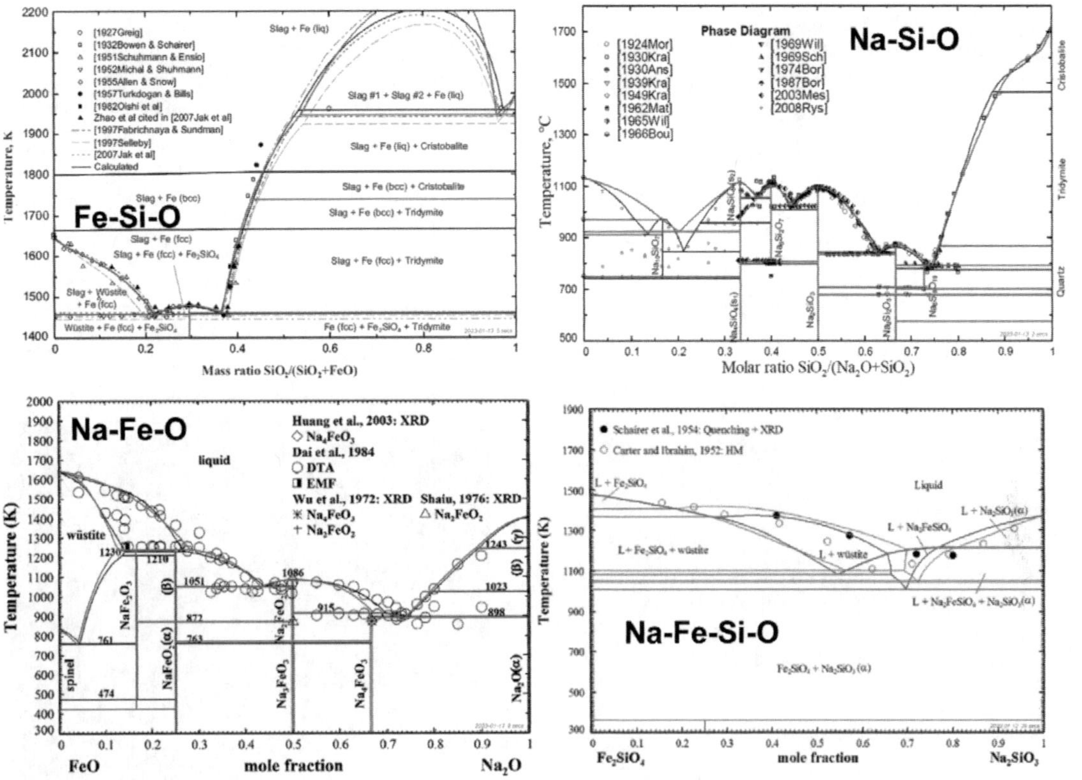

Figure 4. Examples of thermodynamic modelling results for the Fe-Si-O, Na-Si-O, Na-Fe-O and Na-Fe-Si-O systems (slag with predominant Fe^{2+}, in equilibrium with Fe metal). Lines represent thermodynamic calculation using FactSage: red – present study, black – earlier models [16, 19].

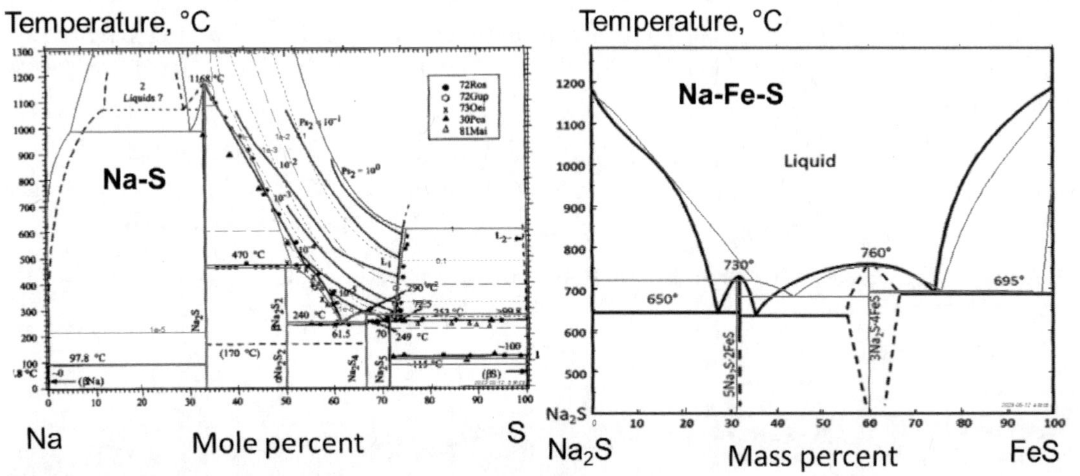

Figure 5. Examples of thermodynamic modelling results for the Na-S [27] and Na-Fe-S [28] systems based on data collected in literature. Red lines represent thermodynamic calculation using FactSage using model parameters of present study.

3.3. Fe-Sb-Si-O system in equilibrium with Fe-Sb metal

No literature data were available for the Fe-Sb-Si-O system, which is not only important for lead acid battery recycling, but also for pyrometallurgical production of antimony [29]. The summary of produced experimental results and example of the microstructure of are shown in **Figure 6**. These data were used to fix the Fe^{2+}-Sb^{3+} interactions in slag, as well as ternary parameters $g^{ijk}_{Fe^{2+},Si^{4+}(Sb^{3+})//O}$ in slag. A new solid compound $FeSb_2O_4$ (schafarzikite) was introduced in the database.

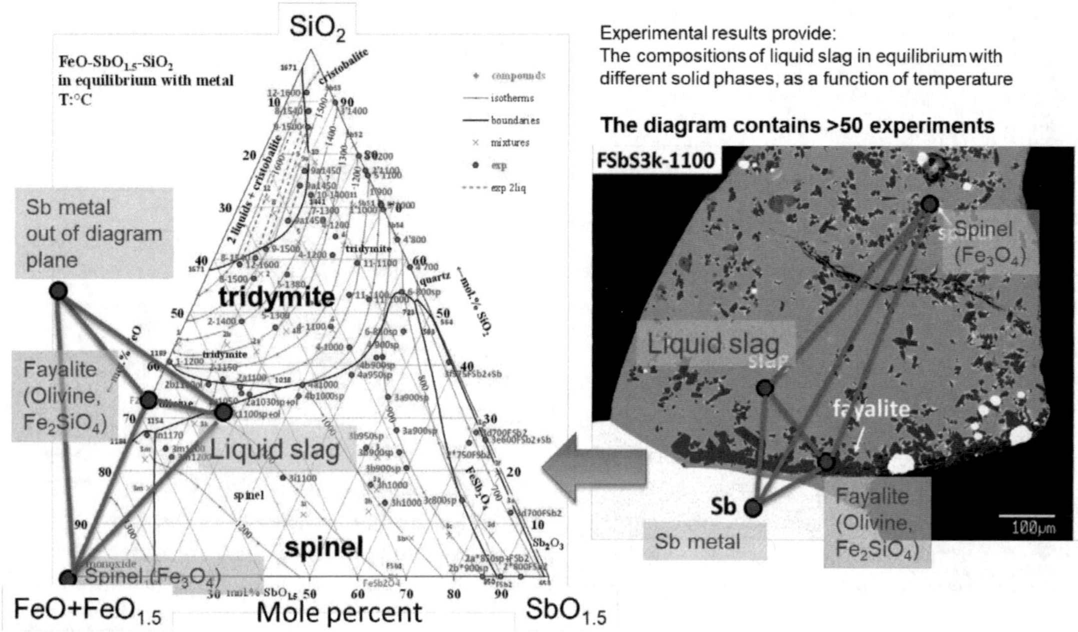

Figure 6. Example of experimental results and model predictions in the Fe-Sb-Si-O system. Lines represent thermodynamic calculation using FactSage, symbols on the diagram are experimental data obtained in 2022-2023.

3.4. Na-Sn-Si-O system, equilibrium with Sn metal and in air

No literature data were available for the Na-Sn-Si-O system. Two experimental series were designed to fix the $g^{ijk}_{Na^{+},Si^{4+}(Sn^{2+})//O}$ and $g^{ijk}_{Na^{+},Si^{4+}(Sn^{4+})//O}$ parameters in slag (**Figure 7**). New compounds were introduced in the database: Na_2SnO_3, Na_4SnO_4, $Na_4Sn_3O_8$, $Na_4Sn_2Si_3O_{12}$. Liquidus of cassiterite in air agrees with literature [30].

Journal of Physics: Conference Series **2738** (2024) 012039 doi:10.1088/1742-6596/2738/1/012039

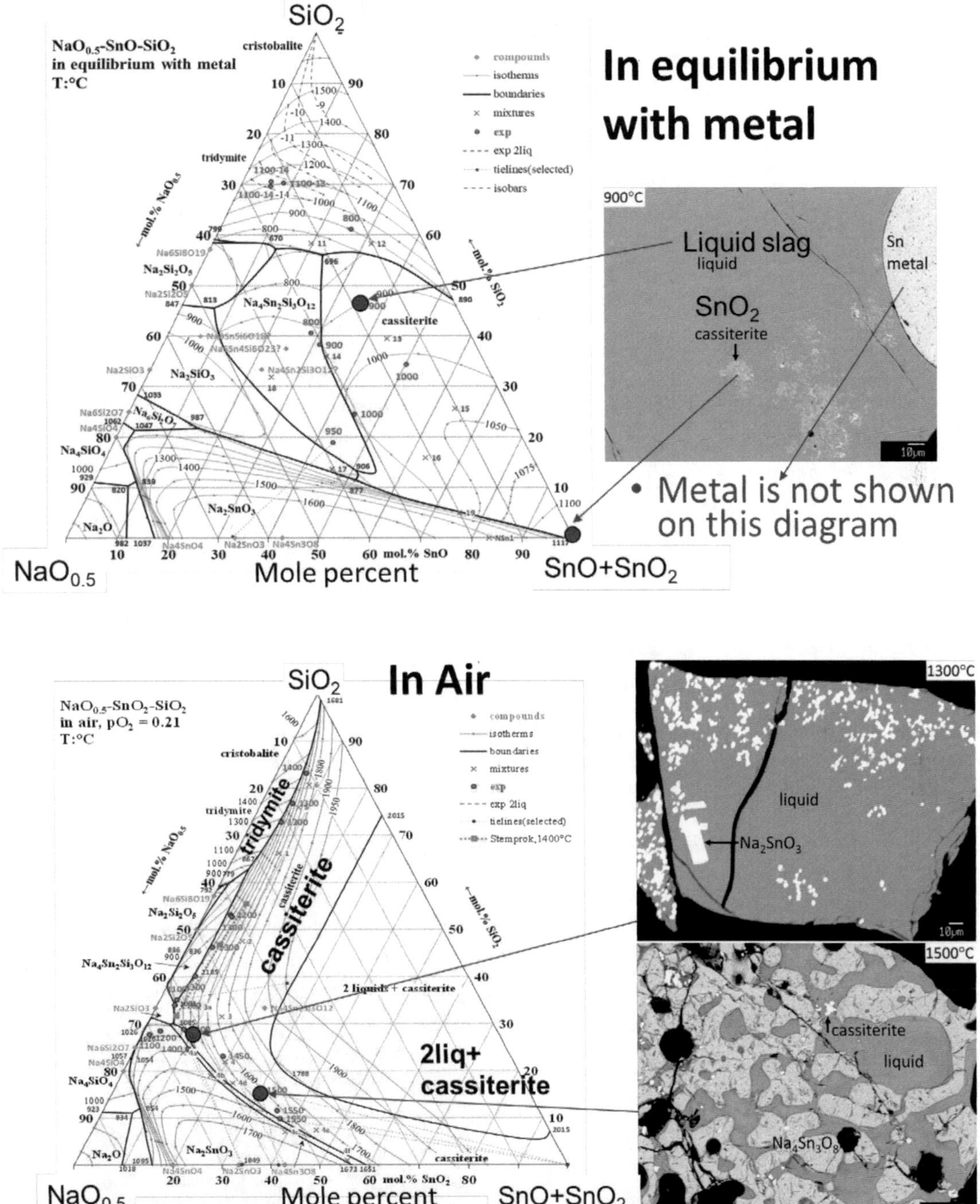

Figure 7. Example of experimental results and model predictions in the Na-Sn-Si-O system. Lines represent thermodynamic calculation using FactSage, symbols on diagrams are experimental data obtained in 2022-2023.

Journal of Physics: Conference Series **2738** (2024) 012039 doi:10.1088/1742-6596/2738/1/012039

3.5. Fe-Sb-Sn-O system in equilibrium with Fe-Sb-Sn metal

First experimental results were obtained for the Fe-Sb-Sn-O system. No new ternary parameters were introduced in the slag, but a Sb^{3+}-Sn^{4+} interaction parameter in slag was optimized using the data in **Figure 8**. Also, the predicted distribution of Sb/Sn/Fe between slag and metal, and distribution of Sn between slag and spinel were tested. Significant solubility of Sn^{4+} in spinel was observed.

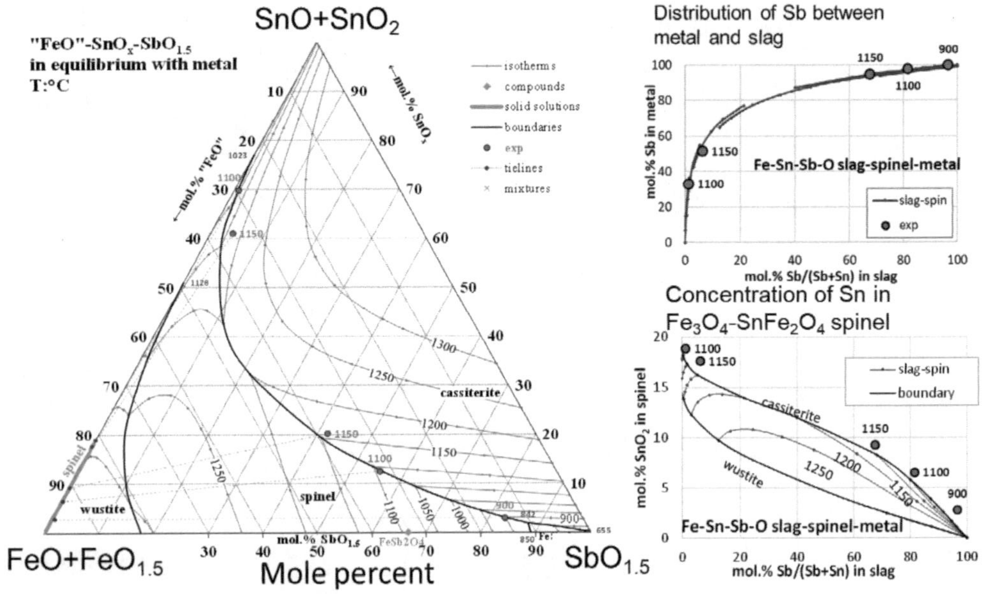

Figure 8. Example of experimental results and model predictions in the Fe-Sb-Sn-O system (slag-metal-solid equilibria). Lines represent thermodynamic calculation using FactSage, symbols on diagrams are experimental data obtained in 2022-2023.

3.6. Fe-As-Si-O system in equilibrium with Fe-As metal (speiss)

The solubility of arsenic in slag can be important if speiss-containing secondary materials are introduced in the process. The solubility of arsenic in the slag was experimentally measured and compared with thermodynamic database predictions. The results are shown in **Figure 9**.

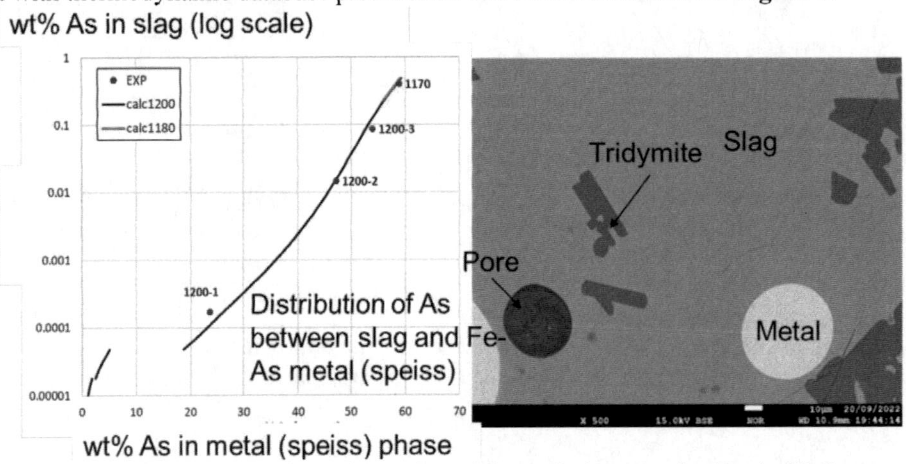

Figure 9. Example of experimental results and model predictions in the Fe-As-Si-O system (slag-metal-tridymite equilibria). Lines represent thermodynamic calculation using FactSage, symbols on diagrams are experimental data obtained in 2022-2023.

3.7. Na-Fe-Si-O-S slag/matte/metal system

To test the newly developed model for Na-Fe-S matte, an experimental series was designed targeting the distribution of sulfur, oxygen and sodium between slag and matte (coexisting with Fe metal and tridymite). The results are shown in **Figure 10**. The agreement is reasonable and allows expanding this direction to analyze the distribution of impurity metals (Sb, Sn, Pb, and As) between slag and matte.

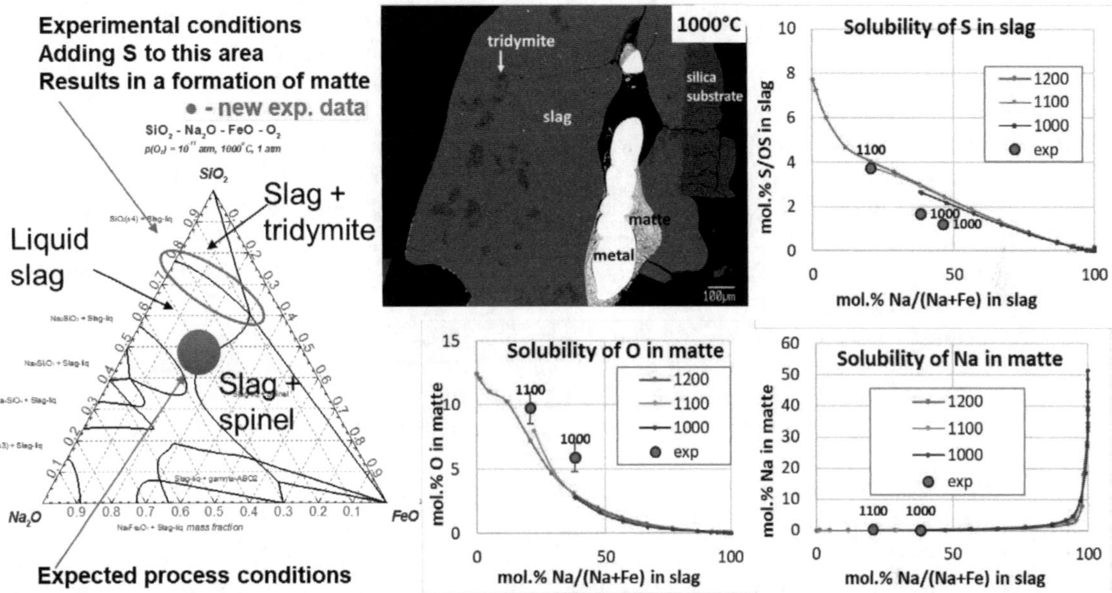

Figure 10. Example of experimental results and model predictions in the Na-Fe-Si-S-O system (slag/matte/Fe metal equilibria). Lines represent thermodynamic calculation using FactSage, symbols on diagrams are experimental data obtained in 2022-2023.

Conclusions

The fundamental research project was initiated for better understanding the thermodynamics of S, Sb, Sn, Pb, and As in Na-Si-Fe-O-based slags at reducing conditions. The existing thermodynamics database has been expanded to include Na in slag and in liquid matte/speiss/metal. Many new stoichiometric compounds were added. The following systems have been studied experimentally: Fe-Sb-Si-O, Na-Sn-Si-O, Fe-Sn-Sb-O, Pb-As-Si-O, Na-Sb-Si-O, Fe-As-Si-O and Na-Fe-Si-O-S. Experimental studies were used to optimize the parameters of thermodynamic models improving the predictive power. The resulting model have been used to analyze the options for pyrometallurgical slag cleaning, but these predictions cannot be disclosed.

Acknowledgments

The funding for the study support is provided by Gopher Resource and US Department of Defense “Materials Recovery Technology for Defense Supply Resiliency” program. The EPMA analysis was conducted using the equipment and technical support by the Microscopy Australia Facility at the Centre for Microscopy and Microanalysis (CMM), The University of Queensland.

Journal of Physics: Conference Series **2738** (2024) 012039 doi:10.1088/1742-6596/2738/1/012039

References

[1] Prengaman PD, Mirza AH. Recycling concepts for lead-acid batteries. In: Lead-Acid Batteries for Future Automobiles. Elsevier B.V.; 2017. 575-598,https://doi.org/10.1016/b978-0-444-63700-0.00020-9.

[2] Errington B, Hawkins P, Lim A. ISASMELT for lead recycling. PbZn 2010. 2010: 709-720.

[3] Blanpain B, Arnout S, Chintinne M, Swinbourne DR. Lead Recycling. In: Worrell E, Reuter M, eds. *Handbook of Recycling*. 2014:95-111.

[4] Anderson A, Kumar V, Rao VM, Grogan J. A Review of Computational Capabilities and Requirements in High-Resolution Simulation of Nonferrous Pyrometallurgical Furnaces. JOM. 2022; 74: 1543-1567, https://doi.org/10.1007/s11837-022-05169-4.

[5] Vanparys RR (2022) Reduction Kinetics between Coke and Slag in the Lead Blast Furnace. PhD thesis, Swinburne University of Technology

[6] Shishin D, Shevchenko M, Starykh R, Sineva S, Prostakova V, Jak E. Development and application of matte/speiss/metal thermodynamic database for optimization of processing of drosses, dusts and reverts from lead, zinc and copper production. In: 10th International Conference on Lead and Zinc Processing, Chansha, China. 2023.

[7] Pelton AD, Decterov SA, Eriksson G, Robelin C, Dessureault Y. The Modified Quasichemical Model. I - Binary Solutions. Metall. Mater. Trans. B. 2000; 31: 651-659, https://doi.org/10.1007/s11663-000-0103-2.

[8] Petersen S, Hack K. The thermochemistry library ChemApp and its applications. Int. J. Mater. Res. 2007; 98: 935-945, https://doi.org/10.3139/146.101551.

[9] Shishin D (2013) Development of a thermodynamic database for copper smelting and converting. Ph. D. thesis, Ecole Polytechnique of Montreal

[10] Pelton AD, Chartrand P. The Modified Quasichemical Model. II - Multicomponent Solutions. Metall. Mater. Trans. A. 2001; 32: 1355-1360, https://doi.org/10.1007/s11661-001-0226-3.

[11] Abdeyazdan H, Shevchenko M, Hayes PC, Jak E. Integrated experimental and thermodynamic modeling study of phase equilibria in the '$CuO_{0.5}$'-MgO-SiO_2 system in equilibrium with liquid Cu metal for characterizing refractory-slag interactions. Ceram. Int. 2022: Ahead of Print, https://doi.org/10.1016/j.ceramint.2022.03.183.

[12] Chartrand P, Pelton AD. On the Choice of "Geometric" Thermodynamic Models. J. Phase Equilib. 2000; 21: 141-147.

[13] Pelton AD. A General "Geometric" Thermodynamic Model for Multicomponent Solutions. CALPHAD. 2001; 25: 319-328.

[14] Shevchenko M, Shishin D, Jak E. Integrated phase equilibria experimental study and thermodynamic modeling of the Cr-Si-O, Fe-Cr-O and Fe-Cr-Si-O systems. Ceram. Int. 2022, https://doi.org/10.1016/j.ceramint.2022.07.286.

[15] Starykh RV, Shevchenko M, Shishin D, Jak E. Experimental study and thermodynamic modeling of the Cu-Sn-Si-O system and sub-systems. CALPHAD. 2021; 74: 102312.

[16] Hidayat T, Shishin D, Decterov SA, Jak E. Experimental Study and Thermodynamic Re-evaluation of the FeO-Fe_2O_3-SiO_2 System. J. Phase Equilib. Diffus. 2017; 38: 477-492, https://doi.org/10.1007/s11669-017-0535-x.

[17] Cheng S, Shevchenko M, Hayes PC, Jak E. Experimental Phase Equilibria Studies in the FeO-Fe_2O_3-CaO-SiO_2 System and the Subsystems CaO-SiO_2, FeO-Fe_2O_3-SiO_2 in Air. Metall. Mater. Trans. B. 2021; 52: 1891-1914, https://doi.org/10.1007/s11663-021-02159-w.

[18] Wu P, Eriksson G, Pelton AD. Optimization of the Thermodynamic Properties and Phase Diagrams of the Na_2O-SiO_2 and K_2O-SiO_2. J. Am. Ceram. Soc. 1993; 76: 2059-2064.

[19] Nekhoroshev E (2019) Thermodynamic Optimization Of The Na_2O-K_2O-Al_2O_3-CaO-MgO-B_2O_3-SiO_2 System. Ph. D. thesis, Polytechnique Montréal

[20] Moosavi-Khoonsari E, Jung I-H. Critical Evaluation and Thermodynamic Optimization of the Na_2O-FeO-Fe_2O_3 System. Metall. Mater. Trans. B. 2016; 47: 576-594, https://doi.org/10.1007/s11663-015-0517-5.

[21] Hidayat T, Shishin D, Jak E, Decterov S. Thermodynamic Reevaluation of the Fe-O System. CALPHAD. 2015; 48: 131-144, https://doi.org/10.1016/j.calphad.2014.12.005.

[22] Shishin D, Prostakova V, Jak E, Decterov S. Critical assessment and thermodynamic modeling of the Al–Fe–O system. Metall. Mater. Trans. B. 2016; 47: 397-424, https://doi.org/10.1007/s11663-015-0493-9.

[23] Moosavi-Khoonsari E, Jung I-H. Critical Evaluation and Thermodynamic Optimization of the Na_2O-FeO-Fe_2O_3-SiO_2 System. Metall. Mater. Trans. B. 2016; 47: 291-308, https://doi.org/10.1007/s11663-015-0504-x.

[24] Waldner P, Pelton AD. Thermodynamic modeling of the Fe-S system. J. Phase Equilib. Diff. 2005; 26: 23-28.

[25] Prostakova V, Shishin D, Jak E. Thermodynamic optimization of the As-S system. Calphad. 2021; 72: 102235, https://doi.org/10.1016/j.calphad.2020.102235.

[26] Shishin D, Chen J, Jak E. Thermodynamic Modeling of the Pb-S and Cu-Pb-S Systems with Focus on Lead Refining Conditions. J. Phase Equilib. Diffus. 2020; 41: 218-233, https://doi.org/10.1007/s11669-020-00811-7.

[27] Sangster J, Pelton AD. The Na-S (Sodium-Sulfur) system. Journal of Phase Equilibria. 1997; 18: 89-96, https://doi.org/10.1007/BF02646762.

[28] Kopylov NI, Novoselov SS. System Cu_2S-FeS-Na_2S. Russ. J. Inorg. Chem. 1964; 9: 1038-1043.

[29] Moosavi-Khoonsari E, Mostaghel S, Siegmund A, Cloutier J-P. A Review on Pyrometallurgical Extraction of Antimony from Primary Resources: Current Practices and Evolving Processes. Processes. 2022; 10: 1590, https://doi.org/10.3390/pr10081590.

[30] Stemprok M, Voldan J. Solubility of tin(IV) oxide in sodium silicate glasses. Silikaty. 1975; 19: 211-221.

Challenges to Treat Complex Zinc Concentrate and Latest Technical Development

Pengfu Tan

Glencore International AG, Switzerland
tanpengfu@yahoo.com

Abstract

The complex zinc concentrate containing high Pb, Cu, SiO2, Na and K, including the Zn-Pb bulk concentrate, cannot be treated in the fluidized bed roaster. The Imperial Smelting Process (ISP) is the main technology to treat this type of concentrate. However, ISP consumes a large amount of coke, which is not only expensive, but also has high CO2 emission.

Several metallurgical processes, which can treat the type of complex zinc concentrate and have low CO2 emission, have been reviewed and compared.

Recently, a new technology "Flash Burner Roasting Process" has been tested successfully at the industrial scale suspension roaster, and the tests demonstrate that this technology can treat the complex zinc concentrate containing high contents of Pb, Cu and SiO2. The thermodynamic modelling for this process has been done. The industrial test results, the modelling results and the further development are presented in this paper.

1. Existing technologies

The fluidized bed roaster cannot treat the complex zinc concentrate containing high Pb, Cu and SiO2 due to the accretion formation in the roaster. There are several technologies that could treat those complex zinc concentrates, such as Imperial Smelting Process, Outotec atmospheric direct leaching process, Albion process, PBOX process, pressure leaching process, and pelletised concentrate fluidized bed roasting process.

1.1. Imperial smelting process (ISP)

Imperial Smelting Processes Ltd (ISP) began development of a new smelting process for zinc in 1943 in UK. In 1947, an experimental blast furnace was built and by 1950, reach a capacity of 2 t per day [1]. The first breakthrough was the gaseous zinc produced at the smelting temperature, could be recovered by an absorption in a liquid lead spray. One major problem was that some of the zinc produced, re-oxidized to ZnO, making the process inefficient. The second breakthrough was to increase furnace top temperature to prevent this re-oxidization. The third breakthrough for this process was that this Zn blast furnace could simultaneously produce lead and zinc metals.

Since then, 20 ISP plants have been built and commissioned. The typical ISP flowsheet is shown below in Figure 1. However, due to high coke costs, most of ISP plants were shut down in the early 2000, and only 6 of them are still in operation. They are Baiyin, Shaoguan, Dongling, Huladao in China, Hachinohe in Japan, and Miasteczko in Poland.

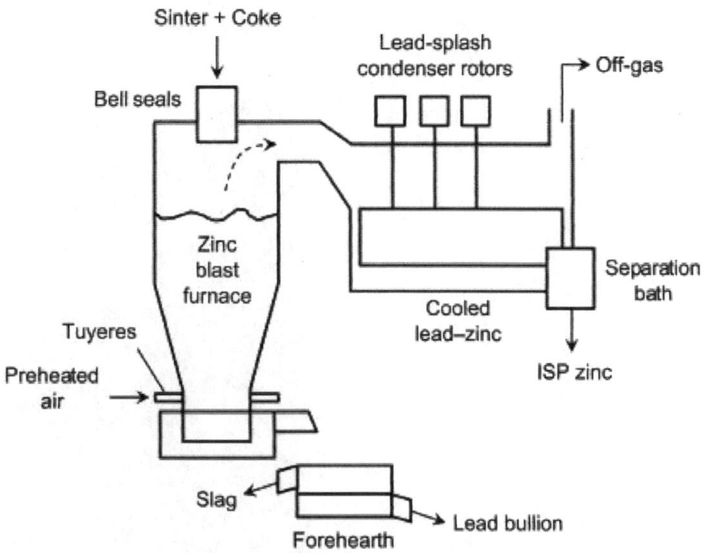

Figure 1. Imperial Smelting Process (ISP) flowsheet

1.2. Outotec atmospheric direct leaching process

In the 1990s, Outokumpu's zinc and technology companies developed an atmospheric zinc concentrate direct leaching process to be able to use a wide variety of sulfidic concentrates in zinc production.

The first laboratory tests before the piloting were performed at Outotec (former Outokumpu) Research Centre (ORC) in Pori with several zinc concentrates used at Kokkola zinc plant [2]. The grain size for most of the zinc concentrates was in the range of 15 to 25 μm. After these laboratory tests the piloting in 1991 with a 10 m high pilot reactor did not give sufficient leaching recovery but showed that the direction of choosing the atmospheric leaching alternative was the best one. So, the piloting was continued at Kokkola zinc plant in a 20 m high pilot reactor, which gave good leaching recovery.

Commercial-scale start-up took place at Kokkola zinc plant in 1998 and three years later another expansion was started using the same technology [2]. The modernization of Boliden Odda plant was finished in autumn 2004 and this technology has also been applied there. The Outotec Zinc Concentrate Direct Leaching (DL) Process equipped with atmospheric leaching reactors of about 900 m^3 having similar height as the big pilot reactor into the existing operations. Schematic picture of the Outotec Direct Leaching reactor is shown below in Figure 2.

1.3. Albion process

The Albion Process technology was developed by Glencore in 1994 and is patented worldwide [3]. It is a combination of ultrafine grinding and oxidative leaching at atmospheric pressure. The feed to the Albion Process™ is a concentrate containing base or precious metals, and the Albion Process is used to oxidise the sulphide minerals in the concentrate and liberate these metals for recovery by conventional means.

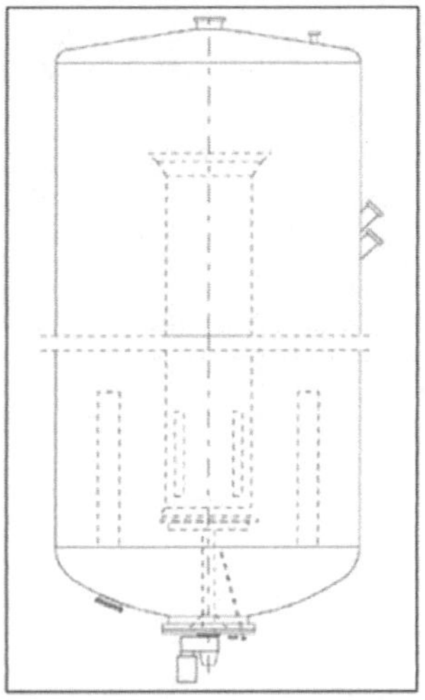

Figure 2. Outotec Direct Leaching Reactor

Figure 3. Albion Oxidative Leach Reactor at Nordenham

Two plants treat a zinc sulphide concentrate and are located at San Juan de Neiva in Spain (4,000 tpa zinc metal), and at Nordenham in Germany (18,000 t/a zinc metal). A summary of the key operating data for the two Glencore Albion Process^{TM} plants is presented in Table 1.

Table 1. Industrial Data for the Albion Process Plants in Spain and Germany

Parameter	San Juan de Neiva [3]	Nordenham
Feed Rate, tph		
MRM Concentrate	1	4
Neutral Leach Residue	0.25	1
Concentrate Composition		
Zinc, %	47.2	41 - 44
Lead, %	8.17	10 - 13
Iron, %	5.4	4.4 - 6.2
SiO_2, %	4.3	4 - 6
Copper, %	0.8	0.8
Sulphur, %	27.2	27 - 29
Oxidative Leach Reactor Size, m^3	280	800, 280
Zinc Recovery in Leach, %, w/w	98.6	98.8
Conversion to elemental sulphur	>95	>95
Leach acid demand, kg/tonne	488	470
Leach oxygen demand, kg/tonne	209	214
Residue Composition		
Sulphur, %	43	40 - 50
Zinc, %	1.3	1 - 2
Lead, %	15.5	17 - 23
Iron, %	8.4	4 - 5
SiO_2, %	7.5	7.1
Copper, %	0.20	0.20

Following the success of the first Albion Process™ plant at San Juan de Neiva, Glencore then constructed a second Albion Process plant at the Nordenham zinc refinery. This plant was commissioned in March 2011. The feed to the plant is also finely ground lead/zinc bulk concentrate from the McArthur River Mine (MRM), and the process flowsheet is similar to that employed at San Juan de Neiva.

The oxidative leach consists of an 800 m^3 and a 280 m^3 reactor in series. The 280 m^3 reactor is similar in geometry to the San Juan de Neiva leach reactor, and the 800 m^3 has a live height of 11.5 m and an aspect ratio of 1.5. Oxygen is injected into the base of the both reactors using HyperSparge supersonic oxygen injection lances, and the slurry is again agitated by a centrally mounted dual hydrofoil impeller. An image of the Nordenham Albion Process™ oxidative leach is shown in Figure 3.

The lead residue from the Nordenham Albion Process™ plant is also sold locally to secondary lead producers.

1.4. PBOX process

The PBOX Process is developed by Glencore as well [4]. The first PBOX process plant at McArthur River Mine (MRM) is used for rapid and selective oxidation of galena, prior to flotation. Oxidation of the galena mineralogy to lead sulphate allows selective flotation of the remaining sphalerite in a conventional flotation circuit, to produce a higher grade sphalerite concentrate. The lead sulphate residue then remains in the flotation tailings, and is recovered to produce lead bullion through the Mt Isa lead smelter. This process, termed PbOX, uses six 800 m^3 tanks constructed from SAF904 steel, shown in Figure 4. The plant was commissioned in 2014 and can process 150,000 to 180,000 tonnes per annum of bulk concentrate, producing up to 150,0000 tonnes per annum of saleable zinc concentrate.

Figure 4. PBOX Reactor in McArthur River Mines

1.5. Pressure leaching process

The Sherritt Zinc Pressure Leach Process was first commercialized at the Trail, British Columbia, zinc refinery of Cominco Limited in early 1981 [5]. Subsequently, Kidd Creek Mines (now Falconbridge Limited, Kidd Creek Division), in Timmins, Ontario and Ruhr Zink GmbH, Datteln,

Germany, have employed the Sherritt Zinc Pressure Leach Process. In each case, the zinc production capacity of existing refineries, using roasting – leaching - electrowinning technology, was expanded without a requirement for increased sulphuric acid production. In the case of the expansions at Cominco and Kidd Creek, conventional zinc concentrates are supplied to the pressure leach circuits. The expansion at Ruhr Zink was designed to treat a blend of conventional zinc concentrates and bulk concentrate. A typical zinc concentrate pressure leaching plant flowsheet is shown in Figure 5.

The application of the Sherritt Zinc Pressure Leach Process to the treatment of lead zinc bulk concentrates was examined in 1977. Bulk flotation concentrate obtained from the treatment of New Brunswick complex sulfide ores was used for the study. A single stage pressure leach process, with upgrading of the oxidic leach residue, was examined.

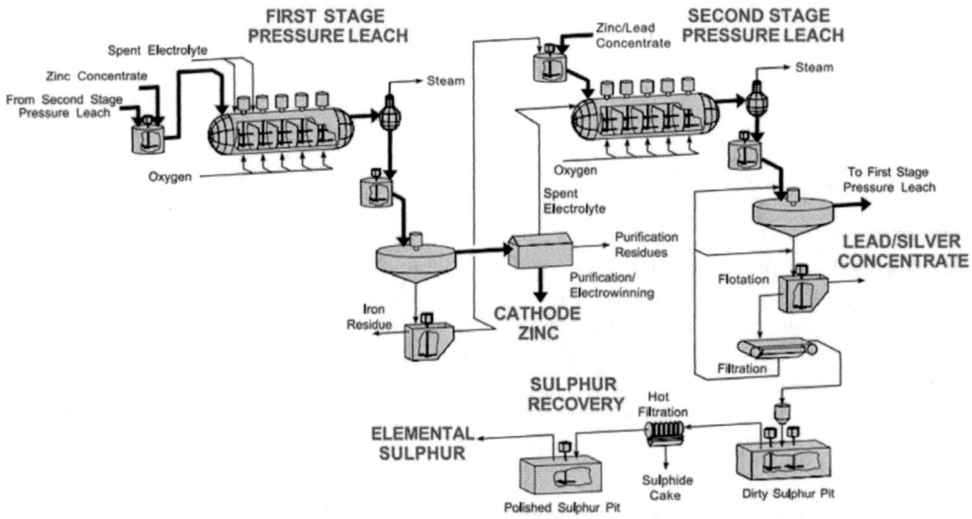

Figure 5. Flowsheet of zinc concentrate pressure leaching

1.6. Pelletised concentrate fluidized bed roasting process

Metallurgie Hoboken-Overpelt in Belgium developed a special fluidized bed roaster to treat the Zn concentrate containing high Cu and high Pb contents in 1970's [6].

In this process, the feed is pelletized and dried, zinc sulphate acting as the binder for the pelletising. The size of the pellets was fixed at 0.5 – 4 mm. The pellets are fed into the rectangular shape roaster, shown below in Figure 6.

This roaster consists mainly of a vertical refractory lined shaft. Over 85% of the length, the roaster bottom consists of an air distribution grate. The rest of the bottom is the aperture, feeding the finishing bin for cooling the calcine and completing the roasting by the secondary air.

As the particle size of the pellets is up to 4 mm, the ascension speed of the gas is very high, and the space rate is higher than 2.5 m/s, which is about 3-4 times higher than that in the normal fluidized bed roaster. Also, the roasting temperature is very high and often exceeds 1000 °C due to the big particle.

It was reported that this roaster treated zinc concentrate from Kipushi with 6% Cu and Tsumeb with 10% Pb and 3.5% Cu during long campaigns without causing accretions on the roaster walls.

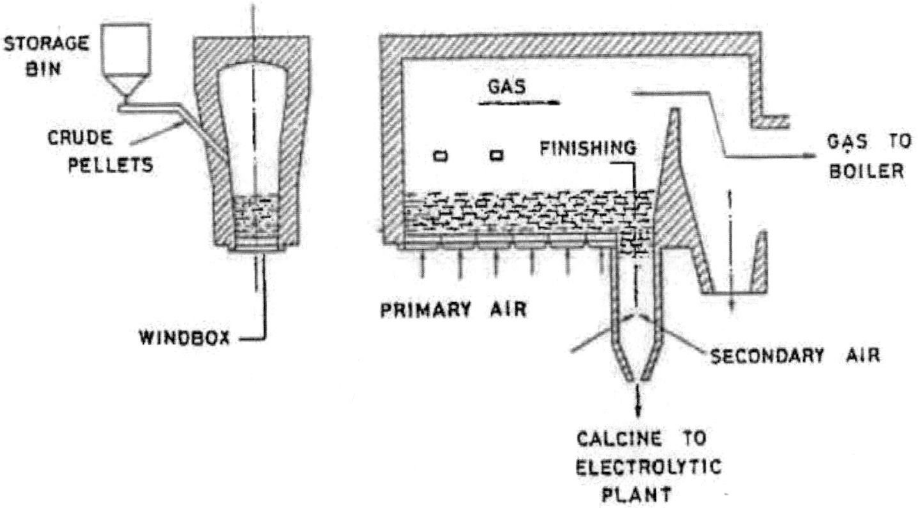

Figure 6. Pelletised concentrate fluidized bed roaster

2. Challenges of complex zinc concentrate treated in fluidized bed roaster

The most typical issue of the complex zinc concentrates is the high Pb content, especially Zn-Pb bulk concentrate. The Pb behaviour during zinc concentrate roasting was studied below.

MRM zinc concentrate has 3.76% Pb, and 5-6% SiO2. The total content of (Pb+Cu+SiO2) is 10–11%. This concentrate is ultra-fine, and the particle size P80 is 7 μm. This MRM zinc concentrate cannot be treated in the fluidized bed roaster alone and needs to be blended with other zinc concentrate to control the total content of (Pb+Cu+SiO2) in the blended feed.

During the fluid-bed roasting of zinc concentrates, agglomeration/accretion and fusion of the bed sometimes occur when the Pb content in the zinc concentrate is high. The lead silicate has long been suspected as the source of the trouble. Recently I investigated this and summarized it below.

Jorgensen from CSIRO in Australia analysed the accretion from an industrial zinc fluidized bed roaster, and their investigations had not found the presence of lead silicate in fused bed accretions from an industrial fluidized bed roasting plant [7].

Jorgensen and his team in CSIRO conducted further lab scale tests which the galena was added to the zinc calcine under the fluidized bed roasting condition, and found that the Pb mineralogy in the accretion are mainly monobasic lead sulphate (PbO.PbSO4) under fluidized bed roasting condition of zinc concentrates, but the lead silicate was not found in the accretion/agglomerates [7].

The Pb compounds in the accretions are shown below: The main one is monobasic lead sulphate ($PbO.PbSO_4$), the minor ones are anglesite ($PbSO_4$) and dibasic and tetrabasic lead sulphate ($2PbO.PbSO_4$ and $4PbO.PbSO_4$), no lead silicate ($PbO.SiO_2$) in the accretion/agglomerates were detected. Those lead sulphate compounds high density, and are difficult to be removed out by air in the fluidized bed roasters, so they form accretion in the roaster bottom to block the tuyeres.

Recently, the test results conducted in XPS, a Glencore company, confirm that Pb is carried in Pb sulphate, Pb sulphide (galena), and fine textures composed of Zn sulphide/sulphate and Pb sulphide/sulphate.

In both these roasting tests of Kidd Creek zinc concentrate and MRM zinc concentrate, Pb exists in the calcine and the sinter/accretion as Pb sulphate, and there is no lead silicate detected.

The Gibbs energies of formation of $PbSO_4$, and $PbO.SiO_2$ (or $PbSiO_3$) are compared as below:

Table 2. Gibbs energy of formation

T, °C	Gibbs energy of formation, J	
	$PbO + SO_2 + 1/2O_2 = PbSO_4$	$PbO + SiO_2 = PbSiO_3$
600	-173798	-20441
700	-148013	-20421
800	-122688	-20436
866	-106289	-20454
886	-101652	-20425
900	-98072	-20091
1000	-73037	-17699
1100	-48432	-15252

The Gibbs energy of formation for $PbSO_4$ is much lower than that for $PbSiO_3$ under the temperature from 600 °C to 1100 °C. Therefore, this confirms that $PbSO_4$ is more stable than lead silicate $PbSiO_3$ thermodynamically and the Pb prefer to form $PbSO_4$ under the roasting temperature range.

The thermodynamic modelling for the MRM concentrate roasting process was done using FactSage. The modelling shows that the $PbSO_4$ is formed, but the $PbSiO_3$ is not formed. This modelling validates the XPS test results.

3. Latest technical developments

3.1. Zn-Pb bulk concentrate tested in pressure leaching process

The pressure leach process was also tested for MRM Zn-Pb bulk concentrate at the lab scale in BGRIMM in 2020.

3.1.1. Pressure leaching tests:

Under the condition of leaching temperature: 150°C, acid zinc ratio: 0.9:1, oxygen partial pressure: 0.5MPa; stirring speed: 550rpm; leaching time 2h, the zinc leaching rate can reach up to 97.84%, The iron leaching rate is 27-35%; the end point acidity is about 28g/l.

Table 3. Pressure leaching tests

	H2SO4/Zn	Lignin, %	Residue Percentage, %	Filtrate, (g/L)			Residue composition, %			Leaching percentage, %	
				Zn	Fe	H_2SO_4	Zn	Fe	S	Zn	Fe
Test 1	0.90	0.2	52.55	75.19	2.55	13.11	8.6	6.68	47.9	89.97	28.26
Test 2	1.10	0.2	48.60	88.91	3.68	30.88	2	5.97	50.23	97.84	40.71
Test 3	1.33	0.2	48.30	86.87		63.87	2.32	3.32	48.23	97.51	67.23

3.1.2. Flotation tests of pressure leaching residue (or sulphur concentrate):

The bubble performance of the leaching pulp is good, the flotation effect is good, the concentration of the control pulp is about 0%, the flotation time is 6~8min, at room temperature or 80°C. The medicament can obtain sulfur concentrate with 80% sulfur.

During the flotation process, the sulfur element is mainly distributed in the concentrate, and the lead element is distributed in the tailings.

Distribution of sulfur: sulfur is mainly distributed in the concentrate, while the zinc and iron are also mostly distributed in the concentrate.

Distribution of lead: ~34% of lead is distributed in sulfide concentrates, and ~66% of lead is distributed in middle mines and tailings.

Distribution of silver: silver is mainly found in flotation concentrates and flotation tailings.

Table 4. Leach residue flotation tests

	Weight, g	Yield, %	Element content, %					Recovery rate, %				
			S	Fe	Zn	Pb	Ag*	S	Fe	Zn	Pb	Ag
Concentrate	23.44	55.52	80.26	2.5	0.41	5.58	216	87.09	37.97	48.93	14.66	43.98
Middle	4.98	11.80	17.6	5.7	0.49	35.06	364	4.06	18.39	12.42	19.57	15.75
Tailings	13.8	32.69	13.85	4.88	0.55	42.51	336	8.85	43.64	38.64	65.76	40.28
Sum	42.22	100.0	51.16	3.66	0.47	21.13	272.68	100.0	100.0	100.0	100.0	100.0

*The unit of Ag content is g/t.

3.2. Sintering – leaching process

This process is proposed to treat the complex zinc concentrate that the fluidized bed roaster cannot treat. The Zn-Pb bulk concentrate is desulphated in sinter machine, similar to that in the sinter machine in the ISP plant. The differences are mainly:

a) The limestone and silica are not added in the feed as flux for the SiO2/CaO and SiO2/Fe ratios control, as the sinter is not treated in the ISP, but is crushed, grinded, and leached.

b) It is also not needed to control the ratio of Zn/Pb in the feed, which is different for ISP.

c) The sinter in this process does not require big size, or the strength is not required as strong as that for ISP.

d) The sulphur content in the sinter should be below 0.5%, so that the leaching process can handle this sinter easily.

3.3. Suspension roasting – leaching process

The industrial scale tests were conducted in a suspension roaster in Glencore plant in Spain in 2023, shown in Figure 7. The flowsheet of this plant is shown in Figure 8.

The concentrate from the feed bins is fed by the conveyor and paddle feeder onto the bottom of the two drying hearths. The concentrate is rabbled across the first hearth into a centre drop-hole, across a second drying hearth and out to a storage bin. Drying air is controlled to 100 -300 °C by mixing hot gas from the roasting hearths with tempering ambient air. The final moisture content of fee after drying is less than 0.5%.

The dried concentrate passes to a ball mill where it is ground to 99% - 150 mesh. The combined air and feed stream is blown into the upper chamber of the roaster. Upon entering the roaster, the concentrate undergoes very rapid spontaneous combustion of its sulphides. The temperature at this point is in the order of 950 – 1200 °C.

Approximately 40% of the calcine falls onto the first of two roasting hearths where retention time and the rabbling action of rotating arms are used to reduce the sulphide content of the calcine.

The remaining 60% of the calcine is carried out of the roaster with the combustion gas and enters the waste heat boiler. Calcine remaining in the gas stream after boiler dropout and hot cyclone separate is removed by ESP. The calcine recovered from the gas cleaning system is returned to the roasting hearths.

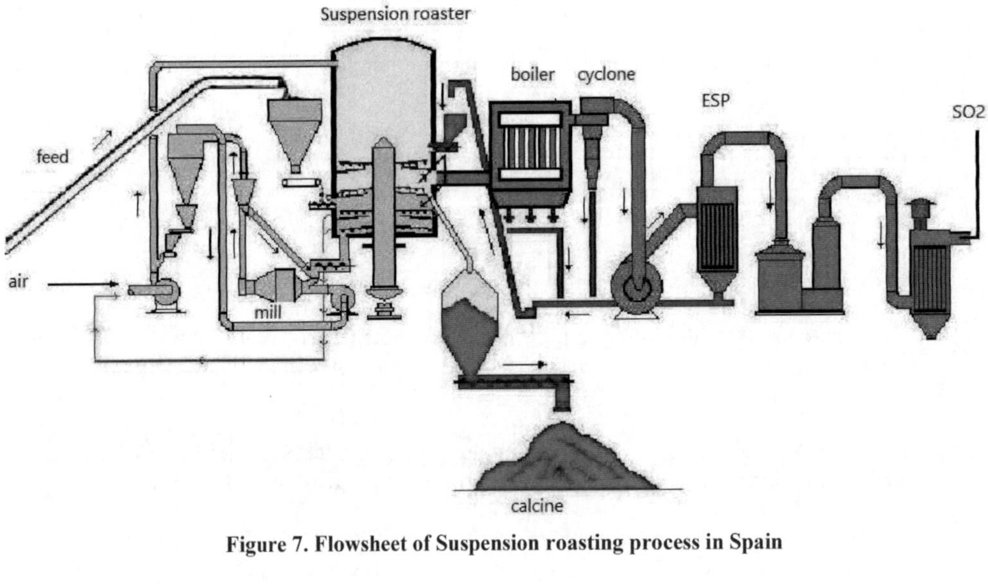

Figure 7. Flowsheet of Suspension roasting process in Spain

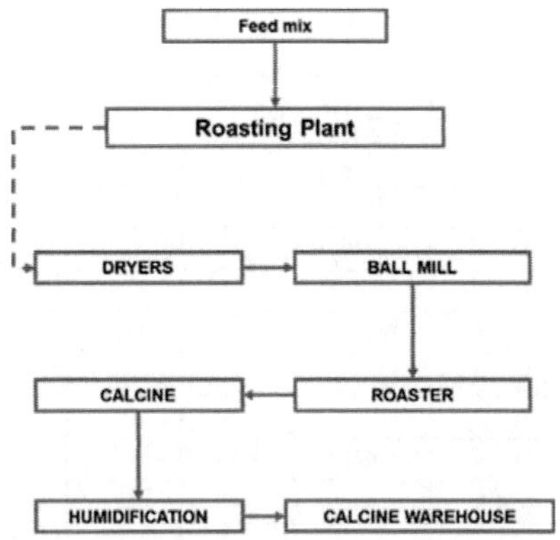

Figure 8. Flowsheet of Suspension Roasting Process in Spain

The roasting tests with MRM zinc concentrate trial at the suspension roaster have been conducted successfully.

The assays of MRM zinc concentrate and particle size are shown in Tables 4 and 5, respectively. MRM zinc concentrate has 3.76% Pb, and 5 - 6% SiO2. The total content of (Pb+Cu+SiO2) is 10 – 11%. This concentrate is ultra-fine, and the particle size P80 is 7 μm.

In the period of 18/June/2023 to 21/June/2023, the suspension roaster treated 90 - 92% MRM zinc concentrate and 8-10% Waelz oxides. The Waelz oxides were introduced in the feed blend to keep the high production but decrease the temperature from 1100 °C to 850 - 900°C. The sulphide content in calcine was <0.3%. The content of (Pb+Cu+SiO2) in the feed reached to 9 - 10%.

Since 22/June/2023, the suspension roaster treated 100% MRM zinc concentrate for 12 days. The roaster operated very stable during those 12 days without any issue observed. The content of (Pb+Cu+SiO2) in the feed reached to 10 % - 11%. The roasting temperature was controlled between 850 - 900°C. The sulphide content in the calcine was well controlled below 0.3 - 0.4%. Feed bypassing the ball mill was set around 25% of the total feed rate.

The industrial test results in the suspension roaster are shown in Table 7.

Table 5. Assays of MRM zinc concentrate

Assays	Zn	Pb	Cu	Fe	S total	SiO2	CaO	MgO	Al	Cd	Ni	Co	Sb	As	Mn	Cl	F	C	Hg	Se	Ag	Tl
Unit	%	%	%	%	%	%	%	%	%	%	%	%	%	%	%	%	%	%	g/t	g/t	g/t	g/t
%	46.30	3.76	0.84	4.99	31.59	5.68	0.16	0.11	0.21	0.10	0.000	0.0048	0.01	0.11	0.00	0.02	0.009	0.42	<10	10	122	138

Table 6. Particle size distribution of MRM zinc concentrate

Particle size	> 100 μm	> 45μ sieve	> 45 μm laser	> 20 μm	> 10 μm	> 2 μm
Unit	μm	μm	μm	μm	μm	μm
%	0	1	1	5.97	21.6	79.18

Journal of Physics: Conference Series **2738** (2024) 012040 doi:10.1088/1742-6596/2738/1/012040

Table 7. Industrial scale test results

Date	Feed to the roaster	Produced Calcine	Roaster Temp	S% in Calcine	Main Impurities (%)				
	(DMT)	(DMT)	(°C, Daily Avg)	S (%)	% Pb	% Cu	% SiO2	% S	% (Pb+Cu+SiO2)
15.06.23	120	110	878	0.26	3.48	0.75	4.90	30.24	9.12
16.06.23	112	103	858	0.32	3.54	0.76	5.00	30.27	9.31
17.06.23	104	95	873	0.26	3.54	0.76	5.00	30.27	9.31
18.06.23	108	98	863	0.26	3.70	0.78	5.30	29.06	9.77
19.06.23	104	95	866	0.26	3.70	0.78	5.30	29.06	9.77
20.06.23	96	87	873	0.26	3.70	0.78	5.30	29.06	9.77
21.06.23	94	85	864	0.22	3.70	0.78	5.30	29.06	9.77
22.06.23	94	85	867	0.22	3.76	0.84	5.68	31.59	10.28
23.06.23	106	96	878	0.26	3.76	0.84	5.68	31.59	10.28
24.06.23	106	96	871	0.26	3.76	0.84	5.68	31.59	10.28
25.06.23	94	85	866	0.26	3.76	0.84	5.68	31.59	10.28
26.06.23	94	85	859	0.29	3.76	0.84	5.68	31.59	10.28
27.06.23	94	85	857	0,32	3.76	0.84	5.68	31.59	10.28
28.06.23	100	91	887	0,38	3.76	0.84	5.68	31.59	10.28
29.06.23	100	91	900	0,42	3.76	0.84	5.68	31.59	10.28
30.06.23	106	98	911	0,46	3.76	0.84	5.68	31.59	10.28
01.07.23	105	97	884	0.26	3.76	0.84	5.68	31.59	10.28
02.07.23	105	97	872	0.29	3.76	0.84	5.68	31.59	10.28

4. Comparison of Different Technologies

All the above technologies can treat the complex zinc concentrate containing high Pb, Cu and SiO2. However, each of them has some advantages and some disadvantages:

a) ISP requires large amount of expensive coke. This was the main reason for the shutdown of most ISP plants in early 2000. Also, the sinter machine operation has a long flowsheet and has serious dust emission. It is also difficult to capture the SO2 from the sinter machine.

b) Outotec atmospheric direct leaching process generates elemental sulphur residue that can't be hot filtered to produce elemental sulphur. Therefore, the whole elemental sulphur residue can't be damped easily and needs to be re-processed in furnaces or roasters. This requires significant Capex and Opex.

c) Albion process requires ultra-grinding. The Albion reactor is cheaper than the Outotec DL reactor, however, the elemental sulphur residue from the Albion process also needs to be re-processed as well.

d) PBOX process just separates Pb and Zn from the Zn-Pb bulk concentrate. The zinc concentrate from PBOX still needs to be treated in the fluidized bed roaster or other processes. Also, the concentrate requires ultra-grinding before it is fed into the PBOX reactors.

e) The pressure leaching process has high Capex, through the elemental sulphur can usually be hot filtered, except some Zn concentrates.

f) The pelletised concentrate fluidized bed roasting process was shut down for decades. Now it is difficult to re-install such roaster. Also, the pelletising requires acid as binder, and the working environment is difficult due to this.

g) The sintering – leaching process has some issues that is similar to ISP, such as the dust emission issue and the SO2 capture etc.

h) The suspension roaster has small capacity due to the rotating arms for calcine removal physically and the pipeline feeding system. This suspension roaster has 10 t/h of feed rate only.

5. Future development

Due to the above disadvantages of those technologies, we are currently focusing on the development of a new technology: Flash Burner Roasting Process, which does not require the expensive coke, does not have elemental sulphur residue issue. The calcine from this flash burner roaster can be treated as the same as the calcine from the fluidized bed roaster.

The concept of the flash burner roaster is based on the industrial tests in 2023, which is presented above, in the suspension roaster in Spain.

The suspension roaster in Spain can treat the complex zinc concentrate, but there are significant disadvantages that limits the scale and capacity.

a) The rotating arms have limited capacity due to the length. This determines the capacity of the suspension roaster.

b) In the suspension roaster, the dried and milled concentrate, and the air are blown into the upper chamber of the roaster via the two pipelines, not via flash burners. Therefore, the mixture efficiency of the concentrate and the air in the pipelines is limited. The flash burners can provide much better mixing between the feed and the air.

In order to solve the above issues, a new flash burner roaster is currently studying and under constructure. The two features are listed below:

a) The cylinder shape is changed to rectangle, so that the calcine can be removed out from the roaster bottom by one of the following two methods:

- Scraper chain conveyor, similar to that for the dust transportation system under the waste heat boiler

- Bed conveyor, similar to the sinter machine bed, but the grates have no holes or gaps, in order to reduce the air leakage into the roaster.

b) The pipeline feeding system is changed to 2 flash burners, similar to the Kivcet.

The sketch of the flash burner roaster is shown in Figures 9 and 10 below:

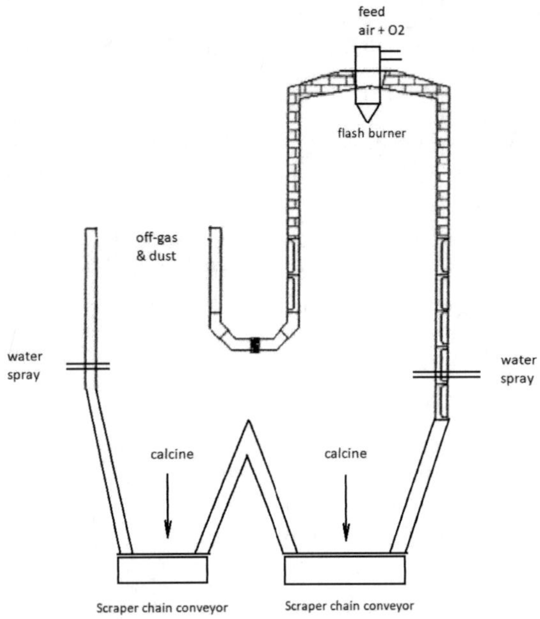

Figure 9. Flash burner roaster with scraper chain conveyor

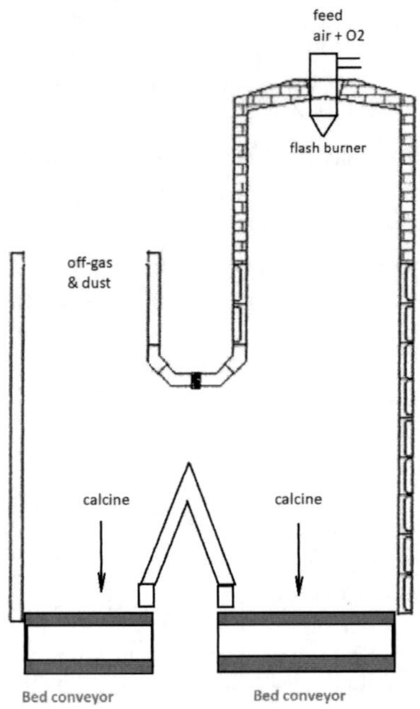

Figure 10. Flash burner roaster with bed conveyor

6. Conclusions

The 6 different technologies for the complex zinc concentrate treatment are discussed and compared.

In order to address the disadvantages of those technologies, a new technology "Flash Burner Roasting Process" concept was tested in an industrial suspension roaster successfully. The industrial tests results showed that the flash roasting process can treat the complex zinc concentrate, and the content of (Pb+Cu+SiO2) in the complex zinc concentrate was up to 11% during the tests. Such complex zinc concentrate is not able to be treated in the fluidized bed roaster.

The flash burner roaster is currently under development. The sketch of the flash burner roaster is presented as well.

References

[1] Dave F. ISP Technology, Lead/Zinc 2010, Vancouver, Oct 2nd, 2010
[2] Kurt S. Outotec Atmospheric Zinc Concentrate Direct Leaching Process – Past, Present and Future. World of Metallurgy – Zrzmetall, 2010, Vol 63, No. 3, pp 136 – 141.
[3] Mike H. Commercialisation of the Albion Process. Website: www.albionprocess.com
[4] Pat B. Zinc Concentrate Processing. Website: www.albionprocess.com

[5] Sherritt technologies – Zinc Technology. Website: www.sherritt.com
[6] Denoiseux R., Winand R. etc. Metallurgie Hoboken Process for Roasting Zinc Concentrates in a Fuild Bed. Lead-Zinc-Tin 80: Proceedings of the World Sysposium on Metallurgy and Environmental Control. 1979, pp. 69 – 84.
[7] Condina V.K., Jorgensen F.R.A. etc. Basic Lead Sulfates as Agglomerating Agents During Fluid-Bed Roasting of Zinc Concentrates. Metallurgical Transaction B. Vol 11B, 1980. pp 637 – 638.

Journal of Physics: Conference Series **2738** (2024) 012041 doi:10.1088/1742-6596/2738/1/012041

The phase structure, transitions, and non-isothermal crystallization kinetics of $Zn_2V_2O_7$ glass

Zhuoyang Li[1,2], Guishang Pei[1,2,*], Mengjiao Jiao[1,2], Yongda Li[1,2], Ningyu Zhang[1,2], and Xuewei Lv[1,2,3,*]

[1] Chongqing Key Laboratory of Vanadium-Titanium Metallurgy and Advanced Materials, Chongqing University, No. 174 Shazheng Street, Shapingba District, Chongqing, 400044 China.
[2] College of Materials Science and Engineering, Chongqing University, No. 174 Shazheng Street, Shapingba District, Chongqing, 400044 China.
[3] The State Key Laboratory of Mechanical Transmissions, Chongqing University, No. 174 Shazheng Street, Shapingba District, Chongqing, 400044 China.
Corresponding author
E-mail: peiguishang@163.com(Guishang Pei), lvxuewei@163.com (Xuewei Lv)

Abstract. The $Zn_2V_2O_7$ glass was prepared by solid-state reaction using the initial reactant ZnO and V_2O_5 powder. X-ray diffraction was employed to characterize the phase composition of the as-prepared $Zn_2V_2O_7$ powder. The crystallization kinetics of $Zn_2V_2O_7$ glass prepared by solid-state reaction was studied using differential thermal analysis under non-isothermal conditions. It was established that the crystallization process of $Zn_2V_2O_7$ can be divided into two steps, which is controlled first by a Disc-like with kinetic equation $G(\alpha)=[-ln(1-\alpha)]^{1/3}$ and then by a Fibril-type growth type with kinetic equation $G(\alpha)=[1-(1-\alpha)^{1/3}]^{1/2}$. The apparent activation energy calculated by different methods (Kissinger method, Ozawa method, Tang method, and Starink method) under non-isothermal conditions were similar, varying between 335.6 kJ·mol^{-1} and 371.6 kJ·mol^{-1}, and the average apparent activation energy was equal to 362.2 kJ·mol^{-1}.
Key words: *synthesis, non-isothermal kinetics, crystallization, apparent activation energy*

1. Introduction

Multicomponent transition metal oxides (TMOs) have attracted widespread attention as the most promising candidate materials for advanced technological applications in contrast with their counter-binary oxides [1-4]. The enhancement in the performance is due to the presence of two metals in these double oxide species, which render additional redox reactions during the device operation.

Journal of Physics: Conference Series **2738** (2024) 012041 doi:10.1088/1742-6596/2738/1/012041

Among them, as to the most important transition metal oxides (TMOs), $Zn_2V_2O_7$ has received considerable attention recently as a material for solid oxide fuel cells, optical, hydrogen storage [5], electrochemical devices materials [6], and white light-emitting diodes [7, 8]. The $Zn_2V_2O_7$ compound has two polymorph forms, namely α-$Zn_2V_2O_7$ at low temperature, and β-$Zn_2V_2O_7$ at high temperature, its state phase transition temperature was studied by Kurzawa *et al.* [9], Makarov *et al.* [10], and Clark *et al.* [11], and its congruent melting was determined by Makarov *et al.* [10] and Clark *et al.* [11], respectively. The cold crystallization process is very important to understand the property of multicomponent oxides, knowledge of the cold crystallization kinetics parameters, such as the apparent activation energy, nucleation type, and crystal growth mechanism of $Zn_2V_2O_7$, are essential for its preparation and the interpretation of its physical properties. In comparison, the crystallization mechanism of $Zn_2V_2O_7$ was less reported in previous studies. Differential scanning calorimetry (DSC), which depends on the relationship between heat flow and temperature was intensively applied in the crystallization process to understand the reaction mechanism [12-15].

In the present study, the $Zn_2V_2O_7$ was prepared by solid-state reaction using ZnO and V_2O_5 powder as raw materials, and X-ray diffraction (XRD) was conducted to analyze and characterize the $Zn_2V_2O_7$ compound. Above all, we investigated the temperature change of two kinds of phase transition of $Zn_2V_2O_7$ using TG and DSC treatment. In addition, we aimed to present a detailed analysis of the non-isothermal crystallization kinetics based on the various DSC curves, mechanism analysis was deduced by using the Avrami method, and the apparent activation energy was calculated and compared by using various methods, such as Kissinger method, Ozawa Method, Tang method, and Starink method.

2. Theoretical basis

2.1. Non-isothermal crystallization mechanism analysis

2.1.1. Avrami model. Non-isothermal kinetic analysis based on the linear heating or cooling process is widely applied with the rapid development of thermal analysis technology [16, 17], various methods were deduced to explore the reaction mechanism and calculate the apparent activation energy during the non-isothermal crystallization process. As the most common method, the Avrami model [18] and Ozawa model [19] have been extensively employed in previous studies, which was derived from the isothermal system and extended to the non-isothermal system by substituting reaction time with temperature. Ding et al [20] studied the non-isothermal kinetics of the $2CaO\cdot Fe_2O_3$ and $CaO\cdot Fe_2O_3$ in the CaO-Fe_2O_3 system by employing the Avrami model, indicating this model is much more suitable for studying its crystallization.

The degree of conversion, α, is the volume fraction of the sample transformed into the crystalline phase at temperature T_1 during crystallization and can be obtained from the DSC curves using the following Eq. (1):

$$\alpha_{(T)} = \frac{S_{T_0}^{T}}{S_{T_0}^{T_1}} \tag{1}$$

where $\alpha_{(T)}$ is the conversion ratio that corresponds to the temperature period in the temperature range T_0 to T_1. $S_{T_0}^{T}$ is the area integrated under the heat flow peak and the baseline from the initial crystallization temperature T_0 to a specific crystallization temperature T. $S_{T_0}^{T_1}$ is the total area from the initial temperature T_0 to the final temperature T_1.

During the non-isothermal crystallization process for the various cooling rates β, the conversion ratio corresponding to temperature can be obtained directly, while the Avrami model was formulated using the conversion ratio as a function of time, the relationship between temperature and time can be defined as shown in Eq. (2):

$$t = \frac{T_0 - T}{\beta} \tag{2}$$

In the present study, the relationship between conversion ratio and time can be determined by substituting the temperature with time, which can be further described as shown in Eq. (3):

$$1-\alpha_{(t)} = \exp(-Kt^n) \tag{3}$$

where $\alpha_{(t)}$ is the conversion ratio at a given time t corresponding to a specific temperature, K is the rate constant of crystallization change, and n is the Avrami exponent, which is determined by using time t and the conversion ratio α by plotting $ln[-ln(1-\alpha)]$ against lnt, as shown in Eq. (4):

$$\ln\left[-\ln(1-\alpha_{(t)})\right] = \ln K + n\ln t \tag{4}$$

where the Avrami model exponent n represents a type of nucleation and growth mechanism [21]. **Table 1** shows the crystallization growth type based on different n values.

Table. 1 Correspondence between growth types and n

n	Growth type	Growth dimensions
3-4	Spherulitic-type	Three-dimensional
2-3	Disc-like	Two-dimensional
1-2	Fibril-like	One-dimensional

The Avrami model derived from the non-isothermal process should be modified to include the parameter K of the formula when extending to the non-isothermal process. Jeziorny [22] amended K as shown in Eq. (5):

$$\ln K_c = \frac{\ln K}{|\beta|} \tag{5}$$

where *Kc* is the modified rate constant based on the non-isothermal crystallization. According to the Avrami model, *Kc* and *n* can be determined by the intercept and slope of plotting $ln[-ln(1-\alpha)]$ against *lnt*.

2.2. Calculation of activation energy using various methods

Besides the mechanism analysis for the non-isothermal crystallization of $Zn_2V_2O_7$, the apparent activation energy is another critical parameter, which indicates the additional energy for an atom to create an activated cluster. The apparent activation energy was determined and compared using various methods, such as the Kissinger method, Ozawa Method, Tang method, and Starink method in the present study.

2.2.1. Kissinger method. The most frequently used method of apparent activation energy evaluation—the Kissinger method[23], is applicable only for non-isothermal experiments where the temperature corresponding to the maximum peak (T_p) shifts with the heating rate (β). The slope of the line, which plots the $ln\ (\beta/T_P^2)$ dependence on $1/T_p$, is equal to the $-E/R$, as shown in Eq. (6).

$$\ln\left(\frac{\beta}{T_\alpha^2}\right) = \ln\frac{AR}{E} - \frac{E}{RT_\alpha} \tag{6}$$

Where T_p is the peak temperature of heat flow curves at a fixing heating rate, *A* is the pre-exponential coefficient, and *E* is the apparent activation energy of the non-isothermal crystallization.

The Kissinger method can be further converted to Eq. (7):

$$\frac{d\left(\ln\frac{|\beta|}{T_\alpha^2}\right)}{d\left(\frac{1}{T_\alpha}\right)} = -\frac{E}{R} \tag{7}$$

Apparent activation energy can be calculated using the slope of the plot of $ln\ (|\beta|/T_P^2)$ against $1/T_P$.

According to the Kissinger method [23], the peak temperature in the DSC curves at various heating rates must be obtained first. After fitting the plots of $ln\ (|\beta|/T_P^2)$ against $1/T_P$, Thus, the apparent activation energy was calculated from the slope of the fitted line.

2.2.2. Ozawa method. Very similar to Kissinger's Method [23], Ozawa [19] simplified it by using Dolye's approximation [24] and proposed a new method to calculate the activation energy under the non-isothermal kinetic analysis of crystallization, as shown in Eq. (8)

$$\ln(\beta) = \ln\frac{AR}{E} - 1.052\frac{E}{RT_\alpha} \tag{8}$$

where T_p is the peak temperature at a specific heating or cooling rate, *A* is the pre-exponential coefficient, and *E* is the apparent activation energy of crystallization.

Apparent activation energy *E* can be calculated using the slope of the plot of $ln\ (\beta)$ against $1/T_P$.

2.2.3. Tang method. The Flynn-Wall-Ozawa [25, 26] and Kissinger-Akahira-Sunose [23] methods can be expressed in the general formula, as shown in Eq. (9)

$$\ln\left(\frac{\beta}{T_{\alpha}^{s}}\right) = -\frac{AE}{RT_{\alpha}} + constant \tag{9}$$

where T_p is the peak temperature, Donghua and Wanjun [27] presented a method in which the parameters s and A were treated as the constant values, namely s = 1.895 and A = 1.00145. Then, Eq. (10) is rearranged as

$$\ln\left(\frac{\beta}{T_{\alpha}^{1.895}}\right) = -\frac{1.00145E}{RT_{\alpha}} + constant \tag{10}$$

Apparent activation energy E can be calculated using the slope of the plot of $ln\ (|\beta|/T^{1.895})$ against $1/T_P$.

2.2.4. Starink method. Modifications were further done based on linear regression and mathematical analysis by Starink [28], and they deduced a new formula in which s = 1.92 and A = 1.0008, Then, Eq. (11) is rearranged as

$$\ln\left(\frac{\beta}{T_{\alpha}^{1.92}}\right) = -\frac{1.0008E}{RT_{\alpha}} + constant \tag{11}$$

where T_p is the peak temperature, Apparent activation energy E can be obtained using the slope of the plot of $ln\ (|\beta|/T^{1.92})$ against $1/T_P$.

3. Experimental process

The raw materials ZnO (99.99%, purity) and V_2O_5 (99.99%, purity) were provided by Shanghai Jingjinle Industry Co., LTD. Before preparing the desired sample, the powder ZnO and V_2O_5 were first separate pre-roasting at 473 K for approximately 3 h under the constant flow of argon (50 ml·min^{-1}) to remove any absorbed moisture. The $Zn_2V_2O_7$ sample was then prepared by the repeated roasting and crushing method, for the first step, the starting materials ZnO and V_2O_5 were thoroughly mixed to obtain a homogenous mixture, according to the specific molar ratio (ZnO: V_2O_5 = 2:1). The mixture was compacted into some pellets with 12 mm diameter utilizing hydraulic press under 10 Mpa, and the pellets was subsequently heated from room temperature to target temperature of 873 K with a heat rate of 10 K/min and maintained for 10 h to complete solid-state reaction, the calcinated samples cooled down to room temperature within furnace and immediately pulverized in a vibration mill. Repeating the above procedure twice until the high-purity and homogenous $Zn_2V_2O_7$ sample was finally obtained.

The phase composition was checked using X-ray diffraction (XRD, Model D/max2500/PC; Cu *Kα*) with an angular range of 10°-90° and a scan rate of 3°/min. Qualitative and quantitative analyses of XRD patterns compared with standard one (**NO #70-1523**) were performed to verify the prepared $Zn_2V_2O_7$ sample after several roasting and cooling processes.

Journal of Physics: Conference Series **2738** (2024) 012041 doi:10.1088/1742-6596/2738/1/012041

Differential scanning calorimetry was carried out using a Netzsch thermal analyzer (Model STA 449C). The $Zn_2V_2O_7$ sample (approximately 12 mg) was tiled in a Pt crucible, with an Al_2O_3 crucible as a reference. The samples were heated from room temperature to 900°C and then cooled to room temperature at the specific cooling rates of 10 °C/min, 15 °C /min, and 20 °C /min in an argon atmosphere with a gas flow rate of 50 ml/min. The various heat flow curves can be obtained to analyze the non-isothermal crystallization process and calculate the apparent activation energy. The heat flow curve was obtained to analyze the polycrystalline transition temperature and enthalpy change.

4. Results and discussion

4.1. Phase structure and transitions of $Zn_2V_2O_7$ glass

The $Zn_2V_2O_7$ compound has two polymorphs forms, namely *α*-$Zn_2V_2O_7$ at low temperature, and *β*-$Zn_2V_2O_7$ at high temperature, **Figure 1** shows the XRD patterns of the prepared sample at room temperature along with a standard PDF card, indicating typical reflections of $Zn_2V_2O_7$ phase **(PDF#70-1523)**. No additional reflection of other phases, e.g., initial reactants, were observed. Narrow and strong reflection confirmed the good crystallinity of the as-prepared sample. The refinement of the cell parameters shows that the $Zn_2V_2O_7$ crystallizes in *monoclinic system*, belonging to space group *C2/c*, with the lattice parameters *a*=7.429 Å, *b*=8.340 Å, *c*=10.098 Å, which was very close to the reference one **(PDF#70-1523)** and previous study by Gopal *et al.* [29] and Chen *et al.* [30]. The results were presented and compared as shown in **Table 2** Quantitative analysis by Jade 6.0® indicated that the $Zn_2V_2O_7$ sample mainly comprises $Zn_2V_2O_7$ with approximately 95%.

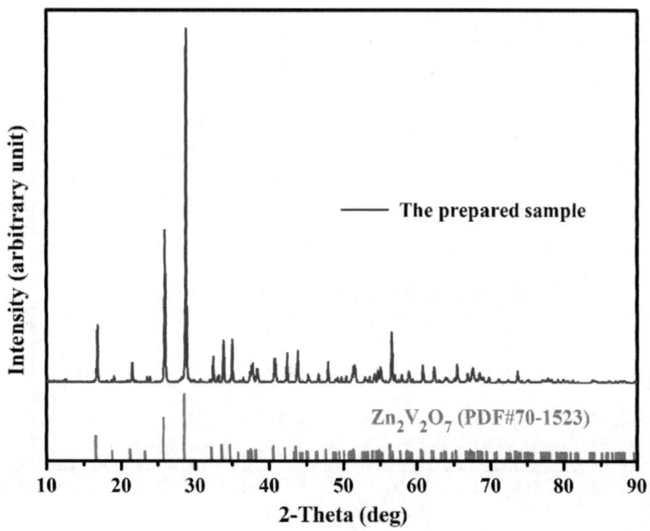

Fig. 1 XRD patterns of $Zn_2V_2O_7$ and standard PDF card.

Table. 2 Crystallographic data of $Zn_2V_2O_7$ compound.

Crystal types	Crystal system	Space group	Lattice parameters			Ref.
			a	*b*	*c*	
α-$Zn_2V_2O_7$	monoclinic	*C2/c*	7.429(5)	8.340(3)	10.098(3)	[29]
	monoclinic	*C2/c*	7.4301(4)	8.3238(1)	10.0928(3)	[30]
	monoclinic	*C2/c*	7.429	8.340	10.098	This study
β-$Zn_2V_2O_7$	monoclinic	*C2/m*	6.9324(2)	8.4394(2)	5.0326(1)	[31]

TG and DSC investigation were both carried out in the present study to determine the temperature change of two kinds of phase transition with a heat rate of 10°C/min up to 900 °C as shown in **Figure 2**, no significant mass loss occurs in the temperature ranging from 150-950 °C in the TG curve. The DSC curve exists in two obvious endothermic peaks corresponding to polymorphic transition and melt fusion. The temperature of polymorphic transition and melt fusion is determined on the onset of heat flow peak, which was presented and compared with the previous studies as shown in **Table 3**, indicating that the values excellently agree with previous work Kurzawa M *et al.* [9], Makarov VA *et al.* [10], and Clark GM *et al.* [11].

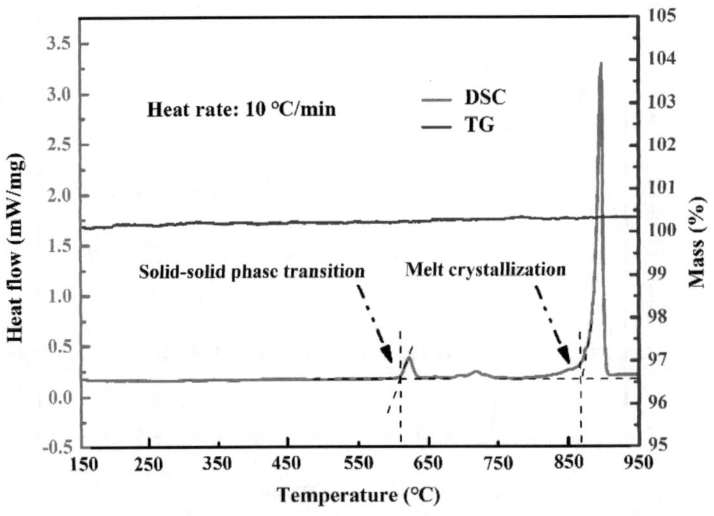

Fig. 2 TG and DSC curves at a heating rate of $10°C \cdot min^{-1}$ in Ar atmosphere.

Table. 3 The phase transition temperature of $Zn_2V_2O_7$ glass.

Transition type	Temperature	Method	Reference
	890°C	DSC	[10]
L↔α-$Zn_2V_2O_7$	872°C	DTA	[11]
	869°C	DSC	This study
	590°C±5°C	DTA	[9]
α-$Zn_2V_2O_7$↔β-$Zn_2V_2O_7$	620°C	DSC	[10]
	615°C	DTA	[11]
	610°C	DSC	This study

4.2. Non-isothermal crystallization kinetics of $Zn_2V_2O_7$ glass

In this study, many curves corresponding to different cooling rates of 10 °C/min, 15 °C/min, and 20 °C/min were obtained during the non-isothermal crystallization process using DSC investigations as shown in **Figure 3** those curves can be applied to analyze the reaction mechanism and calculate apparent activation for non-isothermal crystallization process of $Zn_2V_2O_7$. The positions and intensities of these heat flow peaks change with the cooling rate, the positions of the heat flow peaks are shifted to lower temperatures with increasing cooling rate, which indicates that cold crystallization was activated. The intensities, corresponding to the enthalpy changes during the non-isothermal crystallization process; increase with raising the cooling rates, indicating that increased cooling rates can reinforce the thermal hysteresis phenomenon during the non-isothermal crystallization process, this conforms to the basic physical-chemical rules. For the shape of each heat flow peak, the temperature of heat flow is obvious and can be explained by basic physical-chemical knowledge, the freezing point of a pure substance is constant and only changes with rising cooling rates. See other articles for more on the principles of analysis.[32, 33]

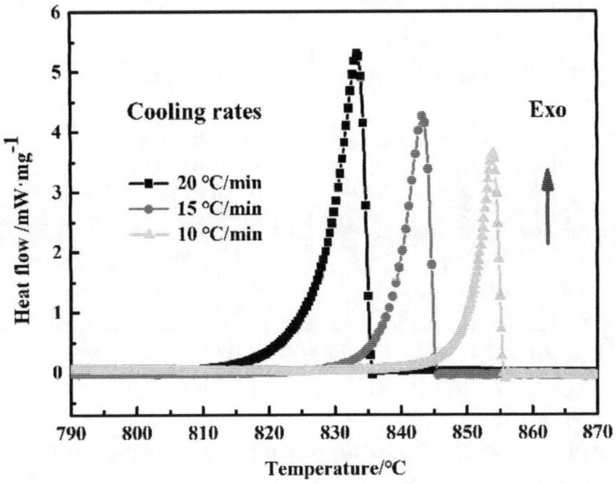

Fig. 3 Heat flow vs. temperature during non-isothermal crystallization at different cooling rates.

4.2.1. Crystal mechanism analysis. According to the Avrami model, the conversion ratio corresponding to temperature and time for non-isothermal melt and cold crystallization is obtained using the original DSC curves as shown in **Figure 4** and **Figure 5**, the sigmoid shape of the temperature dependence of the degree of conversion indicated bulk crystallization and excludes the possibility of surface crystallization. For all the cooling rates, these curves can be separated into two characteristic regions corresponding to the two physical stages of non-isothermal crystallization: a rapid initial period in which nucleation is dominant, followed by a process of slow growth of nuclei. It indicates that raising the heating and cooling rate shortens crystallization time. In addition, by drawing the plots of $ln[-ln(1-\alpha)]$ vs. lnt for the prepared $Zn_2V_2O_7$ compound, the slope of those plots can be divided obviously into two stages with reasonable linear fits, namely Stage 1 and Stage 2, the Avrami kinetics parameter Kc and n of melt and cold crystallization are determined as shown in **Figure 6**, respectively.

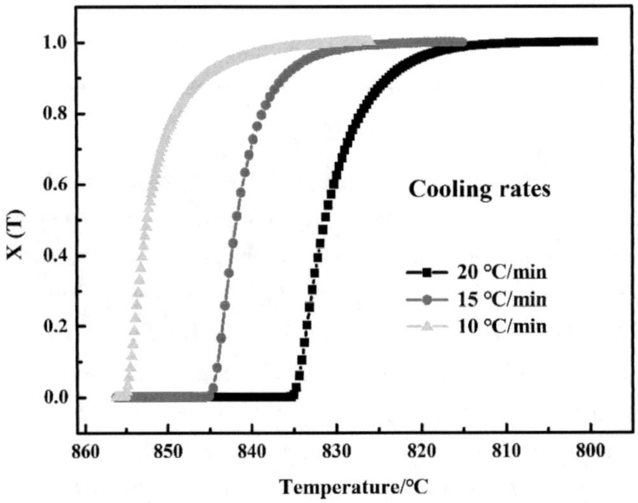

Fig. 4 Conversion ratio corresponding to temperature for non-isothermal crystallization.

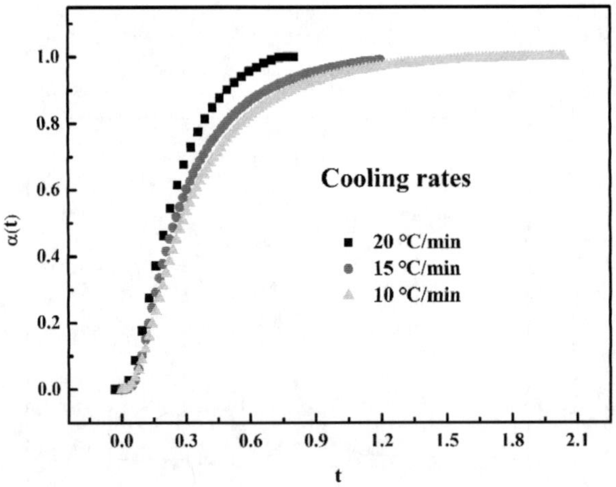

Fig. 5 Conversion ratio corresponding to time for non-isothermal crystallization.

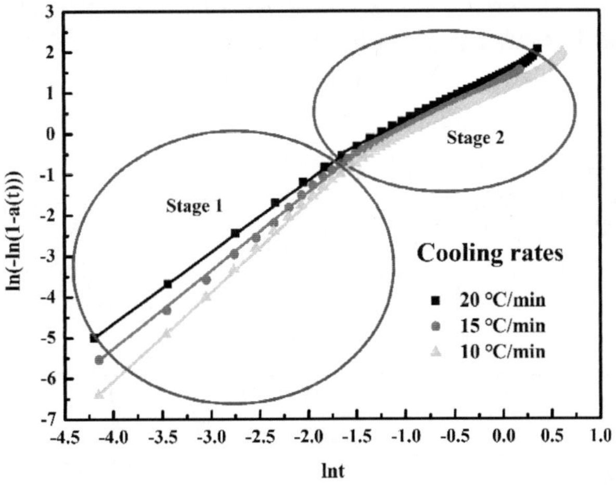

Fig. 6 Plots of $ln[-ln(1-\alpha_{(t)})]$ against *lnt* based on the Avrami model for non-isothermal cold crystallization.

Table. 4 Avrami parameters for non-isothermal cold crystallization.

β	Stage 1			Stage 2		
(K·min^{-1})	n	lnK	R^2	n	lnK	R^2
20	1.7605	2.387	0.99975	1.1515	1.4626	0.9988
15	1.9676	2.4997	0.9977	1.1630	1.2860	0.9978
10	2.210	2.764	0.9993	1.1476	1.0997	0.9943

The non-isothermal cold crystallization of $Zn_2V_2O_7$ included two stages based on the Avrami model, which were discussed in this study. As shown in **Table 4**, *Kc* increased with increasing heating rates both in Stage 1 and Stage 2, indicating that increasing the heating rate increases the cold crystallization rate; this conclusion is consistent with the actual situation. In addition, the *n* values of modification change of $Zn_2V_2O_7$ are approximately between 2 and 3 for Stage 1 and between 1 and 2 for Stage 2. This finding indicates that the cold crystallization process of $Zn_2V_2O_7$ is controlled first by a Disc-like and then by a Fibril-type growth type, which was consistent with the DSC curves.

4.2.2. Calculation of apparent activation energy using various methods. In the present study, the peak temperature, corresponding to the different cooling rates, as shown in **Table 5**, was frequently used to calculate the apparent activation energy during the non-isothermal cool crystallization process [34]. To obtain reasonable and credible results, lots of different methods have been applied to explore the apparent activation energy were applied and compared with each other in this study. After drawing the plots of $ln(\beta/T_\alpha^2)$ against $1000/T_\alpha$ (based on Kissinger method), $ln(\beta)$ against $1000/T_\alpha$ (based on Ozawa method), $ln(\beta/T_\alpha^{1.895})$ against $1000/T_\alpha$ (based on Tang method) and $ln(\beta/T_\alpha^{1.92})$ against $1000/T_\alpha$ (based on

Starink method) for the non-isothermal cool crystallization process with reasonable linear fits, the corresponding apparent activation energy can be obtained from the slopes of those fitted lines and the values are shown in **Table 6** and **Figure 7**, respectively. The apparent activation energy was extremely consistent with each other using the Kissinger method, Tang method, and Starink method, while the value obtained by the Ozawa method is a little bit lower than other methods as shown in **Table 6**, which may be caused by the method itself and the difference is within the normal range. The average apparent activation energy is 362.2 kJ·mol^{-1}, indicating the obtained values were correct and reasonable.

Table. 5 The peak temperature corresponds to different cooling rates during the non-isothermal crystallization process.

Cooling rates (K·min^{-1})	10	15	20
Peak temperature (K)	1126.9845	1116.3185	1106.5932

Table. 6 The calculations of the apparent activation energy during the non-isothermal crystallization process using the Kissinger method, Ozawa method, Tang method, and Starink method.

Types	Kissinger method	Ozawa method	Tang method	Straink method	Average Value
Apparent activation energy (kJ·mol^{-1})	371.6	335.6	370.8	370.9	362.2

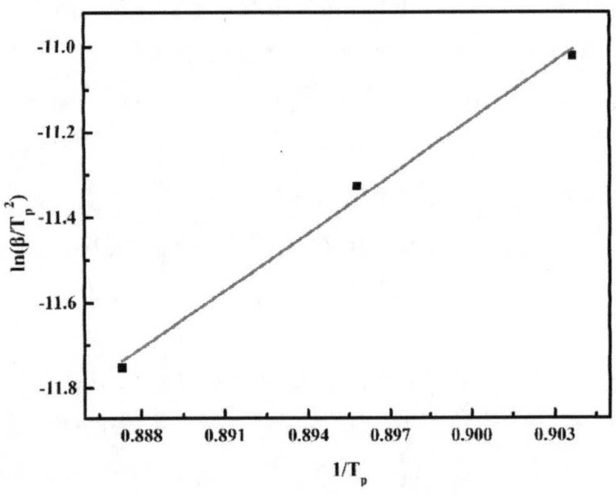

(1) Kissinger method

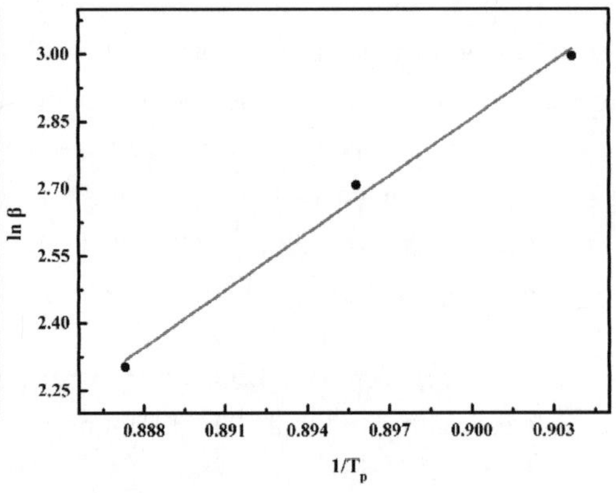

(2) Ozawa method

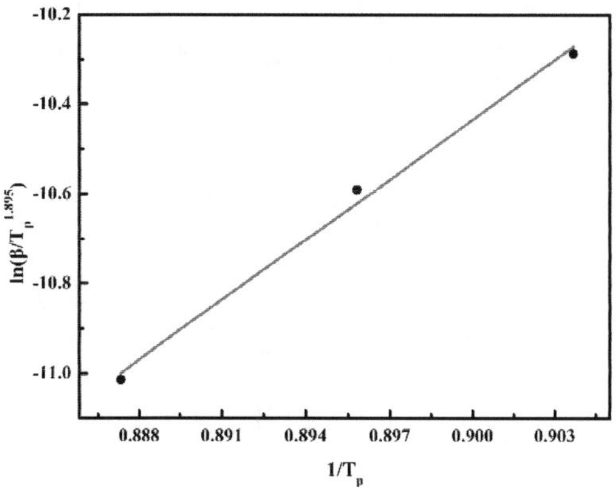

(3) Tang method

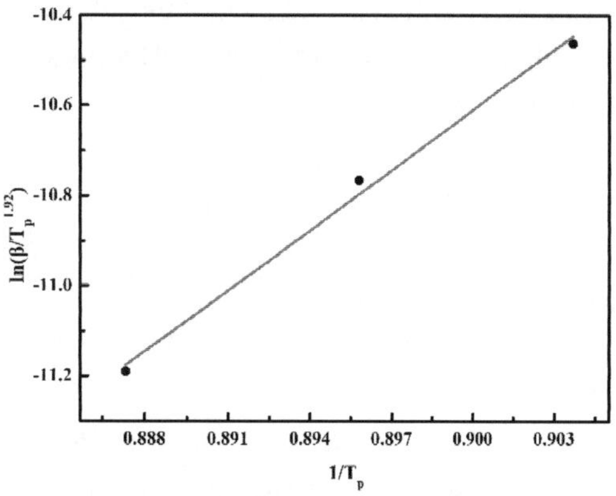

(4) Starink method

Fig. 7 The calculations of the apparent activation energy during the non-isothermal crystallization process using various methods: (1) Kissinger method, (2) Ozawa method, (3) Tang method, and (4) Starink method.

5 Conclusions

The $Zn_2V_2O_7$ glass was synthesized by solid-state reaction and characterized using XRD, TG, and DSC. Thermal analysis was applied to the crystallization kinetics of the $Zn_2V_2O_7$ glass under non-isothermal conditions. It was established that the crystallization process can be divided into two steps, cold crystallization process of $Zn_2V_2O_7$ glass is controlled first by a Disc-like with kinetic equation $G(\alpha)=[-ln(1-\alpha)]^{1/3}$ and then by a Fibril-type growth type with kinetic equation $G(\alpha)=[1-(1-\alpha)^{1/3}]^{1/2}$. The values of apparent activation energy of $Zn_2V_2O_7$ glass calculated by different methods (Kissinger method, Ozawa method, Tang method, and Starink method) under non-isothermal conditions were close to each other and varied between 335.6 $kJ\cdot mol^{-1}$ - 371.6 $kJ\cdot mol^{-1}$, the average apparent activation energy is equal to 362.2 $kJ\cdot mol^{-1}$.

Reference:

[1] Feng, B.Y., et al., *Optimal V_2O_3 extraction by sustainable vanadate electrolysis in molten salts.* Tungsten, 2023.

[2] Olayiwola, A.U., et al., *Cleaner production of ammonium paratungstate by membrane electrolysis-precipitation of sodium tungstate solution.* Tungsten, 2023. **5**(1): p. 145-159.

[3] Chen, Y.-L., et al., *Tungsten and arsenic substance flow analysis of a hydrometallurgical process for tungsten extracting from wolframite.* Tungsten, 2021. **3**(3).

[4] Pei, G.S., et al., *Thermodynamic properties of magnesium orthovanadate $Mg_3(VO_4)_2$ at high temperatures (298.15-1473 K).* Calphad-Computer Coupling of Phase Diagrams and Thermochemistry, 2021. **74**.

[5] Butt, F.K., et al., *Novel $Zn_2V_2O_7$ hierarchical nanostructures: Optical and hydrogen storage properties.* International Journal of Hydrogen Energy, 2015. **40**(30): p. 9359-9364.

[6] Chen, Z., et al., *Hydrothermal synthesis and electrochemical properties of crystalline $Zn_2V_2O_7$ nanorods.* Materials Letters, 2013. **107**: p. 35-38.

[7] Kuang, S., et al., *A new self-activated yellow-emitting phosphor $Zn_2V_2O_7$ for white LED.* Optik, 2013. **124**(22): p. 5517-5519.

[8] Guan, Y., et al., *Spectral Conversion From Ultraviolet to Near Infrared in Yb^{3+}-Doped Pyrovanadate $Zn_2V_2O_7$ Particles.* Journal of the American Ceramic Society, 2014. **97**(10): p. 3202-3207.

[9] Kurzawa, M., et al., *Reinvestigation of Phase Equilibria in the V_2O_5–ZnO System.* Journal of Thermal Analysis & Calorimetry, 2001. **64**(3): p. 1113-1119.

[10] Makarov, V.A., A.A. Fotiev, and L.N. Serebryakova, *PHASE DIAGRAM FOR THE SYSTEM V_2O_5--ZnO.* 1971.

[11] Clark, G.M. and A.N. Pick, *Dta study of the reactions of V_2O_5 with metal (II) oxides.* Journal of Thermal Analysis and Calorimetry, 1975. **7**(2): p. 289-300.

[12] Ding, C., et al., *Non-isothermal crystallization kinetics for CaO–Fe_2O_3 system.* Journal of Thermal Analysis and Calorimetry, 2016. **124**(1): p. 509-518.

[13] Rodrigues, A.M., et al., *Isothermal and non-isothermal crystallization of a fresnoite glass.* Journal of Non-crystalline Solids, 2013. **362**: p. 114-119.

[14] Milicevic, B., M. Marinoviccincovic, and M.D. Dramicanin, *Non-isothermal crystallization kinetics of $Y_2Ti_2O_7$.* Powder Technology, 2017. **310**: p. 67-73.

[15] Pei, G.S., et al., *The phase structure, thermodynamic properties, and dissolution behavior of $Ca_5Mg_4V_6O_{24}$.* Journal of the American Ceramic Society, 2023. **106**(4): p. 2529-2538.

[16. Flynn, J.H., *Thermal analysis kinetics — past, present and future.* Thermochimica Acta, 1992. **203**(2): p. 519-526.

[17] Pei, G.S., et al., *High-temperature heat capacity and phase transformation kinetics of $NaVO_3$.* Journal of Alloys and Compounds, 2019. **794**: p. 465-472.

[18] Henderson, D.W., *Thermal analysis of non-isothermal crystallization kinetics in glass forming liquids.* Journal of Non-Crystalline Solids, 1979. **30**(3): p. 301-315.

[19] Ozawa, T., *Kinetics of non-isothermal crystallization.* Polymer, 1971. **12**(3): p. 150-158.

[20] Ding, C., et al., *Crystallization Kinetics of 2CaO·Fe_2O_3 and CaO·Fe_2O_3 in the CaO–Fe_2O_3 System.* Isij International, 2016. **56**(7): p. 1157-1163.

[21] Balamurugan, G.P. and S.N. Maiti, *Nonisothermal crystallization kinetics of polyamide 6 and ethylene- co -butyl acrylate blends.* Journal of Applied Polymer Science, 2010. **107**(4): p. 2414-2435.

[22] Jeziorny, A., *Parameters characterizing the kinetics of the non-isothermal crystallization of poly(ethylene terephthalate) determined by d.s.c.* Polymer, 1978. **19**(10): p. 1142-1144.

[23] Kissinger, H.E., *Reaction Kinetics in Differential Thermal Analysis.* Analytical Chemistry, 1957. **29**(11): p. 1702-1706.

[24] Doyle, C.D., *Series Approximations to the Equation of Thermogravimetric Data.* Nature, 1965. **207**(4994): p. 290-291.

[25] Ozawa, T., *A New Method of Analyzing Thermogravimetric Data.* Bull.chem.soc.jpn, 1965. **38**(11): p. 1881-1886.

[26] Flynn, J.H. and L.A. Wall, *A quick, direct method for the determination of activation energy from thermogravimetric data.* Journal of Polymer Science Part C Polymer Letters, 1966. **4**(5): p. 323–328.

[27] Wyrzykowski, D., et al., *Thermal analysis of quinolinium tetrachloroferrate(III).* Thermochimica Acta, 2006. **443**(1): p. 72-77.

[28] Starink, M.J., *A new method for the derivation of activation energies from experiments performed at constant heating rate.* Thermochimica Acta, 1996. **288**(1): p. 97-104.

[29] Gopal, R. and C. Calvo, *Crystal Structure of α-$Zn_2V_2O_7$.* Canadian Journal of Chemistry, 1973. **51**(7): p. 1004-1009.

[30] Ye-Qing, C., et al., *Phase relations in the ZnO-V_2O_5-K_2O system.* Chinese Physics B, 2011. **20**(7).

[31] Krasnenko, T., et al., *Crystal structure of β'-$Zn_2V_2O_7$.* Crystallography Reports, 2003. **48**: p. 35-38.

[32] Pei, G., et al., *A clean process of preparing VO as LIBs anode materials via the reduction of V_2O_3 powder in a H-2 atmosphere: Thermodynamic assessment, isothermal kinetic analysis, and electrochemistry performance evaluation.* Journal of Alloys and Compounds, 2020. **845**.

[33] Pei, G., et al., *Isothermal reduction of V_2O_5 powder using H_2 as oxygen carrier: Thermodynamic evaluation, reaction sequence, and kinetic analysis.* Powder Technology, 2021. **378**: p. 785-794.

[34] Atiqullah, M., et al., *Effects of supported ((BuCp)-Bu-n)$_2$ZrCl$_2$ catalyst active center multiplicity on crystallization kinetics of ethylene homo- and copolymers.* Journal of the Taiwan Institute of Chemical Engineers, 2014. **45**(4): p. 1982-1991.

AUTHOR INDEX